VOLUME SEVENTY TWO

ADVANCES IN INORGANIC CHEMISTRY

Materials for Sustainable Energy

ADVISORY BOARD

VOLUME SEVENTY TWO

ADVANCES IN INORGANIC CHEMISTRY

Materials for Sustainable Energy

Edited by

RUDI van ELDIK
University of Erlangen-Nuremberg, Germany
Jagiellonian University, Kraków, Poland

WOJCIECH MACYK
Jagiellonian University, Kraków, Poland

ELSEVIER

ACADEMIC PRESS
An imprint of Elsevier

Academic Press is an imprint of Elsevier
125 London Wall, London EC2Y 5AS, United Kingdom
525 B Street, Suite 1650, San Diego, CA 92101, United States
50 Hampshire Street, 5th Floor, Cambridge, MA 02139, United States
The Boulevard, Langford Lane, Kidlington, Oxford OX5 1GB, United Kingdom

First edition 2018

ISBN: 978-0-12-815077-1
ISSN: 0898-8838

Publisher: Zoe Kruze
Acquisition Editor: Sam Mahfoudh
Editorial Project Manager: Shellie Bryant
Production Project Manager: Denny Mansingh
Cover Designer: Victoria Pearson

Typeset by TNQ Technologies

CONTENTS

CONTRIBUTORS

M. Shaheer Akhtar
New & Renewable Energy Material Development Center (NewREC), Chonbuk National University, Jeonbuk, Republic of Korea

Sadia Ameen
Energy Materials & Surface Science Laboratory, Solar Energy Research Center, School of Chemical Engineering, Chonbuk National University, Jeonju, Republic of Korea

Masakazu Anpo
State Key Laboratory of Photocatalysis on Energy and Environment, College of Chemistry, Fuzhou University, Fuzhou, People's Republic of China

Armand J. Atanacio
Centre for Accelerator Science, Australian Nuclear Science and Technology Organisation, Kirrawee DC, NSW, Australia

Tadeusz Bak
Solar Energy Technologies, School of Computing, Engineering and Mathematics, Western Sydney University, Penrith, NSW, Australia

Monika Bakierska
Faculty of Chemistry, Jagiellonian University, Kraków, Poland

Lucjan Chmielarz
Faculty of Chemistry, Jagiellonian University, Kraków, Poland

Krystian Chudzik
Faculty of Chemistry, Jagiellonian University, Kraków, Poland

Krzysztof Fic
Institute of Chemistry and Technical Electrochemistry, Poznań University of Technology, Poznań, Poland

Elzbieta Frackowiak
Institute of Chemistry and Technical Electrochemistry, Poznań University of Technology, Poznań, Poland

Rafał Knura
Faculty of Chemistry, Jagiellonian University, Kraków, Poland

Marcin Kobielusz
Faculty of Chemistry, Jagiellonian University, Kraków, Poland

Tomasz Kondratowicz
Faculty of Chemistry, Jagiellonian University, Kraków, Poland

Andrzej Kowalczyk
Faculty of Chemistry, Jagiellonian University, Kraków, Poland

Anna Kusior
Faculty of Materials Science and Ceramics, AGH University of Science and Technology, al. A. Mickiewicza 30, 30-059 Kraków, Poland

Piotr Kuśtrowski
Faculty of Chemistry, Jagiellonian University, Kraków, Poland

Marcelina Lis
Faculty of Chemistry, Jagiellonian University, Kraków, Poland

Wojciech Macyk
Faculty of Chemistry, Jagiellonian University, Kraków, Poland

Weronika Marszałowicz
Faculty of Chemistry, Jagiellonian University, Kraków, Poland

Paweł Mikrut
Faculty of Chemistry, Jagiellonian University, Kraków, Poland

Marcin Molenda
Faculty of Chemistry, Jagiellonian University, Kraków, Poland; MarCelLi Adv Tech Sp. z o.o., Kraków, Poland

Mohammad Khaja Nazeeruddin
Group for Molecular Engineering of Functional Materials (GMF), Institute of Chemical Science and Engineering, Faculty of Basic Science, Ecole Polytechnique Federale de Lausanne (EPFL), Sion, Switzerland

Janusz Nowotny
Solar Energy Technologies, School of Computing, Engineering and Mathematics, Western Sydney University, Penrith, NSW, Australia

Joanna Pacek
Faculty of Chemistry, Jagiellonian University, Kraków, Poland; MarCelLi Adv Tech Sp. z o.o., Kraków, Poland

Marta Radecka
Faculty of Materials Science and Ceramics, AGH University of Science and Technology, al. A. Mickiewicza 30, 30-059 Kraków, Poland

Kazi A. Rahman
Solar Energy Technologies, School of Computing, Engineering and Mathematics, Western Sydney University, Penrith, NSW, Australia

Anna Rokicińska
Faculty of Chemistry, Jagiellonian University, Kraków, Poland

Małgorzata Rutkowska
Faculty of Chemistry, Jagiellonian University, Kraków, Poland

Hyung-Shik Shin
Energy Materials & Surface Science Laboratory, Solar Energy Research Center, School of Chemical Engineering, Chonbuk National University, Jeonju, Republic of Korea

Adam Slesinski
Institute of Chemistry and Technical Electrochemistry, Poznań University of Technology, Poznań, Poland

Michał Świętosławski
Faculty of Chemistry, Jagiellonian University, Kraków, Poland

Anita Trenczek-Zajac
Faculty of Materials Science and Ceramics, AGH University of Science and Technology, al. A. Mickiewicza 30, 30-059 Kraków, Poland

Bo Wang
State Key Laboratory of Photocatalysis on Energy and Environment, College of Chemistry, Fuzhou University, Fuzhou, People's Republic of China

Xinchen Wang
State Key Laboratory of Photocatalysis on Energy and Environment, College of Chemistry, Fuzhou University, Fuzhou, People's Republic of China

Katarzyna Zakrzewska
Faculty of Computer Science, Electronics and Telecommunications, AGH University of Science and Technology, al. A. Mickiewicza 30, 30-059 Kraków, Poland

PREFACE

When the use of fossil fuels significantly affects our planet, and their availability decreases in contrast to the growing energy consumption, the interest in sustainable energy continues to grow. The use of sustainable energy sources, mainly solar, requires efficient technologies for energy conversion to electricity or fuels, their transport and storage. These technologies are based on devices and functional materials optimized toward their optimal performance, durability, and sustainability. This volume of *Advances in Inorganic Chemistry*, devoted to *Materials for Sustainable Energy*, describes the state of the art in solar to chemical and solar to electrical energy conversion systems, as well as the development of electricity storage systems and catalytic materials for the depollution of air.

Solar fuels, i.e., fuels produced as the result of solar to chemical energy conversion, can be produced using photocatalytic or photoelectrocatalytic approaches. The first one, described in the chapter by Janusz Nowotny et al., deals with the most commonly studied titanium dioxide materials. The authors show that the photocatalytic properties of TiO_2 and its solid solutions are determined by surface versus bulk defect disorder. Therefore, defect engineering may become a powerful tool for imposing optimized performance of semiconductor photocatalysts. Layered, 2D materials (graphitic carbon nitride, boron carbon nitride), described by Masakazu Anpo and coworkers, constitute an alternative group of photocatalysts capable of water splitting (hydrogen production) and carbon dioxide reduction. These materials offer visible light–induced activity, particular charge separation properties, and the highest specific surface area. Photocatalytic carbon dioxide reduction can be used not only in the production of solar fuels but also for a green synthesis of fine chemicals. This and other examples of photocatalytic synthesis of organic compounds driven by solar energy are discussed in the chapter by Wojciech Macyk et al. The authors classified these reactions according to their mechanisms, involving various types of reduction and oxidation half-reactions. Finally, solar hydrogen can be produced using a photoelectrochemical approach, as discussed by Marta Radecka et al. In their chapter the authors stress the influence of dimensionality of the catalysts (1-3D), as well as the formation of heterostructures on the performance of photoelectrodes.

Solar energy can be converted directly to electricity. Photoelectrochemical solar cells evolved from dye-sensitized solar cells and organic cells to perovskite cells, which attract growing attention. These devices and challenges related to the design of electron and hole transporting materials are described in the chapter by Mohammad Khaja Nazeeruddin and coworkers. The produced electricity can be stored in electrochemical capacitors, described by Elżbieta Frąckowiak et al. Recent progress in the development of electrochemical capacitors, often called *supercapacitors*, show a huge application potential for these devices. Also lithium ion batteries (chapter by Marcin Molenda et al.) offer a high capacity for storage of electrical energy. The development of electrode materials, mainly cathodes, is crucial for their further improvement.

Greener energy–related technologies are also associated with the reduction of air pollution. Elimination of NO_x and volatile organic compounds can be achieved by application of catalysts, as described in the last two chapters, authored by Lucjan Chmielarz, Piotr Kuśtrowski, and their coworkers. Efficient catalysts can be applied together with conventional systems (combustion of fuels) to make these technologies eco-friendly.

All described technologies are based on novel inorganic materials. Their development, features, and drawbacks are presented in nine chapters of this volume of *Advances in Inorganic Chemistry*. Acknowledging the excellent contributions of all authors to this volume, we are convinced that their work will not only be a source of knowledge but also an inspiration to readers of the Inorganic Chemistry community.

Rudi van Eldik
Editor of Advances in Inorganic Chemistry
Emeritus Professor of Inorganic Chemistry
University of Erlangen-Nuremberg, Germany
Professor of Inorganic Chemistry
Jagiellonian University
Kraków, Poland
May 2018

Wojciech Macyk
Co-Editor of this volume
Professor of Inorganic Chemistry
Jagiellonian University
Kraków, Poland

CHAPTER ONE

Defect Engineering of Photosensitive Oxide Materials. Example of TiO_2 Solid Solutions

Armand J. Atanacio*, Tadeusz Bak[§], Kazi A. Rahman[§] and Janusz Nowotny[§,1]

*Centre for Accelerator Science, Australian Nuclear Science and Technology Organisation, Kirrawee DC, NSW, Australia

[§]Solar Energy Technologies, School of Computing, Engineering and Mathematics, Western Sydney University, Penrith, NSW, Australia

[1]Corresponding author: E-mail: J.Nowotny@westernsydney.edu.au

Contents

Advances in Inorganic Chemistry, Volume 72
ISSN 0898-8838
https://doi.org/10.1016/bs.adioch.2018.05.006

Abstract

The imperative to protect the environment from increasingly apparent climate change imposes the urgent need to reduce the emissions of greenhouse gases to the atmosphere. This, consequently, results in intensification of research in the development of new materials and devices for the generation of energy that is environmentally clean. This work considers photosensitive oxide semiconductors for solar energy conversion by light-induced water oxidation. It has been documented that the performance of oxide semiconductors for solar-to-chemical energy conversion is determined by a range of defect-related properties, including the concentration of surface active sites, Fermi level, charge transport, electronic structure, and alignment of band edges with the energy level of the redox couple. The present work considers the research strategy in processing TiO_2-based semiconductors, which are the promising candidates for a new generation of solar materials. It is shown that the performance-related properties of TiO_2 and its solid solutions are determined by surface versus bulk defect disorder and the associated semiconducting properties. Therefore, the development of TiO_2-based materials with enhanced performance could be based on using defect engineering for imposing optimized bulk versus surface properties. In this work, we discuss a range of defect-related properties of TiO_2 and its solid solutions, such as electrical and optical properties and the related photocatalytic performance. We show that the phenomenon of segregation may be used as the technology for imposition of controlled surface versus bulk defect disorder that is required for processing the systems with optimized properties.

1. INTRODUCTION

It becomes increasingly apparent that human activities result in deterioration of the environment.[1] The main cause of environmental damage is related to the extensive use of fossil fuels in energy generation.

The associated increase in greenhouse gas emission leads to climate changes. Therefore, there is an increasingly urgent need to replace fossil fuel-based energy technologies with renewable-types of energy that are environmentally friendly.

The present chapter considers the research strategy in the development of high-performance oxide semiconductors for energy conversion devices. This work, which is focused on the application of defect engineering in processing TiO_2-based semiconductors with enhanced performance in solar-to-chemical energy conversion, considers the following approaches:

- the formation of basic oxide systems
- imposition of surface versus bulk defect-related properties that are required to achieve the desired performance
- characterization of basic properties
- determination of light-induced reactivity of the oxide systems with water
- derivation of theoretical models able to explain the effect of defect disorder on the performance
- verification of the theoretical models.

Increasing efforts in reducing climate changes impose the need to develop new types of energy technologies. The related approaches need new materials with specific properties required for construction of new energy conversion devices with enhanced performance. This, consequently, requires intensification of research addressing the problems that are critical in the engineering of energy materials for the conversion of renewable energy into operational types of energy, such as electrical energy, as well as new types of energy carriers, such as hydrogen.

The major player in renewable energy is solar energy that is the most widely available type of clean energy. The research, which aims at developing solar cells for the conversion of solar energy into electrical energy, is mainly based on silicon. The cost of photovoltaic (PV) electricity already becomes competitive with the costs of energy produced by power plants based on coal and natural gas.[2]

While the technology of silicon solar cells for production of electricity is well advanced, interest is growing in the development of solar cells for the conversion of solar energy into chemical energy. The common approach in this type of conversion is based on light-assisted water oxidation, including:

i. total water oxidation leading to water splitting and the formation of solar hydrogen fuel, and

ii. partial water oxidation leading to removal of toxic contaminants from water.

The present chapter, which considers the research strategy in processing photosensitive oxide semiconductors for water oxidation, is focused on TiO_2-based oxide semiconductors that are promising materials for a new generation of solar materials.[3] The related concepts of defect engineering derived for TiO_2 are applicable for alternative oxide materials as well.

2. PRESENT STATE AND FUTURE CHALLENGES

2.1 Solar Hydrogen Fuel

The intensive research aims at developing solar cells for the conversion of solar energy into chemical energy in the form of solar fuels. The ultimate goal is to develop photoelectrochemical cells (PECs) for water splitting into hydrogen, with oxygen as the by-product.[3–8] This technology is attractive for the following reasons:

1. The ultimate product is hydrogen fuel that is an environmentally clean fuel during all the stages of its generation and combustion.
2. The only driving force of the process is solar energy, which is a renewable energy.
3. The raw material for this technology is water, which is a renewable material.
4. The technology may be used for the production of fuel in remote country areas.

There is an international consensus in place to introduce hydrogen as the main energy carrier along with electricity on the global scale.[9,10] The related international partnerships are aimed at the development of the related technologies, such as hydrogen generation, hydrogen storage, hydrogen transport, and hydrogen distribution. The main advantage of hydrogen as a fuel is that its combustion leads to the formation of water vapor that is environmentally harmless. The technology of solar hydrogen is expected to replace the technology of hydrogen production from natural gas that is not environmentally clean since its by-product is CO_2, which is a greenhouse gas.

The impact of hydrogen generation technologies on the economy and the environment may be considered in terms of the direct costs of distribution and indirect costs associated with environmental damage. The related costs are represented in Fig. 1 in terms of the following components:

1. hydrogen production using the steam reforming of methane (SRM)[10–12]
2. hydrogen production using PECs with solar-to-chemical energy conversion efficiency (ECE) of 5%[11,12]

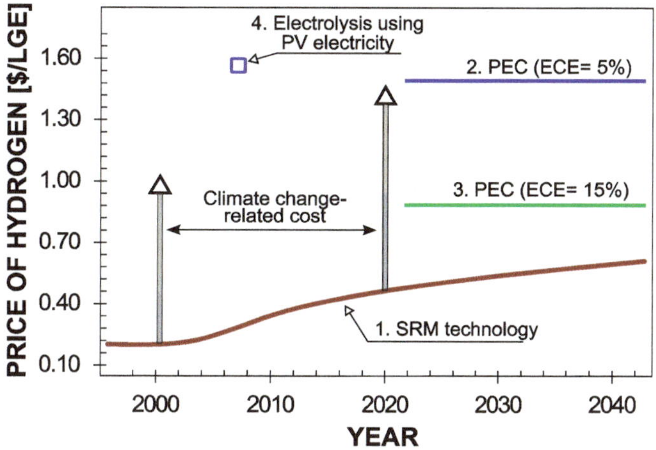

Figure 1 Comparison of hydrogen prices generated using: 1. steam reforming of methane (SRM),[10–12] 2. photoelectrochemical cell (PEC) operating at the energy conversion efficiency (ECE) equal to 5%),[11,12] 3. PEC operating at ECE=15%,[11,12] and 4. water electrolysis by PV electricity.[12] The vertical vectors represent the climate change-related costs according to Bockris[13] (where LGE denotes "liter of gasoline equivalent").

3. hydrogen production using PECs with the hypothetical ECE of 15%[11,12]
4. hydrogen production by water electrolysis using PV electricity[12]
5. the damage to the economy by climate changes.[13]

So far, the costs of hydrogen generation using SRM are the lowest in distribution. However, taking into account the substantial costs associated with the resulting climate change,[13] the PEC technology is a clear winner because it does not include the cost component related to environmental damage.

Intensive research aims at the development of materials for PECs, utilising one and two photoelectrodes.[8] The routes toward the development of high-performance PEC involve novel materials for (i) photoanodes, (ii) cathodes and photocathodes, as well as (iii) internal PV systems boosting the conversion efficiency.

The future challenges aim at better understanding the mechanism of light-induced reactivity at the solid/liquid interfaces for the anode and cathode of PECs.

2.2 Water Purification

The research route toward solar-to-chemical energy conversion involves the development of solar cells for partial water oxidation. The ultimate goal is the development of technology for water purification by removal of infectious microbial agents and toxic organic compounds from water using sunlight as the only driving force of the process.[14–18]

Shortage of clean drinking water on the global scale is a major humanitarian problem since waterborne diseases kill some 2 million people annually.[17,18] Water purification that aims at removal of microbiological agents from water requires energy that is too costly for developing countries. The technology of solar cells for partial water oxidation is the ultimate solution that could provide water treatment that is relatively inexpensive. The objectives of future challenges, aiming at its development, include identification of the key performance-related properties (KPPs) and better understanding the light-induced reactivity of anodic and cathodic sites of photocatalytic systems.

The critical reaction of water oxidation is the removal of electrons from water molecules. This requires the development of materials that are able to react with water, leading to either partial or total oxidation when exposed to sunlight. The performance of materials for these technologies is determined by a range of defect-related properties. Therefore, their processing could be based on defect engineering that aims at the imposition of specific defect disorder required to achieve the desired performance. Consequently, the main sections of this paper are preceded by brief definitions of basic terms used in considering the concepts of defect chemistry of TiO_2-based solid solutions. This paper also considers the concept of surface segregation that can be applied as the technology for the imposition of controlled surface versus bulk composition. These concepts and the related effects pave the way for a new engineering discipline, viz. defect engineering, which can be applied for the development of energy materials in general and photosensitive oxide semiconductors in particular.

3. IMPACT OF STRUCTURAL DEFECTS ON PROPERTIES AND PERFORMANCE OF ENERGY MATERIALS

This section considers the microscopic defects termed point defects[19,20] and their impact on properties and performance of oxide materials of nonstoichiometric composition. The concentration of these defects, which are thermodynamically reversible, depends on the temperature as well as the pressure and composition of the surrounding gas phase. In the case of metal oxides, the disorder of point defects can be imposed in a controlled manner at elevated temperatures (corresponding to gas/solid equilibrium) by the gas phase of controlled oxygen activity.

Perfect crystals, which are free of defects, do not exist. Crystals that exhibit a low concentration of defects are essentially not reactive at room temperature.[21] Accordingly, the properties of solids, including their reactivity and related charge transfer, are profoundly influenced by structural imperfections.

The classical approach of defect chemistry considers thermodynamically reversible defects in the bulk phase of inorganic compounds.[22–24] The related disorder models allow prediction and explanation of bulk properties, such as electrical properties. However, understanding the effect of defects on surface-related properties requires expansion of bulk defect chemistry into the surface layer.

Correct understanding of the properties and the performance of energy materials is possible only when their defect disorder is known. In other words, knowledge of defect disorder is essential in understanding reactivity and performance of materials in energy conversion.

The progress of defect chemistry so far is limited to the bulk phase where charge neutrality is required at any point within the periodic lattice. This research aims at the derivation of defect disorder models that may be used for predicting the defect-related properties for a range of nonstoichiometric compounds.[22–24]

4. DEFECT CHEMISTRY OF TITANIUM DIOXIDE

4.1 Basic Defect Equilibria

The classical approach of solid-state chemistry for crystalline solids forming periodic structures considers the properties mainly in terms of both chemical composition and structure. This approach has changed after it has been realized that inorganic solids exhibit nonstoichiometric compositions.[19,20] The discovery of nonstoichiometry resulted in the development of defect chemistry that considers the effect of structural defects on properties of materials, such as electrical conductivity, catalytic activity, and optical properties. Since knowledge of defect disorder is essential in understanding the properties of solids, defect chemistry is a part of modern solid-state science. The present section considers defect disorder of TiO_2 and its solid solutions in terms of defect chemistry.

Some progress has been recently achieved in defect chemistry of TiO_2, which is the promising compound in a range of applications, including solar energy conversion.[25] According to the present state of understanding, defect disorder of pure TiO_2 involves a wide range of intrinsic ionic defects, including oxygen vacancies, titanium vacancies, and titanium interstitials as well as both electrons and electron holes. The related defect disorder has been derived based on a range of defect-related properties, such as electrical conductivity,[26–29] thermoelectric power,[30–32] and thermogravimetry.[33,34] The derived defect disorder is described by defect equilibria (Table 1)[25] and the related defect diagrams.

The basic requirement for any defect disorder is charge neutrality, which for TiO_2 requires that:

$$2\left[V_O^{\bullet\bullet}\right] + 3\left[Ti_i^{\bullet\bullet\bullet}\right] + 4\left[Ti_i^{\bullet\bullet\bullet\bullet}\right] + p = n + 4\left[V_{Ti}^{''''}\right]. \quad (1)$$

The condition expressed by Eq. (1) includes only the predominant ionic defects. For TiO_2 involving extrinsic ions (donors and acceptors), condition (1) assumes the form:

$$2\left[V_O^{\bullet\bullet}\right] + 3\left[Ti_i^{\bullet\bullet\bullet}\right] + 4\left[Ti_i^{\bullet\bullet\bullet\bullet}\right] + \left[D^{\bullet}\right] + p = n + 4\left[V_{Ti}^{''''}\right] + [A'] \quad (2)$$

where $\left[D^{\bullet}\right]$ and $[A']$ denote the concentrations of ions in the cation sublattice, including singly ionized donor- and acceptor-type extrinsic ions, respectively. Eq. (2) applies for both intentionally formed solid solutions as well as TiO_2 doped unintentionally with impurities.

Since the equilibria (1)–(4) in Table 1 involve $p(O_2)$ as the parameter, combination of the equilibrium constants and condition (2) allows combining the concentration of quasi-free electrons and oxygen activity:

$$n^5 + An^4 - K_i n^3 - 2K_1 P^{1/2} n^2 - 3K_2 P^{-1} n - 4K_3 P^{-1} = 0 \quad (3)$$

where K_1, K_2, K_3, and K_i are the equilibrium constants that are defined in Table 1; A is the effective concentration of acceptors:

$$A = 4\left[V_{Ti}^{''''}\right] + [A'] - \left[D^{\bullet}\right] \quad (4)$$

and

$$P = p(O_2) \quad (5)$$

Expression (3) involves the concentration of both intrinsic and extrinsic defects as well as oxygen activity. Therefore, Eq. (3) may be used for derivation of defect disorder representing the effect of oxygen activity on the concentration of intrinsic defects for pure TiO_2 and its solid solutions with donors and acceptors.

Table 1 Defect Equilibria in Pure TiO_2 and the Related Equilibrium Constants According to the Kröger-Vink Notation,[35]

	Defect Reaction	Equilibrium Constant	ΔH° [kJ/mol]	ΔS° [J/(mol.K)]
1	$O_O^x \Leftrightarrow V_O^{\bullet\bullet} + 2e' + 1/2O_2$	$K_1 = [V_O^{\bullet\bullet}]n^2p(O_2)^{1/2}$	493.1	106.5
2	$Ti_{Ti}^x + 2O_O^x \Leftrightarrow Ti_i^{\bullet\bullet\bullet} + 3e' + O_2$	$K_2 = [Ti_i^{\bullet\bullet\bullet}]n^3p(O_2)$	879.2	190.8
3	$Ti_{Ti}^x + 2O_O^x \Leftrightarrow Ti_i^{\bullet\bullet\bullet\bullet} + 4e' + O_2$	$K_3 = [Ti_i^{\bullet\bullet\bullet\bullet}]n^4p(O_2)$	1025.8	238.3
4	$O_2 \Leftrightarrow V_{Ti}^{''''} + 4h^{\bullet} + 2O_O^x$	$K_4 = [V_{Ti}^{''''}]p^4p(O_2)^{-1}$	354.5	−202.1
5	$nil \Leftrightarrow e' + h^{\bullet}$	$K_i = np$	222.1	−44.6
	$\ln K = [(\Delta S^o)/R] - [(\Delta H^o)/RT]$			

Where n and p Denote the Concentration of Electrons and Electron Holes, Respectively; $p(O_2)$ is Oxygen Activity; Square Brackets Denote the Concentration of the Species Within the Brackets; ΔH° and ΔS° Denote the Enthalpy and Entropy Terms, Respectively; R is the gas Constant; and T is the Absolute Temperature

Eq. (3) allows the determination of defect disorder at constant concentration of titanium vacancies that is defined by Eq. (4). This formula applies when TiO_2 is equilibrated during a relatively short period of time—commonly limited to 1 day. Such equilibration leads to the imposition of the *operational equilibrium* rather than true gas/solid equilibrium.[25]

The gas/solid equilibrium, which requires prolonged equilibration (lasting several months at elevated temperatures) allows the imposition of the concentration of titanium vacancies corresponding to equilibrium. Then the effect of oxygen activity on the concentration of electrons may be described by the following expression:

$$4K_4K_1^{-4}P\mathrm{n}^8 + \mathrm{n}^5 - [D^{\bullet}] - [A']\mathrm{n}^4 - K_i\mathrm{n}^3 - 2K_1P^{-1/2}\mathrm{n}^2 - 3K_2P^{-1}\mathrm{n} - 4K_3P^{-1} = 0 \quad (6)$$

Eq. (6) may be used for derivation of defect disorder corresponding to gas/solid equilibrium in respect to all defects, including titanium vacancies.

The *operational equilibrium* corresponds to a stationary state established after the concentration of fast defects, including oxygen vacancies and titanium interstitials, reach their equilibrium values. However, the concentration of titanium vacancies in the operational equilibrium is usually remote from its equilibrium state owing to their extremely low mobility.

4.2 Defect Diagrams

This section considers defect disorder diagrams for pure TiO_2 that may guide the processing of TiO_2 with controlled concentration of intrinsic defects and the related semiconducting properties. The specimens may be processed by annealing in the gas phase of controlled oxygen activity and subsequent fast quenching. Fig. 2 shows the defect disorder diagram for pure TiO_2 at 1273 K in the range 10^{-12} Pa < $p(O_2)$ < 10^5 Pa, which is derived from Eq. (6).

As seen, the predominant ionic defects in pure TiO_2 in the entire range of oxygen activities are oxygen vacancies, which are compensated by electrons and titanium vacancies at low and high oxygen activities, respectively. The interstitial titanium ions, which represent minor-type defects, have a tangible effect on properties mainly in strongly reducing conditions. As also seen, the n-p transition point, corresponding to the intersection of the log n versus log $p(O_2)$ and log p versus log $p(O_2)$ lines, is at $p(O_2) \approx 10^4$ Pa.

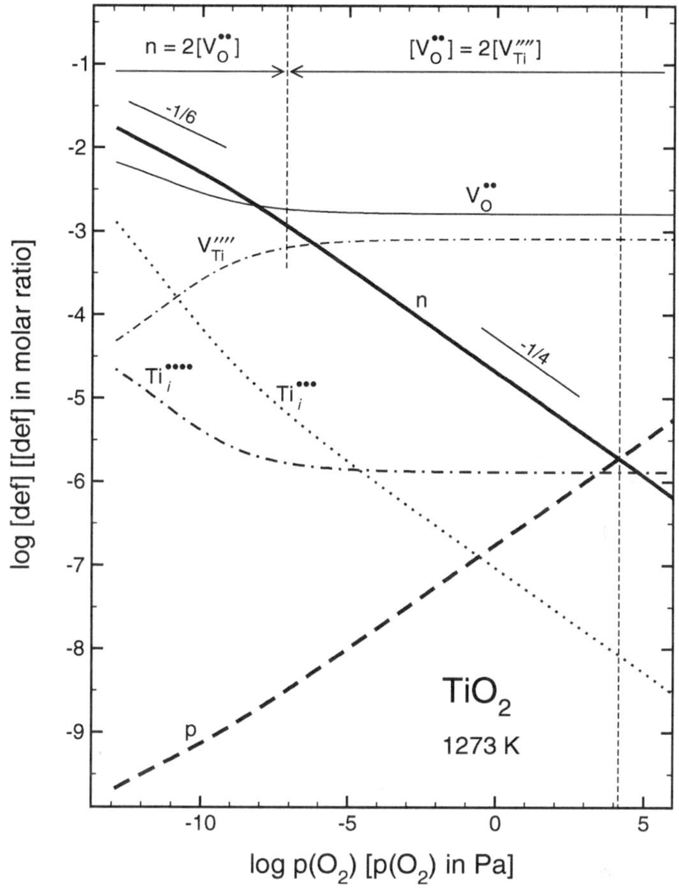

Figure 2 Defect disorder diagram for pure TiO_2 annealed at 1273 K.

According to the diagram in Fig. 2, the semiconducting properties of pure TiO_2 may be imposed in a controlled manner by annealing the TiO_2 specimen in the gas phase of well-defined oxygen activity. As seen, annealing in the gas phase of an oxygen activity that is lower or higher than $p(O_2) = 10^4$ Pa and subsequent quenching results in the formation of n- or p-type TiO_2, respectively. Consequently, oxygen activity in the gas phase may be used as a variable for imposition of desired semiconducting properties.

The effect of oxygen activity on the concentration of defects depends on the temperature of annealing. As seen in Fig. 3, annealing TiO_2 at 973 K results in shifting the n-p transition point toward the level of $p(O_2) \approx 10$ Pa.

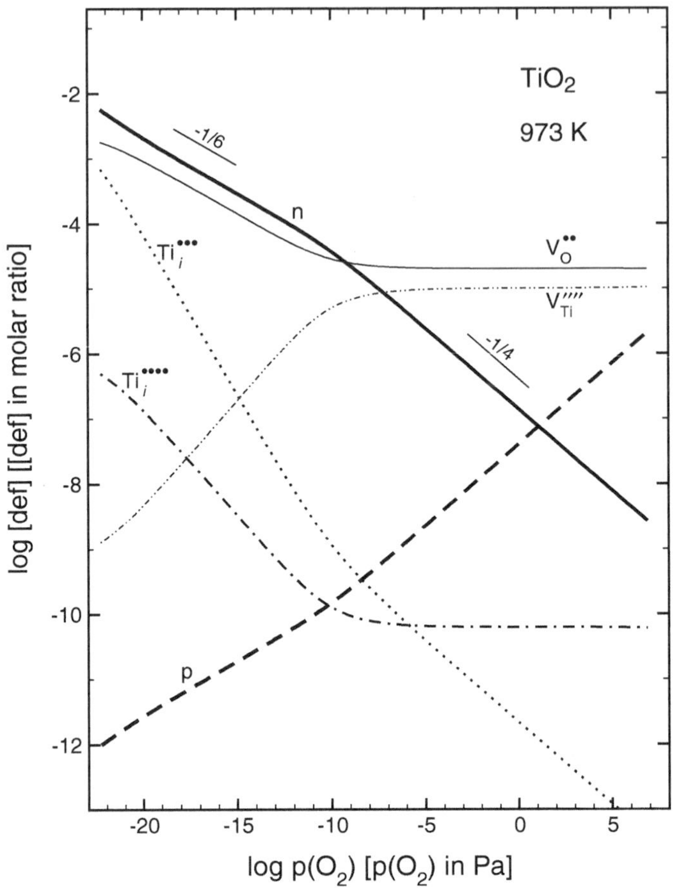

Figure 3 Defect disorder diagram of pure TiO_2 annealed at 973 K.

According to equilibria (1)—(5) in Table 1, both oxygen and titanium sublattices of TiO_2 are defected. Moreover, some of the titanium ions are present in interstitial sites. Therefore, the real chemical composition of pure TiO_2 may be considered in terms of the modules representing the titanium sublattice, the interstitial species, and the oxygen sublattice:

$$A_{\mathrm{Ti}}B_{\mathrm{i}}C_{\mathrm{O}} \tag{7}$$

where the modules A_{Ti}, B_{i}, and C_{O} are represented by the following specific formulas:

$$A_{\mathrm{Ti}} = \left[\left(\mathrm{Ti}_{\mathrm{Ti}}^{x}\right)_{a}\left(\mathrm{Ti}_{\mathrm{Ti}}'\right)_{b}\left(V_{\mathrm{Ti}}''''\right)_{c}\right] \tag{8}$$

$$B_i = \left[\left(\mathrm{Ti}_i^{\bullet\bullet\bullet}\right)_d\left(\mathrm{Ti}_i^{\bullet\bullet\bullet\bullet}\right)_e\right] \tag{9}$$

$$C_O = \left[\left(\mathrm{O}_O^x\right)_f\left(\mathrm{V}_O^{\bullet\bullet}\right)_g\left(\mathrm{O}_O^{\bullet}\right)_h\right] \tag{10}$$

Since each combination of the lattice species must satisfy (1) the charge neutrality condition and (2) the law of conservation of the lattice sites ratio, the concentrations of the lattice species are interdependent. Consequently, the real chemical formula of pure TiO_2, including the basic lattice species in both titanium and oxygen sublattices as well as the structural defects (involving eight different species), is rather complex. It has been documented that the properties of TiO_2 can be properly explained only when all these species are taken into account.[25]

4.3 Charge Transport

4.3.1 Electronic Charge Carriers

Charge transport may be represented by the electrical conductivity that is the product of the mobility and concentration terms for all charge carriers:

$$\sigma = e\left(\mathrm{n}\mu_n + \mathrm{p}\mu_p\right) + \sigma_i \tag{11}$$

where e is an elementary charge, μ is the mobility, σ_i is the ionic conductivity component, and the subscripts n and p are related to electrons and holes, respectively. Since the mobility of electronic charge carriers is markedly larger than that of ionic species, the electrical conductivity of metal oxides at room temperature is practically determined by the first term in Eq. (11).

The mobility of electrons and electron holes for single crystals and polycrystalline specimens of pure TiO_2 at elevated temperatures (1073−1323 K) are represented in Fig. 4.[36] Since the mobility of electrons and holes differ, TiO_2 is considered to be an asymmetric semiconductor.

The mobility of electrons of TiO_2 at room temperature, reported by Cronemayer,[37] is approximately $\mu_n = 10^{-4}$ m^2/V s.

4.3.2 Transport During Equilibration

When oxygen activity or temperature is changed over an initially equilibrated oxide specimen, the gas/solid system has a tendency to assume a new equilibrium state. The equilibration, which involves propagation of

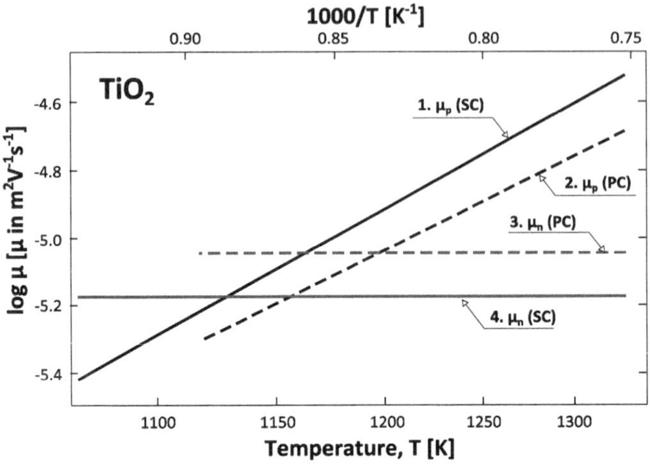

Figure 4 Effect of temperature on the mobility of electrons (μ_n) and electron holes (μ_p) for both single crystal (*SC*) and polycrystalline specimens (*PC*) of pure TiO_2,[36] where n and p in subscripts are related to electrons and holes, respectively.

new oxygen activity from the surface toward the bulk phase, leads to removal of the imposed chemical concentration gradients of defects. The equilibration is determined by chemical diffusion. The chemical diffusion coefficient, D_{chem}, is the rate constant of this process that is a parameter in the diffusion equation expressed by Fick's second law. The solution of this equation for the case of the brick-shaped specimen has been derived by Neuman.[38]

D_{chem} of metal oxides combines the transport of defects in both cation and oxygen sublattices.[22,24,25] The physical meaning of D_{chem} is well defined when one type of defect is involved, as this is the case for NiO.[22] However, for multidefect oxides, such as TiO_2, the physical sense of D_{chem} is complex for the following reasons:

1. D_{chem} reflects the transport of several lattice species taking place simultaneously with different mobilities.
2. The nonstoichiometry is not well defined because the concentration of titanium vacancies frequently corresponds to the operational equilibrium rather than the gas/solid equilibrium.

4.3.2.1 Simultaneous Transport of Several Lattice Species

Chemical diffusion involves the transport of several mobile species. The effect of defects on the experimentally determined D_{chem} depends on their concentrations and mobilities. The picture is relatively simple when the transport is related to two species of entirely different mobilities. Then the operational equilibration is determined by the faster species and the gas/solid equilibration is rate controlled by the slower species.

The progress of chemical diffusion may be monitored using the measurements of defect-related properties, such as electrical conductivity. When the oxide crystal involves two kinds of defects of very different mobilities, the electrical conductivity that changes during the re-equilibration exhibits two-step kinetics involving two stages, as it is the case for pure TiO_2 (Fig. 5).

The equilibration kinetics for pure TiO_2 during oxidation between 42 and 75 kPa at 1323 K is represented in Fig. 5.[39]

As seen, oxidation of TiO_2 at 75 kPa initially exhibits fast kinetics, leading to an operational equilibrium within 1 h, and subsequent slow kinetics, leading ultimately to a gas/solid equilibrium that is achieved after prolonged oxidation within 600—800 h. The fast and slow kinetics are related to the

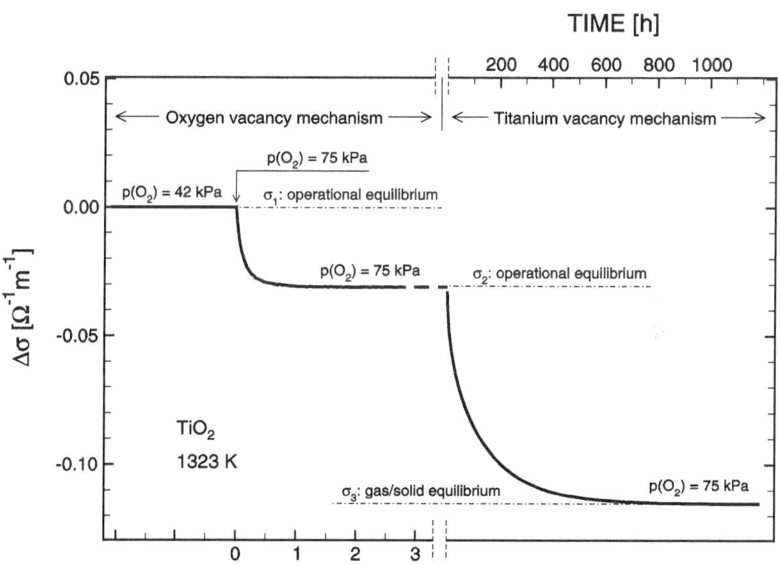

Figure 5 Changes in the electrical conductivity during oxidation-induced double-stage equilibration of pure TiO_2 single crystal, involving the initial fast changes and subsequent slow kinetics while $p(O_2)$ is increased from 42 to 75 kPa at 1323 K, where σ_1, σ_2, and σ_3 denote the electrical conductivity before oxidation, at operational equilibrium, and at the gas/solid equilibrium, respectively.[39]

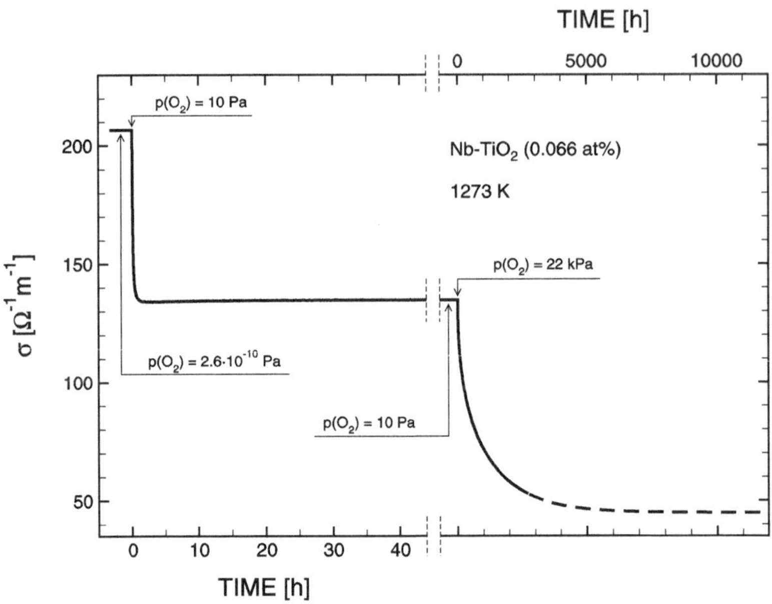

Figure 6 Changes in the electrical conductivity during oxidation-induced equilibration of Nb-doped TiO_2 single crystals, involving the initial fast changes when $p(O_2)$ is increased from 2.6×10^{-10} Pa to 10 Pa, leading to the operational equilibrium, and subsequent slow changes when $p(O_2)$ is increased from 10 Pa to 22 kPa at 1273 K, resulting in the gas/solid thermodynamic equilibrium.[41]

transport determined by oxygen vacancies and titanium vacancies, respectively.[39] Similar two-step kinetics has been reported by Lee and Yoo[40] for the equilibration of TiO_2 in the ranges determined by oxygen vacancies and titanium interstitials.

The consecutive equilibration kinetics during oxidation of Nb-doped TiO_2 at 1273 K, involving the initial fast kinetics, leading to the operational equilibrium at low $p(O_2)$ and the subsequent slow kinetics at higher $p(O)_2$, leading ultimately to the thermodynamic gas/solid equilibrium is represented in Fig. 6.[41]

As seen, oxidation at $p(O_2) = 10$ Pa results in a decrease in the electrical conductivity that is reflective of vanishing oxygen vacancies to the level of their equilibrium concentration at this oxygen activity. The subsequent oxidation at $p(O_2) = 22$ kPa, which is rate controlled by the transport of titanium vacancies, results in the imposition of the gas/solid equilibrium, which corresponds to the following charge neutrality:

$$4\left[V_{Ti}^{''''}\right] + n = \left[Nb_{Ti}^{\bullet}\right] + 2\left[V_{O}^{\bullet\bullet}\right] + 3\left[Ti_{i}^{\bullet\bullet\bullet}\right] + 4\left[Ti_{i}^{\bullet\bullet\bullet\bullet}\right] + p \qquad (12)$$

4.3.2.2 Impact of Titanium Vacancies on Nonstoichiometry

The nonstoichiometry of metal oxides, such as TiO_2, is well defined at elevated temperatures when the system is in gas/solid equilibrium. Then the concentration of all defects and the related nonstoichiometry is well defined by the equilibrium. However, the picture in relation to titanium vacancies is more complex for the following reasons:

1. The commonly reported oxidation\reduction experiments, lasting several hours, allow imposition of the operational equilibrium when the concentration of fast defects (oxygen vacancies and titanium interstitials) corresponds to the thermodynamic equilibrium. Under these conditions, however, the thermodynamic equilibrium for reaction (4) in Table 1 is not reached.
2. The prolonged oxidation may lead to the imposition of the equilibrium concentration of titanium vacancies. However, the determination of the time needed to reach equilibrium is difficult since the changes in concentration versus time are very slow.

As a consequence of (1) and (2), the reported data for TiO_2 and its solid solutions are not reproducible.

4.3.2.3 Experimental Data on Chemical Diffusion for TiO_2

The reported data on the chemical diffusion coefficient for pure and niobium-doped TiO_2, including single crystals and polycrystalline specimens, can be grouped according to the predominant transport mechanism, including (1) oxygen vacancy mechanism and (2) titanium vacancy mechanism. The effect of temperature on the D_{chem} is represented in Fig. 7.[41]

The oxygen vacancy mechanism involves removal of concentration gradients in the TiO_2 lattice through diffusion of oxygen within the oxygen sublattice. The transport of oxygen vacancies, which are fast moving species,[25] leads to removal of oxygen concentration gradients in the oxygen sublattice. The titanium vacancy mechanism consists of the removal of concentration gradients in the TiO_2 lattice through diffusion of cations within the titanium sublattice. However, since the Schottky-type defect disorder and the related charge neutrality condition must be respected within the crystal:

$$nil \leftrightarrow V_{Ti}^{''''} + 2V_{O}^{\bullet\bullet} \tag{13}$$

$$K_S = \left[V_{Ti}^{''''}\right]\left[V_{O}^{\bullet\bullet}\right]^2 \tag{14}$$

the transport results in the removal of concentration gradients in both sublattices.

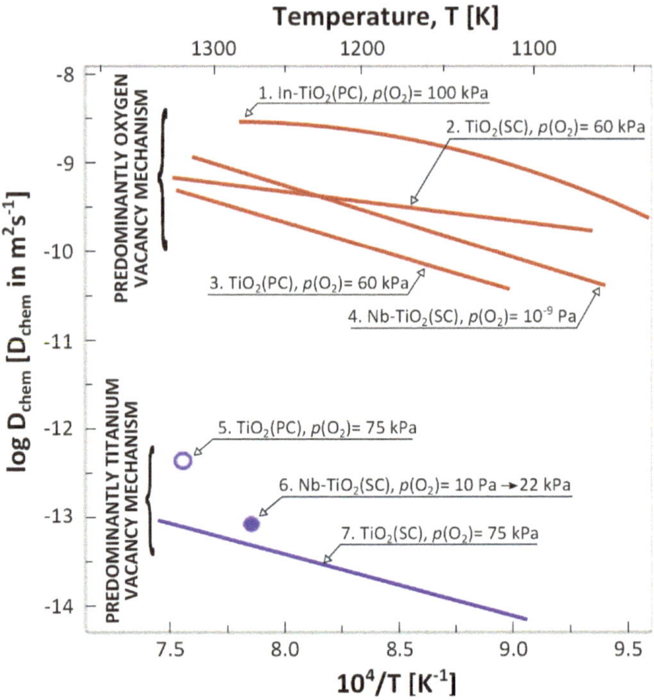

Figure 7 The chemical diffusion for single crystals and polycrystalline specimens of pure, indium- and niobium-doped TiO_2 in the gas phase of controlled oxygen activity, where $p(O_2)$ is oxygen activity, and *PC* and *SC* denote polycrystalline specimen and single crystal, respectively.

The diffusion coefficients for each group are scattered within 1 order of magnitude; however, the difference between these groups is approximately 4 orders of magnitude. The diffusion data in Fig. 7 are discussed below.

4.3.2.4 In-Doped TiO_2

The D_{chem} for In-doped TiO_2 is the highest in the range. This is the consequence of indium incorporation leading to the formation of oxygen vacancies:[42]

$$In_2O_3 \leftrightarrow 2In'_{Ti} + V_O^{\bullet\bullet} + 3O_O^x \quad (15)$$

As a consequence, In-doped TiO_2 exhibits a higher concentration of oxygen vacancies than pure TiO_2 annealed in comparable $p(O_2)$.

4.3.2.5 Pure TiO_2

The D_{chem} for pure TiO_2 single crystal (line 2) and the polycrystalline specimen (line 3), both annealed at $p(O_2) = 60$ kPa, are similar. These results were obtained from the two-step kinetics in the regime determined by oxygen vacancies. The results represented by line 7 and point 5 were determined under almost identical conditions. In this case, however, the kinetics is rate controlled by titanium vacancies.

4.3.2.6 Nb-Doped TiO_2

The D_{chem} for Nb—TiO_2, which are represented by line 4 and point 6, is determined from the subsequent experimental runs that are reflective of the oxygen vacancy and titanium vacancy mechanisms, respectively.[41]

The set of data in Fig. 7 is reflective of a substantial effect of the predominant defect disorder imposed by different processing conditions on the diffusion kinetics. These results indicate that defect engineering may be applied in the modification of the chemical diffusion data to the extent of the imposition of the predominant diffusion mechanism.

5. EFFECT OF OXYGEN ACTIVITY ON SEMICONDUCTING PROPERTIES OF TiO_2

5.1 Intrinsic Defects

The most commonly measured defect-related properties, which can be used in the determination of defect diagrams, are the electrical properties. These can be applied to in situ monitoring of the change in defect disorder at elevated temperatures, while the studied specimen is approaching the gas/solid equilibrium. The related data can be used in the determination of the defect disorder in equilibrium. Additional information can be obtained from simultaneous measurements of both electrical conductivity and thermoelectric power, which provide complementary insight on the concentration and the mobility terms. The effect of oxygen activity on the electrical conductivity and thermoelectric power of amphoteric oxide semiconductors, such as TiO_2, is represented in Fig. 8.

The slope $\partial \log\sigma / \partial \log p(O_2)$ is a defect-related quantity. The experimentally determined slope can serve as verification of theoretical defect disorder

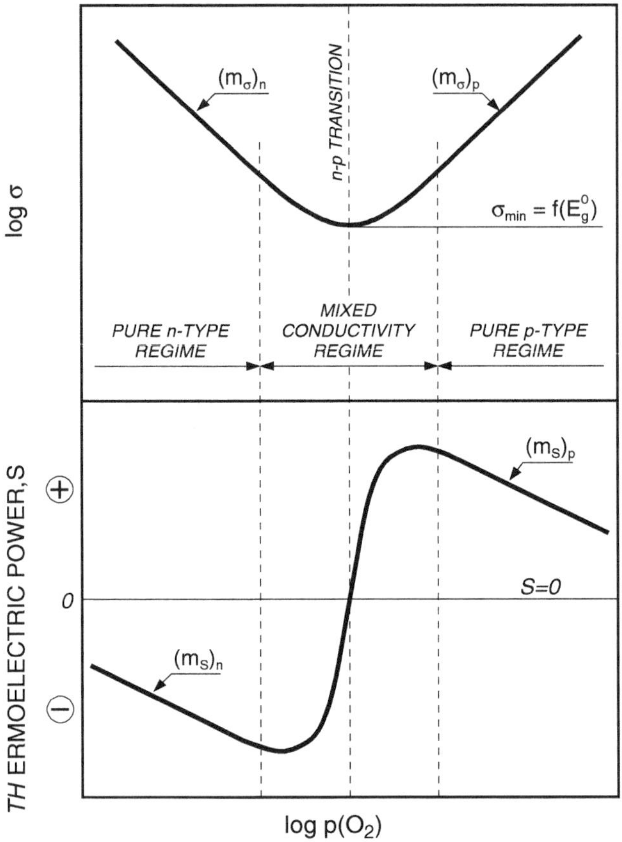

Figure 8 Schematic representation of the effect of oxygen activity on simultaneous changes of the electrical conductivity and thermoelectric power for an amphoteric oxide semiconductor. The parameters $(m_\sigma)_n$, $(m_\sigma)_p$, σ_{min}, $(m_S)_n$, and $(m_S)_p$ are defined by Eqs. (16)–(18).

models. The effect of $p(O_2)$ on the electrical conductivity, σ, of n- and p-type semiconductors may be expressed by the respective expressions:

$$\frac{1}{(m_\sigma)_n} = \frac{\partial \log \sigma}{\partial \log p(O_2)} = \frac{\partial \log n}{\partial \log p(O_2)} + \frac{\partial \log \mu_n}{\partial \log p(O_2)} \tag{16}$$

$$\frac{1}{(m_\sigma)_p} = \frac{\partial \log \sigma}{\partial \log p(O_2)} = \frac{\partial \log p}{\partial \log p(O_2)} + \frac{\partial \log \mu_p}{\partial \log p(O_2)} \tag{17}$$

where the mobility terms μ_n and μ_p are generally assumed as independent of $p(O_2)$. Therefore, the respective derivatives can be neglected. Similar slopes may be considered for the effect of $p(O_2)$ on thermoelectric power, S (Fig. 8):

$$\frac{1}{m_S} = \frac{k}{e} \frac{\partial S}{\partial \ln p(O_2)} \tag{18}$$

where k is the Boltzmann constant.

The slopes of $1/m_\sigma$ and $1/m_S$ may be used for derivation of defect disorder models only within pure n- or p-type regimes where the electrical properties are determined by one type of electronic charge carrier.

According to the diagrams in Figs. 2 and 3, the total concentration of electronic charge carriers (electrons and electron holes) is the lowest at the n-p transition point. Then, at comparable mobility of charge carriers, the electrical conductivity assumes a minimum value.

Becker and Frederikse[43] reported the relationship between the minimum of the electrical conductivity and the bandgap for amphoteric semiconductors, such as TiO_2:

$$\sigma_{\min} = 2e\left(\mu_n \mu_p N_n N_p\right)^{1/2} \exp\frac{\beta}{2k} \exp\left[-\frac{E_g^o}{2kT}\right] \tag{19}$$

where

$$E_g = E_g^o - \beta T \tag{20}$$

where β is the temperature coefficient of the bandgap.

Using simultaneous measurements of the electrical conductivity and thermoelectric power, Chen et al.[44] determined the activation energy for the mobility of electronic charge carriers by measuring the electrical conductivity at $S =$ const. Then the effect of temperature on the electrical conductivity may be represented by the expression:

$$\left(\frac{\partial \ln \sigma}{\partial 1/T}\right)_S = \left(\frac{\partial \ln \mu_n}{\partial 1/T}\right)_{N_p/p} = T - \frac{\Delta H_m}{k} \tag{21}$$

where ΔH_m is the activation energy of the mobility term. They have shown that the component $\partial \ln\sigma/\partial(1/T)$ at constant S is equal to zero when the mobility term is determined by the scatter on phonons (band mechanism). However, the negative value of the component $\partial \ln\sigma/\partial(1/T)$ indicates that charge transport takes place according to the hopping mechanism via small polarons and the related ΔH_m is the activation energy for hopping. This approach may be used as a test on the charge transport mechanism.

The electrical conductivity of amphoteric semiconductors, such as TiO_2, at elevated temperatures is influenced by all charge carriers, including electrons, electron holes, and ions. Then we have:

$$\sigma = \sigma_n + \sigma_p + \sigma_i \tag{22}$$

Taking into account Eqs. (16), (17), and (22), we obtain:

$$\sigma = \sigma_n^o p(O_2)^{-1/m_\sigma} + \sigma_p^o p(O_2)^{1/m_p} + \sigma_i \tag{23}$$

The individual electrical conductivity components may be determined by fitting the parameters of Eq. (23) to the experimental data of the electrical conductivity as a function of oxygen activity.

The components of the electrical conductivity of pure TiO_2 single crystals at 1073 and 1323 K are shown in Fig. 9.[45]

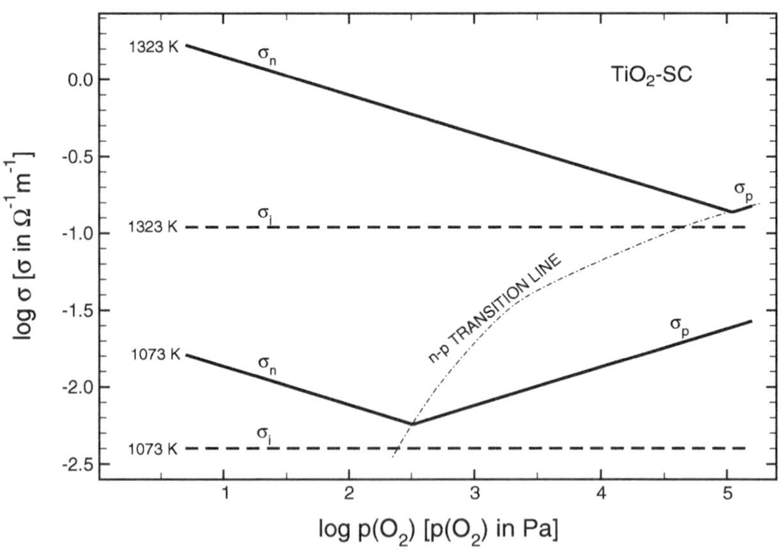

Figure 9 Effect of oxygen activity on the electrical conductivity components for TiO_2 single crystal (*SC*) at 1073 and 1323 K.[45]

As seen, the ionic conductivity component assumes a comparable value with the electronic conductivity components at the n-p transition point. The electrical conductivity at elevated temperatures can be considered as materials-related data. It has been shown that the electrical conductivity assumes different values for single crystals and polycrystalline specimens under the same thermodynamic conditions as shown in Fig. 10.

As seen, the slope $\partial \log\sigma / \partial \log p(O_2)$ assumes $-1/4$ and $-1/6$ under oxidizing and reducing conditions, respectively, which is consistent with the theoretical model shown in Fig. 2. As also seen, the electrical conductivity values in the n-type regime are higher for the polycrystalline specimen

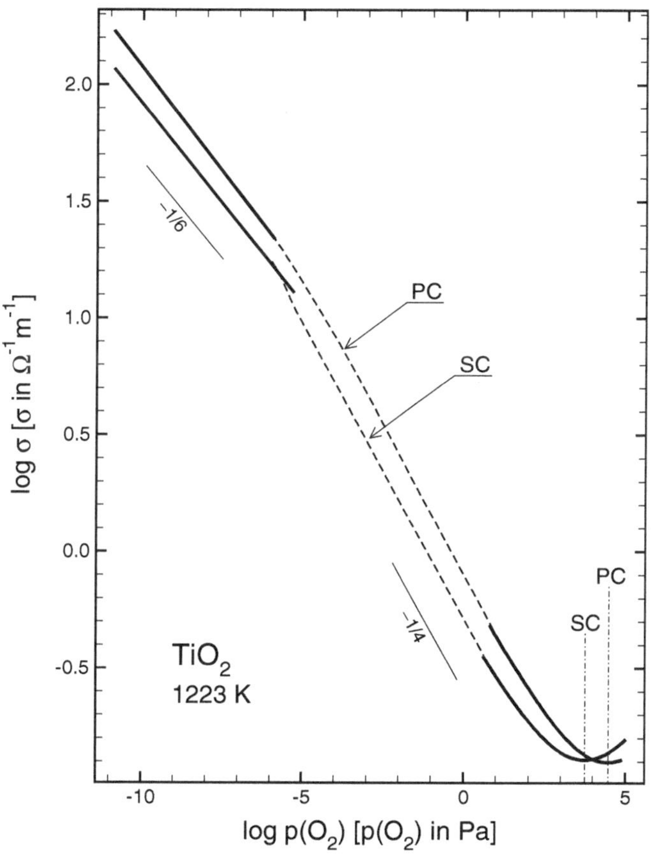

Figure 10 Effect of oxygen activity on the electrical conductivity for pure TiO_2, including single crystal (*SC*) and polycrystalline (*PC*) specimens at 1223 K.

than for the single crystal. This effect indicates that the grain boundaries of the polycrystalline specimens are enriched in donor-type defects, such as oxygen vacancies or titanium interstitials or both, leading in consequence to higher concentrations of electrons and higher electrical conductivity in the n-type regime.

5.2 Extrinsic Defects

The mechanism of incorporation of extrinsic ions, forming donors and acceptors, is profoundly influenced by the oxygen activity of the gas phase during annealing at elevated temperatures. For example, the mechanism of niobium (donor) incorporation into the TiO_2 lattice in reducing and oxidizing atmosphere may be represented by the following reactions, respectively:[25]

$$Nb_2O_5 \rightarrow 2Nb^{\bullet}_{Ti} + 2e' + 4O^{x}_{O} + \frac{1}{2}O_2 \quad (24)$$

$$2Nb_2O_5 \rightarrow 4Nb^{\bullet}_{Ti} + V^{''''}_{Ti} + 10O^{x}_{O} \quad (25)$$

The related simplified lattice charge neutrality conditions can be expressed by the following forms, respectively:

$$\left[Nb^{\bullet}_{Ti}\right] = n \quad (26)$$

$$\left[Nb^{\bullet}_{Ti}\right] = 4\left[V^{''''}_{Ti}\right] \quad (27)$$

As seen, the incorporation mechanism, the related charge neutrality, and the resulting effect on the semiconducting properties depend on the oxygen activity.

The effect of oxygen activity on the mechanism of chromium (acceptor) incorporation under reducing and oxidizing conditions may be represented by the following reactions, respectively:

$$Cr_2O_3 \leftrightarrow 2Cr'_{Ti} + 3O^{x}_{O} + V^{\bullet\bullet}_{O} \quad (28)$$

$$Cr_2O_3 + \frac{1}{2}O_2 \leftrightarrow 2Cr'_{Ti} + 4O^{x}_{O} + 2h^{\bullet} \quad (29)$$

The related charge neutrality must satisfy the following conditions, respectively:

$$\left[Cr'_{Ti}\right] = 2\left[V^{\bullet\bullet}_{O}\right] \quad (30)$$

$$p = \left[Cr'_{Ti}\right] \tag{31}$$

The effect of singly ionized donors and acceptors on the concentration of electrons in TiO_2 after annealing in the gas phase of controlled oxygen activity is represented in Fig. 11. Derivation of the effect of extrinsic donors and acceptors requires taking into account the concentration of titanium vacancies, which are practically immobile intrinsic defects.

As seen, donors and acceptors lead to an increase and decrease in the concentration of electrons, respectively. The effect of doping diminishes with decreasing $p(O_2)$.

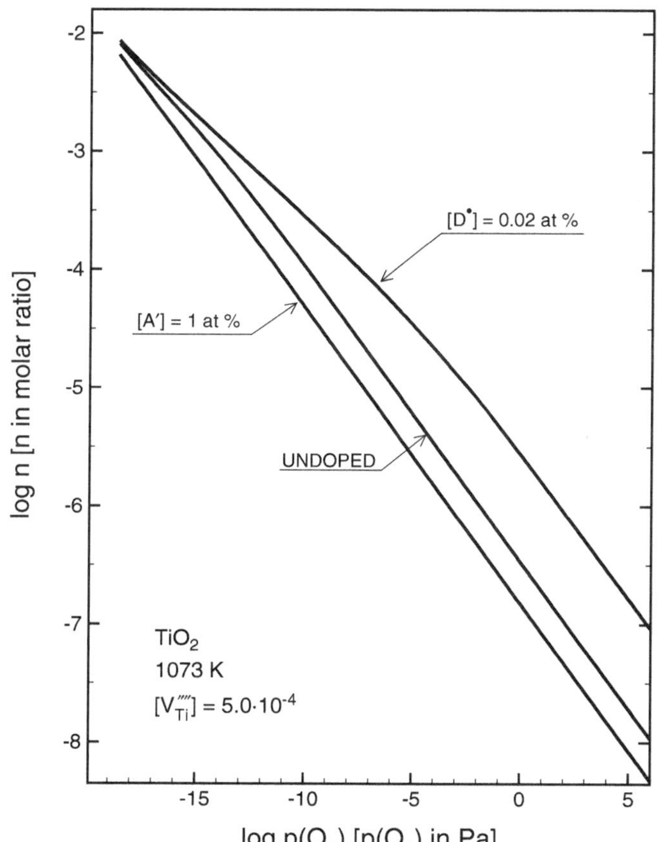

Figure 11 Effect of oxygen activity on the concentration of electrons for pure TiO_2 and its solid solutions with donors ($D^{\bullet}$) and acceptors (A') at 1073 K.

The effect of the annealing temperature on the n-p transition point of TiO_2 is shown in Fig. 12 (the effective concentration of acceptors, A, is defined by Eq. (4)).

As seen in Fig. 12, the increase in the apparent concentration of acceptors at constant temperature results in a shift of the n-p transition point to lower oxygen activity.

The extrinsic defects may be introduced into the host lattice intentionally by doping and unintentionally during processing. In the latter case, the extrinsic defects are termed contaminants. These may have an effect on properties when their concentrations are larger than a certain critical value. Knowledge of the critical level is essential for proper interpretation of the effect of doping on properties.

Fig. 13 represents the effect of the concentration of singly ionized acceptor-type ions on the electronic component of the electrical conductivity and the resulting data of their critical concentrations A_C.

The critical concentration of acceptors, A_C, has been determined from the onset of the dependence of electrical conductivity as a function of concentration at different oxygen activities in the temperature range 873–1273 K. The data in Fig. 13B,D,F indicate that A_C, determined from the onset, increases with annealing temperature from several ppm

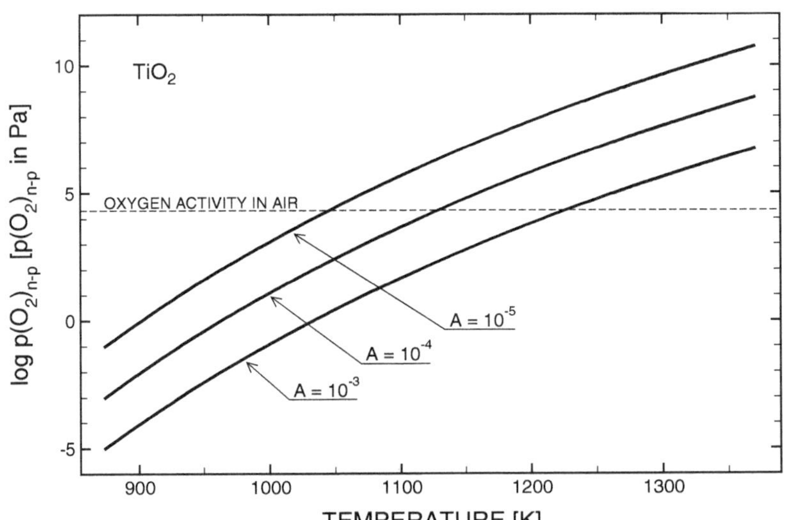

Figure 12 Effect of annealing temperature on the n-p transition point for TiO_2 and its solid solutions with acceptors as defined by the parameter A (Eq. 4).

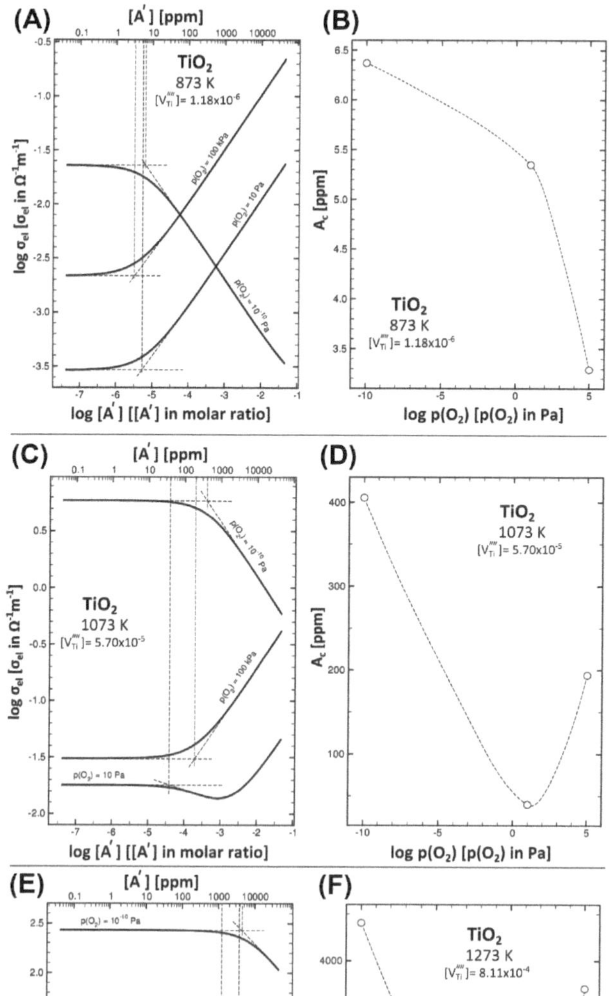

Figure 13 The electronic component of the electrical conductivity of pure TiO_2, annealed under controlled oxygen activity, as a function of the concentration of singly ionized acceptors at (A) 873 K, (B) 1073 K, and (C) 1273 K, as well as the related critical concentration of acceptors, A_C, as a function of oxygen activity at (D) 873 K, (E) 1073 K, and (F) 1273 K. The vertical *dashed lines* in (A), (C), and (E) represent the related critical concentrations of acceptors.

at 873 K to several thousand ppm at 1273 K. Moreover, A_C depends markedly on the oxygen activity. The A_C data are reflective of the level of acceptor-type impurities below which their effect on the performance can be ignored.

6. JONKER FORMALISM

Jonker[46] proposed a quantitative analysis of both electrical conductivity and thermoelectric power determined simultaneously that can be applied for amphoteric semiconductors, such as TiO_2, in equilibrium. The analysis allows the determination of the semiconducting properties, such as the bandgap and the mobility terms. The Jonker formalism involves the following set of equations:

$$S = \pm B\sqrt{1 - \frac{\sigma_{\min}^2}{\sigma^2} - \frac{k}{e}\ln\frac{\sigma}{\sigma_{\min}}\left(1 - \pm\sqrt{1 - \frac{\sigma_{\min}^2}{\sigma^2}}\right)} + D \quad (32)$$

where

$$B = \frac{k}{2e}\left(\frac{E_g}{kT} + A_n + A_p\right) \quad (33)$$

$$D = \frac{k}{2e}\ln\frac{\mu_p N_p e^{A_p}}{\mu_n N_p e^{A_n}} \quad (34)$$

where E_g is the bandgap; A_n and A_p represent the kinetic terms of electrons and holes, respectively; N_n and N_p are the densities of states, respectively; μ_n and μ_p denote the mobility terms, respectively; k is the Boltzmann constant; and e is an elementary charge. Eq. (32), which may be represented graphically on the S-log σ coordinates, exhibits a pear-like shape (Fig. 14).

As seen, the critical quantities of the Jonker formalism may be determined graphically. The parameters B and D may be used for the determination of semiconducting quantities, such as bandgap and the mobility terms. When the minimum of the electrical conductivity is known, the Jonker Eq. (32) may also be represented in a linear form.[47]

The characteristic feature of the Jonker plot is the slope of S versus log σ dependence corresponding to the electrical properties that are remote from the n-p transition. As seen in Fig. 14, the experimental data for the dependence of S versus log σ for TiO_2 annealed in the gas phase, involving

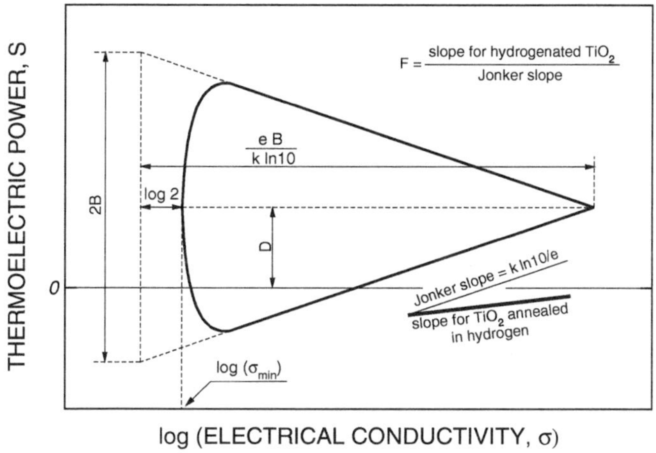

Figure 14 Graphical representation of the Jonker formalism. Factor F represents the ratio between the original Jonker slope of $\partial S/\partial \log\sigma = k\ln 10/e$ and the slope for hydrogenated TiO_2.[47,48]

hydrogen, exhibits a deviation from the Jonker slope by the parameter *F*, which for pure TiO_2 single crystals annealed in the gas phase involving hydrogen, is 0.7[47] and 0.5 for polycrystalline specimen of In-doped TiO_2.[48] The observed deviation from the Jonker slope is reflective of a departure from the classical nondegenerated semiconductors behavior where electrons are described by the Maxwell–Boltzmann statistics. The studies for In-doped TiO_2 annealed in the gas phase including hydrogen were considered in terms of the effect of protons on the formation of structural domains in rutile. It appears that indium ions play a crucial role in the formation of larger defect aggregates, leading to the appearance of structural domains that exhibits a high density of electrons resulting in dynamic correlations within the electron gas.

The Jonker analysis indicates that at elevated temperatures, the bandgap of single-crystal TiO_2 is lower than that for polycrystalline specimens by approximately 0.3 eV.[47] The difference arises from disparate temperature coefficient of the bandgap.

7. SURFACE SEGREGATION

7.1 Regular Solution Approach

The chemical composition of the surface layer differs from that of the bulk phase as a result of segregation. The driving force of segregation is the excess of surface energy. Depending on the effect of composition on surface energy, segregation leads to either an increase in surface concentration of specific species or its depletion.

The effect of segregation on surface composition should be considered in terms of both thermodynamic and kinetic factors. Essentially, when the temperature is high enough to allow fast segregation, the segregation-induced surface composition may be considered in terms of the thermodynamic factor.

A regular solution approach for grain boundary segregation in metallic alloys has been proposed by MacLean.[49] For a binary system, this approach may be expressed as:

$$\frac{X_2^s}{X_1^s} = \frac{X_2^b}{X_1^b} \exp\left(\frac{-\Delta H_{seg}}{kT}\right) \tag{35}$$

where X_1^s and X_1^b denote the surface and bulk molar fraction of the solute at the surface and in the bulk, respectively, X_2^s and X_2^b denote the surface and bulk molar fraction of the solvent, ΔH_{seg} is the enthalpy term, k is the Boltzmann constant, and T is the absolute temperature. MacLean applied this model to the system involving a segregation-modified monolayer and the bulk phase of uniform composition. The related enthalpy change represents the heat associated with the exchange of solute species in the bulk with the solvent species at the surface.[50,51] The enthalpy term involves the following components:

1. the strain energy contribution that depends on the surface energy of the molar area of the solvent
2. the binary interaction contribution that is determined by the heat of mixing of the solid solution
3. the solute elastic strain energy contribution that is related to the mismatch between the species of the solute and the solvent
4. the electrostatic contribution, which depends on the electric field imposed by the surface layer that is electrically charged and results in the relocation of charged species of aliovalent ions within this field. Minimization of the free energy for ionic crystals is the subject of charge neutrality condition that must satisfy the Poisson equation.

The predominant driving force of segregation in oxide materials, such as TiO_2, is the solute elastic strain contribution and electrostatic contribution.[52–56]

The performance of inorganic materials in energy conversion is profoundly influenced by the surface composition and the related defect disorder. Therefore, the phenomenon of segregation may be considered as the technology for surface processing that aims at the imposition of optimized properties.

7.2 Single Solute Segregation. Example of Cr-Doped TiO_2

Analysis of the reported experimental data on chromium segregation in Cr-doped TiO_2 indicates that the segregation-induced surface chromium concentration for the rutile phase is substantially larger than that for the anatase structure (Fig. 15).[52]

The solid line in Fig. 15 shows the chromium concentration that is related to equilibrium segregation. As seen, the surface content for equilibrium segregation in rutile, which above 2 at% Cr remains independent of bulk composition, indicates that segregation-induced enrichment results in the formation of a low-dimensional surface structure (LDSS).[53] The picture for the anatase phase is similar.

7.3 Competitive Segregation. Example of TiO_2 Co-doped With Niobium and Indium

The phenomenon of segregation for an oxide solvent, such as TiO_2, involving a single solute, such as chromium, may be considered in terms of the interaction energy between the host lattice and the solute ions. The picture is more complicated when segregation involves two or more solute ions. Then segregation must be considered in terms of interactions

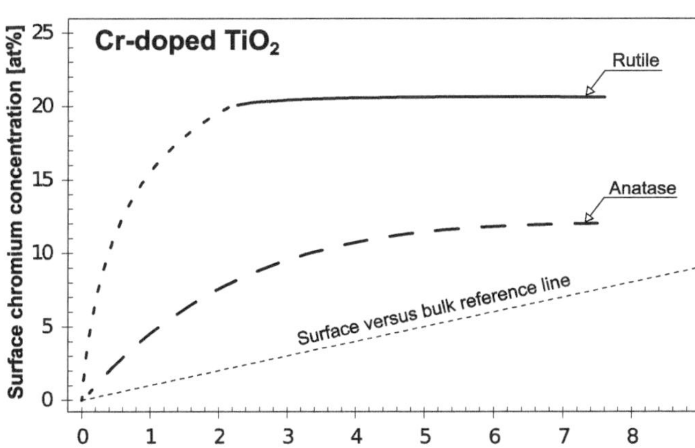

Figure 15 Effect of chromium segregation on surface versus bulk composition for both rutile and anatase forms of Cr-doped TiO_2[52].

between the segregating species, in addition to the interactions between the solute species and the lattice of the solvent.

This section considers the effect of segregation of niobium and indium as a single solute in Nb-doped TiO_2[54] and In-doped TiO_2,[55] respectively, as well as co-segregation of both niobium and indium in TiO_2 co-doped with both ions,[56] after annealing in the gas phase of controlled oxygen activity. The related segregation-induced depth profiles may be considered in terms of the enrichment factor that is defined as the ratio of surface concentration to the bulk concentration of the specific lattice species. Assuming that segregation in the bulk phase can be ignored, the enrichment factor is represented in Fig. 16A[54,55] and Fig. 16B,[56] respectively.

The depth profiles in Fig. 16 allow the following conclusions to be made concerning the effect of $p(O_2)$ on segregation in In-doped TiO_2, Nb-doped TiO_2, and TiO_2 co-doped with indium and niobium.

7.3.1 Indium-Doped TiO_2

Segregation of indium in In-doped TiO_2 results in surface enrichment in indium. The related segregation-induced enrichment factor is $F = 8.7$ and

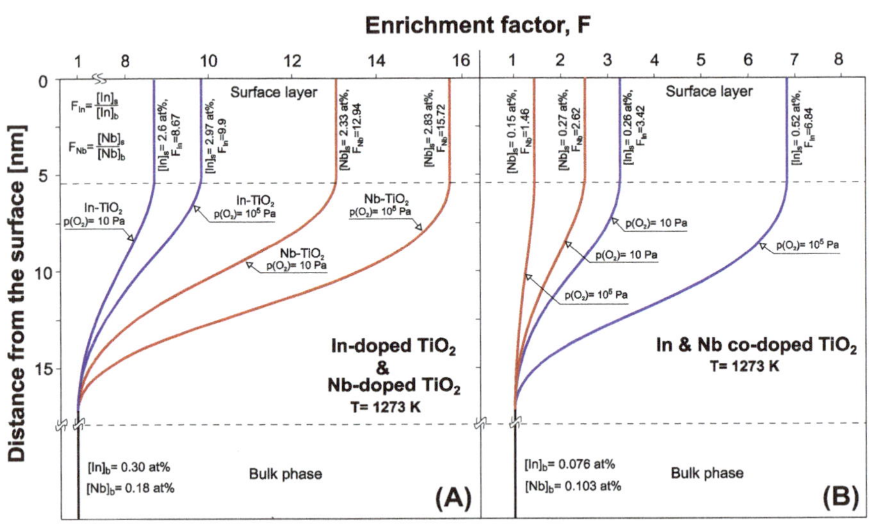

Figure 16 Semiquantitative representation of the depth profiles for the segregation-induced enrichment factor, F, for TiO_2 solid solutions after annealing in the gas phase of controlled oxygen activity, including (A) TiO_2 doped either with niobium or indium[51,52] and (B) TiO_2 co-doped with niobium and indium,[56] where square brackets denote concentration, $F_{Nb} = [Nb_s]/[Nb_b]$, $F_{In} = [In_s]/[In_b]$, and the subscripts s and b are related to the surface layer and the bulk phase, respectively.

F = 9.3 at $p(O_2) = 10$ Pa and $p(O_2) = 10^5$ Pa, respectively. These results are consistent with the theoretical model assuming that positively charged indium ions located in interstitial sites migrate to the surface under the electric field imposed by negatively charged titanium vacancies. In oxidizing atmospheres, the surface of TiO_2-based solids has a tendency to be enriched in titanium vacancies, where they are generated according to reaction (4) in Table 1, but do not migrate into the bulk due to low mobility. As a result, the segregating indium ions react with titanium vacancies leading to the formation of a LDSS:[53,54]

$$4\mathrm{In}_i^{\bullet\bullet\bullet} + 3V_{Ti}^{''''} \leftrightarrow \left\{ \left(\mathrm{In}_i^{\bullet\bullet\bullet}\right)_4 \left(V_{Ti}^{''''}\right)_3 \right\}^x \tag{36}$$

The increase of segregation-induced enrichment with $p(O_2)$ is consistent with this model assuming that the predominant driving force of segregation is the electric field formed by negatively charged surface layer enriched in titanium vacancies.

7.3.2 Niobium-Doped TiO_2

Segregation of niobium in Nb-doped TiO_2 results in surface enrichment with niobium.[51] The resulting enrichment factor is F = 13.05 and F = 15.72 at $p(O_2) = 10$ Pa and $p(O_2) = 10^5$ Pa, respectively. The segregation of niobium, in this case, results in the formation of an LDSS according to the following reaction:

$$4\mathrm{Nb}_{Ti}^{\bullet} + V_{Ti}^{''''} \leftrightarrow \left\{ \left(\mathrm{Nb}_{Ti}^{\bullet}\right)_4 V_{Ti}^{''''} \right\}^x \tag{37}$$

The observed effect of $p(O_2)$ on niobium segregation is consistent with the results reported by Atanacio et al. for Nb-doped TiO_2.[54]

7.3.3 TiO_2 Co-doped With Indium and Niobium

Simultaneous segregation of positively charged indium and niobium in TiO_2 co-doped with both ions results in preferential segregation of indium—the resulting enrichment factor is F = 3.28 and F = 6.84 at $p(O_2) = 10$ Pa and $p(O_2) = 10^5$ Pa, respectively—while the segregation-induced enrichment of niobium is reduced to F = 2.52 and F = 1.45 at $p(O_2) = 10$ Pa and $p(O_2) = 10^5$ Pa, respectively.[56] The latter results indicate that segregation of niobium is suppressed by indium compared to Nb-doped TiO_2. Moreover, the increase of $p(O_2)$ results in further suppression of niobium segregation that indicates the LDSS expressed by Formula (36) is more stable than that represented by formula (37). In summary, the driving force of segregation for both indium and niobium in co-doped TiO_2 is the same as that in the individual solid solutions; however, indium is a clear winner of the race in surface segregation.

8. OPTICAL PROPERTIES

Reflectance or transmission spectroscopy may be applied for the determination of the electronic structure of semiconductors, including the bandgap and the defect-related energy levels within the bandgap. With the exception of thin films, the optical properties of polycrystalline specimens, which are opaque, may be studied using reflectance spectroscopy.

The reflectance (R) can be transformed into light absorption coefficient (α) using the Kubelka–Munk function $f(R)$:[57]

$$f(R) = (1 - R)^2/2R = \frac{\alpha}{s} \tag{38}$$

While in most cases the scattering coefficient, s, is not known, assuming that s is independent of the wavelength of the incident light, the function $f(R)$ is directly proportional to α. The absorption coefficient of an ideal material that exhibits the parabolic dependence of electron energy versus electron wave vector grows with photon energy according to the Tauc equation:[58]

$$\alpha h\nu = A\left(h\nu - E_g\right)^n \tag{39}$$

where A is a proportionality constant, E_g is the optical bandgap, and $n = 1/2$, $3/2$, 2, and 3 for direct allowed transitions, direct forbidden, indirect allowed, and indirect forbidden transitions, respectively.

The bandgap can be graphically estimated from the intersection of the extrapolated linear part of the plot of $[f(R)h\nu]^2$ versus $h\nu$ for direct and $[f(R)h\nu]^{1/2}$ versus $h\nu$ for indirect transitions, with the energy axis as shown in Fig. 17A. As also seen, the Tauc plot may be used for the determination of mid-band energy levels within the forbidden gap (Fig. 17B).

Processing of the experimental reflectance spectra resulted in the determination of the effect of several solutes, including tantalum,[59] niobium,[60] and chromium[61] on the bandgap of polycrystalline TiO_2. The results are shown in Fig. 18.

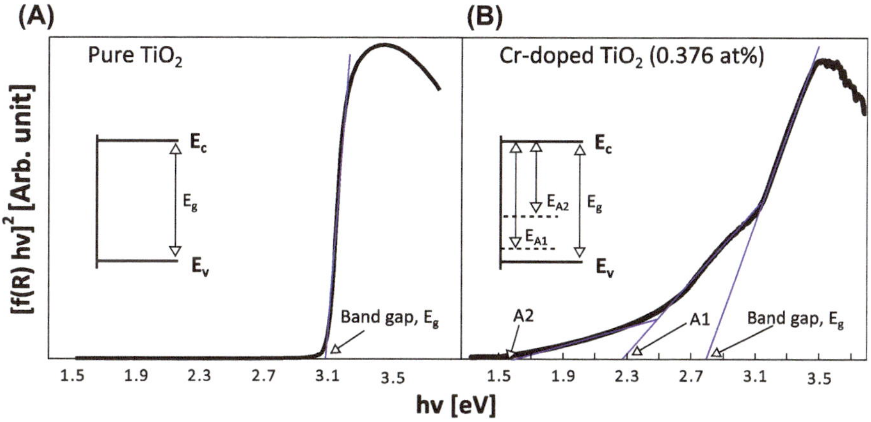

Figure 17 Semiquantitative representation of the Tauc plot showing the method of the determination of (A) the bandgap in pure TiO_2, as well as (B) the bandgap and mid-band energy levels within the bandgap of Cr-doped TiO_2. The related transitions are shown by *arrows* in the *inset*.

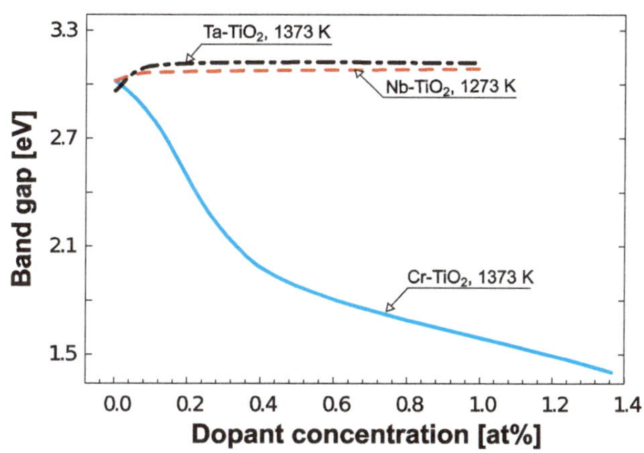

Figure 18 The effect of tantalum,[59] niobium,[60] and chromium[61] on the bandgap of TiO_2 annealed in air (the temperature of annealing is shown in the figure).

As seen, the incorporation of donor-type ions (niobium and tantalum) results in an insignificant increase of the bandgap from 3 to 3.2 eV at 1 at%. As also seen in Fig. 18, the incorporation of chromium into the TiO_2 lattice leads to a substantial reduction of the bandgap to the level below 1.5 eV.[58] The X-ray photoelectron spectroscopy studies indicate that chromium incorporation results in the formation of tri- or/and hexavalent chromium ions incorporated according to the reactions outlined in Eqs. (40)–(47):[58]

$$\frac{1}{2}O_2 + Cr_2O_3 \leftrightarrow 2Cr_i^{\bullet\bullet\bullet} + 2V_{Ti}^{''''} + 2h^{\bullet} + 4O_O^x \tag{40}$$

$$2Cr_2O_3 \rightarrow 4Cr_i^{\bullet\bullet\bullet} + 3V_{Ti}^{''''} + 6O_O^x \tag{41}$$

These reactions are governed by the following charge compensations, respectively:

$$p + 3\left[Cr_i^{\bullet\bullet\bullet}\right] = 4\left[V_{Ti}^{''''}\right] \tag{42}$$

$$3\left[Cr_i^{\bullet\bullet\bullet}\right] = 4\left[V_{Ti}^{''''}\right] \tag{43}$$

The alternative reactions involve the incorporation of hexavalent chromium ions under oxidizing and reducing conditions and may be represented by the following reactions, respectively:

$$2CrO_3 \rightarrow 2Cr_{Ti}^{\bullet\bullet} + V_{Ti}^{''''} + 6O_O^x \tag{44}$$

$$\frac{1}{2}O_2 + CrO_3 \rightarrow Cr_i^{\bullet\bullet\bullet\bullet\bullet\bullet} + 2V_{Ti}^{''''} + 2h^{\bullet} + 4O_O^x \tag{45}$$

where the charge neutrality requires, respectively, that:

$$2\left[V_{Ti}^{''''}\right] = \left[Cr_{Ti}^{\bullet\bullet}\right] \tag{46}$$

$$4\left[V_{Ti}^{''''}\right] = 6\left[Cr_i^{\bullet\bullet\bullet\bullet\bullet\bullet}\right] + p \tag{47}$$

Reactions (40)–(47) include the options associated with chromium incorporation leading to the formation of positively charged donor-type defects that are compensated by negatively charged titanium vacancies.

The analysis of the optical spectra led to the determination of the energy level of titanium vacancies that for Cr-doped TiO_2 is 1.16 eV. This value, which is determined using the optical method,[58] is consistent with the energy level of these defects in TiO_2 determined theoretically by He et al.[62] and Hossain et al.[63] (Table 2).

Table 2 Ionization Energy of Intrinsic Defects in TiO_2 According to Theoretical Calculations Using the Density Functional Theory Method

Oxygen Vacancies		Titanium Interstitials		Titanium Vacancies		
					Energy [eV]	
Ionization	Energy [eV]	Ionization	Energy [eV]	Ionization	Theory	Experiment
$V_O^x/V_O^{\bullet}$	2.50[a]	$Ti_i^x/Ti_i^{\bullet}$	3.08[a]	V_{Ti}^x/V_{Ti}'	−0.39[a]	
$V_O^{\bullet}/V_O^{\bullet\bullet}$	2.11[a]	$Ti_i^{\bullet}/Ti_i^{\bullet\bullet}$	2.48[a]	V_{Ti}'/V_{Ti}''	0.22[a]	
		$Ti_i^{\bullet\bullet}/Ti_i^{\bullet\bullet\bullet}$	2.33[a]	V_{Ti}''/V_{Ti}'''	0.82[a]	
		$Ti_i^{\bullet\bullet\bullet}/Ti_i^{\bullet\bullet\bullet\bullet}$	1.71[a]	V_{Ti}'''/V_{Ti}''''	1.15[b]	1.16[c]
					1.44[a]	

[a]He et al.[62]
[b]Hossain et al.[63]
[c]The experimental value from the optical method[52] (all energy data are related to the energy of the valence band).

9. LIGHT-INDUCED WATER OXIDATION

9.1 Total and Partial Water Oxidation

The applied aspects of the light-induced water oxidation can be considered in terms of either total or partial water oxidation, which can be represented by the following respective chemical reactions:

$$2H_2O + 4h^{\bullet} \leftrightarrow 4H^+ + O_2 \quad (48)$$

$$H_2O + h^{\bullet} \leftrightarrow H^+ + HO^{\bullet} \quad (49)$$

The reaction represented by equilibrium (48) is related to water oxidation with 2 holes per water molecule. Its final products are elemental water components in the form of protons and oxygen gas. The reaction (49), which represents partial water oxidation, involves only one electron hole per water molecule. Both reactions take place at the surface of oxide semiconductors and can be induced by light leading to ionization over the bandgap. The reaction (48) takes place at the anodic surface site that exhibits high affinity to electrons. However, charge neutrality requires that the excess of electrons formed in the semiconductor as a result of reaction (48) be removed, and this tendency is the driving force of the following reaction:

$$4H^+ + 4e^- \leftrightarrow H_2 \quad (50)$$

Consequently, Reactions (48) and (50) taking place at the surface of a semiconductor results in the formation of the gas mixture of both oxygen and hydrogen. Separation of these gases requires energy. Therefore, the total water oxidation usually takes place in PECs where hydrogen and oxygen are evolved on separate electrodes. Instead, the partial water oxidation aims at maximum efficiency of the Reaction (49) leading to the formation of radicals that play a key role in water purification.

The main approach in solar-to-chemical energy conversion is based on water oxidation, including partial and total oxidation. Total water oxidation leads to water splitting into its elemental components. The process involves several stages, such as (1) light absorption leading to ionization over the bandgap, (2) transport of the generated electronic charge carriers from the site of their generation to the reaction site, and (3) reactions of the light-induced electrons and holes at anodic and cathodic sites, respectively. Effective light conversion requires that recombination of electronic charge carriers is prevented. This, consequently, requires imposition of an electric field for charge separation. It may be achieved by the deposition of surface islets of alternative phases, such as noble metals.[14,17]

The present section considers both partial and total oxidation of water using TiO_2 and its solid solution as photocatalysts and photoelectrodes, respectively.

9.2 Partial Water Oxidation

Partial oxidation consists in selective water oxidation leading to the formation of radical species, such as hydroxyl radicals. The most common primary oxidation reaction, which takes place at anodic sites, occurs between water molecules and electron holes, and may be represented by the anodic reaction (49). The charge neutrality of the system requires that the excess electrons are removed from the system. Consequently, the electrons react with molecular oxygen dissolved in water at cathodic sites leading to the formation of superoxide species:

$$O_2 + e^- \leftrightarrow O_2^- \tag{51}$$

The superoxide species then react with protons to form hydrogen peroxide:

$$O_2^- + 2H^+ + e' \leftrightarrow H_2O_2 \tag{52}$$

Thus generated active radicals HO^*, O_2^-, and H_2O_2 species react with organic contaminations (microbial species and toxic compounds) in water leading to their oxidation. The reaction mechanism at the TiO_2/H_2O interface and the related charge transfer is shown in Fig. 19.

As seen, the water molecule is adsorbed at the anodic site (titanium vacancy) forming an active complex that is decomposed according to reaction (49). At the same time, oxygen is adsorbed on the cathodic site (oxygen vacancy) to form an active complex that leads to the formation of radical species according to reaction (51). Charge neutrality requires that reactions (49) and (51) take place simultaneously.

The commonly applied research strategy in the development of high-performance photocatalysts for water purification is based on reduction of the bandgap.[64–67] This approach assumes that the solar-to-chemical ECE is determined by the bandgap and the associated amount of light absorption. Recent progress in photocatalysis indicates, however, that the performance of oxide semiconductors in solar-to-chemical energy conversion is determined by a range of defect-related properties, including the bandgap, concentration of surface active sites, Fermi level, and charge transport.

The rate constant for the photocatalytic reaction, tested by oxidation of methylene blue by pure TiO_2 and its solid solutions with niobium, tantalum, and chromium is shown in Fig. 20.

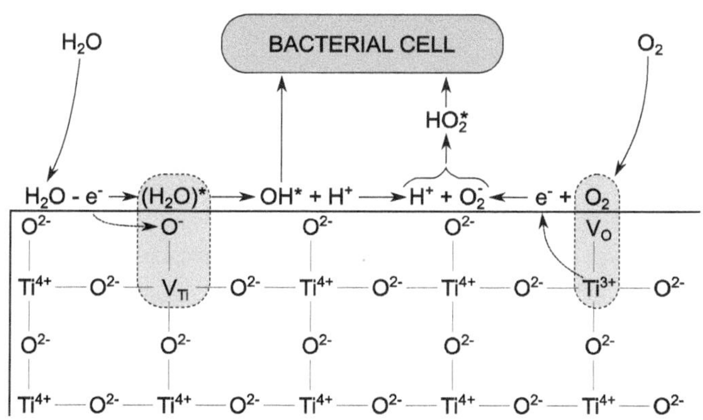

Figure 19 Theoretical model representing the reactions at anodic and cathodic sites formed by titanium vacancies and oxygen vacancies, respectively.

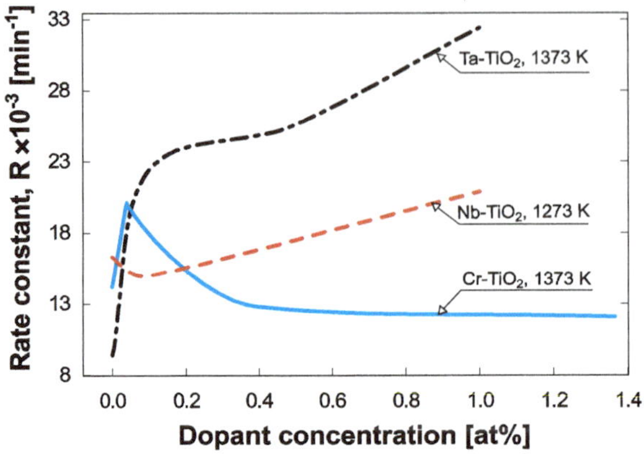

Figure 20 Effect of composition on the rate constant of photocatalytic activity for TiO_2 doped with tantalum,[59] niobium,[60] and chromium.[61]

The data in Fig. 20 may be considered in terms of the effect of selected extrinsic ions on the TiO_2 defect disorder and the associated KPPs.

9.2.1 Nb-doped TiO_2

Increasing the content of niobium results in enhanced performance within the entire range of concentrations (except very dilute solid solutions). The observed experimental relationship indicates that the performance of Nb-doped TiO_2 is related to the effect of niobium on the formation of titanium vacancies (reaction (25)) acting as surface active sites, and enhanced charge transport.

> *At the same time, the effect of niobium on the increase of the Fermi level (not desired) and the bandgap (not desired as well) has a minor influence.*

9.2.2 Ta-doped TiO_2

Increasing the content of tantalum results in enhanced performance. The effect, which is more substantial than that observed for niobium, indicates that the performance is determined by the increasing concentration of titanium vacancies, acting as surface active sites, and enhanced charge transport.

9.2.3 Cr-doped TiO_2

The incorporation of chromium leads to enhanced performance only within dilute solid solutions (up to 0.04 at% Cr). Apparently, chromium promotes the formation of anodic surface active sites (titanium vacancies) mainly within this range. The effect of chromium in this range may also be considered in terms of the decrease of the Fermi level. The deteriorating performance above 0.04 at% chromium indicates that the beneficial effect of bandgap reduction is compensated by the negative effects related to the decrease in charge transport.

9.3 Total Water Oxidation

Total water oxidation, leading to water splitting, may be expressed by the respective anodic and cathodic reactions:

$$H_2O + 2h^{\bullet} \leftrightarrow 2H^+ + \frac{1}{2}O_2 \tag{53}$$

$$2H^+ + 2e^- \leftrightarrow H_2 \tag{54}$$

The related electrochemical chain is shown in Fig. 21.

The process of water oxidation involves the following steps:

1. adsorption of water molecules on the anode formed by n-type TiO_2
2. exposure to light results in the reaction of water with TiO_2, leading to the formation of protons, oxygen gas, and electrons at the anode
3. the excess of electrons is transported through the external circuit to the cathode

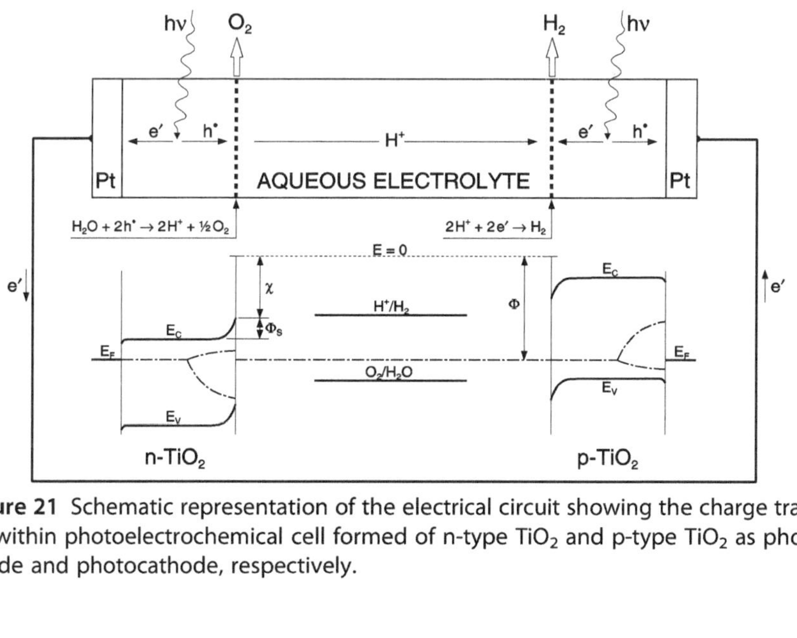

Figure 21 Schematic representation of the electrical circuit showing the charge transfer within photoelectrochemical cell formed of n-type TiO_2 and p-type TiO_2 as photoanode and photocathode, respectively.

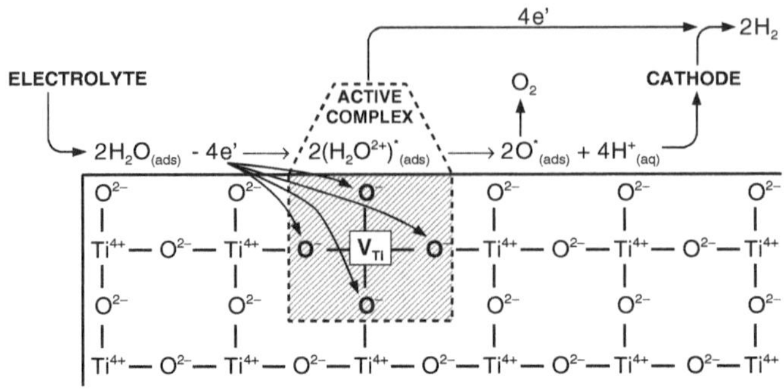

Figure 22 The model representing the reactivity of TiO_2 with water leading to total water oxidation (water splitting).

4. the protons migrate to the cathode via the electrolyte
5. both electrons and protons react at the cathode and form hydrogen gas.

Schematic representation of the reactivity of TiO_2 with water, leading to total water oxidation and the related charge transfer is in Fig. 22.

The performance of the PEC may be modified by a change of defect disorder and the related reactivity of the anode and the cathode.

10. RESEARCH STRATEGY IN PROCESSING TiO_2-BASED PHOTOSENSITIVE SEMICONDUCTORS

The research strategy in the development of TiO_2-based semiconductors with enhanced performance in solar-to-chemical energy conversion depends on a range of factors. The latter involves light intensity, surface area, pH of the solution, and the structure of the photoreactor. This section considers the research strategy in the use of defect engineering in processing oxide semiconductors with optimized performance-related properties, including the concentration of surface active sites, Fermi level, charge transport, electronic structure, and proper alignment of the bandgap edges to energy levels of the redox couple.

10.1 Concentration of Surface Active Sites

The first step of light-induced water oxidation involves adsorption of water molecules on the semiconductor surface. The most active surface anodic sites for water oxidation are acceptor-type defects, such as titanium vacancies. These defects are formed as a result of either oxygen incorporation into the TiO_2 lattice or doping TiO_2 with donor-type defects, such as niobium and tantalum.[25] Adsorption of water molecule on these defects leads to the formation of an active complex and its subsequent decomposition into water oxidation products and regeneration of the TiO_2 lattice site by light-induced electron holes.[25]

10.1.1 Fermi Level

The Fermi level, E_F, is the chemical potential of electrons that is related to the concentration of electronic charge carriers by the following expressions for n- and p-type semiconductors, respectively:

$$E_F = const + kT \ln \frac{\mathrm{n}}{N_n} \tag{55}$$

$$E_F = const - kT \ln \frac{\mathrm{p}}{N_p} \tag{56}$$

Accordingly, E_F, the position of the Fermi level depends on the defect disorder and the related concentration of electronic charge carriers. The E_F may be lowered and elevated by the incorporation of acceptor- and donors-type ions. The E_F may have an effect on photocatalytic properties only when it concerns the surface layer that is directly involved in photoreactivity and the related charge transfer.

The incorporation of acceptors leads to enhanced charge transfer at the anodic sites. It is important to note here that the mechanism of incorporation in the bulk and at the surface may differ.

10.1.2 Charge Transport

Electrical conductivity, which is the product of the mobility and concentration of charge carriers, can be considered as a measure of charge transport. Therefore, the modification of charge transport may be achieved by a change in these terms. A photocatalytic process involves the formation of light-induced charge carriers within the light penetration distance beneath the surface and subsequent charge migration toward the sites of the reaction. Therefore, charge transport is critical for the performance.

10.1.3 Electronic Structure

The electronic structure, mainly the bandgap and the energy levels within the bandgap, is of crucial importance for the light-induced charge generation. The gap between the valence band and the conduction band, termed the bandgap, is the energy gap through which the light-induced ionization takes place. While the bandgap is one of the critical performance-related

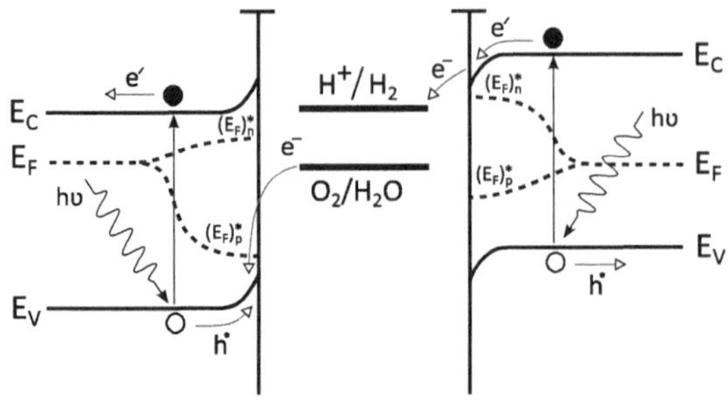

Figure 23 Schematic representation of the optimal alignment of band edges with respect to energy levels of the redox couple in light-induced charge transfer at anodic and cathodic sites (*left* and *right*, respectively).

quantities of semiconductors, the recent progress in photocatalysis indicates that the bandgap is only one of several KPPs that are competitively influencing the conversion efficiency of solar energy into chemical energy.[17] A range of approaches have been reported for the imposition of an optimized bandgap that is required for maximized light absorption of the surface layer. However, it has been documented that in some cases the bandgap has little effect on the performance.[61,68]

10.1.4 Alignment of Band Edges With Energy Levels of Redox Couple

Proper alignment of the band edges with the energy levels of the redox couple in the solution is essential for spontaneous charge transfer at the solid/liquid interface as it is represented schematically in Fig. 23.

The alignment may be modified by shifting the band edges up and down through appropriate doping with aliovalent ions and imposition of the optimal level of segregation.

11. CONCLUSIONS

The properties and performance of inorganic nonstoichiometric compounds may be modified in a controlled manner by a change in the concentration of the structural imperfections (point defects), which are thermodynamically reversible. The present work considers the application of defect chemistry in processing photosensitive oxide semiconductors with optimized properties that are required for imposition of desired properties. In this work, we have concentrated on the research strategy in processing TiO_2-based oxide semiconductors with enhanced performance in solar-to-chemical energy conversion. We show that defect engineering in the processing of pure (undoped) TiO_2 consists of the imposition of controlled intrinsic defect disorder by annealing under controlled oxygen activity. The additional modification of defect disorder may be achieved by the incorporation of extrinsic ions (donors and acceptors). It is shown that the mechanism of doping is determined by the oxygen activity. Proper understanding of the effect of doping on properties requires the determination of the effect of segregation on surface versus bulk composition that is profoundly influenced by the oxygen activity.

HIGHLIGHTS

- TiO_2-based semiconductors are promising candidates for solar energy conversion.
- Performance of TiO_2-based materials is determined by structural imperfections.
- Oxygen activity is the key parameter that has a profound effect on defect disorder of oxide semiconductors.
- The phenomenon of segregation may be used as the technology in surface processing.
- The defect-related properties determining photocatalytic performance, such as bandgap, Fermi level, and charge transport, are competitive.

REFERENCES

1. Future Earth. *2025 Vision, Future Earth Interim Secretariat;* International Council of Science: Paris, 2013.
2. SunShot Vision Study, US DOE, February 2012, Chapter 4. https://energy.gov/eere/solar/sunshot-vision-study.
3. Fujishima, A.; Fujishima, A.; Honda, K. *Nature* **1972,** *238,* 37–38.
4. Khan, S. U.; Al-Shahry, M.; Ingler, W. B. *Science* **2002,** *297,* 2243–2245.
5. Park, J. H.; Kim, S.; Bard, A. J. *Nano Letters* **2006,** *6,* 24–28.
6. Nowotny, J.; Bak, T.; Li, W. In *Handbook of Hydrogen Energy;* Sherif, S. A., Goswami, D. Y., Stefankos, E. K., Steinfeld, A., Eds.; CRC: Boca Raton, 2014; pp 445–535.
7. Abdi, F. F.; Han, L.; Smets, A. H.; Zeman, M.; Dam, B.; Van De Krol, R. *Nat. Commun.* **2013,** *4,* 2195.
8. Nowotny, J.; Atanacio, A.; Bak, T.; Belova, I.; Fiechter, S.; Ikuma, Y.; Ionescu, M.; Kennedy, B.; Majewski, P.; Murch, G. *Int. Mater. Rev.* **2014,** *59,* 449–478.
9. Sherif, S. A.; Barbir, F.; Veziroglu, T. N. In *Handbook of Hydrogen Energy;* Sherif, S. A., Goswami, D. Y., Stefankos, E. K., Steinfeld, A., Eds.; CRC: Boca Raiton, 2014; pp 1–16.
10. Nowotny, J.; Bak, T.; Chu, D.; Fiechter, S.; Murch, G. E.; Veziroglu, T. N. *Int. J. Hydrogen. Energ.* **2014,** *39,* 4151–4157.
11. DOE hydrogen and fuel cell program: www.hydrogen.gov/h2a_prod_studies.html.

12. Schoots, K.; Ferioli, F.; Kramer, G. J.; Van der Zwaan, B. C. C. *Int. J. Hydrogen. Energ.* **2008,** *33,* 2630−2645.
13. Bockris, J.O.. Private Communication to Nowotny, J. on June 6 , 2002.
14. Carp, O.; Huisman, C. L.; Reller, A. *Prog. Solid State Ch.* **2004,** *32,* 33−177.
15. Choi, J.; Park, H.; Hoffmann, M. R. *J. Phys. Chem. C* **2010,** *114,* 783−792.
16. Asahi, R. Y. O. J. I.; Morikawa, T. A. K. E. S. H. I.; Ohwaki, T.; Aoki, K.; Taga, Y. *Science* **2001,** *293,* 269−271.
17. Bak, T.; Nowotny, J.; Sucher, N.; Wachsman, E. *Adv. Appl. Ceram.* **2012,** *111,* 4−15.
18. Sucher, N.; Carles, M.; Nowotny, J.; Bak, T. *Adv. Appl. Ceram.* **2012,** *111,* 16−33.
19. Frenkel, J. *Z. Phys.* **1926,** *35,* 652−669.
20. Wagner, C.; Schottky, W. *Zeitschr. Phys. Chem.* **1930,** *11,* 163−171.
21. Henrich, V. E. In *Surface and Near-surface Chemistry of Oxide Materials;* Nowotny, J., Dufour, L. C., Eds.; Elsevier: Amsterdam, 1988; pp 23−60.
22. Kofstad, P. *Nonstoichiometry, Diffusion and Electrical Conductivity of Binary Metal Oxides;* Wiley: New York, 1972.
23. Kroger, F. A. *The Chemistry of Imperfect Crystals;* North Holland: Amsterdam, 1974.
24. Mrowec, S. *Defects and Diffusion in Solids: An Introduction;* Elsevier Scientific Pub. Co, 1980.
25. Nowotny, J. *Oxide Semiconductors for Solar Energy Conversion: Titanium Dioxide;* CRC Press: Boca Raton, 2012.
26. Tani, E.; Baumard, J. F. *J. Solid State Chem.* **1980,** *32,* 105−113.
27. Baumard, J. F.; Tani, E. *J. Chem. Phys.* **1977,** *67,* 857−860.
28. Radecka, M.; Rekas, M. *In Solid State Phenomena;* Trans Tech Publications, 1994; pp 135−138.
29. Carpentier, J. L.; Lebrun, A.; Perdu, F. *J. Phys. Chem. Solids* **1989,** *50,* 145−151.
30. Baumard, J. F.; Tani, E. *Physica Status Solidi A* **1977,** *39,* 373−382.
31. Itakura, M.; Niizeki, N.; Toyoda, H.; Iwasaki, H. *Jpn. J. Appl. Phys.* **1967,** *6,* 311.
32. Nowotny, M. K.; Bak, T.; Nowotny, J. *J. Phys. Chem. B* **2006,** *110,* 16283−16291.
33. Eror, N.; Wagner, J. *J. Phys. Chem. Solids.* **1968,** *29,* 1597−1611.
34. Kofstad, P. *J. Phys. Chem. Solids* **1962,** *23,* 1579−1586.
35. Kröger, F.; Vink, H. *Solid State Phys.* **1956,** *3,* 307−435.
36. Bak, T.; Nowotny, M.; Sheppard, L.; Nowotny, J. *J. Phys. Chem. C* **2008,** *112,* 12981−12987.
37. Cronemeyer; Charles, D. *Phys. Rev.* **1952,** *87,* 876.
38. Newman; Broadus, A. *Trans. Am. Inst. Chem. Eng.* **1931,** *27,* 203−220.
39. Nowotny, J.; Bak, T.; Burg, T. *Ionics* **2007,** *13,* 79−82.
40. Lee, D.-K.; Yoo, H.-I. *Solid State Ionics* **2006,** *177,* 1−9.
41. Nowotny, J.; Bak, T.; Dickey, E. C.; Sigmund, W.; Alim, M. A. *J. Phys. Chem. A.* **2016,** *120,* 6822−6837.
42. Nowotny, J.; Malik, A.; Alim, M. A.; Bak, T.; Atanacio, A. J. *ECS J. Solid State Sci. Tech.* **2014,** *3,* P330−P339.
43. Becker, J. H.; Frederikse, H. P. R. *J. Appl. Phys.* **1962,** *33,* 447−455.
44. Chen, C.; Garstein, E.; Mason, T. O. *J. Phys. Chem. Solids* **1982,** *43,* 991.
45. Nowotny, J.; Bak, T.; Burg, T. *Phys. Status Solidi B* **2007,** *244,* 2037−2054.
46. Jonker, G. *Philips Res. Rep.* **1968,** *23,* 131−138.
47. Nowotny, J. *J. Phys. Chem. C* **2011,** *115,* 18316−18326.
48. Nowotny, J.; Bak, T.; Ionescu, M.; Alim, M. A. *J. Phys. Chem. A* **2015,** *119,* 4032−4040.
49. MacLean, D. *Grain Boundaries in Metals;* Clarendon Press: London, 1957.
50. Wynblatt, P.; McCune, R. *Surface and Near-surface Chemistry of Oxide Materials,* 1988; pp 247−279.

51. Burggraaf, A.; Winnubst, A. *Surface and Near-surface Chemistry of Oxide Materials,* 1988; pp 448–477.
52. Rahman, K. A.; Bak, T.; Atanacio, A.; Ionescu, M.; Liu, R.; Nowotny, J. *Ionics* **2017,** 1–9. https://doi.org/10.1007/s11581-017-2326-0.
53. Atanacio, A. J.; Bak, T.; Chu, D.; Ionescu, M.; Nowotny, J. In *Handbook of Nanomaterials Properties;* Springer Berlin: Heidelberg, 2014; pp 891–910.
54. Atanacio, A. J.; Bak, T.; Nowotny, J. *J. Phys. Chem. C* **2014,** *118,* 11174–11185.
55. Atanacio, A.; Bak, T.; Nowotny, J. *ACS Appl. Mater. Interfaces* **2012,** *4,* 6626–6634.
56. Atanacio, A. J.; Alim, M. A.; Bak, T.; Ionescu, M.; Nowotny, J. *J. Am. Ceram. Soc.* **2017,** *100,* 419–428.
57. Tandon, S.; Gupta, J. *Phys. Status Solidi B* **1970,** *38,* 363–367.
58. Murphy, A. *Sol. Energy Mater. Sol. Cells* **2007,** *91,* 1326–1337.
59. Alim, M. A.; Bak, T.; Atanacio, A. J.; Ionescu, M.; Kennedy, B.; Price, W. S.; Du Plessis, J.; Pourmahdavi, M.; Zhou, M.; Torres, A.; Nowotny, J. *Ionics* **2017,** *23,* 3517–3531.
60. Li, W.; Bak, T.; Atanacio, A.; Nowotny, J. *Appl. Catal. B* **2016,** *198,* 243–253.
61. Rahman, K. A.; Bak, T.; Atanacio, A.; Ionescu, M.; Nowotny, J. *Ionics* **2017**. https://doi.org/10.1007/s11581-017-2369-2.
62. He, J.; Behera, R. K.; Finnis, M. W.; Li, X.; Dickey, E. C.; Phillpot, S. R.; Sinnott, S. B. *Acta Mater.* **2007,** *55,* 4325–4337.
63. Hossain, F. M.; Sheppard, L.; Nowotny, J.; Murch, G. E. *J. Phys. Chem. Solids* **2008,** *69,* 1820–1828.
64. Lee, J. S.; You, K. H.; Park, C. B. *Adv. Mater* **2012,** *24,* 1084–1088.
65. Chen, D.; Jiang, Z.; Geng, J.; Wang, Q.; Yang, D. *Ind. Eng. Chem. Res.* **2007,** *46,* 2741–2746.
66. Diaz-Uribe, C.; Vallejo, W.; Ramos, W. *Appl. Surf. Sci.* **2014,** *319,* 121–127.
67. Hao, W.; James, P. L. *J. Phys. Conden. Matter* **2005,** *17,* L209.
68. Rahman, K. A.; Bak, T.; Atanacio, A.; Ionescu, M.; Nowotny, J. *Ionics* **2018**. https://doi.org/10.1007/s11581-017-2304-6.

CHAPTER TWO

Visible Light-Responsive Photocatalysts—From TiO_2 to Carbon Nitrides and Boron Carbon Nitride

Bo Wang, Masakazu Anpo[1] and Xinchen Wang
State Key Laboratory of Photocatalysis on Energy and Environment, College of Chemistry, Fuzhou University, Fuzhou, People's Republic of China
[1]Corresponding author: E-mail: anpo@chem.osakafu-u.ac.jp

Contents

Advances in Inorganic Chemistry, Volume 72
ISSN 0898-8838
https://doi.org/10.1016/bs.adioch.2018.05.008

Abstract

It can be said that energy depletion and environmental pollution on a global scale are the most serious and urgent issues facing mankind in recent times. It is, thus, vital to design novel energy production and conversion systems that utilize natural energy and allow sustainable development without environmental destruction or pollution. In line with these objectives, the decomposition of H_2O into H_2 and O_2, as well as the related reactions using visible light-responsive photocatalysts under sunlight irradiation, has been intensively investigated as one of the most promising environmentally benign energy production systems to address these issues. In the past half century, research on various photocatalytic systems using metal oxides has been carried out. However, to achieve higher efficiency in the production of H_2, more innovative breakthroughs in the development of photocatalytic materials are strongly desired.

To address such issues, graphitic carbon nitride (g-C_3N_4) and hexagonal boron carbon nitride (BCN) nano-materials have been investigated as promising visible light-responsive photocatalytic materials. In this chapter, we have highlighted the development of such g-C_3N_4 and hexagonal BCN photocatalytic nano-materials, focusing on their design, construction and optimization, as well as their applications to environmentally benign solar energy conversion systems, for the decomposition of H_2O, reduction and fixation of CO_2, and organic redox reactions. Such visible light-responsive photocatalytic materials are considered to be the most important research for the development of safe and clean energy production technologies for the 21st century and beyond.

1. INTRODUCTION

Humanity is now facing the serious and urgent issues of the lack of natural energy resources and fossil fuel-driven environmental pollution and destruction on a global scale. The increase in world population and rampant unregulated industrial growth have all led to accelerated energy consumption and the unabated release of toxic agents and industrial wastes into the air and waterways, leading to pollution-related diseases and abnormal climate changes such as global warming. In addition, the earthquake—tsunami disaster on March 11, 2011, which destroyed the nuclear reactors of the nuclear power plants in Fukushima, Japan, has raised serious concerns over the supply of electric power, which seemed to reduce the dependency on fossil fuels but actually requires coolants polluted with radioactive materials and creates highly radioactive nuclear waste as a by-product.

While our industries are constantly providing a variety of new products and materials based on innovative new technologies, it is becoming imperative to focus on the recycling and treatment of waste materials as well.

Moreover, we need to reduce the consumption of our limited natural resources and raise awareness of the great impact that our consumerism has on our environment.

Thus, the greatest challenge facing researchers today is to develop environmentally harmonious, ecologically clean and safe, sustainable and energy-efficient chemical technologies. Photocatalysis, in which the inexhaustible energy of the sun can be harnessed and converted to clean chemical or electrical energy for sustainable, nontoxic, and economically viable technologies, is a major advance in this direction. In this chapter, we will present the progress made so far on UV to visible light-driven photocatalysts.

2. BRIEF HISTORY OF THE PROGRESS IN METAL OXIDES (TIO_2) PHOTOCATALYSTS

Metal oxide semiconducting photocatalysts, such as TiO_2 materials, have been the focus of much attention for their ability to induce efficient and effective reactions at room temperature under sunlight irradiation. In fact, as shown in Fig. 1, in the past half century, many scientists have studied the effect of UV light irradiation on semiconductors[1–4] and, in 1972, Honda and Fujishima reported their groundbreaking work on the sensitization effect of a TiO_2 electrode for the electrolysis of H_2O.[5] Since then, many chemists have further investigated the development of semiconductor photocatalysis to decompose H_2O into H_2 and O_2 using various powdered metal oxide catalysts for various practical applications.[6–15]

As a photocatalytic material, TiO_2 is cost-efficient, nontoxic and has relatively high reactivity and stability even in aqueous solution. Thus, as shown in Fig. 1, it has been applied for the development of highly dispersed single site photocatalysts,[16–23] dye sensitized solar cell by Graetzel in 1991,[24] and the photocatalytic degradation or mineralization of organic pollutants under relatively weak UV light irradiation.[25–28] The photo-induced superhydrophilic properties of TiO_2 surfaces were then established,[29–34] and finally, visible light-responsive TiO_2 photocatalysts were developed by applying a metal ion implantation technique in 1998.[35–44] Then, from 2005, separate evolution of H_2 and O_2 from H_2O using a visible light-responsive TiO_2 thin film photocatalyst was successfully carried out.[45–51] Highly efficient H_2O splitting into H_2 and O_2 mixture using particulate semiconducting powdered materials such as oxynitrides and oxychalcogenides under visible light irradiation has been reported by

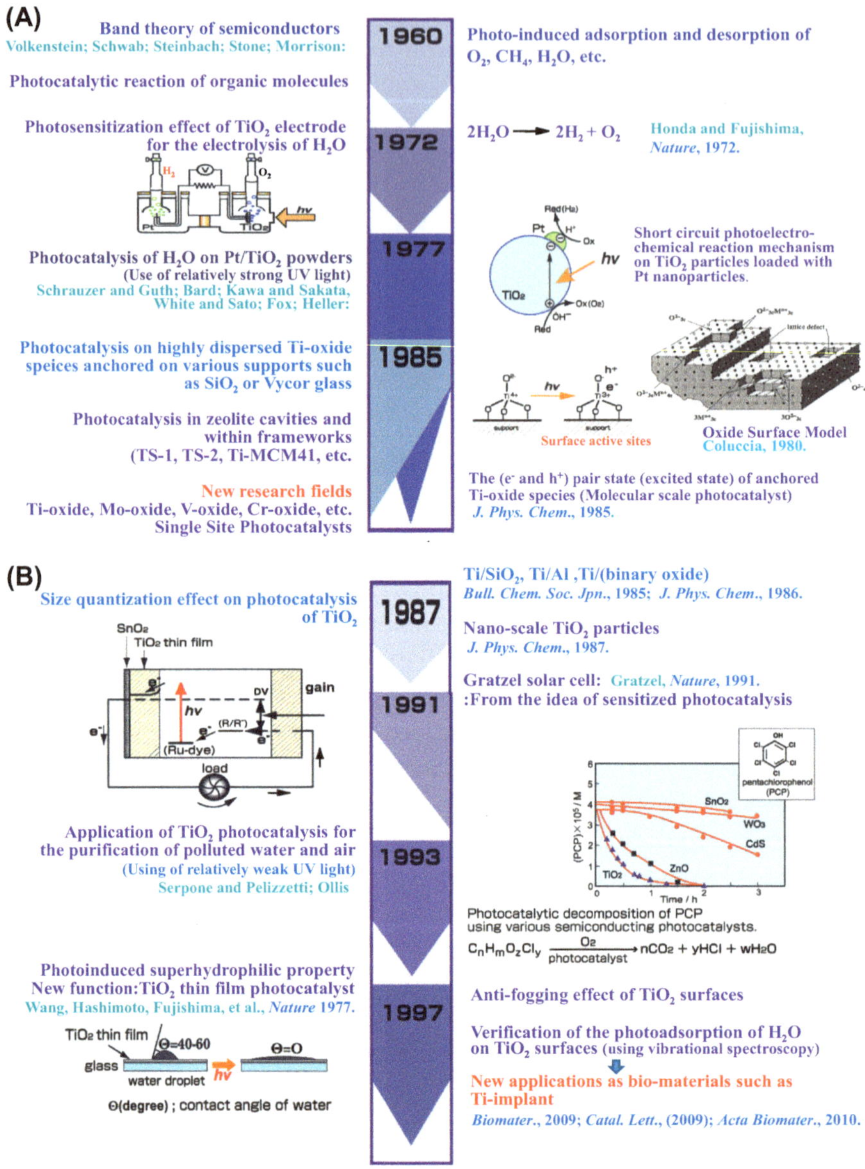

Figure 1 Progress in Research on Metal Oxide-based Photocatalysts and their Practical Applications (past, present, and future). (A) 1960-1986; (B) 1987-1997; (C) 1998-future.

Domen and Kudo.[52–59] Moreover, Mitoraj et al. reported a kind of N-modified titania (TiO_2—N) by calcining a mixture of titania and urea or melamine at 400°C, which exhibited visible light photocatalytic activity. They attributed this mainly to the higher melamine condensation

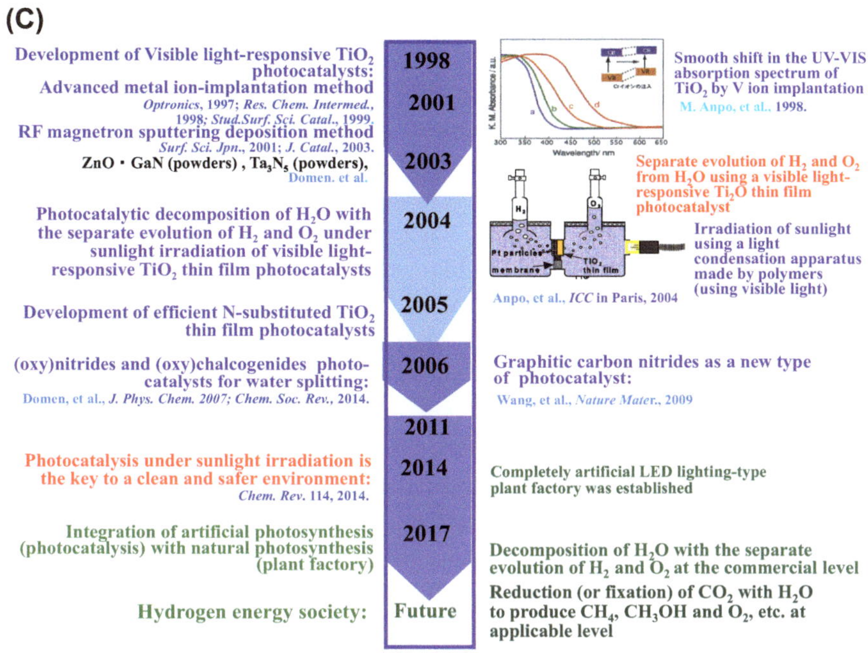

Figure 1 (*continued*).

products (such as melam, melem, and melon), which are precursors of graphitic carbon nitride and act as visible-light sensitizers.[60] These research results suggested that higher melamine condensation products would be a potential visible light-responsive material to sensitize TiO_2, which may create an inspirational image of graphitic carbon nitride photocatalysts.

In 2009, Wang et al. reported the highly efficient production of H_2 from H_2O splitting using visible light-responsive graphitic carbon nitride photocatalysts.[61] As one of the most promising photocatalytic materials over metal oxides, visible light-responsive graphitic carbon nitride (g-C_3N_4)[61–71] has recently been investigated centering on their design, construction, and optimization to establish the high conversion of light energy into chemical energy. In 2015, Huang et al. reported another visible light-responsive ternary boron carbon nitride (BCN), which works as a photocatalyst.[72–76] Such work should contribute to the advancement of photocatalytic research for clean renewable energy technologies and the purification of the environment so that future societies may thrive and prosper.

3. GRAPHITIC CARBON NITRIDE (G-C_3N_4)

Graphitic carbon nitride (known as g-C_3N_4) semiconducting nano-materials contain earth-abundant elements of C, N, and H. In general, carbon nitride exists in several allotropes such as α-C_3N_4, β-C_3N_4, graphitic-C_3N_4, cubic-C_3N_4, and pseudocubic-C_3N_4 with various different properties. Among them, graphitic carbon nitride (g-C_3N_4) is regarded as the most stable allotrope under ambient conditions.[77–79]

g-C_3N_4 nano-material consists of a conjugated polymeric system, being constructed from s-triazine or tri-s-triazine units interconnected via tertiary amines. The atoms in the layers are arranged in honeycomb configurations with strong covalent bonds. The force between the two-dimensional sheets (2D) is weak due to the van der Waals force.[80,81] g-C_3N_4 nano-material is thermally stable even in air up to 600°C, being attributed to its aromatic C—N heterocycles due to the van der Waals interactions between the layers. g-C_3N_4 nano-material is also chemically stable in various solvents such as water, alcohols, N, N-dimethylformamide, tetrahydrofuran, diethyl ether, and toluene, as well as glacial acetic acid and 0.1 M NaOH aqueous solution.[82,83] g-C_3N_4 nano-material is also easily controllable on its surfaces by the molecular level chemical modification[84,85] and surface engineering.[86,87] Moreover, g-C_3N_4 nano-material is flexible to be served as a host matrix of outstanding compatibility to various inorganic nano-particles, which is beneficial for the fabrication of g-C_3N_4—based nano-composites.[88–91] Density functional theory calculations showed that the valence band (VB) edge of g-C_3N_4 is mainly constructed with the N 2p state, whereas the conduction band (CB) edge consists of the hybridization of the N 2p and C 2p states.[61]

As mentioned above, g-C_3N_4 nano-material is only composed of C, N, and H. It is prepared by the thermal condensation of several low-cost nitrogen-rich precursors such as cyanamide,[78,92,93] dicyandiamide,[61,94–96] melamine,[97,98] urea,[85] and thiourea,[99] or mixtures thereof.[100] Normally, g-C_3N_4 is made of the tri-s-triazine (melem) unit (Fig. 2A). The X-ray diffraction (XRD) patterns of g-C_3N_4 are characterized by two strong diffraction peaks at c. 27.4° and c. 13.0°, which are used to determine its phase (Fig. 2B).

g-C_3N_4 nano-material possesses a small band gap of c. 2.7 eV with its CB and VB positions at c. −1.3 V and c. +1.4 V, respectively, versus normal

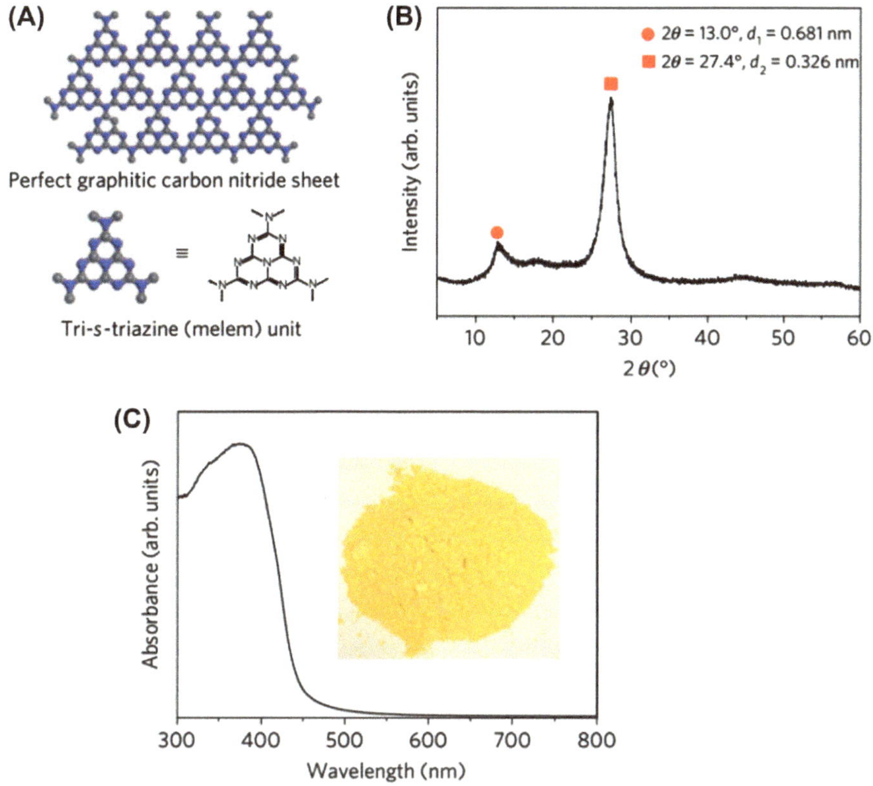

Figure 2 Crystal structure and optical properties of g-C_3N_4 material. (A) Schematic diagram of a perfect graphitic carbon nitride sheet constructed from melem units. (B) XRD pattern of the polymeric carbon nitride, revealing a graphitic structure with an interplanar stacking distance of aromatic units of 0.326 nm. (C) UV-Vis diffuse reflectance spectrum of the polymeric carbon nitride. *Inset: Photograph of the photocatalyst. Copyright © 2009 Springer Nature.*

hydrogen electrode at pH = 7.[101] This nano-material enables theoretically to operate as a visible light-responsive photocatalyst for various reactions such as H_2O splitting.

g-C_3N_4 nano-material was introduced into the field of heterogeneous catalysis in 2006.[102,103] Then, Wang et al. first utilized g-C_3N_4 nano-material as a low-cost, robust metal-free, and visible light-responsive photocatalyst for H_2O splitting to produce H_2 and O_2 in 2009.[61] Since then, great efforts have been carried out in the field of g-C_3N_4–based nano-material photocatalyst. In fact, as shown in Fig. 3, the publications and cross-citations concerning basic and applied studies on the synthesis

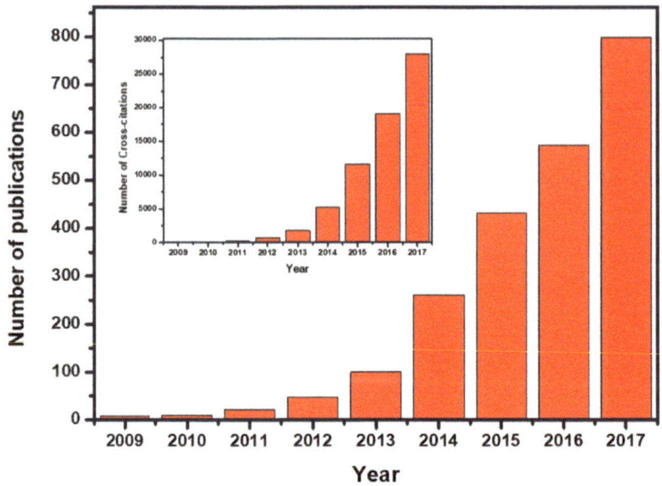

Figure 3 Number of annual publications using "Graphitic carbon nitride" and "photocatalysis" as a topic keyword since 2009. The insert is the cross-citations concerning basic and applied studies on the synthesis and modification of g-C_3N_4 for sustainable photocatalyst. *Adapted from ISI Web of Science, dated October 31, 2017.*

and modification of g-C_3N_4 as a photocatalytic material remarkably increased. Herein, we mainly focus on the latest progresses on the modification and photocatalytic application of g-C_3N_4 (mainly during 2015—17).

3.1 Design and Optimization of g-C_3N_4

3.1.1 Nano-structure Design

Due to the similar layered structure as in graphite, the theoretical specific surface area of an ideal monolayer g-C_3N_4 nano-material is as high as 2500 m^2/g.[66] The emergence of g-C_3N_4 having nano-structured semiconducting property has opened promising and flexible pathways to improve efficient and effective photocatalytic materials.[92,104] By applying a method of a simple and cost-effective liquid exfoliation, Yang et al. prepared those g-C_3N_4 nanosheets with minimal thickness, large aspect ratios, and favorable electronic band gap lead to a high photocatalytic activity for H_2 evolution from H_2O under visible light irradiation.[105]

By using inorganic or organic nano-structures as templates to synthesize nanostructural materials named as a template approach method, including hard-template approach and soft-template approach, is one of the most effective methods to synthesize porous or nanoarchitecture materials.[106,107] The size, morphology, and pore structure are easily modulated by the different selection of templates.[108,109] Various nanostructured g-C_3N_4

nano-materials, such as g-C_3N_4 nanosheets,[105] hollow g-C_3N_4 nanospheres (HCNSs),[110] nano-spherical carbon nitride frameworks,[111] helical g-C_3N_4 nano-rods,[112] have been synthesized by the hard-template approach.

Inspired by the structure of thylakoids in photosynthetic organisms in biological systems, the HCNSs were applied as a light-harvesting platform for the catalytic H_2 evolution from H_2O under visible light irradiation.[110] Moreover, the postannealing treatment of HCNS was found to be effective for reinforcing the interface/surface, leading to the improvement of the polymerization quality and optimizing the texture of HCNS, achieving a high photocatalytic activity for the H_2 evolution reaction.[113]

Zhang fabricated NS-g-C_3N_4 photocatalysts with highly opened up three-dimensional (3D) framework assembled from 2D nanosheets with sharp edge tips by using KCC-1 nano-structures as the template.[111] With excellent structural benefits for the mass transfer and especially for the charge separation, a remarkably improved H_2 evolution with an apparent quantum yield (AQY) of 9.6% at 420 nm light was reported with 3 wt% Pt loaded g-C_3N_4 nanosheet, Pt(0.3/NS-g-C_3N_4).[111]

On the other hand, Zheng synthesized helical g-C_3N_4 rods based on a chiral mesoporous silica template by a nanocasting method.[112] Zheng also found that this nano-material exhibits a photocatalytic activity toward H_2 or O_2 evolution from H_2O and the conversion of CO_2 into CO. For the helical rod-like g-C_3N_4 nano-material, it was found that unequal quantities of left- and right-handed helical nano-rods exhibited unique optical activity to circularly polarized light at the semiconductor absorption edge of g-C_3N_4.[112]

In addition, ionic liquids (ILs) have also been used as soft templates to react with dicyandiamide or urea to modify g-C_3N_4 nanoarchitectures. Lin et al. functionalized g-C_3N_4 as a multifunctional modifier for the self-polymerization of urea to control the texture, surface chemistry, and semiconducting properties of g-C_3N_4 nano-material, by introducing urea polymerized with ILs.[114] They also functionalized g-C_3N_4 nano-material by codoping with B and F elements in a simple one-pot fashion. The obtained g-C_3N_4 nanosheets exhibited an improved photocatalytic performance for H_2 evolution from H_2O as compared with the pure g-C_3N_4 nano-material.[114] The use of ILs as soft modifiers for the preparation of heteroatom-containing g-C_3N_4 nanosheets is known as a general and versatile route for tailoring the chemistry and morphology of the surfaces of conjugated g-C_3N_4 nano-materials.

Furthermore, very recently, a facile and green approach to prepare few-layered polymeric g-C_3N_4 polymeric carbon nitride (PCN) semiconductors nano-material was reported.[115] Bulky g-C_3N_4 nanomaterials obtained from typical precursors (urea, melamine, dicyandiamide, and thiourea) were exfoliated into few-layered nanosheets by a one-step carbon/nitrogen steam-reforming reaction, leading to occupation of the surface carbon vacancies. The unique nanosheet structures with strengthened surface properties endowed g-C_3N_4 nano-material with high active sites, and an increased charge separation efficiency with a prolonged charge lifetime, resulting in a remarkably enhancement in their photoredox performance. An AQY for the H_2 evolution of 11.3% at 405 nm light was recorded for the PCN nanosheets.

3.1.2 Crystalline-Structure Engineering

Pristine g-C_3N_4 synthesized by the traditional thermal-induced polycondensation of N-containing precursors with a lower crystallinity is known to exhibit a moderate photocatalytic activity because of the kinetic problem encountered during the bulk condensation process.[116] In fact, an improvement in crystallinity of the pristine g-C_3H_4 nano-material resulted in an enhancement of the photocatalytic activity due to a minimization of the number of defects, which act as recombination centers for the photogenerated electrons and holes.[117,118]

For the efficient photocatalytic performance for the decomposition of H_2O, there are three important steps as shown in Fig. 4: (1) electron–hole pairs are generated in semiconducting photocatalysts through light absorption with energy equal to or higher than the electronic band gap energy of the photocatalyst; (2) charge carriers (separated electrons and holes) migration to the surface of the photocatalyst on which the redox reaction proceeds or recombine in the bulk of the photocatalyst; (3) photogenerated electrons in the CB reduce H^+ to produce H_2, while photogenerated holes in the VB oxidize OH^- to produce O_2 via the formation of H_2O_2 on the surfaces.

Lin et al. synthesized crystalline g-C_3N_4 semiconductor nano-material with tri-s-triazine subunits using preheated melamine as precursor in combination with KCl/LiCl salts melt method.[116] This synthetic method was found to induce the effects on not only crystallinity of the sample but also enhancement in mobility of photogenerated charge carriers, electrons, and holes, as well as increase in the surface area. The synthesized crystalline g-C_3N_4 sample exhibited a much higher activity for H_2 production from H_2O with an AQY of 50.7% at 405 nm light.[116]

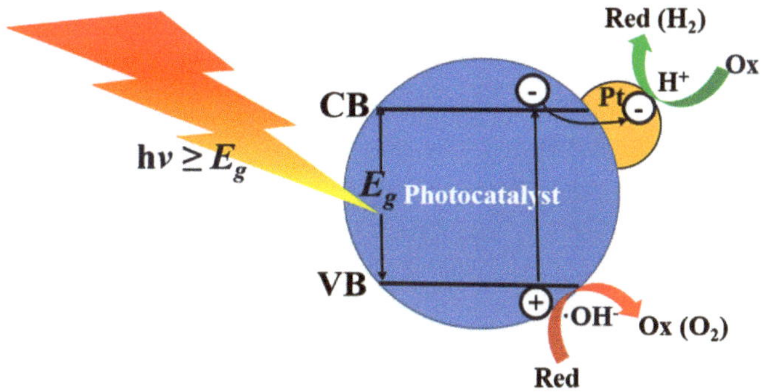

Figure 4 Schematic illustration of photocatalytic overall water splitting over a semiconductor photocatalyst modified with both H_2- and O_2- evolution cocatalysts.

Moreover, Ou et al. prepared tri-s-triazine–based crystalline carbon nitride, C_3N_4, nanosheets (CCNNSs) by applying a conventional and cost-effective sonication–centrifugation process.[119] The tri-s-triazine–based CCNNSs was found to exhibit a significantly enhanced photocatalytic activity for H_2 production from H_2O under visible light irradiation as compared to that of g-C_3N_4, poly (triazine imide)/Li^+ Cl^-, and bulk tri-s-triazine–based crystalline carbon nitrides (Fig. 5), exceeding the values obtained with carbon nitride nanosheets.[119]

The preparation methods of g-C_3N_4 are categorized into top-down and bottom-up approaches.[120] More recently, Zhang et al. reported a facile bottom-up strategy to improve the photocatalytic activity of g-C_3N_4 nano-material to a level in which most of the absorbed photons are really utilized to induce photoredox chemistry by the cocondensation reaction of urea and oxamide (OA) followed by its postcalcination in molten salt (KCl/LiCl).[121] The obtained CN-OA-m was found to produce highly crystalline species with an obviously reduced interlayer stacking distance of heptazine units of 0.292 nm and additional donor–acceptor (D–A) structure, as well as an expanded absorption spectrum up to 500 nm. These properties were also found to improve charge migration in the lateral direction and an exciton dissociation at the interlayers (Fig. 6). The addition of OA resulted in a decrease in the electronic band gap from 2.74 to 2.56 eV, leading to an efficient photocatalytic reaction even under green light irradiation. The AQY for the H_2 evolution reached 57% and 10% at 420 and 525 nm, respectively, which were much higher than those for the current results obtained with many polymeric photocatalysts.[121]

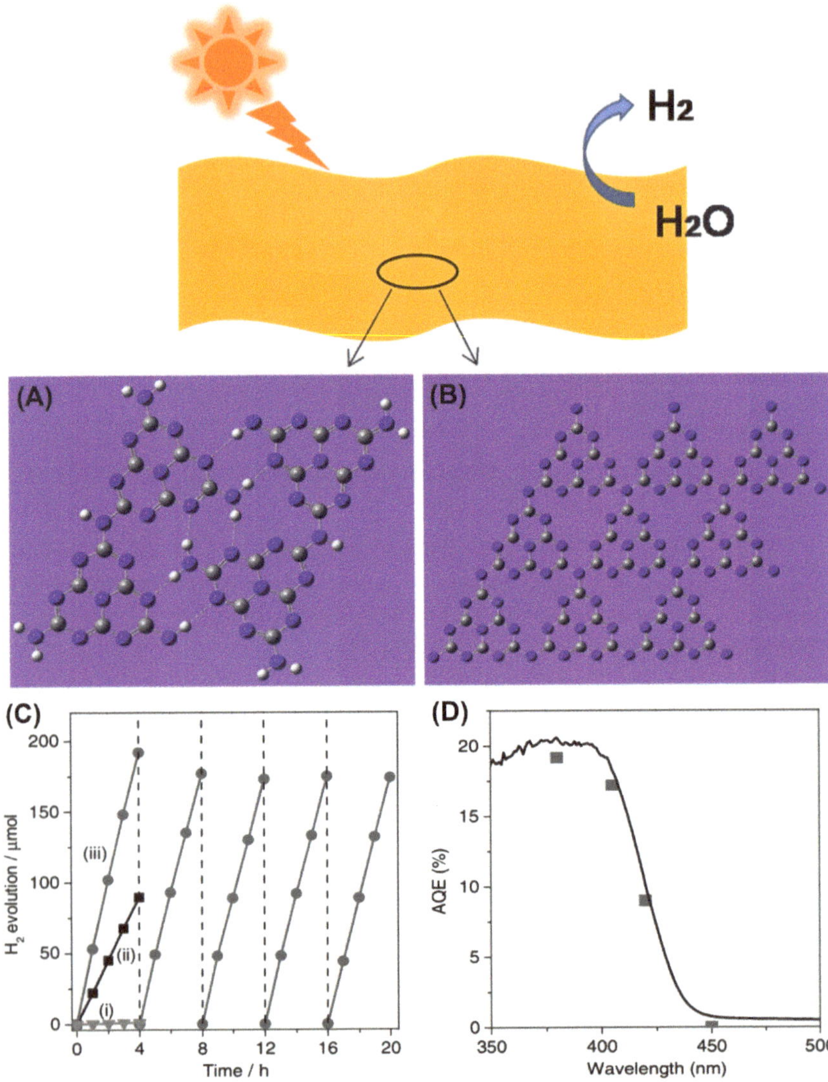

Figure 5 Schematic illustration of carbon nitride nanosheets for hydrogen evolution under visible light. A single layer atomic structure of (A) the pristine carbon nitrides with hydrogen bonds and (B) the crystalline carbon nitrides. H, C, and N atoms are denoted by white, gray, and blue balls, respectively. Photocatalytic activity test of the samples. (C) Photocatalytic activities test of bulk graphitic carbon nitrides (i), bulk crystalline carbon nitrides (ii), and crystalline carbon nitride, C_3N_4, nanosheets (iii) under visible light irradiation ($\lambda > 420$ nm). (D) Wavelength-dependent apparent quantum efficiency (AQE) of H_2 evolution over nanosheets (left axis), UV/Vis light absorption spectra of nanosheets (right axis). *Copyright © 2017 John Wiley and Sons.*

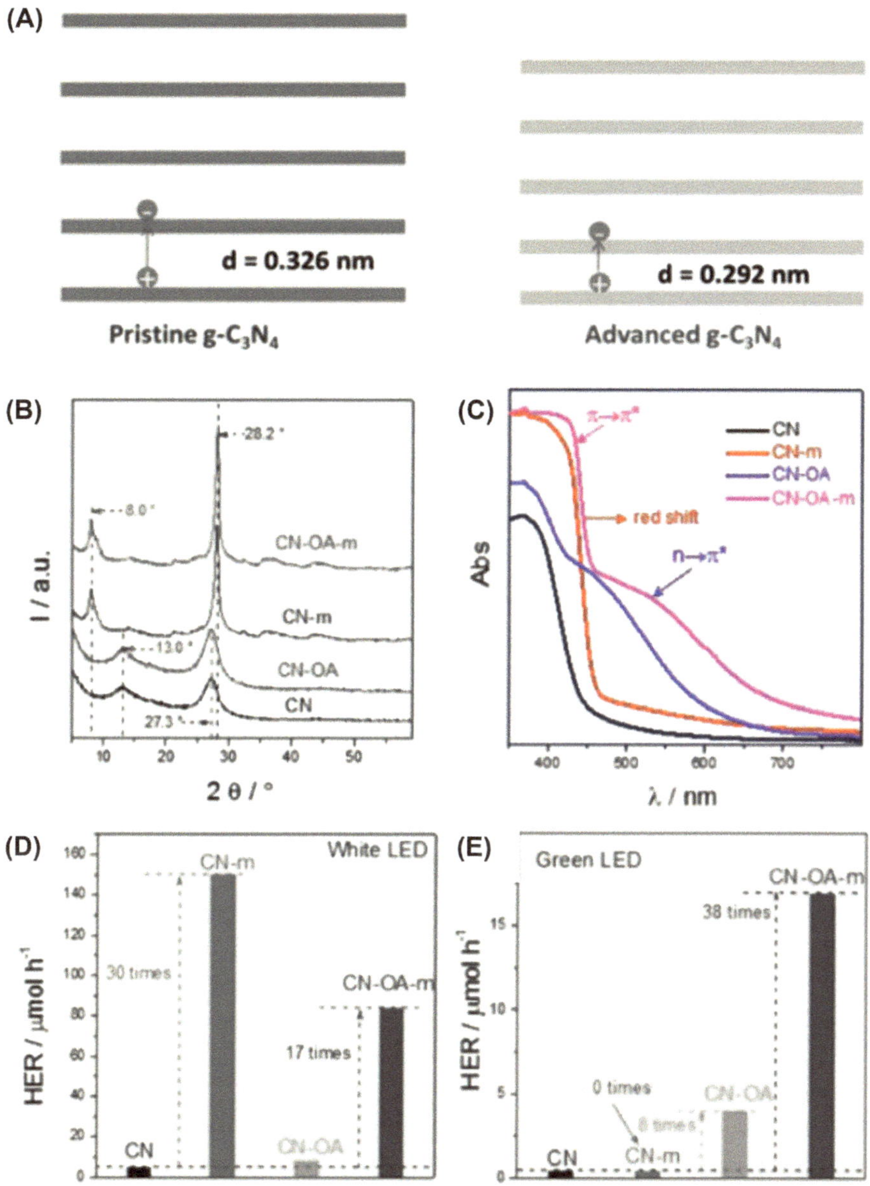

Figure 6 (A) Proposed interlayer charge carrier transfer of (left) pristine CN (d = 0.326 nm) and (right) optimized CN-OA-m with shortened layer stacking distance (d = 0.292 nm); (B) Powder XRD patterns of CN, CN-OA, CN-m, and CN-OA-m; (C) UV/Vis absorption spectra of CN, CN-M, CN-OA, and CN-OA-m; (D) Photocatalytic H_2 evolution activities of the prepared samples under white and (E) green LED light irradiation. *Copyright © 2017 John Wiley and Sons.*

3.1.3 Elemental Doping

Doping of g-C_3N_4 nano-material with nonmetals, such as B,[122] C,[123] O,[124] S,[125] P,[126] F,[127] Br,[128] I,[129] and combinations of these anions into g-C_3N_4,[114,130] can modify its texture, optical, and electronic properties.[65] Such band gap engineering of g-C_3N_4 nano-material by incorporation of anions, cations, or codoping of both elements plays an important role in the modification of the light absorption and redox potentials of g-C_3N_4 nano-material.

For example, Zhang et al. employed dicyandiamide and ammonium iodine to fabricate iodine-doped g-C_3N_4 (I-doped g-C_3N_4) photocatalysts.[129] The intimate interaction between g-C_3N_4 and iodine atoms was found to extend the π-conjugated aromatic heterocycle, resulting in an enhancement of the mobility of free charge carriers, being directly observed by electron paramagnetic resonance (EPR) analysis (Fig. 7A).[129] I-doped g-C_3N_4 exhibited an increased surface area, prolonged optical absorption range, as well as improved charge separation and an excellent high yield for H_2 production as compared with that of the original g-C_3N_4 (Fig. 7B).[129]

A series of modifications of g-C_3N_4 nano-material with Br via one-pot cocondensation of urea with NH_4Br was reported.[128] The incorporation of Br into g-C_3N_4 nano-material was found to modify its texture, optical absorption, conductivity, charge carrier migration rate, and photocatalytic performance, without destroying the major construction and architecture of the g-C_3N_4 nano-material.[128] The optimized CNU–$Br_{0.1}$ nano-composite exhibited an improvement of the photocatalytic activity and the stability for the photocatalytic H_2 evolution from H_2O under visible light irradiation (Fig. 7C and D). Furthermore, the photocatalytic potential of CNU$Br_{0.1}$ for H_2O oxidation was found to be greatly enhanced by loading of different amounts of cobalt oxide (CoO_x) as H_2O oxidation cocatalyst.[128]

3.1.4 Copolymerization

An important challenge in artificial photosynthesis is to develop earth-abundant 2D nanosheets with suitable electronic band structures because such 2D nanosheets have a high surface area and many active surface sites with N atoms.[61] The photocatalytic properties of bulk g-C_3N_4 nano-material originate from its π-conjugated system constructed with sp2 hybridization of C and N.[61] Therefore, it is desirable to extend the delocalization of the π electrons and to change the intrinsic semiconductor properties by a simple one-pot chemical condensation of urea with aromatic groups to form a

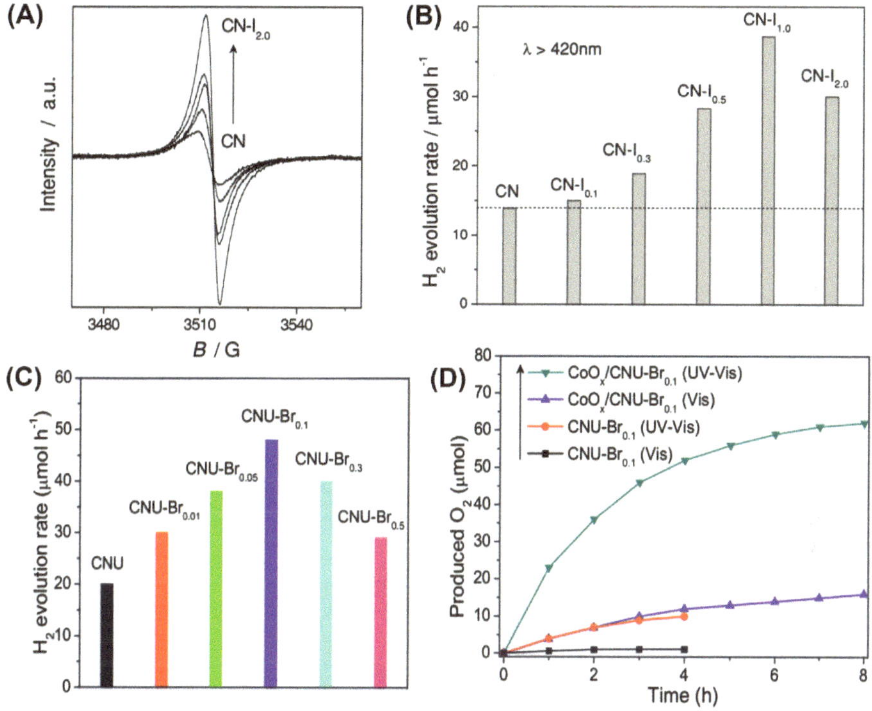

Figure 7 (A) Room temperature EPR spectra for CN and CN-I_x samples, (B) Photocatalytic H_2 evolution rate for the CN-I_x nanomaterials with different iodine doping amounts under visible light irradiation, (C) photocatalytic performance for H_2 evolution of pure g-C_3N_4 and doped g-C_3N_4 with different nonmetal elements, (D) O_2 evolution by CoO_x/CNU−$Br_{0.1}$ and CNU−$Br_{0.1}$ nanomaterials under visible light ($\lambda \geq 420$ nm) or UV−Vis light irradiation. *Copyright © 2016 Elsevier & © 2013 John Wiley and Sons.*

functionalized 2D g-C_3N_4. Such strategy modulates the π-delocalized system of g-C_3N_4 and its electronic band structure, resulting in a red shift of the absorption edge and an improved charge separation in the semiconductor nanomaterial, as well as the photocatalytic performance, leading to an enhancement in the photocatalytic H_2 evolution rate (HER) by a factor of about 9 compared to the sheet-like default sample and a quantum efficiency determined to be 8.8% at 420 nm by CNU−ATCN[131] (Fig. 8).

More recently, Yang et al. applied a mixture of nucleobases and urea to construct layered carbon nitrides. Different nucleobases such as adenine, guanine, cytosine, thymine, and uracil have been applied.[85] These biomolecule-derived binary carbon nitride nano-material enabled the generation of

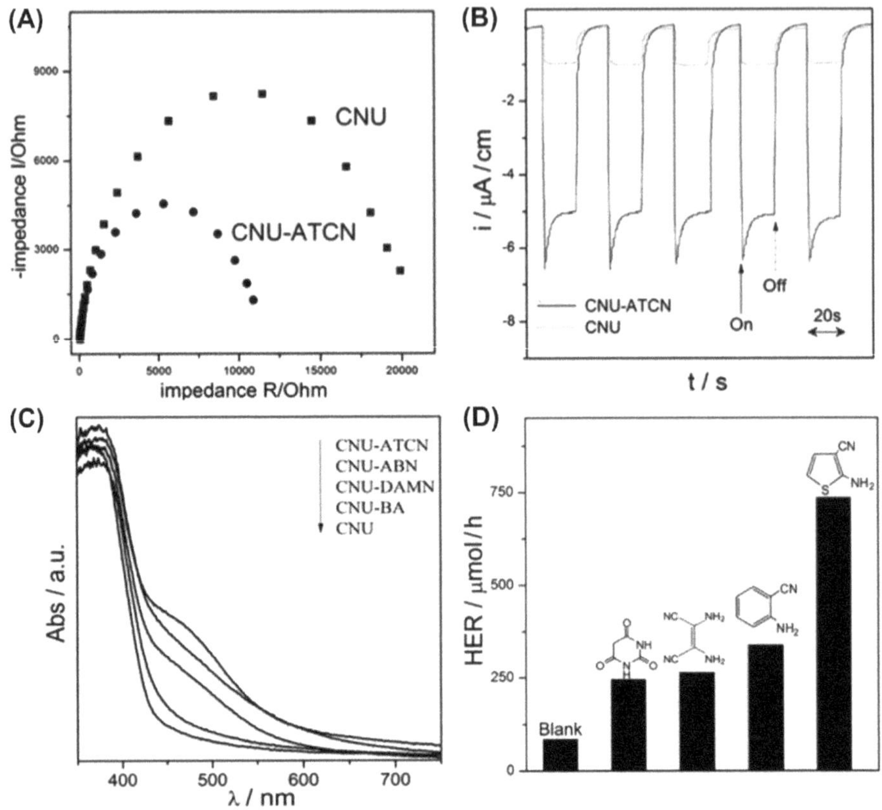

Figure 8 Photoelectrochemical properties of the CNU and CNU–ATCN samples at a −200 mV bias potential versus Ag/AgCl in a 0.2 M Na_2SO_4 aqueous solution (pH = 6.8). (A) Electrochemical impedance spectroscopy Nyquist plots in the dark, and (B) periodic on/off photocurrent response under visible-light irradiation ($\lambda > 420$ nm), (C) UV–Vis Diffuse reflectance spectroscopy spectra of the samples, (D) H_2 evolution rates (HERs) of CNU and modified CNU photocatalysts under visible light irradiation, using 3 wt% Pt particles as cocatalysts. *Copyright © 2014, Royal Society of Chemistry.*

energized charge carriers under visible light irradiation enabling to induce photoredox reactions such as H_2 production, as well as heterogeneous organo synthesis of C–O, C–C, C–N, and N–N bonds.

3.1.5 Construction of Carbon Nitride Colloids

g-C_3N_4 was prepared by thermal-induced self-polymerization of cyanamide at 823 K, which resulted in a condensed solid with poor contact between the large aggregated particles and substrates, as well as abundant boundary defects. These properties have limited the applications of g-C_3N_4 in the field of photochemistry and photoelectrochemistry.[132] Zhang et al. demonstrated a promising method for the production of a stable colloidal suspension of g-C_3N_4 (CN-Sol) by applying chemical protonation with strong oxidizing acids such as HNO_3.[133] However, various problems such as abundant boundary defects and poor contact between particles and substrates were reported in the fabrication of thin-film electrodes. After that, as expected, the thin-film electrodes prepared from the CN-Sol were found to show a remarkable performance.[133] Furthermore, the high flexibility and compatibility of the polymer system of CN-Sol were modified with desired functionalities (for example, carbon nano-tubes, activated carbon, and mesoporous carbon membranes), and then such modifications in unique nitrogen-rich chemical functions opened new ways for the application of g-C_3N_4 for specific tasks.

As bridging nano- and macro-dimensions, the assembly of low-dimensional nano-structures into macroscopic aerogel frameworks has been known to play a major role in the development of advanced functional materials,[134,135] as well as creating new properties originating from the collective interparticle interactions. The assembly of g-C_3N_4 nano-structures into macroscopic aerogel frameworks has rarely been reported because of poor interfacial properties such as very limited active groups at edges, large lateral sizes, peculiar thermal stability, and insolubility in most conventional solvents.[133,136] However, Ou et al. prepared an aerogel structure of g-C_3N_4 by self-assembly without using any cross-linking agents.[137] (Fig. 9) This nano-material is characterized by a large surface area, incorporated functional groups and 3D network structures. They applied this nano-material as a photocatalyst and found that it operated as a good photocatalyst with a high activity for H_2 evolution from H_2O, as well as H_2O_2 production from H_2O in the presence of O_2 under visible light irradiation.

3.2 Construction of g-C_3N_4—Based Nano-composites

The coupling of different semiconductors induces heterojunctions and/or influences on the mobility of electrons and holes, as well as their trapping rates in semiconductors, resulting in differences in the reduction and oxidation processes.[138,139] Often, such coupling procedures lead to a remarkable decrease in their photocatalytic performance because coupling influences various factors, including surface area, defect density, crystallinity, and quantum size effects, etc., which are closely related with the photocatalytic performance.[140]

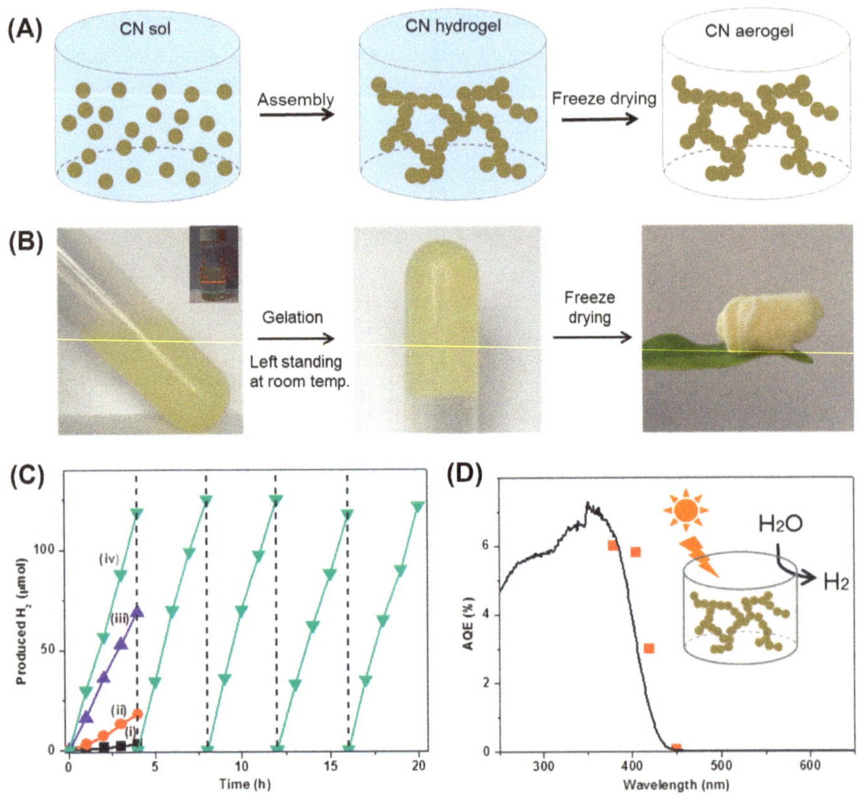

Figure 9 (A) Fabrication processes of CN-aerogel; (B) Prepared colloidal solutions of CN-NPs (inset: Tyndall effect of CN-NPs aqueous solution), CN-hydrogels and CN-aerogels; (C) Photocatalytic H_2 production from H_2O on CN_{450} (i), CN (ii), CN-xerogels (iii), and CN-aerogels (iv) under visible light irradiation ($\lambda > 420$ nm); (D) Wavelength-dependent AQE of H_2 production on CN-aerogels (left axis) and UV/Vis light absorption spectra of CN-aerogels (right axis). *Copyright © 2017 John Wiley and Sons.*

The polymeric structure of g-C_3N_4 nano-material provides high flexibility, which is predominant and suitable for the formation of van der Waals heterojunctions[81] with close interconnection between g-C_3N_4 nano-material and various semiconductors. Typically, g-C_3N_4−based nano-material photocatalysts with heterojunctions are engineered by utilizing g-C_3N_4 and other semiconductor (or cocatalyst) such as TiO_2[141] and CdS.[142,143] The incorporation of g-C_3N_4 nano-material with another semiconductor unit with dissimilar electronic band structures results in the formation of nano-materials with new electronic band structures. In other words, electronic energy band bending is created at the interface of the heterojunction nano-composites as a result of the generation of potential differences between two sides.[144] This, in turn, induces a built-in electric field within space charge regions to enhance the spatial separation and migration of the photogenerated electrons and holes.[145]

An inorganic-polymeric hybrid nano-heterostructure of CdS/g-C_3N_4 nano-composite was prepared by integrating CdS quantum dots (QDs) onto the exterior HCNSs. This nano-composite exhibited an efficient H_2 evolution from H_2O under visible light illumination.[146] An ordered and long-sized assembly of g-C_3N_4 nano-material with many accessible active sites is very important for the application of g-C_3N_4. The unique 3D hollow architectural framework of HCNS was applied as a polymeric scaffold to form intimate interfacial contact with QDs by a self-assembly force, which forms a typical type II heterojunction.[132,132a] This type II heterojunction operated to facilitate surface kinetics of the charge separation of photogenerated electrons and holes, as well as mass transfer of the reactants and products. The schematic representation of the charge transfer process of the photogenerated electrons and holes in CdS—HCNS composite is shown in Fig. 10A.[146] In the CdS/HCNS nano-composite, electrons and holes are effectively photogenerated in both HCNS and CdS QDs under visible light irradiation. Then, as the CB of HCNS is more negative than that of CdS QDs, an internal local electric field is formed, in which electrons in HCNS effectively transfer to the CB of CdS QDs. And then, electrons in CdS QDs move to and accumulate on Pt nano-particles on CdS QDs, on which electrons reduce H^+ to induce H_2 evolution. On the other hand, photogenerated holes in CdS QDs, effectively transfer to the VB of HCNS. Thus, the heterojunction effectively operates to separate photogenerated electrons in CdS QDs and holes in HCNs, respectively, reducing their recombination. These situations lead to a remarkable increase of H_2 evolution (Fig. 10B), as well as a drastic decrease in the photocorrosion of CdS QDs.[146]

Kailasam et al. reported that the WO_3/mpg-C_3N_4 nano-composites of mesoporous polymeric carbon nitride (mpg-C_3N_4) and WO_3 showed very high photocatalytic activity for the evolution of H_2 from H_2O under visible light irradiation in the presence of sacrificial electron donors.[147] The photogenerated electrons in the CB of g-C_3N_4 transfer to the CB of WO_3. Similarly, the transfer of photogenerated holes from the VB of WO_3 to g-C_3N_4 results in a lower oxidation ability of the nano-composite.[147] However, contrary to the above expectations, it was reported that the yield of H_2 evolution is considerably higher than that for the pure g-C_3N_4 system (Fig. 10D), which implies that the electrons in the CB of g-C_3N_4 are effectively utilized

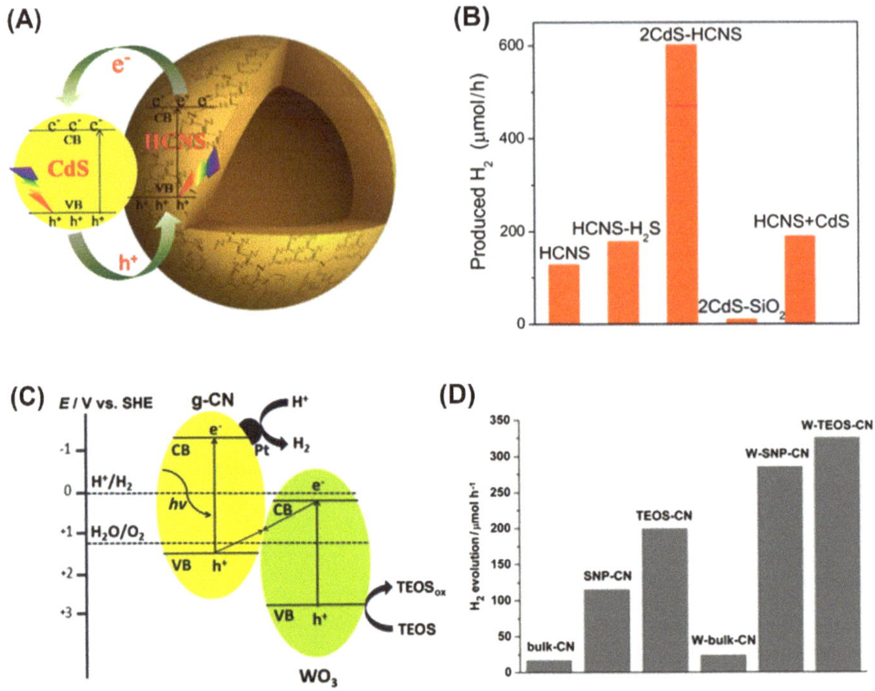

Figure 10 (A) Schematic illustration of the CdS–HCNS heterojunction system; (B) Photocatalytic hydrogen evolution by CdS–HCNS photocatalysts under visible light irradiation ($\lambda > 420$ nm), with some reference samples of HCNS, HCNS–H_2S, 2CdS–SiO_2, physically mixed HCNS and CdS. (C) Schematic diagram of the solid state Z-scheme photocatalytic mechanism in g-C_3N_4/WO_3 composites (SHE = standard hydrogen electrode); (D) Photocatalytic H_2 generation by bulk, mesoporous CNs and the respective CN/WO_3 composites. *HCNS*, hollow g-C_3N_4 nanospheres. *Copyright © 2015 Elsevier; © 2015 John Wiley and Sons.*

to reduce H^+ to produce H_2, whereas the holes in the VB of WO_3 effectively participate in the oxidation of triethanolamine as a sacrificial electron donor.[147] These processes occurring on the WO_3/g-C_3N_4 nano-composite are regarded to be similar to the Z-scheme mechanism occurring in natural photosynthesis.[148]

Apart from heterojunction structures, it has been reported that semiconductors operate as cocatalysts to improve the photocatalytic reaction. Various metal oxides (NiO,[149] CoO_x[150]), metal sulfide (WS_2,[151] MoS_2,[152] and $[Mo_3S_{13}]^{2-}$ clusters[153]), and MOFs (Co-ZIF-9[154]) have been utilized as cocatalysis for g-C_3N_4 photocatalytic systems, which take the place of Pt catalyst, resulting in an enhanced photocatalytic performance. Zhang et al. have successfully engineered a core-shell Ni/NiO architecture on the surface of g-C_3N_4 nano-material by an impregnation method, resulting in an enhancement in the photocatalytic H_2 evolution from H_2O under visible light irradiation.[149] This high photocatalytic activity and durability of this core-shell Ni/NiO/g-C_3N_4 nano-composite as compared with those of g-C_3N_4 nano-materials modified by Ni or NiO are attributed to the well-constructed nano-structure leading to the prevention of charge build-up on the surface of g-C_3N_4. Thus, the enhanced photocatalytic activity of g-C_3N_4 of this nano-composite is attributed to the promoted charge carrier transfer rate and the reduced overpotential for H_2 evolution.

3.3 Photocatalytic Application of g-C_3N_4 Nano-material

The unique characteristics of g-C_3N_4 nano-material make it a promising photocatalyst for various reaction systems. Up to now, a variety of g-C_3N_4—based photocatalytic systems have been reported in versatile redox reactions, including H_2O splitting for H_2 and/or O_2 evolution,[61] reduction of CO_2 into hydrocarbon fuels,[155] pollutants decomposition,[156] bacterial inactivation,[109] photocatalytic organic synthesis,[157] and photoelectronic devices.[64]

3.3.1 Photocatalytic Water Splitting With g-C_3N_4 Nano-material

There are many advantages of g-C_3N_4 photocatalysts to apply to H_2O splitting: (1) the electronic band gap of 2.7 eV is sufficiently large to overcome the endothermic reaction of the H_2O splitting reaction and the optical absorption edge at around 460 nm allows to utilize sunlight more efficiently and effectively.[61] (2) the electronic band structure of g-C_3N_4 enables to cover the electrochemical redox potentials of H_2O, as a result, both reduction of H^+ by electrons to produce H_2 and oxidation of OH^- by holes to liberate O_2 are expected to sufficiently proceed on g-C_3N_4. (3) g-C_3N_4 is chemically and thermally stable and not subjected to photocorrosion in the H_2O splitting reaction unlike other semiconductors such as CdS.[101] (4) g-C_3N_4 nano-material has a specific microstructure and many surface terminations such as defects and N atoms, which function as electron localization or anchoring inorganic/organic functional moieties on the surfaces as active sites.[158]

In the pioneering work, Wang et al. applied g-C_3N_4,[61] which was prepared by heating cyanamide at temperatures in the range of 673—873 K for 4 h, in the photocatalytic reaction to decompose H_2O into H_2 and O_2 under visible light irradiation. The production of H_2 with bare

$g\text{-}C_3N_4$ was observed to fluctuate and demonstrated discrepancies from batch to batch (0.1−4 μmol/h). However, they found that by modifying the sample with 3 wt.% Pt, the amount of H_2 evolution reached 770 μmol after 72 h, which far exceeded the amount of C_6N_8 (tri-s-triazine) units or the deposited Pt nano-particles.[61] They also explored the evolution of O_2 from the surface of $g\text{-}C_3N_4$ loaded with small amounts of RuO_2 as cocatalyst, which is known as a typical oxidation catalyst for O_2 generation, in the presence of $AgNO_3$ as a sacrificial electron acceptor, and found that under UV irradiation ($\lambda > 300$ nm), the total amount of O_2 production reached 53 μmol after 8 h.[61]

3.3.2 Water Splitting to Produce H_2

To investigate the H_2 production half reaction of the overall H_2O splitting reaction, electron donors such as triethanolamine, methanol, and ethanol, etc. have been applied into the aqueous system to accelerate the oxidation reaction, being the rate determining step of the overall reaction.[63,66,80] Furthermore, cocatalysts, such as Pt and RuO_2, play important roles in trapping photogenerated electrons and holes, respectively. This leads to a suppression in the recombination of the electrons and holes, resulting in a prolonging charge carriers' lifetime.[159] Meanwhile, these cocatalysts work as real catalysts to enhance the surface redox reactions by decreasing activation energies (overpotential), facilitating multielectron transfer reactions to improve the photocatalytic performance.[61,160]

As cocatalysts to enhance H_2 evolution in H_2O splitting, noble metals such as Pt,[133,146,161,162] Pd,[163,164] Rh,[78] Au,[165,166] and Ag[167,168] are well known. These noble metal cocatalysts exhibit low overpotential, being close to zero for H_2 evolution; therefore, they exhibit excellent performance to enhance H_2 evolution. Among them, Pt is known to be the most effective cocatalyst for H_2 evolution due to its largest work function and its lowest activation energy for the H_2 production reaction.

Furthermore, the amounts of cocatalysts loaded on the $g\text{-}C_3N_4$ nanomaterial photocatalyst influence the photocatalytic activity.[78] For example, at lower loading contents, the photocatalytic activity increases with increasing the amounts of cocatalyst. In such ranges, the active sites available for the redox reactions increase with increasing amount of cocatalyst. However, further increasing the amounts of cocatalyst, passing through a maximum, leads to a decrease in the number of surface active sites on which reaction molecules adsorb as reactive species. Also, a large amount of cocatalyst on the semiconductor leads to a shielding effect on the incident light and prevents the absorption of light. Even worse, a large amount of loaded cocatalyst leads to induce a function as the charge carrier recombination center on the surfaces, resulting in a low performance due to shortening the lifetime of electrons and holes.

In addition, transition metals such as Fe, Co, and Ni, as well as transition metal compounds such as $Cu(OH)_2$,[169] MoS_2,[152,170] NiO,[150] $Ni(OH)_2$,[171] NiS_x,[172] WS_2,[151,173] and CoP,[174,175] have been applied as cocatalysts to improve the photocatalytic performance of g-C_3N_4 nano-material for the H_2 evolution reaction. By comparison with noble metals, these transition metal complexes have various advantages such as low-cost, low-toxicity, abundant resources and a variety of properties.

Layered MoS_2/g-C_3N_4 nano-material exhibited a remarkable improvement for the photocatalytic H_2 evolution.[170] As is shown in Fig. 11, the

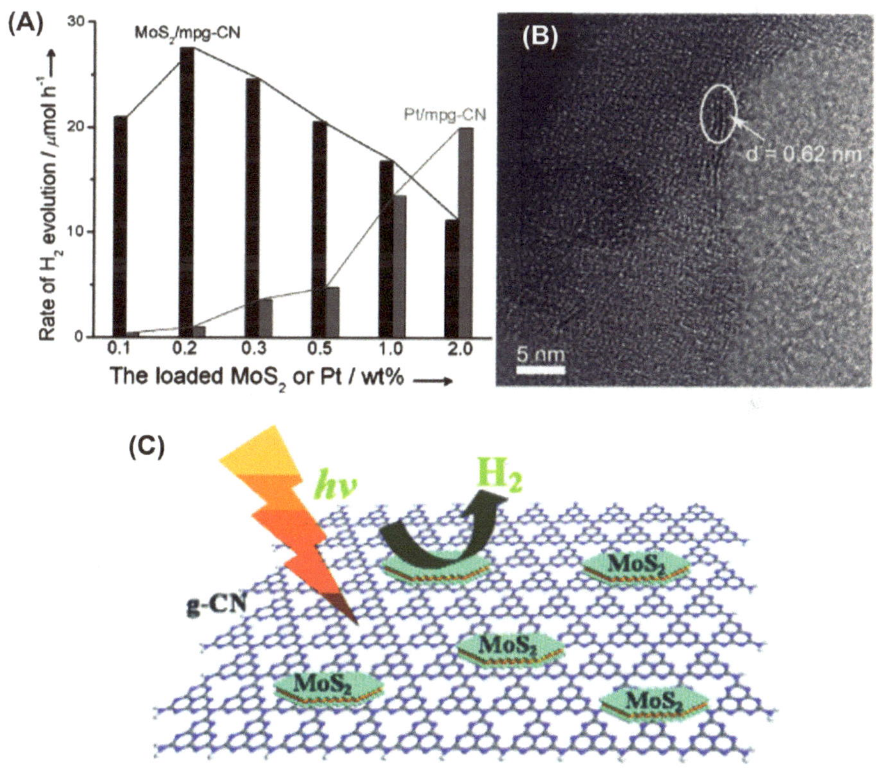

Figure 11 (A) The rate of H_2 production over mpg-C_3N_4 loaded with different amounts of MoS_2 or Pt; (B) TEM images for MoS_2/g-C_3N_4; (C) The procedure for gas-phase sulfidation and the idealized structural model of the resultant MoS_2/g-CN–layered junctions. *Copyright © 2013 John Wiley and Sons.*

MoS_2(1.0 wt%)/g-C_3N_4 nano-material exhibited the higher H_2 evolution activity than that of the Pt/g-C_3N_4 nano-material. This high activity was attributed to the similar layered geometries of MoS_2 and g-C_3N_4, operating to enhance the charge carrier transfer at the interfaces and prolonging the charge carrier lifetime. Future research is focused on how to improve the stability of these materials for practical applications.

It has been reported that by increasing the dimension of conjugated polymers from 0D to 1D, to 2D, and to 3D, mobility of electron becomes higher and the binding energies of the bound electron–hole pairs are reduced, as well as the robust stability of the extended conjugated units is increased.[81,176] 2D graphene nanosheets incorporated with g-C_3N_4 led to a decrease in the charge carriers' recombination rate[177] and also it works as cocatalyst to lead to an increase in the photocatalytic H_2 evolution. The optimum activity of this system for the photocatalytic H_2 evolution was found at the deposition amount of graphene of 1 wt % on g-C_3N_4, its extent exceeding more than 3.1 times as compared with that of pure g-C_3N_4 (Fig. 12A). As shown in Fig. 12B, in such system, the photogenerated electrons in g-C_3N_4 easily transfer to the surface of graphene, cocatalyst, inducing a decrease of the charge carrier recombination, being beneficial for the subsequent H_2O redox reactions.

3.3.3 Water Splitting to Produce O_2

The O_2 production half reaction of the overall H_2O splitting reaction, which involves the four-electron process to break the O–H bond of OH^- and to form an O–O bond. As mentioned above, the O_2 formation reaction is more restrained in comparison with the H_2 formation reaction, being regarded as the key bottleneck, rate determining step, to achieve an efficient and high overall H_2O-splitting reaction rate. Thus, in the H_2O splitting reaction, the key point is how to develop an efficient and sustainable H_2O oxidation cocatalyst.

Co is known as an active cocatalyst for the photo/electrochemical water oxidation reaction. The performance of g-C_3N_4 nano-material photocatalyst relies strongly on the contact and adhesion of Co species with g-C_3N_4 nano-material. The impregnation of Co into g-C_3N_4 frameworks has been investigated by applying a facile soft-chemical method.[150] Interestingly, by impregnation of Co into g-C_3N_4 nano-material, the optical absorption, charge carriers transfer rate, and electronic conductivity of g-C_3N_4 have been improved, whereas the basic graphitic semiconductor

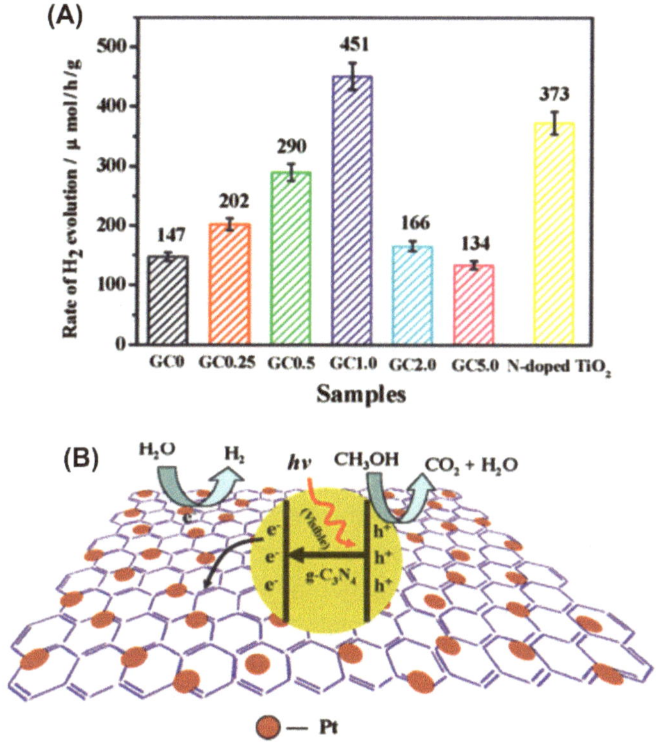

Figure 12 (A) Comparison of the photocatalytic activities of samples with different loading amounts of graphene on g-C_3N_4 polymers, (B) proposed mechanism for the enhanced electron transfer process of the graphene/g-C_3N_4 polymers.

nano-structure of g-C_3N_4 scarcely changed. As a result, the Co/g-C_3N_4 nano-composite photocatalyst enhances the O_2 production reaction as compared with that of the original g-C_3N_4.[178]

The layered $Co(OH)_2$/g-C_3N_4 nano-composite has been prepared by the facile impregnation method and utilized for the photocatalytic O_2 production under UV–Vis and visible light irradiation.[179] An enhancement in the photocatalytic O_2 formation has been attributed to not only an acceleration of the interface transfer rates of the photogenerated charge carriers but also a reduction in the excessive energy barrier for the formation of O_2. The optimum O_2 evolution rates (OERs) from H_2O with $Co(OH)_2$/g-C_3N_4 photocatalyst reached 27.4 and 7.1 μmol/h under UV–Vis ($\lambda > 300$ nm) and visible light ($\lambda > 420$ nm) irradiation, respectively.[179] These values are 5.5 and 7.0 times higher than those with the original g-C_3N_4 nano-material. Similar results have been reported recently for the c-$CoMn_2O_4$/g-C_3N_4 photocatalytic system, in which the OER was enhanced to be four times higher than that of the original g-C_3N_4.[179]

3.3.4 Overall Water Splitting to Produce H_2 and O_2

As mentioned above, the following three steps are important to achieve highly efficient photocatalytic decomposition of H_2O: (1) photogeneration of electron–hole pairs; (2) separation of photogenerated charge carriers migration to the surface; (3) redox reactions by electrons and holes on the surfaces. The first two steps are closely related to the properties of the photocatalytic material such as optical electronic properties (absorption), specific surface area, texture, and microstructure.[180] The third step is the most difficult and rate determining step. As a result, photocatalytic materials, which theoretically enable to proceed the overall H_2O splitting reaction are limited by the following requirements: (1) the electronic band gap energy should be larger than 1.23 eV (to meet the thermodynamic barrier, ideally the band gap should be 2.0 eV after considering the kinetic barriers); (2) the conduction (or valence) band potentials should be sufficiently negative (or positive) for the H_2O reduction (or oxidation) half reactions; (3) it should be stable enough toward light and solution corrosion.[181] Beyond these strict requirements, g-C_3N_4 nano-material is expected to be an excellent H_2O-splitting photocatalyst with a strong absorption in the visible region, an efficient charge separation and fast migration to the interfaces on which redox reactions proceed.[61]

Zhang et al. have demonstrated that the g-C_3N_4–conjugated polymer nano-material induces a one-step H_2O splitting reaction to produce H_2 and O_2 in a good stoichiometry when its surfaces are modified by small amounts of Pt, PtO_x, and CoO_x via the photodeposition method.[182] The optimal (Pt,PtO_x, CoO_x)/g-C_3N_4 nano-composite exhibited a turnover number of 3.1 for the overall H_2O splitting reaction. These nano-composites were stable in H_2O and under light irradiation, and the results clearly show that these kinds of nano-materials work as a new, clean, renewable, and highly efficient photocatalyst under sunlight irradiation.

To avoid the back reaction of H_2 and O_2 into H_2O, an approach to fabricate Pt and Co_3O_4 nano-particles onto the interior and exterior surface, respectively, of the hollow g-C_3N_4 (HCNS) has been achieved.[183] (Fig. 13A) As shown in Fig. 13B, the enhanced photocatalytic activity of this nano-composite system has been attributed to the spatially separated redox reaction active sites for the evolution of H_2 and O_2 and also to the

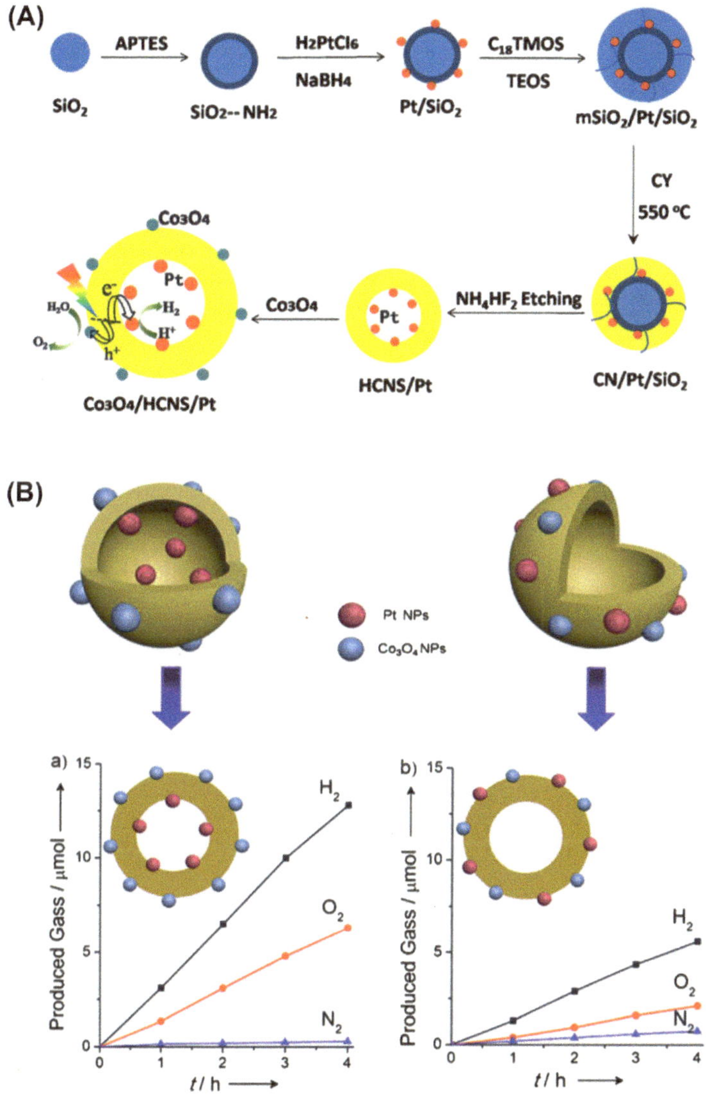

Figure 13 (A) An illustration of the preparation of Co_3O_4/HCNS/Pt composites; (B) Time courses of photocatalytic evolution of H_2 and O_2 by using relevant chemical reactions. (A) Co_3O_4/HCNS/Pt and (B) (Co_3O_4+Pt)/HCNS under UV irradiation ($\lambda > 300$ nm). *HCNS*, hollow g-C_3N_4 nanospheres. *Copyright © 2016 John Wiley and Sons.*

unidirectional migration of the electrons and holes on the Janus surfaces, which decreases the recombination of the electrons and holes.[183]

The so-called "Z-scheme mechanism" (two-step photoexcitation system) (Fig. 14), being inspired by natural photosynthesis in green plants,

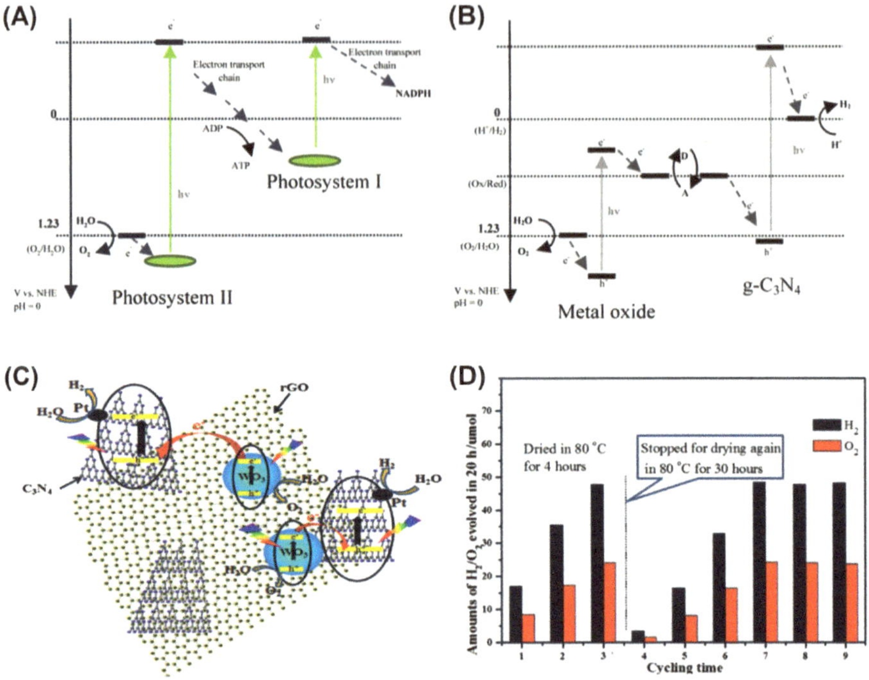

Figure 14 (A) Natural photosynthesis and (B) an artificial analogy used in this study, composed of an organic semiconductor C_3N_4 and a robust metal oxide (C) proposed mechanism for the H_2/O_2 evolution by WO_3/C_3N_4 nano-composite with or without the mediation of reduced graphene oxide (rGO) (D). The cycling performance of the sample (C_3N_4-rGO-150-WO_3 1000–260 loaded with 1 wt% Pt) dried after several photo-test cycles as fresh slurry. *(A–B) Copyright © 2014, American Chemical Society. (C–D) Copyright © 2015 Royal Society of Chemistry.*

has been reported to enhance the overall H_2O splitting.[148] The advantages of the Z-scheme H_2O-splitting system are considered as follows: (1) a wider range of visible light is available, leading to easily overcome the energy required to drive the reaction, (2) the separation of the evolution of H_2 and O_2 is possible, (3) various semiconductors having either the H^+ reduction (water reduction) or OH^- oxidation potential are utilized for one side of the Z-scheme system.[88]

By inspiring the natural photosynthesis system shown in Fig. 14A, two parallel systems with different metal oxides, $BiVO_4$/g-C_3N_4 and WO_3/g-C_3N_4 nano-composites, have been successfully developed for the overall H_2O splitting under visible light irradiation using I^-/IO_3^- as a soluble mediator (Fig. 14B).[88] The evolution rates in both systems have been found to depend strongly on pH and concentration of the redox mediator.

Furthermore, Zhao et al. developed the WO_3/g-C_3N_4 nano-composites with a heterojunction structure in which the reduced graphene oxide (rGO) was introduced for the mediator-free overall H_2O splitting and found that it enhanced the efficiency of H_2O splitting (Fig. 14C).[184] It was reported that under visible light irradiation ($\lambda > 420$ nm) of WO_3/g-C_3N_4 nano-composite loaded with Pt, H_2, and O_2 were evolved with a 2:1 ratio in the absence of any redox mediator. Thus, introducing rGO into the WO_3/g-C_3N_4 nano-composites enhances the efficiency of the overall H_2O splitting reaction, and the highest HERs and OERs of 2.84 and 1.46 μmol/h are observed with a quantum yield of 0.9% (Fig. 14D).[184]

3.3.5 Photocatalytic Fixation of CO_2

The g-C_3N_4 nano-material has been utilized to capture, activate, or reduce CO_2 as photocatalyst in the presence of the sacrificial hydrocarbons.[185] The formation of carbamate species is evidence for the binding of CO_2 to the tris-triazine—based covalent framework of g-C_3N_4 with Lewis basic functions. In this reaction, CO_2 is initially bonded to N atoms to form active carbamate intermediates, being similar to the first step of biological photosynthesis.[186] Lin et al. successfully activated CO_2 to form CO with g-C_3N_4 nano-composite under visible light irradiation and also found that the Co/g-C_3N_4 nano-composite enhances the reaction due to the presence of Co as a redox promoter to accelerate the charge carrier separation and transfer rates.[187]

Moreover, the modified g-C_3N_4 nanosheets with barbituric acid (BA), BA/g-C_3N_4 nanosheets, were found to effectively photocatalyze CO_2 conversion into CO with a 15 times higher rate than that of g-C_3N_4, being attributed to the improvement in the optical absorption and creation of the surface heterojunction at molecular scale promoting the charge separation.[188]

3.3.6 Direct Oxidation of Benzene to Phenol

Phenol is one of the most important intermediates for the industrial production of resins, fibers, dyes, and medicines. Phenol technology is mostly based on the three-step cumene process invented in 1940s and involves the coproduction of phenol and acetone.[189] Therefore, a more economical, environmentally harmonious, sustainable, and coproduct-free reaction system is strongly desired for phenol production.

Light-induced generation of the active oxy-species for the one-step oxidation of benzene to phenol has been proposed as an alternative process.

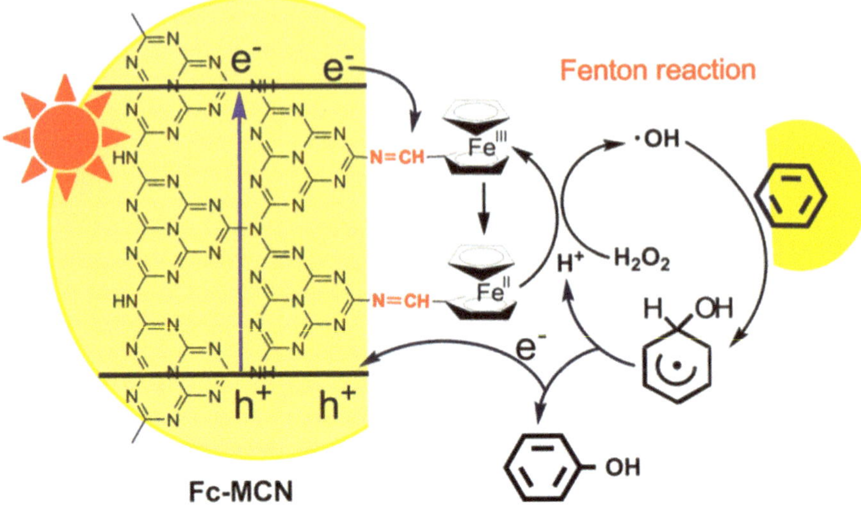

Figure 15 Merging of the redox function of ferrocene with carbon nitride photocatalyst to construct a heterogeneous photo-Fenton system for directing hydroxylation of benzene to phenol. *Copyright © 2014 John Wiley and Sons.*

Chen et al. reported that the Fe/g-C_3N_4 nano-composite was active for the direct oxidation of benzene to phenol in the presence of H_2O_2. Moreover, by applying the Fe/g-C_3N_4 nano-composite onto the SBA-15 support[190] or titanium silicate zeolite (TS-1),[191] the yield of the phenol production was remarkably enhanced under visible light irradiation ($\lambda > 420$ nm).

Ye et al. developed a ferrocene-grafted mesoporous g-C_3N_4 (Fc-M/g-C_3N_4) nano-composite by the Schiff base reaction of the surface-terminating amine moieties of mesoporous g-C_3N_4 and ferrocene moieties such as Fc-CHO or Fc-COOH.[192,193] Furthermore, such mesoporous g-C_3N_4 nano-material surfaces have been modulated at the molecular level using its photo and electronic properties. In these systems, the synergistic donor—acceptor interaction between the carbon nitride matrix and ferrocene groups play an important role in the improvement of the photocatalytic performance leading to the direct synthesis of phenol from benzene in the presence of H_2O_2 under visible light irradiation[192] (Fig. 15).

3.3.7 Photocatalytic Oxidation of Alcohols

The selective oxidation of alcohols to aldehydes is also one of the most relevant chemical reactions.[194,195] Conventional synthetic processes rely on the utilization of ClO^-, Cr(IV), Cl_2, and peroxy-acids as oxygen donors, which are regarded as corrosive, toxic, and stoichiometric chemical agents.[196] As a green oxidizing agent, the utilization of O_2 to achieve oxidation process is desired, though the activation of O_2 is not easy. Su et al. used fluorinated solvents (such as perfluorohexane) to dissolve a larger amount of oxygen. Due to the solvolysis of O—O bonds in polar media, the reaction becomes less surface controlled. Then, the molecular oxygen dissolved in fluorinated solvents is reduced by photogenerated electrons into superoxide radical anion ($\bullet O_2^-$) on the surface of g-C_3N_4 photocatalyst under visible light irradiation. The formed active $\bullet O_2^-$ species lead to the selective oxidation of alcohols to aldehydes.[157] Zhou et al. reported that a highly crystalline g-C_3N_4 nano-material exhibits an excellent performance to facilitate adsorption of benzyl alcohol and desorption of benzaldehyde, and simultaneously, to lower the energy barrier for the activation of O_2.[197] In fact, the highly crystalline g-C_3N_4 nano-material exhibited a roughly 12 times higher performance as compared with that of the normal g-C_3N_4, this remarkable enhancement being attributed to the synergistic effect of an increased electron—hole separation and increased surface reaction kinetics.

3.3.8 Photocatalytic Degradation of Organic Pollutants

g-C_3N_4 nano-material has received interest due to its promising application in the degradation of organic pollutants as a visible light-responsive photocatalyst. In these systems, the photogenerated electrons react with O_2 at the surface to produce superoxide anion radical ($\bullet O_2^-$) species, which could contribute to the formation of $\bullet$OH radicals by the reaction with H^+ and electrons in aqueous systems.[198] At the same time, the photogenerated holes react with OH^- groups on the surfaces to produce $\bullet$OH radicals. Thus, formed $\bullet$OH radicals attack any kind of organic pollutants to lead to the mineralization of these compounds into CO_2 and H_2O.[199] As shown in the reaction mechanisms below:

$$g - C_3N_4 \xrightarrow{h\nu} e^- + h^+$$

$$e^- + O_2 \rightarrow \bullet O_2^-$$

$$\bullet O_2^- + 2H^+ + 2e^- \rightarrow \bullet OH + OH^-$$

$$h^+ + OH^- \rightarrow \bullet OH$$

$$\text{Organic pollutants} \xrightarrow{\bullet OH} CO_2 + H_2O$$

Zheng et al. developed S-doped g-C_3H_4 microrods through the thermal polymerization based on trithiocyanuric acid and melamine supramolecular aggregates. The degradation of Rhodamine B and the reduction of hexavalent chromium Cr(VI) were studied as probe reactions to evaluate the photocatalytic activities of these S-doped g-C_3H_4 microrods.[200]

4. TERNARY BORON CARBON NITRIDE

Hexagonal boron nitride (*h*-BN), referred to as white graphite, is isostructural to graphene and has received extensive attention due to its high specific surface area, high efficiency in adsorption of aromatic molecules and chemical inertness.[201] From their unique properties, *h*-BN nanomaterial is expected to be widely applied in various fields such as hydrogen storage, adsorption agent of organic pollutants and drug delivery systems.[202] However, the poor electron delocalization with a wide electronic energy band gap of 5.96 eV limits the application of *h*-BN as a photocatalyst.[203]

Interestingly, Huang et al. found that the doping of C into *h*-BN produces a ternary BCN alloy nano-material. This was carried out by doping glucose during the preparation of *h*-BN, and the prepared samples were named as BCN-x, where x is the percentage weight content of glucose to boron oxide. The prepared BCN nano-material kept full advantages of graphene and *h*-BN.[72] Thus, the prepared BCN nano-material was successfully characterized by FTIR and XRD, showing similar structure of *h*-BN. In addition, to these techniques, Raman and electron energy loss spectroscopy (EELS) clearly showed the successful synthesis of BCN (Fig. 16A).[72] An additional C signal was clearly observed in the spectra of BCN that was not seen in the pure *h*-BN. EELS showed the K-edges of B, C, and N, and the B and N signals of BCN were consistent with those of *h*-BN, thus indicating the prepared BCN had a layered structure with an sp^2 domain.[72] This delocalized 2D electron system with a C(sp^2) atom incorporated in the *h*-BN lattice allowed the tunable electronic band gap by changing the content of C.[72] These results showed a good agreement with density functional theory (DFT) calculations, in which the details of the electronic band structures of $B_{16}N_{16}$, $B_{13}C_8N_{11}$, and $B_{11}C_{12}N_9$ were simulated (Fig. 16B).[72] SEM and TEM were also used to study the ultrathin layers of the nano-structure (Fig. 16C), while HRTEM and energy-dispersive X-ray spectroscopy were used to study the crystallinity and elemental analysis.[72] X-ray photoelectron spectroscopy (XPS) was used to fully visualize the bond formation in BCN, including B—C, B—N, and N—C bonds.

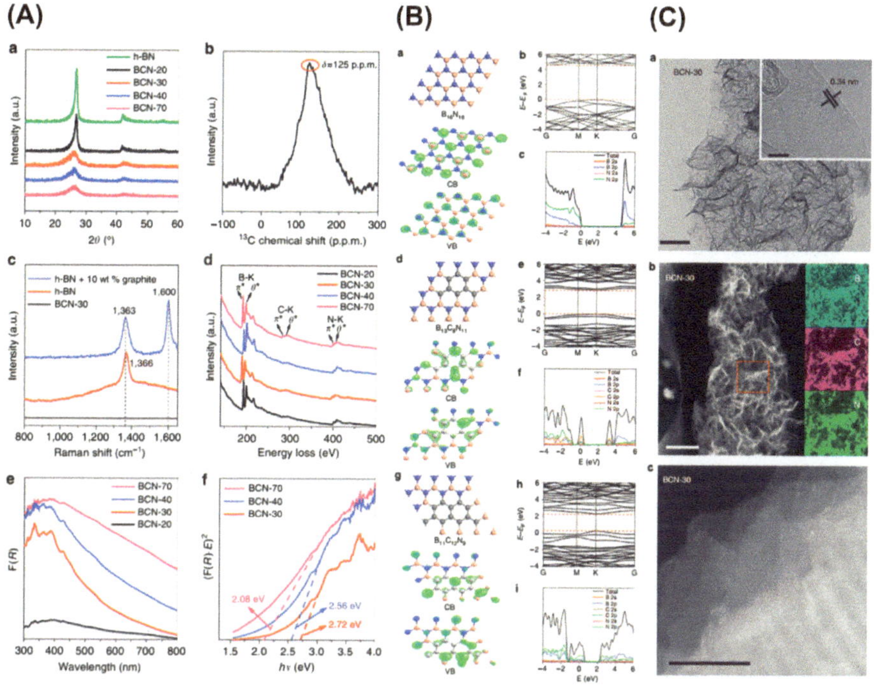

Figure 16 (A) Characterization of *h*-BN and BCN, as well as the corresponding DFT calculations. (a) Powder XRD patterns of BCN; (b) solid-state ^{13}C NMR spectrum of BCN-30; (c) Raman spectra of h-BN and BCN-30; (d) EELS spectra of *h*-BN and BCN; (e) UV–Vis DRS of BCN-x; (f) the corresponding plot to define the band gap. (B) The plane-wave DFT calculations of electronic structure of *h*-BN and BCN. The optimized structure of $B_{16}N_{16}$ (a) $B_{13}C_8N_{11}$ (d) and $B_{11}C_{12}N_9$ (g) with the corresponding valence band (VB)/conduction band (CB), calculated energy band (b,e,h) and corresponding total and ion-decomposed electronic density of states (c,f,i). (C) Surface morphology of BCN-30. (a) HRTEM image of the BCN-30 sample. Scale bar, 100 nm, inset scale bars, 5 nm. (b) Typical TEM dark-field image of BCN-30 sample and the elemental mapping images of B, C, and N of the enlargement of selected area in the picture. Scale bar, 300 nm. (c) High-resolution STEM of the BCN-30 sample along (002) facet. Scale bar, 5 nm. *BCN*, boron carbon nitride; *DRS*, diffuse reflectance spectroscopy; *EELS*, electron energy loss spectroscopy; *h-BN*, hexagonal boron nitride. *Copyright © 2016 Springer Nature.*

At the same time, minor signals for the B—O and N—H bonds were identified and were attributed to residual starting material and terminal edges, respectively.[72]

Photocatalytic studies on the H_2O splitting and CO_2 reduction were carried out using BCN nano-material. Wang et al. reported that among the prepared BCNs with different C contents, BCN-30 showed the highest performance in H_2 production; its photocatalytic activity was sustained for at

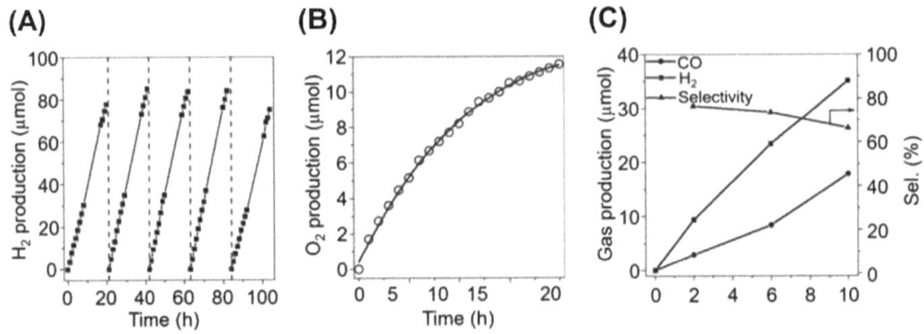

Figure 17 The performance of BCN-30. Time course of the performance of (A) H_2 evolution, (B) O_2 evolution, and (C) CO_2 reduction achieved by BCN-30. *BCN*, boron carbon nitride. *Copyright © 2016 Springer Nature.*

least 104 h, indicating an excellent working stability (Fig. 17A).[72] As a consequence of its appropriate electronic band structure, BCN nano-material was found to operate for the O_2 production by using Ni—Co layered double hydroxide (LDH) as a cocatalyst under visible and UV light irradiation (Fig. 17B).[72] More importantly, the photocatalytic reduction of CO_2 was found to proceed with BCN nano-material under visible light irradiation, and an isotopic labeling experiment clearly showed that the photoformed CO originated from CO_2 (Fig. 17C).[72]

The BCN nano-material with a high specific surface area showed a strong chemical affinity toward metal ions due to the "lop-sided" densities characteristic of the ionic B—N bonding, enabling the creation of metal/BCN hybrid-layered nano-structures with unique properties.[73] For the utilization of BCN nano-material as a photocatalyst, Co^{2+} ions were supported onto the surfaces of BCN (Fig. 18).[73] The monolayered nano-structure of BCN was sufficient to immobilize Co ions in its frameworks, and such designed hybrid nanostructure provided a remarkable increase in the photocatalytic evolution of O_2 from H_2O.[73] Thus, the kinetic barrier for the O_2 evolution with BCN was found to reduce by the use of suitable Co cocatalyst.

5. CONCLUSION AND PERSPECTIVES

Semiconductor-based photocatalysis is a green, feasible, and versatile technology that has been widely used for solving the worldwide energy shortage and environmental pollution issues. Up to now, photocatalysts vary from traditional metal oxides to 2D covalent organic frameworks.

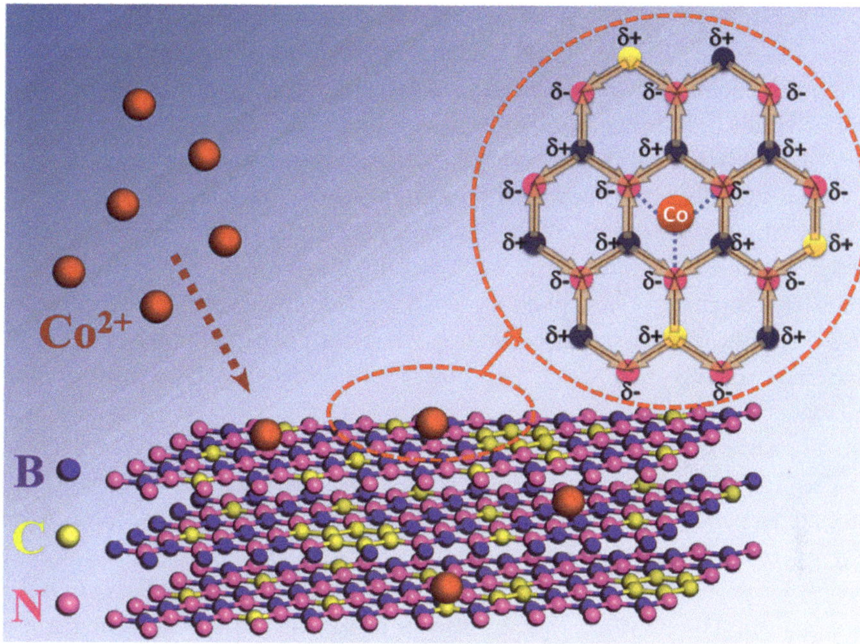

Figure 18 The synthetic process of Co/BCN via chemical adsorption. *BCN*, boron carbon nitride. *Copyright © 2015, Springer Nature.*

Most traditional metal oxide photocatalysts such as TiO_2 function only under UV light, which makes up less than 4%–5% of the solar light reaching the earth. Visible light-responsive TiO_2 photocatalysts, which can effectively and efficiently utilize sunlight more than 25% have been developed by applying an advanced physical ion-implantation method. However, it is rather expensive and unfeasible for large scale production. Thus, innovative yet cost-efficient methods to produce such visible light-responsive photocatalysts are strongly desired.

To develop more efficient visible light-driven photocatalysts, it is necessary to narrow the band gaps to harvest visible light in the longer-wavelength regions and enhance the photogenerated charge separation in photocatalysis. In the recent decade, 2D nanosheet materials, in respect to their ultimate structural anisotropy in nanoscale thickness versus their submicro or micro- scale lateral size, exhibit unique physical properties and promising electronic features, which strongly support their promising applications in both energy- and environmental-related topics.

To date, the development of 2D materials, such as g-C_3N_4 and BCN nanomaterials, have been developed as highly efficient photocatalysts, which can operate even under visible light irradiation. As we mentioned above, a variety of g-C_3N_4–based photocatalysts have been utilized in various redox reactions. The construction of two parallel systems for overall water splitting under visible light irradiation by using g-C_3N_4 with different metal oxides of $BiVO_4$ and WO_3 has been reported.[88] A photosystem for the conversion of CO_2 into CO has been designed by using g-C_3N_4. Maeda et al. successfully utilized the g-C_3N_4 grafted by a Ru-complex on its surface as a photocatalyst to convert CO_2 into formic acid with the AQY of 2%.[155] There are also many reports that g-C_3N_4 photocatalyst can decompose organic compounds as TiO_2 photocatalysts, demonstrating the great potential of g-C_3N_4 nanomaterials for a broad range of solar applications based on photoredox catalysis.[109,204]

The anisotropic 2D geometric morphology, together with the aromatic π-conjugated framework, endows these 2D photocatalytic materials with attractive band gap- and surface-engineered applications ranging from energy-related topics to other new emerging fields. The photocatalytic processes observed with 2D nano-materials focused on the following three key points: (1) extending the absorption bands toward visible light regions (visible light harvesting), (2) promoting photogenerated charge carrier separation (separation and migration to the surfaces), and (3) increasing the amount of the reactant species on the photocatalyst surfaces (number of active sites). With these objectives in mind, the advantages of 2D nano-materials as photocatalysts were discussed focusing on their nanostructure design, crystalline structure engineering, elemental doping, and the construction of nano-composites involving the type II heterostructure and Z-scheme mechanism, and what effect these factors have on enhancing their photocatalytic performance. For the further applications of g-C_3N_4 and BCN as photocatalysts, it is important to enhance the efficiency of their photocatalytic reactivity. To solve this issue, it would be an attractive way to construct heterostructuring of g-C_3N_4 or BCN with other semiconductors such as highly active TiO_2 because such heterostructure would be expected to prolong the charge carrier lifetime to cover the fast recombination of charge carriers in these organic polymer compounds. Furthermore, the fabrication of g-C_3N_4 or BCN hybrids and nano-composites would be expected to lead to the more efficient band gap tuning than the doping or morphology design of these compounds. Finally, 2D nanosheets of g-C_3N_4 and BCN would be more extensively studied for their application because they exhibit high specific surface areas and would be expected to have better electron transport along with the in-plane direction, as well as prolonged lifetimes of the photogenerated charge carriers.

Besides, given that all the enhanced catalytic activity of 2D materials originates from the modification of their electronic structure, van der Waals heterostructures can be formed by the combination of various 2D materials. Although they are rather complex structurally, such stacks of different 2D crystals can be produced by self-assembly. By modifying the surface charge of 2D nano-flakes, one could create strong interactions between different crystals. Such heterostructures would offer even greater control over the electronic properties and have recently been fabricated and investigated, revealing unusual properties and new phenomena.

2D materials and their heterostructures are excellent candidates for both fundamental catalytic research and commercial applications. However, their application in catalysis still faces a number of scientific and technological issues. First of all, 2D materials are often a composite of nanosheets with all kinds of layer thickness, size, edges, and defects. The controllable preparation of 2D materials is of vital importance. Secondly, although there has been rapid development of 2D materials and their heterostructures in catalysis recently, there are still lack of successful examples of an industrial application. Apart from the scientific issues, there are a number of technological problems that need to be solved. It remains challenges to produce them on the large scale.

Further research into the development of such visible light-responsive 2D photocatalytic materials will be vital in establishing artificial photosynthetic systems to convert solar energy effectively and efficiently into useful chemical or electrical energy, thus ensuring a stable energy source, as well as a cleaner and more sustainable environment for future generations.

ACKNOWLEDGMENTS

We acknowledge funding supported by the National Natural Science Foundation of China (21425309 and 21761132002), National Key R&D Program of China (2018YFA0209301) and the 111Project. We also thank Profs. Rudi van Eldik and Wojtek Macyk for their careful reading and valuable comments as well as kind assistance in editing the volume.

REFERENCES

1. Morrison, S. R. *The Chemical Physics of Surfaces;* Plenum Press: New York, 1977. And references therein.
2. Kennedy, D. R.; Ritchie, M.; Mackenzie, J. *Trans. Faraday Soc.* **1958,** *54,* 119.
3. Fujita, Y.; Kwan, T. *Bull. Chem. Soc. Jpn.* **1958,** *31,* 379.
4. Barry, T. I.; Stone, F. S. *Proc. R. Soc. Lond. Ser. A* **1960,** *255,* 124. And references therein.

5. Honda, K.; Fujishima, A. *Nature* **1972,** *238,* 37–38.
6. Schrauzer, G. N.; Guth, T. D. *J. Am. Chem. Soc.* **1977,** *99,* 7189–7193.
7. Frank, S. N.; Bard, A. J. *J. Am. Chem. Soc.* **1977,** *99,* 303–304.
8. Halmann, M. *Nature* **1978,** *275,* 115–116.
9. Yoneyama, H.; Koizumi, M.; Tamura, H. *Bull. Chem. Soc. Jpn.* **1979,** *52,* 3449–3450.
10. Kiwi, J.; Graetzel, M. *J. Am. Chem. Soc.* **1979,** *101,* 7214–7217.
11. Kawai, T.; Sakata, T. *Nature* **1980,** *286,* 474–476.
12. Sato, S.; White, J. M. *J. Am. Chem. Soc.* **1980,** *102,* 7206–7210.
13. Domen, K.; Naito, S.; Soma, M.; Onishi, T.; Tamaru, K. *J. Chem. Soc. Chem. Commun.* **1980,** 543–544.
14. Heller, A.; Shalon, A. A.; Bonner, W. A.; Miller, B. *J. Am. Chem. Soc.* **1982,** *100,* 5985–5992.
15. Fax, M. A. *Acc. Chem. Res.* **1983,** *16,* 314–321.
16. Anpo, M.; Aikawa, N.; Kodama, S.; Kubokawa, Y. *J. Phys. Chem.* **1984,** *84,* 569.
17. Anpo, M.; Aikawa, N.; Kubokawa, Y. *J. Phys. Chem.* **1984,** *88,* 3998–4000.
18. Anpo, M.; Kondo, M.; Coluccia, S.; Louis, C.; Che, M. *J. Am. Chem. Soc.* **1989,** *111,* 8791–8799.
19. Anpo, M. *Sol. Energy Mater. Sol. Cells* **1995,** *38,* 221–238.
20. Anpo, M.; Thomas, J. M. *Chem. Commun.* **2006,** 3238–3273.
21. Anpo, M.; Dzwigai, S.; Che, M. *Adv. Catal.* **2009,** *52,* 10. And references therein.
22. Horiuchi, Y.; Toyao, T.; Takeuchi, M.; Matsuoka, M.; Anpo, M. *Phys. Chem. Chem. Phys.* **2013,** *15,* 13243–13253.
23. Schnider, J.; Matsuoka, M.; Takeuchi, M.; Zhang, J. L.; Anpo, M.; Bahnemann, D. W. *Chem. Rev.* **2014,** *114,* 9919–9986. And references therein.
24. Oregan, B.; Graetzel, M. *Nature* **1991,** *353,* 737–740.
25. Serpone, N.; Pelizzetti, E. *Photocatalysis –Fundamental and Application;* Wiley: New York, 1989. And references therein.
26. Ollis, D. F.; Al-Ekabi, H. *Photocatalytic Purification and Treatment of Water and Air;* Elsevier: Amsterdam, 1991. And references therein.
27. Anpo, M. In *Handbook of Heterogeneous Catalysis;* Ertl, G., Knozinger, H., Weitkamp, J., Eds.; Wiley-VCH: Weinheim, 1997. And references therein.
28. Anpo, M. *Pure Appl. Chem.* **2000,** *72,* 1265–1270.
29. Wang, R.; Hashimoto, K.; Fujishima, A.; Chikuni, M.; Kojima, E.; Kitamura, A.; Shimohigoshi, M.; Watanabe, T. *Nature* **1997,** *388,* 431–432.
30. Wang, R.; Hashimoto, K.; Fujishima, A.; Chikuni, M.; Kojima, E.; Kitamura, A.; Shimohigoshi, M.; Watanabe, T. *Adv. Mater.* **1998,** *10,* 135–138.
31. Takeuchi, M.; Yamashita, H.; Matsuoka, M.; Anpo, M.; Hirao, T.; Iwamoto, N. *Catal. Lett.* **2000,** *66,* 185–187.
32. Takeuchi, M.; Yamashita, H.; Matsuoka, M.; Anpo, M.; Hirao, T.; Iwamoto, N. *Catal. Lett.* **2000,** *67,* 135–137.
33. Takeuchi, M.; Dohshi, S.; Eura, T.; Anpo, M. *J. Phys. Chem. B* **2003,** *107,* 14278–14282.
34. Anpo, M. *Bull. Chem. Soc. Jpn.* **2004,** *77,* 1427–1442. And references therein.
35. Anpo, M.; Yamashita, H.; Ichihashi, Y. *Optronics* **1997,** *186,* 161.
36. Anpo, M. *Catal. Surv. Jpn.* **1997,** *1,* 169.
37. Anpo, M.; Ichihashi, Y.; Takeuchi, M.; Yamashita, H. *Res. Chem. Intermed.* **1998,** *24,* 143–149.
38. Yamashita, H.; Ichihashi, Y.; Takeuchi, M.; Kishiguchi, S.; Anpo, M. *J. Synchrotron Radiat.* **1996,** *6,* 451.
39. Yamashita, H.; Harada, M.; Misaka, J.; Takeuchi, M.; Ichihashi, Y.; Ishida, M.; Sasaki, T.; Anpo, M. *J. Synchrotron Radiat.* **2001,** *8,* 569–571.

40. Yamashita, H.; Harada, M.; Misaka, J.; Takeuchi, M.; Neppolian, B.; Anpo, M. *Catal. Today* **2003,** *84,* 191–196.
41. Anpo, M.; Takeuchi, M. *J. Catal.* **2003,** *216,* 505–516. And references therein.
42. Yamashita, H.; Anpo, M. *Catal. Surv. Asia* **2004,** *8,* 35–45.
43. Anpo, M.; Doshi, S.; Kitano, M.; Hu, Y.; Takeuchi, M.; Matsuoka, M. *Annu. Rev. Mater. Res.* **2005,** *35,* 1–27.
44. Anpo, M.; Kishiguchi, S.; Ichihashi, Y.; Takeuchi, M.; Yamashita, H.; Ikeue, K.; Morin, B.; Davidson, A.; Che, M. *Res. Chem. Intermed.* **2001,** *27,* 459–467.
45. Kitano, M.; Takeuchi, M.; Matsuoka, M.; Thomas, J. M.; Anpo, M. *Chem. Lett.* **2005,** *34,* 616–617.
46. Matsuoka, M.; Kitano, M.; Takeuchi, M.; Anpo, M.; Thomas, J. M. *Top. Catal.* **2005,** *35,* 305–310.
47. Kitano, M.; Tsujimaru, K.; Anpo, M. *Appl. Catal. A Gen.* **2006,** *314,* 179–183.
48. Kitano, M.; Takeuchi, M.; Matsuoka, M.; Thomas, J. M.; Anpo, M. *Catal. Today* **2007,** *120,* 133–138.
49. Kitano, M.; Iyatani, K.; Tsujimaru, K.; Matsuoa, M.; Takeuchi, M.; Ueshima, M.; Thomas, J. M.; Anpo, M. *Top. Catal.* **2008,** *49,* 24–31.
50. Matsuoka, M.; Kitano, M.; Fukumoto, S.; Iyatani, K.; Takeuchi, M.; Anpo, M. *Catal. Today* **2008,** *132,* 159–164.
51. Matsuoka, M.; Ebrahimi, A.; Nakagawa, M.; Kim, T. H.; Kitano, M.; Anpo, M. *Res. Chem. Intermed.* **2009,** *35,* 997–1004.
52. Abe, R.; Takata, T.; Sugihara, H.; Domen, K. *Chem. Commun.* **2005,** 3829–3831.
53. Maeda, K.; Teramura, K.; Lu, D.; Takata, T.; Saito, N.; Inoue, Y.; Domen, K. *Nature* **2006,** *440,* 295.
54. Maeda, K.; Teramura, K.; Lu, D.; Takata, T.; Saito, N.; Inoue, Y.; Domen, K. *Angew. Chem. Int. Ed.* **2006,** *45,* 7806–7809.
55. Maeda, K.; Domen, K. *Chem. Mater.* **2009,** *22,* 612–623.
56. Maeda, K.; Xiong, A.; Yoshinaga, T.; Ikeda, T.; Sakamoto, N.; Hisatomi, T.; Takashima, M.; Hisatomi, T.; Lu, D.; Kanehara, M.; Setoyama, T.; Teranishi, T.; Domen, K. *Angew. Chem. Int. Ed.* **2010,** *49,* 4096–4099.
57. Maeda, K.; Lu, D.; Domen, K. *Chem. Eur. J.* **2013,** *19,* 4986–4991.
58. Hisatomi, T.; Kubota, J.; Domen, K. *Chem. Soc. Rev.* **2014,** *43,* 7520–7535. And references therein.
59. Takata, T.; Domen, K. *Dalton Trans.* **2017,** *46,* 10529–10544.
60. Matoraj, D.; Kisch, H. *Angew. Chem. Int. Ed.* **2008,** *47,* 9975–9978.
61. Wang, X.; Maeda, K.; Thomas, A.; Takanabe, K.; Xin, G.; Carlsson, J. M.; Domen, K.; Antonietti, M. *Nat. Mater.* **2009,** *8,* 76–80.
62. Wang, Y.; Wang, X.; Antonietti, M. *Angew. Chem. Int. Ed.* **2012,** *51,* 68–89.
63. Zheng, Y.; Lin, L.; Wang, B.; Wang, X. *Angew. Chem. Int. Ed.* **2015,** *54,* 12868–12884.
64. Kessler, F. K.; Zheng, Y.; Schwarz, D.; Merschjann, C.; Schnick, W.; Wang, X.; Bojdys, M. J. *Nat. Rev. Mater.* **2017,** *2,* 17030.
65. Ong, W.-J.; Tan, L.-L.; Ng, Y. H.; Yong, S.-T.; Chai, S.-P. *Chem. Rev.* **2016,** *116,* 7159–7329.
66. Cao, S.; Low, J.; Yu, J.; Jaroniec, M. *Adv. Mater.* **2015,** *27,* 2150–2176.
67. Liu, J.; Zhang, T.; Wang, Z.; Dawson, G.; Chen, W. *J. Mater. Chem.* **2011,** *21,* 14398–14401.
68. Zhang, Y.; Liu, J.; Wu, G.; Chen, W. *Nanoscale* **2012,** *4,* 5300–5303.
69. Li, J.; Shen, B.; Hong, Z.; Lin, B.; Gao, B.; Chen, Y. *Chem. Commun.* **2012,** *48,* 12017–12019.
70. Naseri, A.; Samadi, M.; Pourjavadi, A.; Moshfegh, A.; Ramakrishna, S. *J. Mater. Chem. A* **2017,** *5,* 23406–23433. And references therein.

71. Zhou, L.; Zhang, H.; Sun, H.; Liu, S.; Tade, M. O.; Wang, S.; Jin, W. *Catal. Sci. Technol.* **2016,** *6,* 7002–7023.
72. Huang, C.; Chen, C.; Zhang, M.; Lin, L.; Ye, X.; Lin, S.; Antonietti, M.; Wang, X. *Nat. Commun.* **2015,** *6,* 7698.
73. Zhang, M.; Luo, Z.; Zhou, M.; Huang, C.; Wang, X. *Sci. Chin. Mater.* **2015,** *58,* 867–876.
74. Guo, F.; Yang, P.; Cao, X.; Pan, Z.; Xie, Z.; Wang, X. *Angew. Chem. Int. Ed.* **2017,** *56,* 8231–8235.
75. Fang, Y.; Wang, X. *Angew. Chem. Int. Ed.* **2017,** *56,* 15506–15518.
76. Chen, L.; Wang, X. *Chem. Commun.* **2017,** *53,* 11988–11991.
77. Teter, D. M.; Hemley, R. J. *Science* **1996,** *271,* 53–55.
78. Maeda, K.; Wang, X.; Nishihara, Y.; Lu, D.; Antonietti, M.; Domen, K. *J. Phys. Chem. C* **2009,** *113,* 4940–4947.
79. Xu, Y.; Gao, S. *Int. J. Hydrogen Energy* **2012,** *37,* 11072–11080.
80. Zhang, J.; Chen, Y.; Wang, X. *Energy Environ. Sci.* **2015,** *8,* 3092–3108.
81. Geim, A. K.; Grigorieva, I. V. *Nature* **2013,** *499,* 419–425.
82. Wang, X.; Blechert, S.; Antonietti, M. *ACS Catal.* **2012,** *2* (8), 1596–1606.
83. Dong, G.; Zhang, Y.; Pan, Q.; Qiu, J. *J. Photochem. Photobiol. C* **2014,** *20,* 33–50.
84. Zhang, J.; Zhang, G.; Chen, X.; Lin, S.; Mohlmann, L.; Dolega, G.; Lipner, G.; Antonietti, M.; Blechert, S.; Wang, X. *Angew. Chem. Int. Ed.* **2012,** *51,* 3183–3187.
85. Yang, C.; Wang, B.; Zhang, L.; Yin, L.; Wang, X. *Angew. Chem. Int. Ed.* **2017,** *56,* 6627–6631.
86. He, F.; Chen, G.; Yu, Y.; Zhou, Y.; Zheng, Y.; Hao, S. *Chem. Commun.* **2015,** *51,* 425–427.
87. Ong, W.-J.; Tan, L.-L.; Chai, S.-P.; Yong, S.-T. *Chem. Commun.* **2015,** *51,* 858–861.
88. Martin, D. J.; Reardon, P. J. T.; Moniz, S. J. A.; Tang, J. *J. Am. Chem. Soc.* **2014,** *136* (36), 12568–12571.
89. Tian, Y.; Chang, B.; Lu, J.; Fu, J.; Xi, F.; Dong, X. *ACS Appl. Mater. Interfaces* **2013,** *5,* 7079–7085.
90. Miao, G.; Huang, D.; Ren, X.; Li, X.; Li, Z.; Xiao, J. *Appl. Catal. B* **2016,** *192,* 72–79.
91. Wang, S.; Liu, D.; Sun, C.; Yang, Y.; Guan, Y.; He, H. *Appl. Catal. B* **2014,** *144,* 885.
92. Liu, J.; Wang, H.; Chen, Z. P.; Moehwald, H.; Fiechter, S.; van de Krol, R.; Wen, L.; Jiang, L.; Antonietti, M. *Adv. Mater.* **2015,** *27,* 712–718.
93. Shiraishi, Y.; Kofuji, Y.; Sakamoto, H.; Tanaka, S.; Ichikawa, S.; Hirai, T. *ACS Catal.* **2015,** *5,* 3058–3066.
94. Liang, Q.; Li, Z.; Huang, Z.-H.; Kang, F.; Yang, Q.-H. *Adv. Funct. Mater.* **2015,** *25,* 6885–6892.
95. Liu, G.; Wang, T.; Zhang, H.; Meng, X.; Hao, D.; Chang, K.; Li, P.; Kako, T.; Ye, J. *Angew. Chem. Int. Ed.* **2015,** *54,* 13561–13565.
96. Schwinghammer, K.; Tuffy, B.; Mesch, M. B.; Wirnhier, E.; Martineau, C.; Taulelle, F.; Schnick, W.; Senker, J.; Lotsch, B. V. *Angew. Chem. Int. Ed.* **2013,** *52,* 2435–2439.
97. Yan, S. C.; Li, Z. S.; Zou, Z. G. *Langmuir* **2009,** *25,* 10397.
98. Chen, Y.; Wang, B.; Lin, S.; Zhang, Y.; Wang, X. *J. Phys. Chem. C* **2014,** *118,* 29981–29989.
99. Zhang, G.; Zhang, J.; Zhang, M.; Wang, X. *J. Mater. Chem.* **2012,** *22,* 8083–8091.
100. Jorge, A. B.; Martin, D. J.; Dhanoa, M. T. S.; Rahman, A. S.; Makwana, N.; Tang, J.; Sella, A.; Corà, F.; Firth, S.; Darr, J. A.; McMillan, P. F. *J. Phys. Chem. C* **2013,** *117,* 7178–7185.
101. Cui, Y.; Ding, Z.; Liu, P.; Antonietti, M.; Fu, X.; Wang, X. *Phys. Chem. Chem. Phys.* **2012,** *14,* 1455–1462.

102. Goettmann, F.; Fischer, A.; Antonietti, M.; Thomas, A. *Angew. Chem. Int. Ed.* **2006,** *45,* 4467–4471.
103. Goettmann, F.; Fischer, A.; Antonietti, M.; Thomas, A. *Chem. Commun.* **2006,** 4530–4532.
104. Yang, Z.; Zhang, Y.; Schnepp, Z. *J. Mater. Chem. A* **2015,** *3,* 14081–14092.
105. Yang, S.; Gong, Y.; Zhang, J.; Zhan, L.; Ma, L.; Fang, Z.; Vajtai, R.; Wang, X.; Ajayan, P. M. *Adv. Mater.* **2013,** *25,* 2452–2456.
106. Yan, H. *Chem. Commun.* **2012,** *48,* 3430–3432.
107. Li, X.; Masters, A. F.; Maschmeyer, T. *ChemCatChem* **2015,** *7,* 121–126.
108. Kailasam, K.; Epping, J. D.; Thomas, A.; Losse, S.; Junge, H. *Energy Environ. Sci.* **2011,** *4,* 4668–4674.
109. Huang, J.; Ho, W.; Wang, X. *Chem. Commun.* **2014,** *50,* 4338–4340.
110. Sun, J.; Zhang, J.; Zhang, M.; Antonietti, M.; Fu, X.; Wang, X. *Nat. Commun.* **2012,** *3,* 1139.
111. Zhang, J.; Zhang, M.; Yang, C.; Wang, X. *Adv. Mater.* **2014,** *26,* 4121–4126.
112. Zheng, Y.; Lin, L.; Ye, X.; Guo, F.; Wang, X. *Angew. Chem. Int. Ed.* **2014,** *53,* 11926–11930.
113. Zheng, D.; Huang, C.; Wang, X. *Nanoscale* **2015,** *7,* 465–470.
114. Lin, Z.; Wang, X. *ChemSusChem* **2014,** *7,* 1547–1550.
115. Yang, P.; Ou, H.; Fang, Y.; Wang, X. *Angew. Chem. Int. Ed.* **2017,** *56,* 3992–3996.
116. Lin, L.; Ou, H.; Zhang, Y.; Wang, X. *ACS Catal.* **2016,** *6,* 3921–3931.
117. Diebold, U. *Nat. Chem.* **2011,** *3,* 271–272.
118. Maeda, K. *ACS Catal.* **2013,** *3,* 1486–1503.
119. Ou, H.; Lin, L.; Zheng, Y.; Yang, P.; Fang, Y.; Wang, X. *Adv. Mater.* **2017,** *29,* 1700008.
120. Guo, Y.; Xu, L.; Liu, H.; Li, Y.; Che, C.-M.; Li, Y. *Adv. Mater.* **2015,** *27,* 985–1013.
121. Zhang, G.; Li, G.; Lan, Z.-A.; Lin, L.; Savateev, A.; Heil, T.; Zafeiratos, S.; Wang, X.; Antonietti, M. *Angew. Chem. Int. Ed.* **2017,** *56,* 13630–13634.
122. Lu, C.; Chen, R.; Wu, X.; Fan, M.; Liu, Y.; Le, Z.; Jiang, S.; Song, S. *Appl. Surf. Sci.* **2016,** *360,* 1016–1022.
123. Zhang, P.; Li, X.; Shao, C.; Liu, Y. *J. Mater. Chem. A* **2015,** *3,* 3281–3284.
124. Guo, S.; Zhu, Y.; Yan, Y.; Min, Y.; Fan, J.; Xu, Q. *Appl. Catal. B* **2016,** *185,* 315–321.
125. Xu, C.; Han, Q.; Zhao, Y.; Wang, L.; Li, Y.; Qu, L. *J. Mater. Chem. A* **2015,** *3,* 1841–1846.
126. Ma, T.; Ran, J.; Dai, S.; Jaroniec, M.; Qiao, S. *Angew. Chem. Int. Ed.* **2015,** *54,* 4646–4650.
127. Wang, Y.; Di, Y.; Antonietti, M.; Li, H.; Chen, X.; Wang, X. *Chem. Mater.* **2010,** *22,* 5119–5121.
128. Lan, Z.-A.; Zhang, G.; Wang, X. *Appl. Catal. B Environ.* **2016,** *192,* 116–125.
129. Zhang, G.; Zhang, M.; Qiu, X.; Wang, X. *Adv. Mater.* **2014,** *26,* 805–809.
130. Lu, Y.-C.; Chen, J.; Wang, A.-J.; Bao, N.; Feng, J.-J.; Wang, W.; Shao, L. *J. Mater. Chem. C* **2015,** *3,* 73–78.
131. Zhang, M.; Wang, X. *Energy Environ. Sci.* **2014,** *7,* 1902–1906.
132. Zhang, J.; Zhang, M.; Sun, R.-Q.; Wang, X. *Angew. Chem. Int. Ed.* **2012,** *51,* 10145–10149.
132a. Marschall, R. *Adv. Funct. Mater.* **2014,** *24,* 2421–2440.
133. Zhang, J.; Zhang, M.; Lin, L.; Wang, X. *Angew. Chem. Int. Ed.* **2015,** *54,* 6297–6301.
134. Hunt, S. T.; Milina, M.; Alba-Rubio, A. C.; Hendon, C. H.; Dumesic, J. A.; Roman-Leshkov, Y. *Science* **2016,** *352,* 974–978.
135. Singh, A.; Lindquist, B. A.; Ong, G. K.; Jadrich, R. B.; Singh, A.; Ha, H.; Ellison, C. J.; Truskett, T. M.; Milliron, D. J. *Angew. Chem. Int. Ed.* **2015,** *54,* 14840–14844.

136. Zhou, Z.; Wang, J.; Yu, J.; Shen, Y.; Li, Y.; Liu, A.; Liu, S.; Zhang, Y. *J. Am. Chem. Soc.* **2015,** *137,* 2179–2182.
137. Ou, H.; Yang, P.; Lin, L.; Anpo, M.; Wang, X. *Angew. Chem. Int. Ed.* **2017,** *56,* 10905–10910.
138. Vinodgopal, K.; Bedja, I.; Kamat, P. V. *Chem. Mater.* **1996,** *8,* 2180–2187.
139. Shiragami, T.; Fukami, S.; Wada, Y.; Yanaida, S. *J. Phys. Chem.* **1993,** *97,* 12882–12887.
140. Hu, X.; Li, G.; Yu, J. C. *Langmuir* **2010,** *26* (5), 3031–3039.
141. Jiang, Z.; Zhu, C.; Wan, W.; Qian, K.; Xie, J. *J. Mater. Chem. A* **2016,** *4,* 1806–1818.
142. Zhang, J.; Wang, Y.; Jin, J.; Zhang, J.; Lin, Z.; Huang, F.; Yu, J. *ACS Appl. Mater. Interfaces* **2013,** *5,* 10317–10324.
143. Zheng, D.; Pang, C.; Wang, X. *Chem. Commun.* **2015,** *51,* 17467–17470.
144. Wang, H.; Zhang, L.; Chen, Z.; Hu, J.; Li, S.; Wang, Z.; Liu, J.; Wang, X. *Chem. Soc. Rev.* **2014,** *43,* 5234–5244.
145. Wang, Y.; Wang, Q.; Zhan, X.; Wang, F.; Safdar, M.; He, J. *Nanoscale* **2013,** *5,* 8326–8339.
146. Zheng, D.; Zhang, G.; Wang, X. *Appl. Catal. B* **2015,** *179,* 479–488.
147. Kailasam, K.; Fischer, A.; Zhang, G.; Zhang, J.; Schwarze, M.; Schröder, M.; Wang, X.; Schomäcker, R.; Thomas, A. *ChemSusChem* **2015,** *8,* 1404–1410.
148. Zhou, P.; Yu, J.; Jaroniec, M. *Adv. Mater.* **2014,** *26,* 4920–4935.
149. Zhang, G.; Li, G.; Wang, X. *ChemCatChem* **2015,** *7,* 2864–2870.
150. Zhang, G.; Huang, C.; Wang, X. *Small* **2015,** *11,* 1215–1221.
151. Hou, Y.; Zhu, Y.; Xu, Y.; Wang, X. *Appl. Catal. B Environ.* **2014,** *156,* 122–127.
152. Zheng, D.; Zhang, G.; Hou, Y.; Wang, X. *Appl. Catal. A Gen.* **2016,** *521,* 2–8.
153. Guo, F.; Hou, Y.; Asiri, A. M.; Wang, X. *Chem. Commun.* **2017,** *53,* 13221–13224.
154. Wang, S.; Lin, J.; Wang, X. *Phys. Chem. Chem. Phys.* **2014,** *16,* 14656–14660.
155. Kuriki, R.; Sekizawa, K.; Ishitani, O.; Maeda, K. *Angew. Chem. Int. Ed.* **2015,** *54,* 2406–2409.
156. Chang, F.; Xie, Y.; Li, C.; Chen, J.; Luo, J.; Hu, X.; Shen, J. *Appl. Surf. Sci.* **2013,** *280,* 967–974.
157. Su, F.; Mathew, S. C.; Lipner, G.; Fu, X.; Antonietti, M.; Blechert, S.; Wang, X. *J. Am. Chem. Soc.* **2010,** *132* (46), 16299–16301.
158. Cao, S.; Yu, J. *J. Phys. Chem. Lett.* **2014,** *5* (12), 2101–2107.
159. Yang, J.; Wang, D.; Han, H.; Li, C. *Acc. Chem. Res.* **2013,** *46* (8), 1900–1909.
160. Ran, J.; Zhang, J.; Yu, J.; Jaroniec, M.; Qiao, S. *Chem. Soc. Rev.* **2014,** *43,* 7787–7812.
161. Guo, Y.; Li, J.; Yuan, Y.; Li, L.; Zhang, M.; Zhou, C.; Lin, Z. *Angew. Chem. Int. Ed.* **2016,** *55,* 14693–14697.
162. Martin, D.; Qiu, K.; Shevlin, S.; Handoko, A.; Chen, X.; Guo, Z.; Tang, J. *Angew. Chem. Int. Ed.* **2014,** *53,* 9240–9245.
163. Wang, Y.; Yao, J.; Li, H.; Su, D.; Antonietti, M. *J. Am. Chem. Soc.* **2011,** *133,* 2362–2365.
164. Sayed, F. N.; Jayakumar, O. D.; Sasikala, R.; Kadam, R. M.; Bharadwaj, S. R.; Kienle, L.; Schurmann, U.; Kaps, S.; Adelung, R.; Mittal, J. P.; Tyagi, A. K. J. *Phys. Chem. C* **2012,** *116,* 12462–12467.
165. Cheng, N.; Tian, J.; Liu, Q.; Ge, C.; Qusti, A.; Asiri, A.; AlYoubi, A.; Sun, X. *ACS Appl. Mater. Interfaces* **2013,** *5,* 6815–6819.
166. Murdoch, M.; Waterhouse, G. I. N.; Nadeem, M. A.; Metson, J. B.; Keane, M. A.; Howe, R. F.; Llorca, J.; Idriss, H. *Nat. Chem.* **2011,** *3,* 489–492.
167. Xu, Y.; Zhang, W. *ChemCatChem* **2013,** *5,* 2343–2351.
168. Ge, L.; Han, C.; Liu, J.; Li, Y. *Appl. Catal. A* **2011,** *409–410,* 215–222.
169. Yu, J.; Ran, J. *Energy Environ. Sci.* **2011,** *4,* 1364–1371.

170. Hou, Y.; Laursen, A. B.; Zhang, J.; Zhang, G.; Zhu, Y.; Wang, X.; Dahl, S.; Chorkendorff, I. *Angew. Chem. Int. Ed.* **2013,** *52,* 3621–3625.
171. Yu, J.; Wang, S.; Cheng, B.; Lin, Z.; Huang, F. *Catal. Sci. Technol.* **2013,** *3,* 1782–1789.
172. Hong, J.; Wang, Y.; Wang, Y.; Zhang, W.; Xu, R. *ChemSusChem* **2013,** *6,* 2263–2268.
173. Akple, M.; Low, J.; Wageh, S.; Al-Ghamdi, A.; Yu, J.; Zhang, J. *Appl. Surf. Sci.* **2015,** *358,* 196–203.
174. Pan, Z.; Zheng, Y.; Guo, F.; Niu, P.; Wang, X. *ChemSusChem* **2017,** *10,* 87–90.
175. Yi, S.; Yan, J.; Wulan, B.; Li, S.; Liu, K.; Jiang, Q. *Appl. Catal. B Environ.* **2017,** *200,* 477–483.
176. Liu, Q.; Chen, T.; Guo, Y.; Zhang, Z.; Fang, X. *Appl. Catal. B Environ.* **2016,** *193,* 248–258.
177. Xiang, Q.; Yu, J.; Jaroniec, M. *J. Phys. Chem. C* **2011,** *115,* 7355–7363.
178. Zhang, G.; Zang, S.; Lin, L.; Lan, Z.-Z.; Li, G.; Wang, X. *ACS Appl. Mater. Interfaces* **2016,** *8,* 2287–2296.
179. Zhang, G.; Zang, S.; Wang, X. *ACS Catal.* **2015,** *5,* 941–947.
180. Tong, H.; Ouyang, S.; Bi, Y.; Umezawa, N.; Oshikiri, M.; Ye, J. *Adv. Mater.* **2012,** *24,* 229–251.
181. Ma, Y.; Wang, X.; Jia, Y.; Chen, X.; Han, H.; Li, C. *Chem. Rev.* **2014,** *114,* 9987–10043.
182. Zhang, G.; Lan, Z.-A.; Lin, L.; Lin, S.; Wang, X. *Chem. Sci.* **2016,** *7,* 3062–3066.
183. Zheng, D.; Cao, X.; Wang, X. *Angew. Chem. Int. Ed.* **2016,** *55,* 11512–11516.
184. Zhao, G.; Huang, X.; Fina, F.; Zhang, G.; Irvine, J. T. S. *Catal. Sci. Technol.* **2015,** *5,* 3416.
185. Wang, K.; Li, Q.; Liu, B.; Cheng, B.; Ho, W.; Yu, J. *Appl. Catal. B* **2015,** 176–177, 44–52.
186. Chang, X.; Wang, T.; Gong, J. *Energy Environ. Sci.* **2016,** *9,* 2177–2196.
187. Lin, J.; Pan, Z.; Wang, X. *ACS Sustain. Chem. Eng.* **2014,** *2* (3), 353–358.
188. Qin, J.; Wang, S.; Wang, X. *Appl. Catal. B Environ.* **2017,** *209,* 476–482.
189. Tani, M.; Sakamoto, T.; Mita, S.; Sakaguchi, S.; Ishi, Y. *Angew. Chem. Int. Ed.* **2005,** *44,* 2586–2588.
190. Chen, X.; Zhang, J.; Fu, X.; Antonietti, M.; Wang, X. *J. Am. Chem. Soc.* **2009,** *131,* 11658–11659.
191. Ye, X.; Cui, Y.; Qiu, X.; Wang, X. *Appl. Catal. B Environ.* **2014,** *152–153,* 383–389.
192. Ye, X.; Cui, Y.; Wang, X. *ChemSusChem* **2014,** *7,* 738–742.
193. Ye, X.; Zheng, Y.; Wang, X. *Chin. J. Chem.* **2014,** *32,* 498–506.
194. Brink, G. T.; Arends, I. W. C. E.; Sheldon, R. A. *Science* **2000,** *287,* 1636–1639.
195. Enache, D. I.; Edwards, J. K.; Landon, P.; Solsonaespriu, B.; Carley, A. F.; Herzing, A. A.; Watanabe, M.; Kiely, C. J.; Knight, D. W.; Hutchings, G. J. *Science* **2006,** *311,* 362–365.
196. Hoffmann, M. R.; Martin, S. T.; Choi, W.; Bahnemann, D. W. *Chem. Rev.* **1995,** *95,* 69–96.
197. Zhou, M.; Yang, P.; Yuan, R.; Asiri, A. M.; Wakeel, M.; Wang, X. *ChemSusChem* **2017,** *10,* 4451–4456.
198. Ong, W.-J.; Yeong, J.-J.; Tan, L.-L.; Goh, B. T.; Yong, S.-T.; Chai, S.-P. *RSC Adv.* **2014,** *4,* 59676–59685.
199. Su, C.; Ran, X.; Hu, J.; Shao, C. *Environ. Sci. Technol.* **2013,** *47,* 11562–11568.
200. Zheng, Y.; Yu, Z.; Lin, F.; Guo, F.; Alamary, K. A.; Taib, L. A.; Asiri, A. M.; Wang, X. *Molecules* **2017,** *22,* 572.

201. Cai, Q.; Mateti, S.; Yang, W.; Jones, R.; Watanabe, K.; Taniguchi, T.; Huang, S.; Chen, Y.; Li, L.-H. *Angew. Chem. Int. Ed.* **2016,** *55,* 8405–8409.
202. Ma, R.; Bando, Y.; Zhu, H.; Sato, T.; Xu, C.; Wu, D. *J. Am. Chem. Soc.* **2002,** *124,* 7672–7673.
203. Watanabe, K.; Taniguchi, T.; Kanda, H. *Nat. Mater.* **2004,** *3,* 404–409.
204. Niu, P.; Yang, Y.; Yu, J. C.; Liu, G.; Cheng, H. M. *Chem. Commun.* **2014,** *50,* 10837–10840.

CHAPTER THREE

Photocatalytic Synthesis of Chemicals

Marcin Kobielusz, Paweł Mikrut and Wojciech Macyk[1]
Faculty of Chemistry, Jagiellonian University, Kraków, Poland
[1]Corresponding author: E-mail: macyk@chemia.uj.edu.pl

Contents

Advances in Inorganic Chemistry, Volume 72
ISSN 0898-8838
https://doi.org/10.1016/bs.adioch.2018.05.002

Abstract

Although production of solar fuels from water or carbon dioxide seems still too expensive to compete with fossil fuels, application of photocatalysis for synthesis of valuable organic chemicals may appear an interesting approach. The use of a photocatalyst and light as the energy source can be considered as a green, sustainable alternative for other methods of fine chemicals production. In this chapter we review approaches to synthesize organic compounds in heterogeneous photocatalytic systems. The described reactions can be divided into three main groups due to their mechanisms: (1) reactions of organic substrates with photocatalytically activated small molecules; (2) a direct photocatalytic activation of one organic substrate; and (3) reactions involving a direct activation of two organic molecules. The understanding of plausible mechanisms of the photocatalytic reactions and properties of photocatalytic systems favoring the desired reaction should facilitate the design of new photocatalysts and convenient conditions to synthesize new specific chemicals.

1. INTRODUCTION

Photosynthesis is the natural process, which enables synthesis of organic compounds from simple, low-energy substrates (carbon dioxide, water). Because the energy is provided in the form of light, photosynthesis can be considered as the process resulting in solar to chemical energy conversion. As such, it is a perfect example of a sustainable energy production process with chlorophyll and other components of photosynthetic systems being examples of sustainable materials used in energy production. However, photosynthesis can also be considered as a sustainable process of fine chemicals production. Synthesized organic compounds are not only fuels but are often valuable chemicals produced in a "green fashion." Photocatalytic systems mimicking natural photosynthesis should retain the most important features of natural photosynthesis: light as the energy source, mild reaction conditions, and green solvents. Thus, photosynthesis of chemicals, not only fuels, appears a green alternative to other methods of fine chemicals production.

Artificial photosynthetic systems are based on photosensitizers or photocatalysts facilitating, in principle, a photoinduced charge separation. Separated charges are involved in reduction and oxidation reactions. Alternatively, photoexcited systems may transfer energy instead of charges (electrons and holes). Photocatalytic systems can be based on molecular materials, like in natural photosynthesis, or on crystalline solids, as in the case of semiconductor-based heterogeneous photocatalysis. The latter systems often offer a higher stability and photostability, as well as a simpler separation of products. In this work, heterogeneous, semiconductor-based photocatalytic systems for synthesis of organic compounds will be reviewed and discussed. Our goal is to discuss photocatalytic synthesis of organic compounds in terms of the mechanisms of these processes.

Synthesis of organic compounds in the discussed systems results from redox reactions initiated by electrons from the conduction band (CB) and holes from the valence band (VB). However, mechanisms of the reactions may involve a direct electron transfer between an organic substrate and the photocatalyst, or an indirect electron transfer, encompassing activation (oxidation or reduction) of small molecules as initial steps of the overall reaction (Fig. 1).

Water and hydroxyl anions are the main small molecules being oxidized in these processes. Formed hydroxyl radicals are responsible for a consecutive oxidation of organic substrates. $HO^{\bullet}$ is a very reactive, strong oxidant, which shows usually a low selectivity in oxidation reactions. Furthermore, hydroxyl radicals may lead to a complete oxidation of organics to carbon dioxide, water, and simple inorganic anions. Generation of $HO^{\bullet}$ takes place mainly at oxides (TiO_2, ZnO, *etc.*). It is worth mentioning, that differentiation between a direct oxidation of an organic molecule by photogenerated holes and its indirect oxidation, involving photogenerated hydroxyl radicals, may be quite difficult. Such processes will be considered in this paper as those initiated by hydroxyl radicals.

Dioxygen, water, carbon dioxide, and hydrogen peroxide are major small molecules undergoing reduction reactions with electrons from the conduction band. Superoxide, hydrogen, $CO_2^{\bullet-}$, and water are the primary products of these processes, respectively. Also inorganic ions may act as electron acceptors.

Two decades ago Kisch proposed a classification of photocatalytic reactions based on the role of primary reduction and oxidation reactions in the formation of products.[1] According to this classification, *semiconductor photocatalysis type A* encompasses formation of two different products as the result of reactions involving photogenerated holes and electrons. Reactions presented in Fig. 1A–D follow this scheme. As the result of *semiconductor photocatalysis type B*, only one product is formed. In this case, products of primary redox reactions (radicals) couple together, yielding a new stable molecule, as depicted in Fig. 1E. Because photocatalysis type B requires coupling of two radicals, their lifetimes strongly affect the overall efficiency of the reaction.

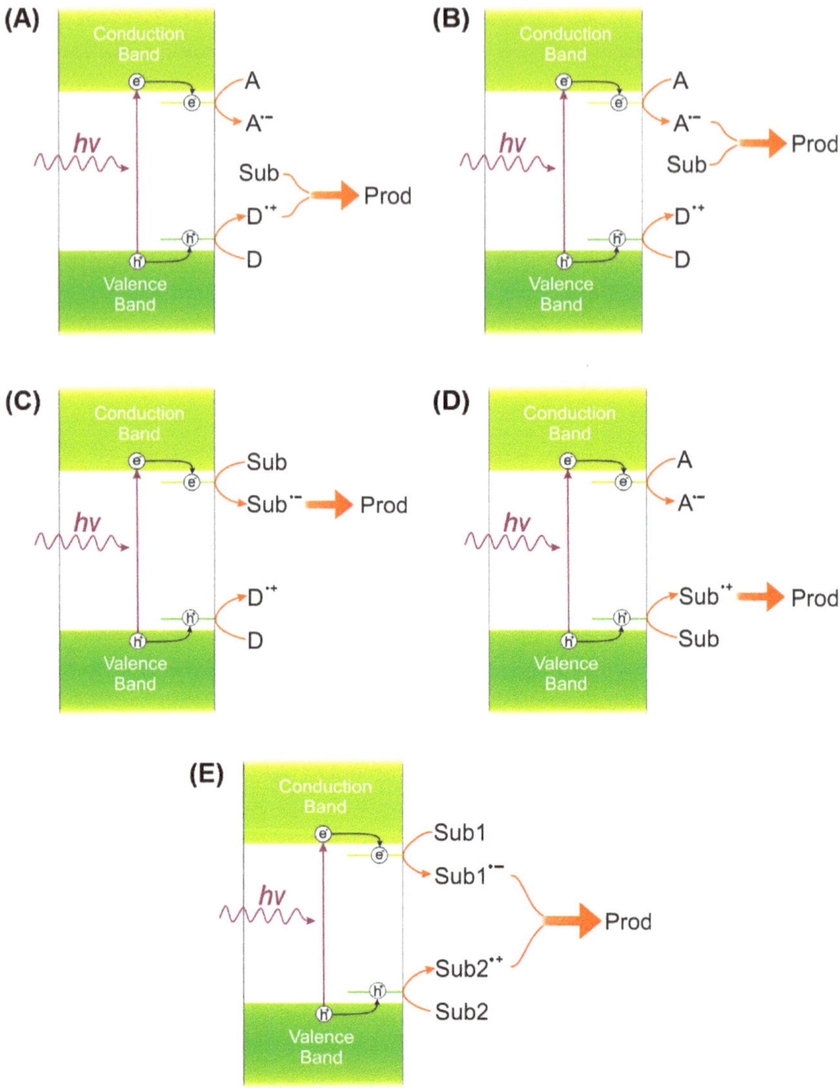

Figure 1 Primary redox reactions leading to photocatalytic synthesis of organic compounds (Prod) from substrates (Sub, Sub1, Sub2) and small molecules acting as electron donors (D) or acceptors (A): (A, B) reaction involving reduction and oxidation of two small molecules (A, D); (C) reaction involving reduction of an organic substrate (Sub) and oxidation of a small molecule (D); (D) reaction involving reduction of a small molecule (A) and oxidation of an organic substrate (Sub); and (E) reaction involving reduction and oxidation of organic substrates (Sub1, Sub2).

In the following chapters photocatalytic synthesis of organic compounds will be described on classification of the reaction into one of the groups presented in Fig. 1.

2. REACTIONS INVOLVING ACTIVATION OF SMALL MOLECULES

2.1 Oxidation

2.1.1 Oxidation of Alcohols

Wang et al.[2] proposed two possible mechanisms for the TiO_2-assisted photocatalytic oxidation of methanol in the gas phase: direct oxidation by photogenerated holes and indirect oxidation via $HO^{\bullet}$ radicals.

$$HO^{\bullet} + RCH_2OH \rightarrow R^{\bullet}CHOH + H_2O \tag{1}$$

$$TiO_2(h^+) + RCH_2OH \rightarrow TiO_2 + R^{\bullet}CHOH + H^+ \tag{2}$$

$$R^{\bullet}CHOH + O_2 \rightarrow RCHO + HO_2^{\bullet} \tag{3}$$

According to Wang's studies, the methanol oxidation mechanism depends on the competitive adsorption between water and methanol at the TiO_2 surface. Oxidation of methanol at TiO_2 in the presence of high water concentrations involves photocatalytically generated $HO^{\bullet}$ radicals. At a lower water concentration (or higher methanol content), methanol can be oxidized directly by holes. Wang estimated that the critical water to methanol molar ratio is *c.* 300, at which both mechanisms may operate.

Pillai et al.[3] studied oxidation of various primary and secondary alcohols to aldehydes and ketones in a gas-phase photochemical reactor using immobilized TiO_2 photocatalysts (Table 1). The influence of different O_2/alcohol ratios on the photooxidation yield was monitored. The amount of oxygen was found to be critical for the photooxidation of alcohols; the conversion yield increased with the increasing oxygen content. A large excess of oxygen (O_2 to alcohol molar ratio $\approx$ 22) enabled optimal formation of products. Pillai studied also the effect of water vapor on the oxidation yield. Relative humidity of the reaction mixture at the level of up to 90% did not affect significantly the alcohol oxidation yield.

The complexity of alcohol oxidation can be illustrated by glycerol oxidation. This reaction can proceed via both h^+- and radical-mediated routes,

Table 1 Gas-Phase Photocatalytic Oxidation of Alcohols at TiO_2[3]

Alcohol	Conv. (%)	Product	Sel. (%)
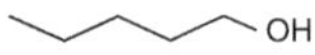	18	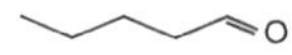	>95
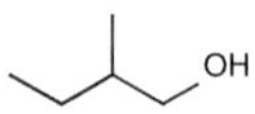	20	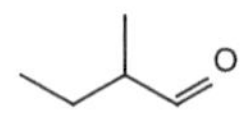	>95
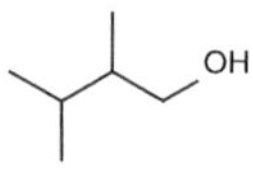	24	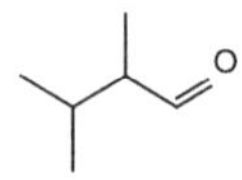	>95
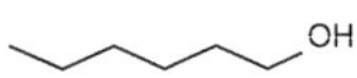	26		>95
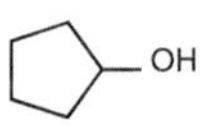	37	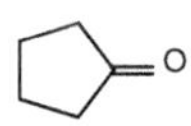	>95
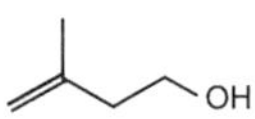	14	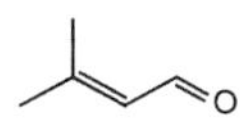	>95
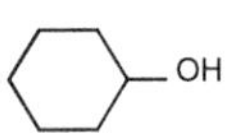	21	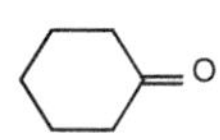	>95
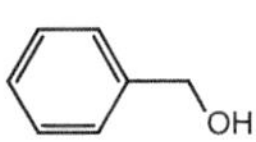	35	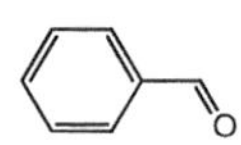	>95
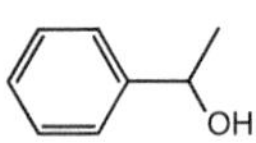	97	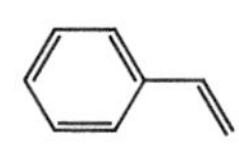	83
		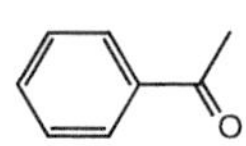	7
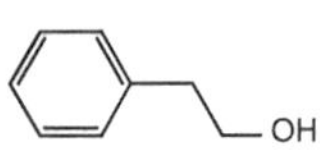	53	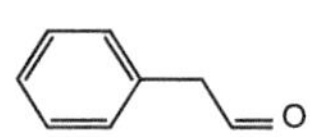	26
		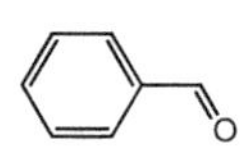	48
		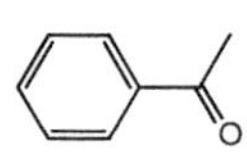	10

Reaction conditions: alcohol (1.43 mmol/min), O_2/alcohol ratio (22), temperature (463 K), and irradiation time (2 h).

with the latter mechanism prevailing. Jedsukontorn et al.[4] investigated the route of glycerol photoconversion at TiO_2 in the presence of H_2O_2. After 8 h of irradiation, the conversion of 0.3 M glycerol to dihydroxyacetone (DHA), glyceraldehyde (GCD), and glyceric acid (GCA; Fig. 2) reached *c.* 19% in the absence of H_2O_2 with a selectivity of 24%, 68%, and 8%, respectively. Conversion of glycerol via the radical-mediated oxidation pathway,

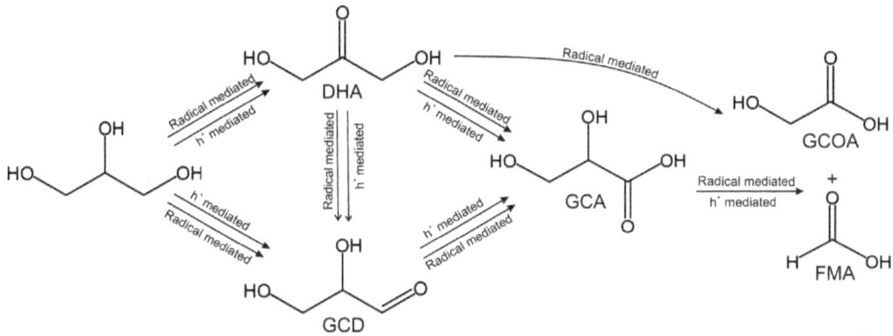

Figure 2 Pathways of photocatalytic oxidation of glycerol in the presence of TiO_2 and H_2O_2.[4]

increased in the presence of hydrogen peroxide to *c.* 71%. The excess of $HO^•$ radicals, resulting from decomposition of H_2O_2, can oxidize organic substrates by decreasing the DHA and GCD concentration and increasing the GCA production. Moreover, further oxidation steps lead to the formation of glycolic acid (GCOA) and formic acid (FMA).

Hermes et al.[5] also investigated photocatalytic oxidation of glycerol at ZnO photocatalyst. The authors studied the influence of the main reaction parameters, i.e., initial pH and temperature, on the yield and selectivity of GCD and DHA production. The best results of glycerol conversion (initial concentration of 3 mM, reaction time 1 h) were achieved for an initial pH = 11 at 50°C. Under these conditions, 64.6% of glycerol was converted, with a selectivity of 35.3% and 14.0% for GCD and DHA, respectively (yields of 22.8 and 9.0 for GCD and DHA, respectively).

2.1.2 Oxidation of Aromatics

Park and Choi[6] studied photocatalytic synthesis of phenol from benzene. Benzene can be directly converted into phenol through the reaction with $HO^•$ radicals. Direct addition of a hydroxyl radical to the benzene ring results in hydroxycyclohexadienyl (HCHD) radical production, which subsequently undergoes an H-atom abstraction to form phenol.

Park et al. indicated that the addition of Fe^{3+}, H_2O_2, and $Fe^{3+} + H_2O_2$ strongly enhanced the phenol production yield. These agents increase the efficiency of $HO^•$ production and act as efficient electron acceptors, hindering HCHD reduction by electrons from the conduction band and suppressing regeneration of benzene. Moreover, addition of polyoxometalates to the TiO_2 suspension, deposition of Pt nanoparticles, and surface fluorination of

Table 2 Results of Photocatalytic Transformation of Benzene to Phenol after 2 h of Irradiation[8]

Photocatalyst	Anatase (%)	Benzene Conversion (%)	Phenol Yield (%)	Phenol Selectivity (%)
TiO_2	100	26	2	8
TiO_2	58[b]	16	1	6
$mTiO_2$	61[b]	23	19	83
$mTiO_2$[a]	61[b]	42	34	81

[a]Irradiation time (6 h).
[b]Other crystalline form does not exist.

TiO_2 increased the phenol production yield. Shiraishi et al.[7] managed to transform benzene to phenol with a very high selectivity (>80%; Table 2). The authors used mesoporous TiO_2 ($mTiO_2$) prepared by surfactant-templating or particle aggregation method, which strongly affected the adsorption of substrate molecules and $HO^{\bullet}$ radical generation. Benzene is less polar than phenol, and hence it adsorbs well on the less polar surface of $mTiO_2$ inside the pores. In contrast, polar phenol molecules are weakly adsorbed inside pores. As a result, $HO^{\bullet}$ radicals generated inside the pores of $mTiO_2$ react efficiently with the well-adsorbed benzene, whereas less polar phenol leaves the reaction site.

Bui et al.[9] studied the photocatalytic oxidation of benzene in aqueous systems in the presence of different kinds of TiO_2 powders (anatase or rutile) and labeled oxygen ($^{18}O_2$) to elucidate the origin of oxygen atoms introduced to the substrate. When anatase powders were used as photocatalysts, 70%–90% of oxygen in phenol molecules originated from water (Fig. 3), pointing at indirect oxidation of benzene by hydroxyl radicals. In contrast, when rutile powders were used, only 20%–40% of the oxygen originated from water, whereas the rest came from molecular oxygen. This indicates a direct oxidation of benzene by holes to the $C_6H_6^{\bullet+}$ radical cation, followed by its reaction with molecular oxygen or photogenerated superoxide anion.

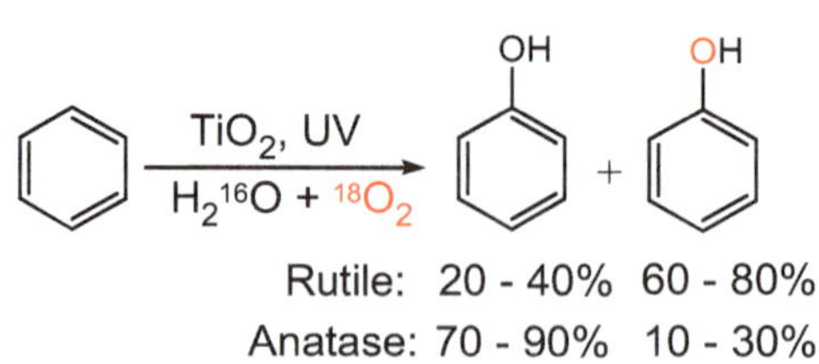

Figure 3 Photocatalytic oxidation of benzene to phenol on TiO_2.[10]

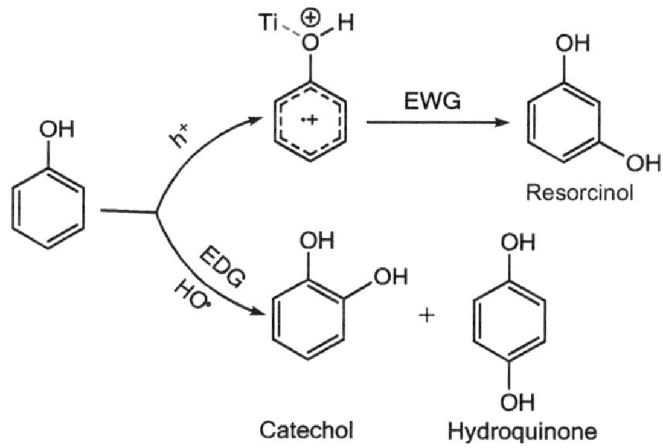

Figure 4 Photocatalytic oxidation of phenol at TiO_2.[12] *EDG*, electron donor group; *EWG*, electron withdrawing group.

The principal photocatalytic oxidation of phenol involves mainly $HO^•$ radicals.[11] Phenol oxidation in the presence of TiO_2 led to the formation of hydroxylated products, i.e., hydroquinone, catechol, and resorcinol, according to Fig. 4. Lv et al.[12] investigated the distribution of hydroxylated products as a function of irradiation time in TiO_2 suspensions. After 1 h of irradiation, almost all phenol was transformed to hydroquinone (the predominant product), catechol, and resorcinol. The highest hydroquinone concentration was observed after 35 min of irradiation, with the product distribution of 1:2:97 for *ortho-*, *meta-*, and *para-*products, respectively. Phenol hydroxylation with $HO^•$ radicals results in *o-* and *p-*dihydroxybenzene production (Fig. 4); however, a direct phenol oxidation with holes leads to production of *m-*dihydroxybenzene, as confirmed in experiments with $HO^•$ quenchers. After 1.5 h of irradiation, nearly all $C_6H_4(OH)_2$ products were decomposed. Recently, we have demonstrated that introduction of additional −OH groups to an aromatic ring results in self-sensitized photocatalytic degradation of organic compounds at TiO_2.[13]

Lv et al.[14] studied effects of anatase and rutile surface modification by fluoride ions on the photocatalytic oxidation of phenol to catechol. Addition of NaF to the solution had two opposite effects on the oxidation rate of phenol, depending on the crystal structure of TiO_2. In the case of anatase, phenol to catechol transformation was more efficient (phenol conversion increased from 20% to 54% after 1 h of irradiation), compared with rutile (phenol conversion decreased from 12% to 2.7% after 6 h of irradiation) under the same reaction conditions. In the presence of a fluorinated surface, the selectivity of catechol formation was enhanced, both for anatase and rutile TiO_2 suspensions.

Palmisano et al.[15] investigated the effect of substituent for a group of benzene derivatives on the formation of monohydroxy derivatives. Organic molecules, containing electron withdrawing groups (EWGs), electron donor groups (EDGs), or a combination of EWG and EDG, were used as substrates. The first group of molecules encompassed four compounds with EWGs: nitrobenzene, cyanobenzene, benzoic acid, and 1-phenylethanone. Phenol, phenylamine, and *N*-phenylacetamide contained EDGs. The compound with a combination of EWG and EDG was represented by 4-chlorophenol and hydroxy-cyanobenzenes. In the case of molecules with EDGs, the $HO^{\bullet}$ radical attack led to formation of *ortho-* and *para-*substituted hydroxy products. In the case of the substrates with EWGs, the $HO^{\bullet}$ attack was unselective, resulting in a mixture of all three possible hydroxy isomers. In the case of 4-chlorophenol and hydroxy-cyanobenzenes, the attack of the hydroxyl radical took place only in the positions activated by the $-OH$ group.

Huixian et al.[16] studied the synthesis of α-naphthol from naphthalene at TiO_2 as a photocatalyst. A direct oxidation of naphthalene by $HO^{\bullet}$ radicals, generated on illuminated TiO_2 nanocrystals, took place according to the reaction scheme shown in Fig. 5. Huixian checked the influence of surface modifications and electron acceptors on the production of α-naphthol. Production yield of α-naphthol reached 22.2% on addition of Fe^{3+} and H_2O_2 to the reactor, while in the absence of these additives, formation of α-naphthol was not detected. The authors showed the enhancement of α-naphthol production yield for all tested surface modifications (La–Eu/TiO_2, La–Y/TiO_2, $H_3PW_{12}O_{40}/TiO_2$, $H_3PMo_{12}O_{40}/TiO_2$, Fe/TiO_2, Ag/TiO_2, Cu/TiO_2, and N/TiO_2) compared to the unmodified TiO_2 photocatalyst. Fe/TiO_2 was the most effective among the surface-modified photocatalysts (conversion of 50% was reached), but the yield was only 5.6% due to the oxidative degradation of the formed product.

1,4-Naphthoquinone is another interesting product of naphthalene oxidation. Karam et al.[17] studied the synthesis of 1,4-naphthoquinone in the process of photocatalytic oxidation of naphthalene by photogenerated $HO^{\bullet}$ radicals. The authors found optimal conditions to run the reaction in a closed reactor system. The yield of 1,4-naphthoquinone production

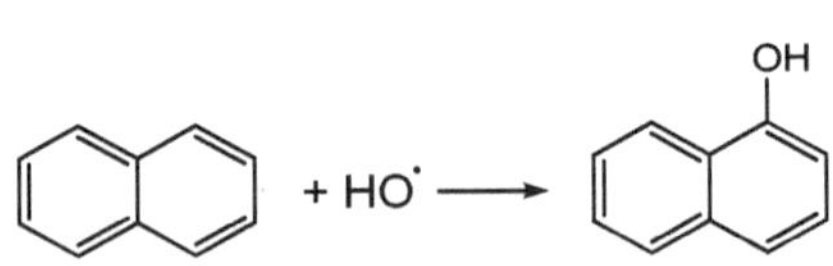

Figure 5 Direct hydroxylation of naphthalene at TiO_2.

reached 57%. Further oxidation led to ring-opening of the product. Ohno et al.[18] demonstrated the photocatalytic oxidation of naphthalene to 2-formylcinnamaldehyde in acetonitrile/water environment with molecular oxygen as the electron acceptor. The authors studied the influence of various anatase to rutile ratios on the oxidation of naphthalene. TiO_2 photocatalysts containing both anatase and rutile phases were prepared either by annealing the anatase powder (Toho Titanium; 99.9% anatase) at different temperatures or by mixing rutile (Toho Titanium; 1.5% anatase) with the anatase powder (Ishihara Sangyo; 100% anatase) in different ratios. Ohno et al. found a synergistic effect for rutile/anatase mixtures in the naphthalene oxidation reaction. In the aforementioned first method, the highest activity (2-formylcinnamaldehyde production: *c.* 5.5 μmol/h) was observed for samples annealed at 800°C, which contained 90% of the rutile phase. In the aforementioned second method, the highest activity (2-formylcinnamaldehyde production: *c.* 6.2 μmol/h) was observed for the sample containing 70% of the rutile phase.

2.1.3 Oxidation of Side Chains

An interesting example of photocatalytic oxidation reactions leading to an efficient production of valuable chemicals was presented by Augugliaro et al.[19] Photocatalytic synthesis of vanillin at room temperature in aqueous medium in the presence of TiO_2 photocatalyst was confirmed to occur from different substrates: ferulic acid, isoeugenol, eugenol, and vanillyl alcohol (Fig. 6). Photocatalytic oxidation of these chemicals resulted in vanillin production with a selectivity ranging from 1.4 to 21 mol% with respect to the converted substrate.

Photocatalytic production of vanillin involving membrane separation in a pervaporation photocatalytic reactor (PVPR) was studied by G. Camera-Roda et al.[20] Due to an efficient separation from the reaction mixture, vanillin produced in the PVPR system did not undergo any further degradation, compared to a high instability of vanillin produced in a traditional photoreactor.

Shiraishi et al.[7] compared photocatalytic selective transformation of organics on $m TiO_2$ and nonporous TiO_2 ($n TiO_2$). The reaction yield on $m TiO_2$ depended on the adsorption of substrates onto the catalyst surface and increased for well-adsorbed molecules (Table 3). The degree of substrate adsorption onto the photocatalyst was defined as the distribution ratio, D [i.e., $D = (C_0 - C_e)/C_e$, where C_0 denotes the initial concentration of the substrate in solution and C_e denotes the equilibrium concentration of the substrate in the solution after addition of the photocatalyst (313 K; 0.5 h)].

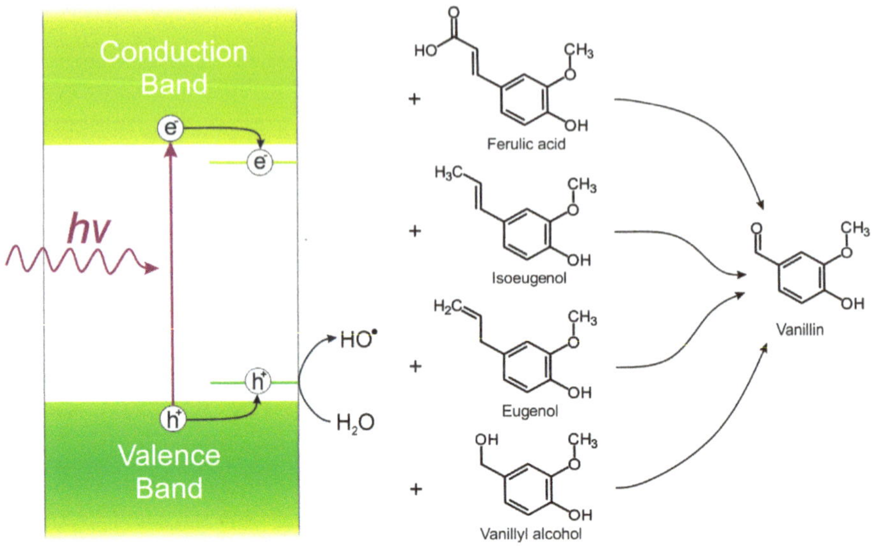

Figure 6 Photocatalytic synthesis of vanillin.[19]

Table 3 Photocatalytic Transformation of Various Substrates[7]

Catalyst(x)	Reactant	D	Conv. (%)	Product	D	Sel. (%)
$nTiO_2(100)$	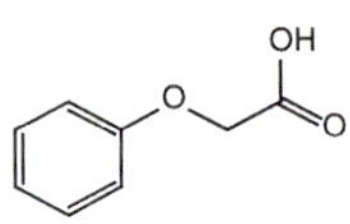	0.17	90	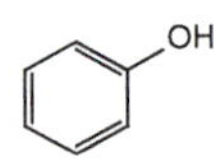	0.01	35
$nTiO_2(58)$		0	82		0	34
$mTiO_2(61)$		0.11	89		0	72
$nTiO_2(100)$	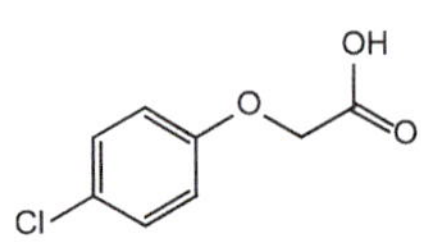	0.10	91	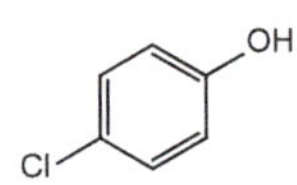	0.01	18
$nTiO_2(58)$		0	87		0	13
$mTiO_2(61)$		0.12	90		0	72
$nTiO_2(100)$	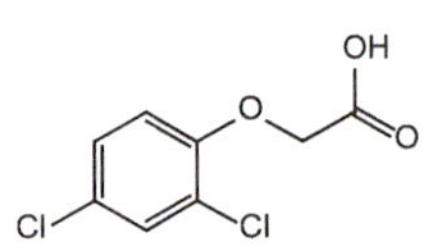	0.05	60	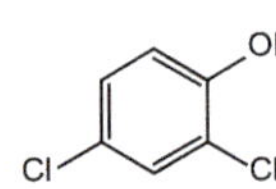	0	23
$nTiO_2(58)$		0	71		0	17
$mTiO_2(61)$		0.09	84		0	89

Reaction conditions: reactants 20 μmol; irradiation time 2 h; catalysts 10 mg; buffered (pH = 7) aqueous solution 10 mL; temperature 313 K; and *x* denotes anatase content/%.

The authors suggested that substrate adsorption occurred mainly inside the pores of $mTiO_2$. Moreover, $HO^•$ production on $mTiO_2$ took place also inside the mesopores. Therefore, higher D values correspond to higher substrate concentrations inside the pores, where oxidation by $HO^•$ was efficient.

2.1.3.1 Oxidation of N-Containing Compounds

Decomposition of *N*-nitrosodimethylamine (NDMA) in water leads to formation of methylamine (MA) and dimethylamine (DMA).[21] Photocatalytic degradation of NDMA is initiated mainly by $HO^•$ radicals rather than VB holes. In acidic suspensions of pure TiO_2, formation of MA rather than DMA was highly favored. On the other hand, in alkaline solutions, all degradation products (MA, DMA, and NH_3) were generated at comparable concentrations. Lee et al.[21] suggested that the product distribution depends on the position of the initial attack of $HO^•$ radicals on NDMA and can follow three different pathways. An enhanced decomposition of NDMA was achieved at silica-loaded TiO_2 and Nafion-coated TiO_2, at which formation of DMA rather than MA was preferred. In the case of surface fluorinated TiO_2, formation of MA over DMA dominated.

2.1.3.2 Oxidation of S-Containing Compounds

Sulfoxidation of alkanes is one of those synthetic processes which can be efficiently photocatalyzed by titanium dioxide. Production of *n*-heptanesulfonic acid, adamantanesulfonic acid, and cyclohexane sulfonic acid was reported by Parrino et al.[22] on visible light irradiation of titania suspension under sulfur dioxide and oxygen atmosphere. Visible light is absorbed by the formed TiO_2-SO_2 surface complex and leads to the charge separation (Fig. 7). The alkyl radical generation is a result of either a direct alkane oxidation by surface $-SO_2^{•+}$ groups or an indirect oxidation involving superoxide anion radicals formed in the reduction of O_2 with electrons from the conduction band. Another possibility of alkyl radicals formation involves the reaction of alkanes with the excited SO_2 ($^3SO_2$) molecule (Fig. 7). The alkyl radicals thus produced are involved in a radical chain reaction, encompassing reaction with sulfur dioxide and molecular oxygen, as summarized in Fig. 7. The formed RSO_2-$OO^•$ radicals react with RH to reproduce $R^•$ radicals and generate RSO_2-OOH, which rearranges to the final RSO_3H product.

Sulfoxides can be also formed on selective oxidation of sulfides. Visible light—active mesoporous graphitic carbon nitride (mpg-C_3N_4) was successfully applied by Zhang et al.[23] as the photocatalyst of this reaction. High yield and selectivity were achieved in the system containing aldehyde and O_2. Isobutyraldehyde (IBA) appeared to increase the yield of reaction most efficiently. It has been suggested that superoxide radical anion $O_2^{•-}$ was created during irradiation, which, in turn, attacked the sulfur atom with the formation of the reactive intermediate persulfoxide or thiadioxirane (Fig. 8). In the presence of aldehyde, the generated $O_2^{•-}$ can oxidize

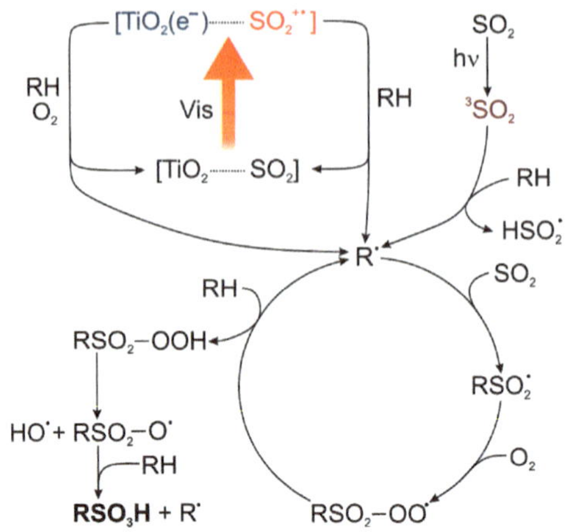

Figure 7 Photocatalytic sulfoxidation of alkanes (according to Ref. 22).

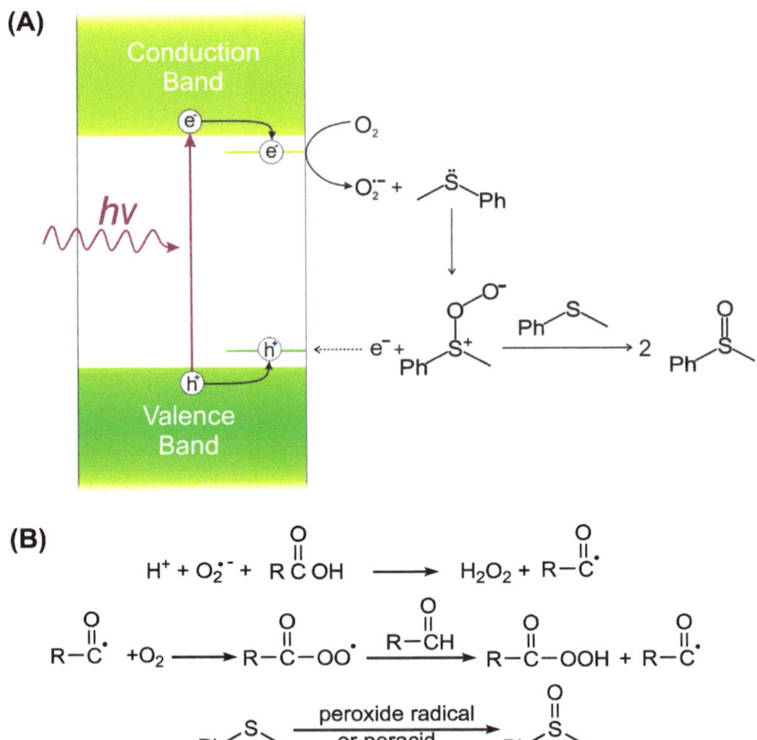

Figure 8 Possible reaction mechanism for the catalytic cycle of the mpg-C_3N_4 (A) and mpg-C_3N_4–aldehyde (B) system.[23]

aldehyde to the corresponding acyl radicals. Acyl radicals react with oxygen to yield a peroxy acid, which can oxidize sulfides to the corresponding sulfoxides. Examples of selectivity and yields of various sulfoxide syntheses are presented in Table 4.

Schoumacker et al.[24] studied photocatalytic synthesis of propan-1-thiol on TiO_2 and CdS photocatalysts. High selectivity of propan-1-thiol formation was achieved. The photocatalytic activity of TiO_2 was three times higher than that of CdS. The photocatalytic synthesis of propan-1-thiol proceeds in the following steps:

1. electron-hole pair generation on TiO_2;
2. formation of HS^-_{ads} by dissociative chemisorption of H_2S at the surface of TiO_2:

$$H_2S + TiO_2 \rightarrow HS^-_{ads} + H^+_{ads} \tag{4}$$

3. formation of $HS^\bullet_{ads}$ by abstraction of e^- from HS^- by photogenerated holes:

$$HS^-_{ads} + h^+ \rightarrow HS^\bullet_{ads} \tag{5}$$

4. reaction of $HS^\bullet$ with olefins:(6)

$$HS^\bullet_{ads} + CH_2 = CH{-}CH_{3ads} \rightarrow HS{-}CH_2{-}{}^\bullet CH{-}CH_{3ads} \tag{6}$$

5. formation of propan-1-thiol:(7)

$$HS{-}CH_2{-}{}^\bullet CH{-}CH_{3\,ads} \overset{H^+e^-}{\rightarrow} HS{-}CH_2{-}CH_2{-}CH_{3\,ads} \tag{7}$$

2.1.3.3 Oxidation of Hydrocarbons

Photocatalytic oxidation of hydrocarbons proceeds according to mechanisms usually involved in photodegradation of organic pollutants. Under some specific conditions it can, however, be used as a synthetic method. Oxidation of cyclohexane at TiO_2, studied by Almquist and Biswas,[25] is a good example of such a process. Oxidation of alkanes with either hydroxyl radicals or holes results in alkyl radical formation:

$$h^+ + RH_{2,ads} \rightarrow {}^\bullet RH + H^+ \tag{8}$$

Table 4 Results of Oxidation of Various Sulfides by the Mesoporous Graphitic Carbon Nitride–Isobutyraldehyde System[23]

Substrate	**Product**	**Time (min)**	**Conv. (%)**	**Sel. (%)**
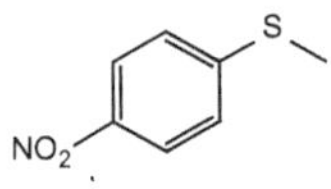	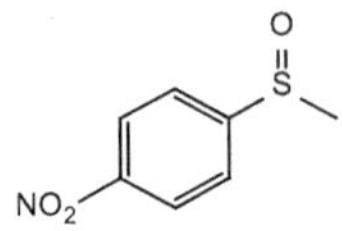	50	98	98
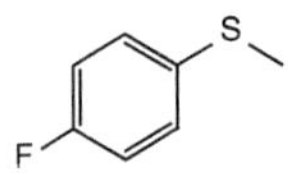	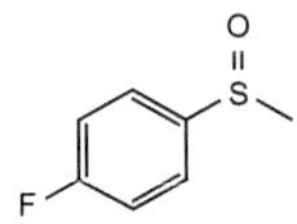	480	76	95
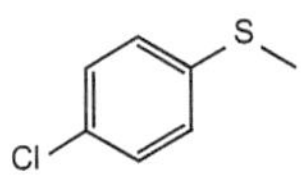	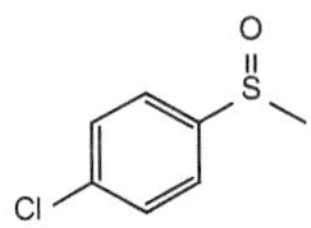	330	96	99
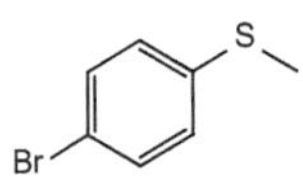	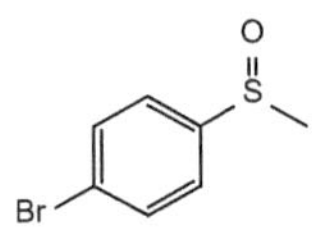	80	98	99
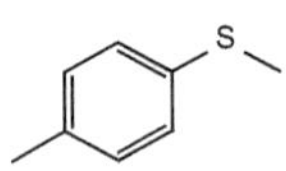	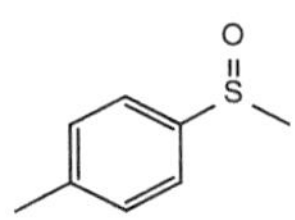	480	92	92
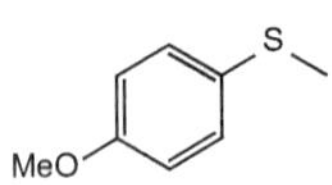	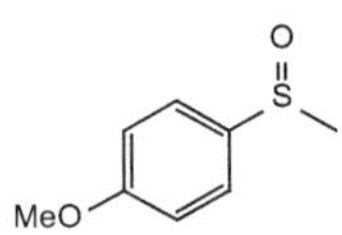	300	96	97
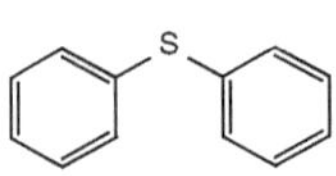	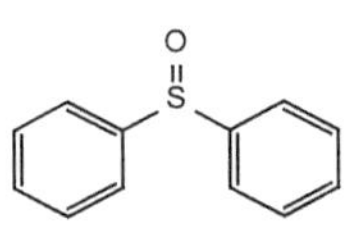	35	98	97
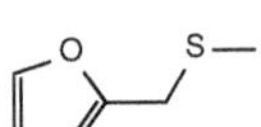	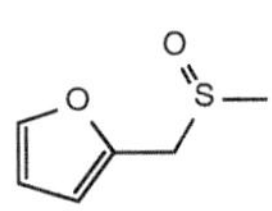	180	89	91
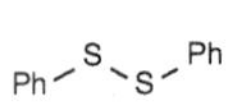	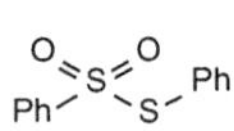	240	39	77

Reaction conditions: substrate 1 mmol, mpg-C_3N_4 50 mg, IBA 2 mmol, CH_3CN 3 mL, O_2 1 atm, 25ºC, and mercury lamp (150 W) with a cutoff filter >420 nm as a visible light source.

$$HO^{\bullet} + RH_{2,ads} \rightarrow {}^{\bullet}RH + H_2O \quad (9)$$

Boarini[26] proposed that cyclohexanol is formed in the reaction:

$${}^{\bullet}RH + HO^{\bullet} \rightarrow RHOH. \quad (10)$$

However, this product can be formed from ${}^{\bullet}RH$ also in other pathways. Cyclohexanol can be further oxidized to form cyclohexanone:

$$RHOH + 2HO^{\bullet} \rightarrow R = O + 2H_2O \quad (11)$$

Studies on the influence of various solvents on photocatalytic oxidation of cyclohexane on TiO_2 showed that the products' formation rates and distribution were significantly affected by the relative strengths of adsorption of the substrate, solvent, and oxidation products on titanium dioxide.[25] In nonpolar solvents, the selectivity of cyclohexanol production was very low, due to its strong adsorption on the TiO_2 surface, which facilitated a complete mineralization to carbon dioxide and water. However, in polar solvents, selectivity of the cyclohexanol synthesis was significantly higher.

Cyclohexane oxidation can also follow another mechanism, involving $O_2^{\bullet -}$ radicals.[27] Shirashi et al.[28] demonstrated the oxidation of cyclohexane by $O_2^{\bullet -}$ radicals under visible light irradiation at platinum-loaded WO_3 with *c.* 93% selectivity (Fig. 9).

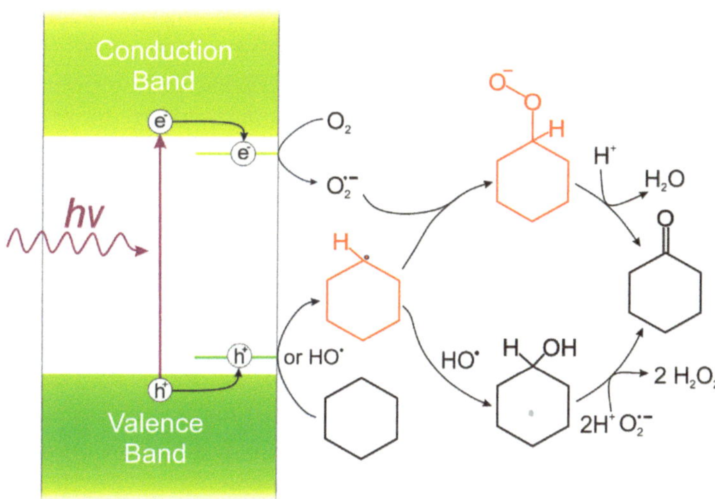

Figure 9 Mechanism of photocatalytic oxidation of cyclohexane to cyclohexanol and cyclohexanone at TiO_2.[29]

2.1.3.4 Free-Radical Polymerization Initiated by Heterogeneous Photocatalysts

Free-radical polymerization of methyl methacrylate (MMA) photoinitiated by TiO_2 (Degussa P25) in aqueous environment encompasses photogeneration of $HO^•$ radicals or h^+, which initiates the chain reaction.[30] The dependence of polymerization rate on concentration of TiO_2 and monomer, incident light intensity, and pH was demonstrated. Polymerization rate and the average molecular weight of the synthesized polymer increased readily with the monomer concentration. Moreover, the quantum yield decreased, and the average molecular weight increased with increasing incident light intensity. The rate of polymerization at $pH < pH_{pzc}$ was two times higher than at $pH > pH_{pzc}$. The yield of $HO^•$ radical generation measured at $pH > pH_{pzc}$ was *c.* two times higher than that at $pH < pH_{pzc}$. At low pH, polymerization was initiated by reaction with the photogenerated hole, whereas with increasing pH, the role of $HO^•$ radicals in polymerization was more and more pronounced.

Photoinduced free-radical polymerization of vinyl monomers was achieved by Dadashi-Silab et al. in the system containing ZnO and Fe^{3+}-doped ZnO (Fe/ZnO) nanoparticles.[31] Polymerization of acrylamide (AAm), as a water-soluble monomer in aqueous solution, and MMA, as an oil-soluble monomer in organic media, was observed on UV light irradiation and the absence of oxygen proved a versatility of these photocatalysts to initiate polymerization. $HO^•$ radicals were responsible for initiation of polymerization of AAm in aqueous solution. In organic media, triethylamine (TEA) and diphenyliodonium hexafluorophosphate ($Ph_2I^+PF_6^-$) were used as coinitiators. Initiating radicals in MMA photopolymerization were formed by hydrogen abstraction or reduction processes when tertiary amines or iodonium salts were used, respectively.

2.2 Reduction

Hydrogenation of furan to tetrahydrofuran (THF) in alcoholic suspension of metal-loaded TiO_2 was demonstrated by Nakanishi et al.[32] Various metals, acting as cocatalysts, were deposited on TiO_2: Au, Ag, Cu, Pd, and Pt. The highest efficiency of THF formation (37%) was recorded for Pd-loaded TiO_2. As a side reaction to furan hydrogenation over Pd/TiO_2, H_2 evolution was observed. The proposed reaction mechanism encompasses reduction of protons (water) on metal nanoparticles, followed by C=C bond hydrogenation by active hydrogen species or H_2 formation (Fig. 10).

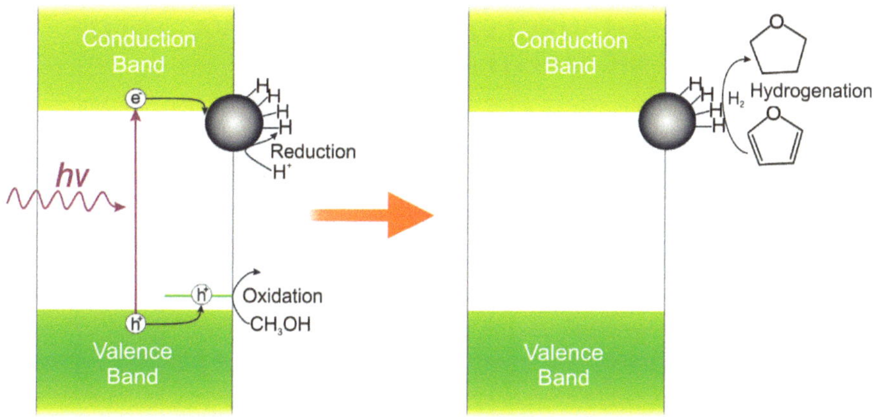

Figure 10 Expected reaction process for photocatalytic hydrogenation of furan to THF on $Pd{-}TiO_2$.[32]

The same research group studied also the hydrogenation of aromatic compounds in aqueous suspensions of Rh-loaded TiO_2 in the presence of oxalic acid as a sacrificial reagent.[33] Under H_2-free conditions, aromatics were very efficiently hydrogenated to the corresponding cyclohexane derivatives. For instance, cyclohexanecarboxylic acid was produced by hydrogenation of benzoic acid (Fig. 11). The yield of this reaction reached 16% after 15 min irradiation and 53% after 60 min irradiation at 360 nm.

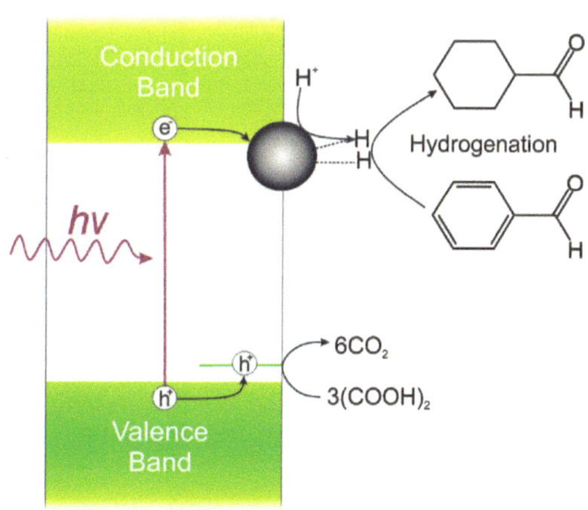

Figure 11 The mechanism of photocatalytic hydrogenation of benzoic acid to cyclohexanecarboxylic acid on $Rh{-}TiO_2$.[33]

3. REACTIONS INVOLVING DIRECT ACTIVATION OF ONE ORGANIC SUBSTRATE

In the previous section large groups of oxidation and reduction reactions of various organic compounds were discussed, encompassing photogeneration of radicals originating from water ($HO^\bullet$), oxygen ($O_2^{\bullet-}$), or protons ($H^\bullet$). Although these reactions are very efficient, they usually do not offer a high selectivity. In the case of oxidation with hydroxyl radicals, although a desired oxidation product is achieved, there are often further steps leading finally to the total mineralization of the products.

3.1 Reduction

3.1.1 Carbon Dioxide Reduction

An example of a direct reduction leading to formation of an organic product is the reduction of carbon dioxide. This process attracts more and more attention due to a significant increase in CO_2 concentration in the atmosphere, hereby causing the global climate change. Photocatalytic reduction of carbon dioxide to valuable chemicals or renewable fuels seems to be a particularly interesting approach utilizing solar energy. The subject of these studies is very extensive; therefore, only the most important aspects of CO_2 reduction will be presented here.

Photocatalytic reduction of CO_2 is a difficult process. One electron reduction causes bending of the linear molecule, which affects the LUMO level by increasing its energy.[34] Therefore, the one-electron CO_2 reduction reaction takes place at a highly negative potential of −1.9 V versus NHE.[35,36] A multielectron CO_2 reduction requires much milder conditions. Eqs. (12)−(17) summarize the possible reactions of multielectron CO_2 reduction and their potential values (at $pH = 7$ in aqueous solution, relative to NHE):[34,37]

$$CO_2 + 2H^+ + 2e^- \rightarrow HCOOH \quad E = -0.61\ V \tag{12}$$

$$CO_2 + 2H^+ + 2e^- \rightarrow CO + H_2O \quad E = -0.53\ V \tag{13}$$

$$CO_2 + 4H^+ + 4e^- \rightarrow C + 2H_2O \quad E = -0.20\ V \tag{14}$$

$$CO_2 + 4H^+ + 4e^- \rightarrow HCHO + H_2O \quad E = -0.48\ V \tag{15}$$

$$CO_2 + 6H^+ + 6e^- \rightarrow CH_3OH + H_2O \quad E = -0.38\ V \tag{16}$$

$$CO_2 + 8H^+ + 8e^- \rightarrow CH_4 + 2H_2O \quad E = -0.24\ V \tag{17}$$

However, involvement of several electrons in these reactions is associated with the difficulty to collect them simultaneously at the CO_2 adsorption site. These difficulties are responsible for the still unsatisfactory yields of such processes. Approaches to increase the efficiency of photocatalytic reduction comprise, but are not limited to, the use of photocatalysts characterized by a relatively low, negative potential of the conduction band edge,[38,39] formation of various types of composites with reduced recombination efficiency,[39,40] increase of the adsorption of CO_2,[41] and the use of cocatalysts to support the process.[42–44]

3.1.2 Reduction of Aldehydes and Ketones

The first report on photocatalytic reduction of aldehydes, to our knowledge, dealt with the reduction of acetaldehyde on zinc sulfide.[45] Ethanol and hydrogen were the main reaction products. By-products such as acetic acid and various products of partial oxidation and reduction of reagents were also observed.

Photocatalytic reduction of aldehydes with a much higher selectivity was proposed by Joyce-Pruden et al.[46] They studied photocatalytic reduction of aromatic aldehydes to the corresponding alcohols, in the absence of oxygen using titanium dioxide as the photocatalyst. Photocatalytic tests were carried out in alcohols (methanol, ethanol, or 2-propanol), which played the role of efficient scavengers of holes and hydroxyl radicals, thus reducing the chances of unfavorable oxidation of aromatic alcohols. The most efficient reduction (*c.* 80% yield in the case of benzaldehyde (BzAD) to benzyl alcohol (BzA) transformation, Fig. 12) was observed when ethanol was used as solvent.

Matsushita et al. performed this photocatalytic reduction in a microreaction system.[47] Also Fukui et al. reported on the reduction of BzAD derivatives, having other reducible functional groups, to the corresponding BzA analogues with a high selectivity.[48,49] Almost quantitative conversion of aldehydes was achieved, both in the presence of titanium dioxide excited

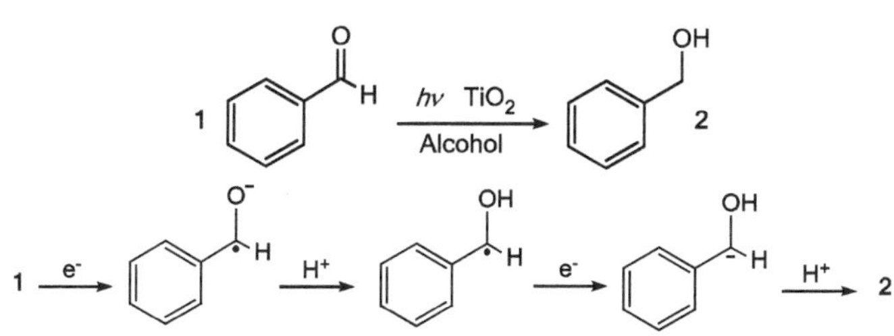

Figure 12 Reaction pathway of photocatalytic reduction of benzaldehyde.

Table 5 Photocatalytic Selective Reduction of Aromatic Aldehydes with Various Functional Groups in Acetonitrile Suspension of 2,3-Dihydroxynaphthalene@TiO_2 Under Visible Light Irradiation (as Reported in 48)

Substrate	Product	Time/h	Conv./%	Sel./%
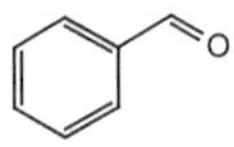	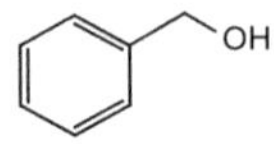	40	>99	>99
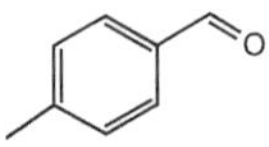	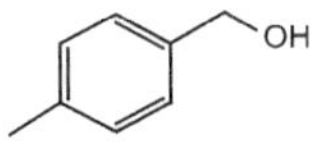	60	81	81
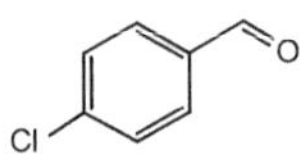	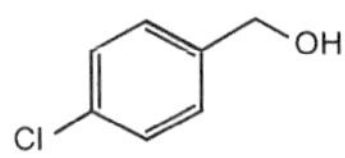	30	>99	>99
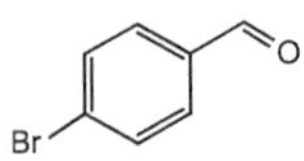	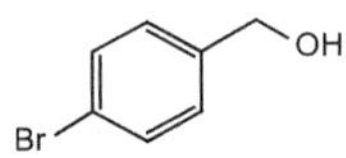	24	>99	>99
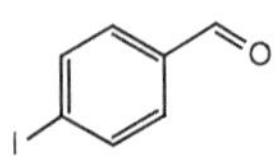	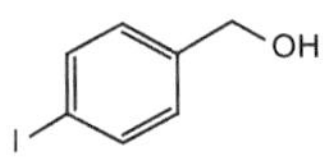	40	>99	93
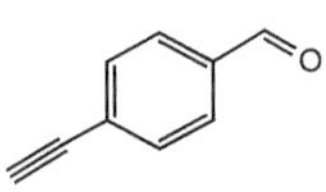	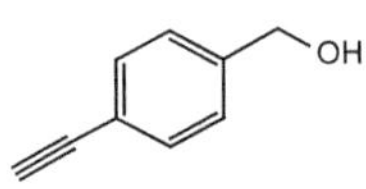	40	>99	>99
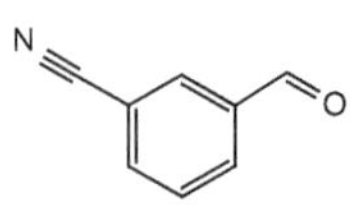	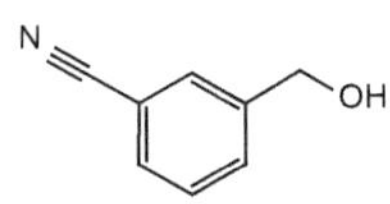	20	>99	>99
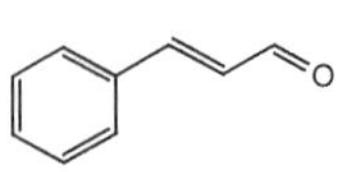	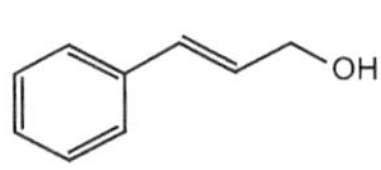	46	78	43

with ultraviolet light[49] and in the photosensitized system involving titanium dioxide with organic modifiers excited with visible light (Table 5).[48]

Reduction (hydrogenation) of ketones is a more often reported process. Park et al. studied photoreduction of 1,2-diketones in the presence of TiO_2 (P25).[50] Under these conditions, photoreduction leads to the formation of α-hydroxyketones. In the case of reduction of camphorquinone, a series of isomers was obtained, including 2-endo-, 3-endo-, 2-exo-, and 3-exo-hydroxycamphors. 2-Endo- and 3-endo-products were obtained with higher efficiencies (35% and 46%, respectively) when a methanol/water mixture was employed as solvent. However, the selectivity between the 2 and 3 positions is poor. In contrast, reduction of 1-phenyl-1,2-propanedione also gave the desired α-hydroxyketones as main products, but

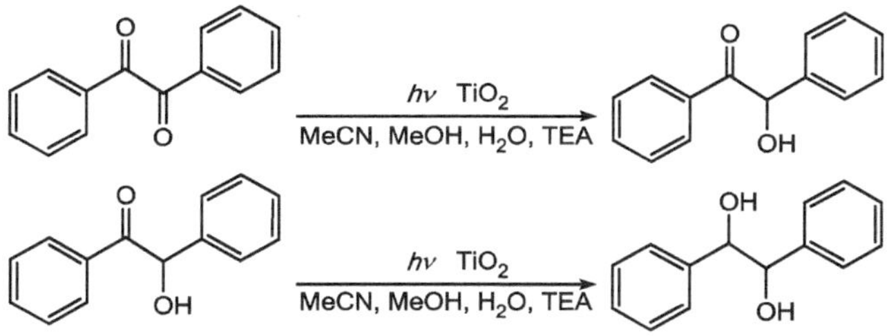

Figure 13 Photocatalytic reduction of benzil in the presence of TiO_2 (P25).

1-hydroxy-1-phenyl-2-propanone significantly predominates over 2-hydroxy-1-phenyl-1-propanone. In methanol containing 0.5 M TEA as a sacrificial electron donor, the yields of these products reached 56% and 4%, respectively.

Also in the case of benzil reduction (Fig. 13) catalyzed by titanium dioxide, the selectivity was not satisfactory.[51] In the reaction mixture composed of MeCN/MeOH/H_2O/TEA (88/7/2/3; 0.105 g of benzil in 50 mL of solvents), the yield of benzoin production was as high as 85% at 50% conversion of benzil. Nevertheless, the competitive reduction of benzoin into hydrobenzoin was also observed. Under the same conditions, in an independent experiment, benzoin was reduced to hydrobenzoin with a yield of 45%.

Extensive research on the photocatalytic reduction of acetophenone derivatives and diaryl ketones (Fig. 14) was carried out by Kohtani et al.[52] The photocatalytic tests were carried out under UV light irradiation (>340 nm) in oxygen-free atmosphere with titanium dioxide photocatalyst (P25) and alcohol (methanol, ethanol, or isopropanol) as a sacrificial hole scavenger. In most cases high product yields were observed. Oxygen inhibited the reaction. A correlation between the reduction potential of ketones and the efficiency of the reaction was observed. In the case of aliphatic ketones the reaction did not occur under the applied conditions due to too negative redox potentials that characterize the reduction of these ketones. Low yields were also observed for ketones with sterical hindrances.

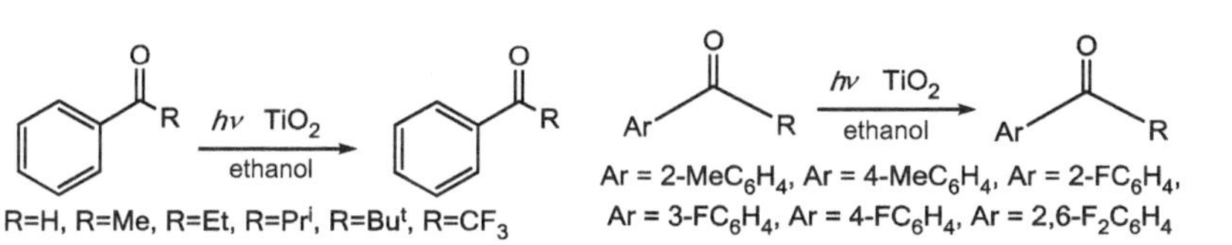

Figure 14 TiO_2-catalyzed reduction of acetophenone derivatives studied by Kohtani et al.[52]

The same group studied also the influence of adsorption of acetophenone derivatives on the kinetics of their photocatalytic reduction at TiO_2 (P25).[53,54] Surface defects (Ti sites) play a crucial role in the reduction process. They act as adsorption sites and electron traps that participate in the reduction of acetophenone derivatives. The reaction rate of this process is strongly affected also by the possibility to form ketal and hemiketal intermediates from some acetophenone derivatives and ethanol.[53]

3.1.3 Reduction of Nitroaromatics

Functionalized aniline derivatives are important substrates for the industry, where they are used for the synthesis of drugs, dyes, herbicides, *etc.*[55] Usually, suitable hydrogen donors, such as hydrazine, borohydride, and other reducing agents, are used for reduction of nitroaromatic compounds. They are, however, not ideal feedstocks for these reactions because their application generates a large amount of chemical wastes. The catalytic reduction of nitro compounds with hydrogen is a much safer alternative; however, it is also an imperfect approach. The catalytic reduction offers a relatively low selectivity, and often other functional groups are also reduced leading to undesirable side-products.[56] Therefore, the emphasis is on processes that offer an increase in reduction selectivity. Photocatalysis may appear to be an interesting approach here. On one hand, the selectivity of photocatalytic reduction can be advantageous, and on the other hand, it appears as a more environmental friendly approach.

Many attempts have been made to reduce nitroaromatic compounds photocatalytically. Most often, titanium dioxide (P25) was used as a photocatalyst. The reduction of nitroaromatics was carried out in the presence of electron donors, usually alcohols, such as methanol, ethanol, and isopropanol,[57,58] or acids (oxalic acid).[59] The reduction process is a multistep reaction involving 6 electrons:

$$TiO_2 + h\nu \rightarrow e^- + h^+ \tag{18}$$

$$CH_3CH_2OH \xrightarrow{2h^+} CH_3CHO + 2H^+ \tag{19}$$

$$Ar - NO_2 \xrightarrow{2e^-,\ 2H^+} Ar - NO + H_2O \tag{20}$$

$$Ar{-}NO \xrightarrow{e^-,\ H^+} [Ar{-}NOH^\bullet] \xrightarrow{e^-,\ H^+} Ar{-}NHOH \tag{21}$$

$$Ar{-}NHOH \xrightarrow{e^-,\ H^+} [Ar{-}NH^\bullet] + H_2O \xrightarrow{e^-,\ H^+} Ar{-}NH2 \tag{22}$$

It was assumed that the reduction of the nitro group may occur not only involving electrons from the conduction band but also by α-hydroxyalkyl radicals generated during the one-electron oxidation of alcoholic donors (see Chapter 2.1.1).[60] Compared with electrons from the conduction bands of TiO_2 and ZnO, α-hydroxyalkyl radicals, such as $^{\bullet}CH_2OH$, appeared to be a stronger reductant. However, most authors agree that the main role of alcohol in this reaction is the efficient consumption of photogenerated holes.[61,62]

Brezova et al.[57] studied the influence of the solvent on the photocatalytic reduction of 4-nitrophenol on TiO_2 (P25). 4-Nitrophenol was successfully reduced to 4-aminophenol in the presence of various alcohols, such as methanol, ethanol, 1-propanol, 2-propanol, 1-butanol, and 2-butanol. The authors suggested that polarity affects the reaction rate, which increased with increasing polarity of the solvent.

An important issue in the reduction of nitro compounds is the selectivity of this reaction. As concluded from Eqs. (21) and (22), during multielectron reduction of nitro compounds, reactive radicals are generated ([Ar−NOH$^{\bullet}$] and [Ar−NH$^{\bullet}$]).[63] These intermediates can dimerize to form azoaromatic and azoxyaromatic compounds. Due to the possibility of a side reaction between these two radical intermediates, their rapid reduction is desired. Nevertheless, the reduction of nitroaromatics proceeds usually with a very high selectivity and efficiency, reaching in some cases more than 99%.[64–68] The high selectivity of photocatalytic reduction of nitroaromatics is maintained also for derivatives possessing other functional groups, which might be reduced with conventional reductants, but are preserved under photocatalytic conditions. Reduction of the nitro group requires a much milder potential compared with many other functional groups, as shown in Fig. 15. Therefore, the use of photocatalysts offering mild reduction properties, such as zinc oxide,[69] titanium dioxide,[62,64,65,70] and cadmium sulphide[66,67,71] to reduce nitroaromatics, does not lead to the reduction of other functional groups. Interestingly, particularly, high selectivity and yields of nitroaromatics reduction were observed for the rutile polymorph of titanium dioxide.[64] This advantage of rutile over anatase very likely results from a lower potential (higher energy) of the conduction band edge of rutile-TiO_2, as evidenced in our previous studies.[72,73] Another, redox active moiety that could potentially affect other functional groups, is photogenerated hydroxyl radicals (see Chapter II.A.1). Nevertheless, alcohols present in the systems act as efficient scavengers of $HO^{\bullet}$ radicals (Table 6).

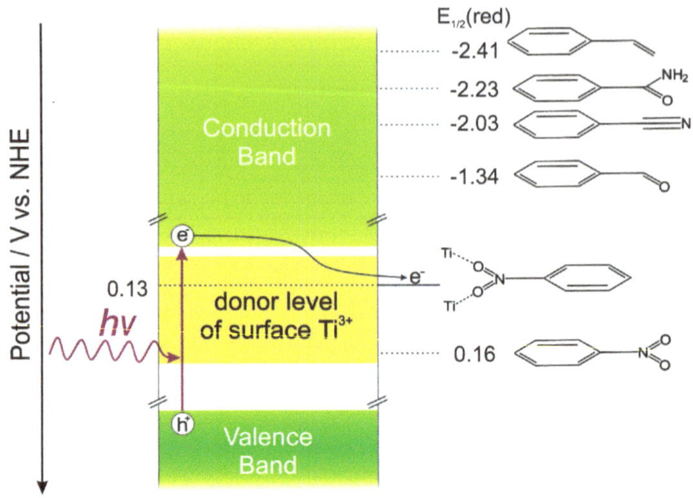

Figure 15 Potentials of TiO_2 (rutile) bandgap edges and half-wave reduction potentials of selected functional groups attached to the benzene ring (according to Ref. 64).

3.1.4 Reduction of Nitroaromatic Compounds to Azoxyaromatics and Azoaromatics

The reduction of nitroaromatics to the corresponding amines is an important industrial process. Nevertheless, the reduction of nitroaromatics to azoaromatic and azoxyaromatic compounds provides important and valuable precursors for the chemical industry. These types of products can be formed as by-products of the N—N coupling of radical intermediates. Although this type of reaction is marginal for the previously mentioned photocatalytic systems, there are also reports on materials that are able to photocatalyze this process with a high selectivity (Fig. 16).

Pal et al. irradiated ($\lambda = 436$ nm) a deaerated isopropanol/water solution containing nitrobenzene and rhodium-loaded silica-coated cadmium sulfide nanoparticles.[75] Azoxybenzene was the main product formed with a relatively high selectivity (68%). Moreover, Naya et al. synthesized azobenzenes from nitrobenzenes at room temperature on visible light irradiation ($\lambda > 430$ nm; TiO_2-supported Au nanoparticles with a bimodal size distribution) with the yield exceeding 95% and the selectivity higher than 99%.[76] It is worth mentioning that other authors using similar photocatalysts, i.e., Au@TiO_2[77] and Ag@TiO_2,[78] observed a high selectivity toward amine formation. A plausible reason of this difference is a different spectrum of the applied light (UV vs. visible light).

Table 6 The Highest Performances of Photocatalytic Reduction of Nitroaromatics (with Various Functional Groups) to the Corresponding Amines

Substrate	Photocatalyst	Product	Product Yield/%
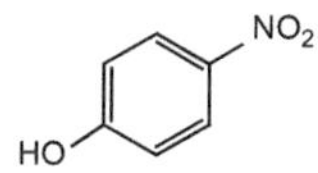	N-doped TiO_2 and KI[65]	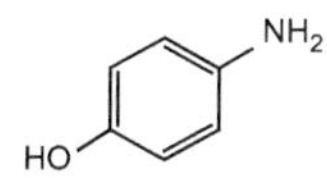	99.37
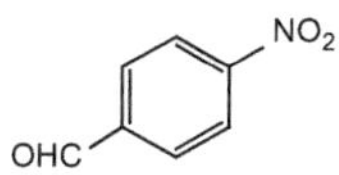	$PbBiO_2Br$[74]	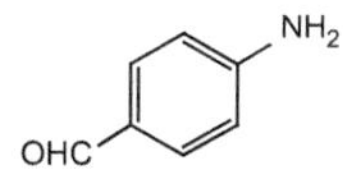	>99
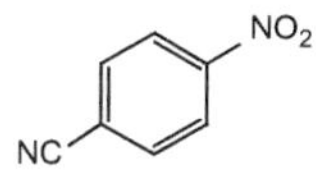	TiO_2 (rutile)[64]	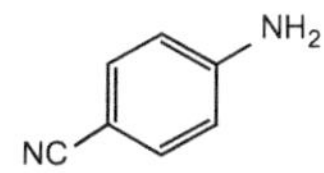	97
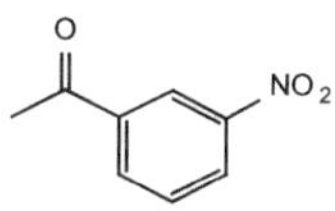	CdS nanostructure[67]	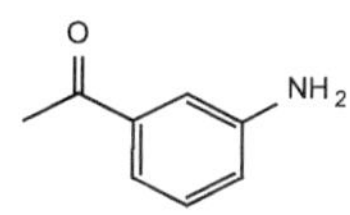	97
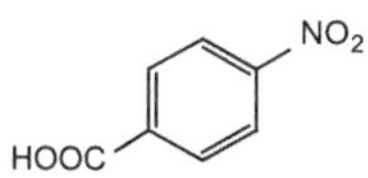	N-doped TiO_2 and KI[64]	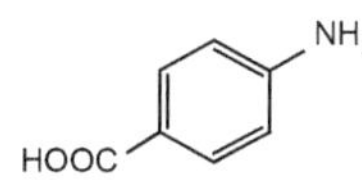	99.39
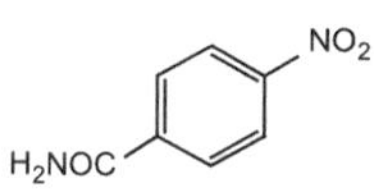	TiO_2 (rutile)[64]	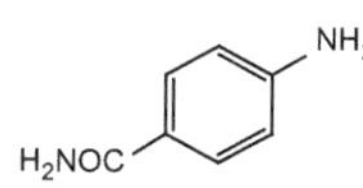	94
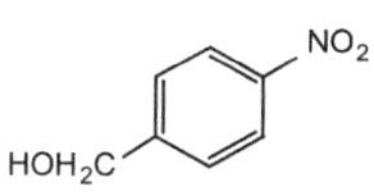	N-doped TiO_2 and KI[65]	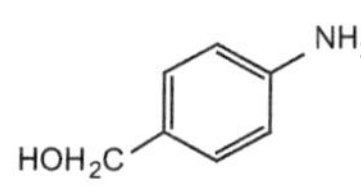	99
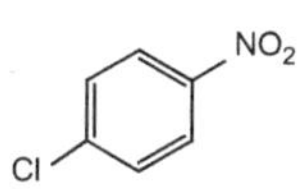	TiO_2 (rutile)[64]	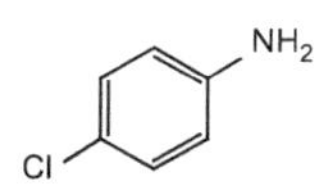	99.47
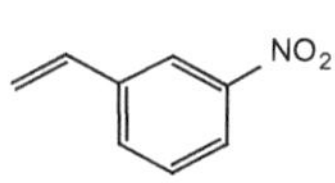	TiO_2 (rutile)[64]	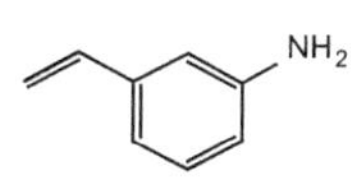	98
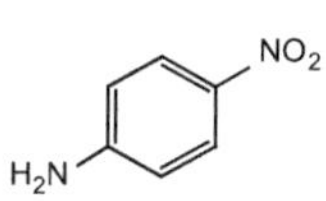	N-doped TiO_2 and KI[65] CdS-TiO_2-NS-60[66]	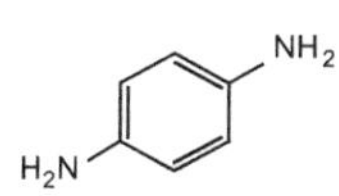	98.15 98

Dai et al. demonstrated that inexpensive graphitic C_3N_4 can act as an efficient photocatalyst for selective synthesis of a series of azoaromatic and azoxyaromatic compounds from their corresponding nitroaromatics.[79] The selectivity of this reaction could be simply controlled by the photon

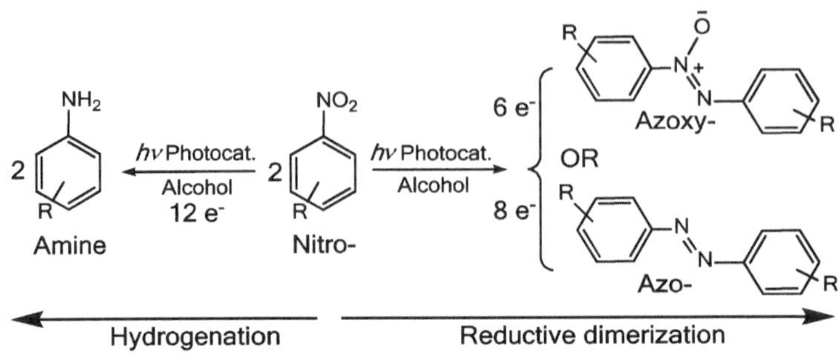

Figure 16 Plausible products formed on photocatalytic reduction of nitroaromatic compounds as the result of hydrogenation and reductive dimerization.

energy. Under purple irradiation (410 nm), azobenzene was gradually formed in a two-step reduction reaction; in the first step, nitrobenzene was converted to azoxybenzene, which in turn was subsequently reduced to azobenzene. On the other hand, under blue light (450 nm) excitation, azoxybenzene was found to be the main product. Regardless of the color of light, the nitrobenzene reduction reached a high conversion (>95%) and a high selectivity (>94%) with a negligible formation of aniline. The authors attributed such high selectivity of the formation of azoaromatic and azoxyaromatic compounds, to a weak binding of photogenerated hydrogen atoms, consecutive H_2 evolution, and in consequence, coupling of radical intermediates with the formation of N—N bond (compare Fig. 16, hydrogenation vs. reductive dimerization).

3.2 Oxidation

3.2.1 Dehydrodimerization

In the systems discussed so far, condensation of the generated radicals was rather a side, unfavorable reaction. However, these types of reactions can be performed intentionally. Kisch et al. investigated an anaerobic photocatalytic oxidation of cyclic unsaturated ethers at sulfides (ZnS or platinized CdS). In the case of 2,5-dihydrofuran (2,5-DHF), apart from dihydrogen, dehydrodimers were generated.[80–83] The mechanism encompasses oxidation of 2,5-DHF with photogenerated holes to the dihydrofuryl radical. The radical products in the next step undergo dimerization. In the reductive part of the reaction, hydrogen is photogenerated from the reduction of water by electrons from the conduction band (Fig. 17). The analogous dehydrodimers were also observed when 3,4-dihydropyran, 3-methyl-2,3-dihydropyran, and cyclohexene were used as substrates.

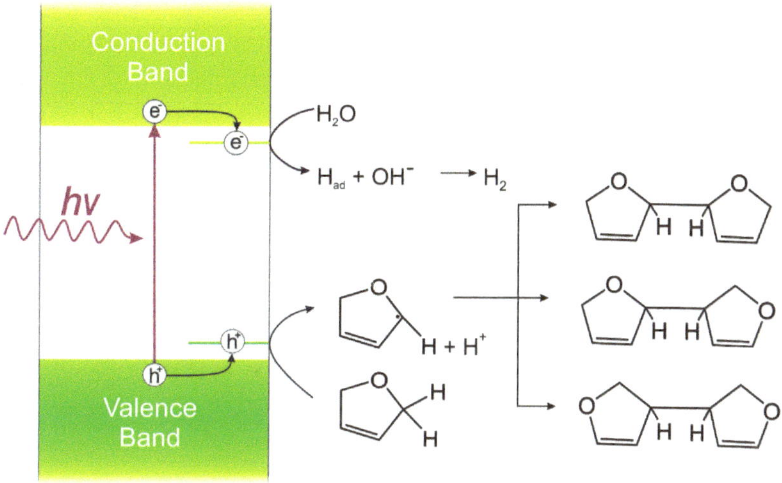

Figure 17 Anaerobic dehydrodimerization of 2,5-dihydrofuran in water (according to Ref. 84).

Oxidation of BzA derivatives (3; Fig. 18) is another example of dehydrodimerization. One-electron oxidation of alcohols, as discussed previously, leads to the formation of the α-hydroxyalkyl radical, which can be oxidized further to aldehyde. Mitkina et al. demonstrated that irradiation of a BzA in acetonitrile in an oxygen-free atmosphere in the presence of cadmium sulfide leads to the formation of three main products: BzAD (4), hydrobenzoin (5; as a mixture of diastereoisomers), and benzoin (6).[85]

The authors proposed a mechanism in which holes from the valence band oxidize BzA to benzyl radicals and protons (Eq. 23). Highly reactive

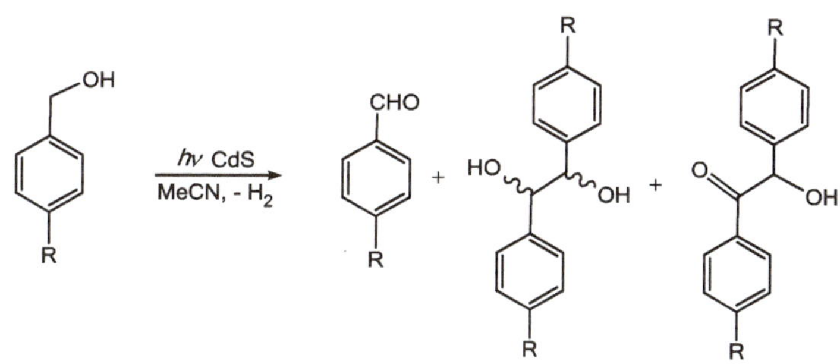

3 4 5 6

Figure 18 Photocatalytic oxidation of benzyl alcohols in MeCN in the presence of cadmium sulfide. R = H, OMe, Cl, Br, I, COOMe (according to Ref. 85).

benzyl radicals may undergo a homocoupling reaction to give hydrobenzoin (5; Eq. 24) or BzAD (4) as the result of the second oxidation step that involves oxidation with a hole (Eq. 25) or injection of an electron into the conduction band (Eq. 26). The authors suggest that benzoin is most likely a product of oxidation of BzAD and may be further oxidized to benzil.

$$ArCH_2 - OH + h^+ \rightarrow Ar\dot{C}H - OH + H^+ \quad (23)$$

$$2Ar\dot{C}H - OH \rightarrow Ar(OH)CH - CH(OH)Ar \quad (24)$$

$$2Ar\dot{C}H - OH \rightarrow ArCHO + ArCH_2OH \quad (25)$$

$$Ar\dot{C}H - OH \rightarrow ArCHO + e^- + H^+ \quad (26)$$

$$Ar\dot{C}H - OH + h^+ \rightarrow ArCHO + H^+ \quad (27)$$

Because benzoin as an important feedstock extensively used in chemical industry, new photocatalysts are required to provide a higher selectivity. Sun et al. used an alkaline modified g-C_3N_4 photocatalyst for oxidative coupling of BzA to yield benzoin.[86] They achieved a high selectivity ($\sim$97%) and conversion ($\sim$90.2%) when K^+-modified g-C_3N_4 was applied. A different mechanism of product formation operates here. In the first step, BzA is oxidized to BzAD. The second step encompasses oxidative C—C coupling of two BzAD molecules to yield benzoin.

Mitkina et al. noticed that substrates with an amino group are converted to imines (8) as the main products, and not to benzoin.[85] Su et al. reported the aerobic oxidation of amines into imines with high yields using mpg-C_3N_4 as the photocatalyst (Fig. 19).[87] It is noteworthy that also a high selectivity and conversion efficiency (both over 99%) were achieved for heterocyclic amines containing nitrogen and sulfur atoms, which usually poison most metal catalysts.

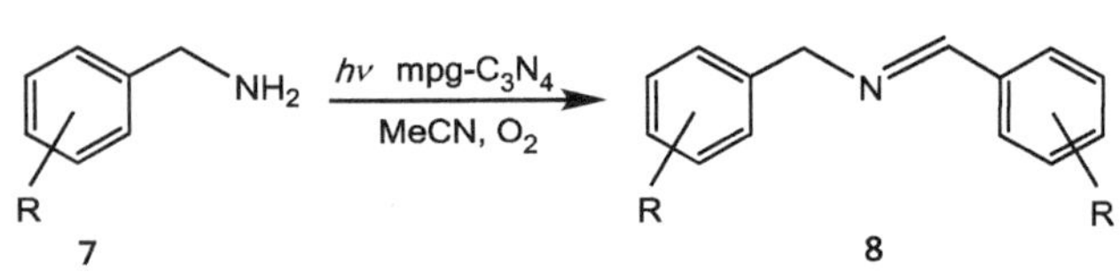

Figure 19 Oxidation of various amines using mpg-C_3N_4. R = H, *p*-Me, *p*-OMe, *p*-Cl, *p*-CF_3, *o*-OMe, *m*-OMe (according to Ref. 87).

4. REACTIONS INVOLVING ACTIVATION OF TWO ORGANIC SUBSTRATES

So far, systems in which the main reactant was reduced or oxidized were discussed, while the counter redox processes were of minor importance for the produced chemicals. In this section we will focus on products formed as the result of direct photocatalytic oxidation and reduction of two organic substrates or one organic substrate and CO_2 as the second reactant. These processes lead to C—C or C—N bond formation and offer a possibility to synthesize more complex compounds. In some cases the synthesis of various stereoisomers of the product can be obtained. The reactions described here can be considered as a *photocatalysis type B*, according to the classification proposed by Kisch (compare *Introduction* and Fig. 1E).

4.1 Photocatalytic Carboxylation of Organic Compounds

Increasing attention focused on the photocatalytic reduction of carbon dioxide as reflected in numerous studies on CO_2 to fuel conversion (see Section 3). The main products of this reaction are simple and cheap C_1 and C_2 molecules (e.g., CO, formic acid, formaldehyde, methanol, methane, oxalate, acetic acid). The cost-effectiveness of such processes is questionable due to low quantum yields, unsatisfactory photostability of the systems, several possible side reactions (e.g., reoxidation of the products), difficulties with the counterpart oxidation reaction (e.g., water oxidation), and relatively low commercial value of the formed products.

Recently, we have proposed another approach to photocatalytic carbon dioxide fixation (Fig. 20).[88] Electrons from the conduction band of a photocatalyst should enable one-electron reduction of CO_2 to form $CO_2^{\bullet-}$ radicals. This is a crucial step of the process because the reduction takes place at a highly negative potential of −1.9 V versus NHE.[35,36] Therefore, the photocatalyst should be carefully selected from those offering such low redox potential of the conduction band bottom. Using EPR spectroscopy and spin traps we have demonstrated that zinc sulfide decorated with ruthenium nanoparticles enables $CO_2^{\bullet-}$ generation.[88] On the other hand, photogenerated holes oxidize organic substrates (DH) to organic radicals (Fig. 20). Formed $D^{\bullet}$ and $CO_2^{\bullet-}$ radicals couple together to yield the corresponding carboxylic acid (DCOOH) identified by ^{13}C-NMR and IR techniques. To increase the efficiency of product formation, the DH substrate should facilitate formation of possibly stable radicals. Therefore, carboxylation of

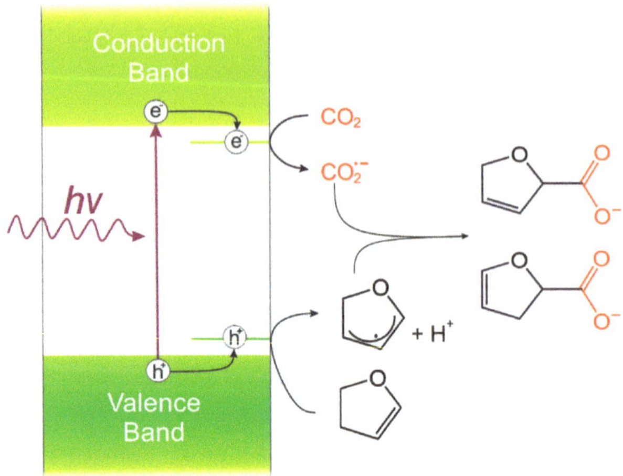

Figure 20 Photocatalytic carboxylation with carbon dioxide.[88] 2,3-DHF represents an example of an organic substrate yielding allyl radicals.[89]

acetylacetone[88] and 2,3-DHF (2,3-dihydrofuran)[89] was tested. The reported yields for photocatalytic carboxylation reactions are small, up to a few percent of the substrate conversion, but further work on stabilization of the formed radicals (e.g., spin-trapping approach) may result in a significant enhancement of the overall reaction efficiency.

Zinc sulfide offers highly reductive electrons able to reduce CO_2, but at the same time, holes can be considered as relatively mild oxidants. They can oxidize some adsorbed organic compounds, but they are not able to oxidize water to hydroxyl radicals. Therefore, an uncontrolled oxidation of reactants is limited. Recently, we have proposed to consider mainly *p*-type semiconductors as prospective photocatalysts for CO_2 reduction.[38] In general, such materials offer electrons and holes characterized by energies more suitable for carbon dioxide reduction and less prone to reoxidation of the formed products.[90] The main disadvantage of such materials is their limited photostability compared to typical oxides.

4.2 Photocatalytic C—C Coupling

Coupling of two photogenerated radicals was studied and described by Kisch et al.[91–93] In this reaction, zinc sulfide or cadmium sulfide photocatalysts oxidize an organic substrate, which can form a stabilized radical (e.g., allyl radicals; Fig. 21). Released protons can be involved in the reduction process in which imine is reduced to the corresponding

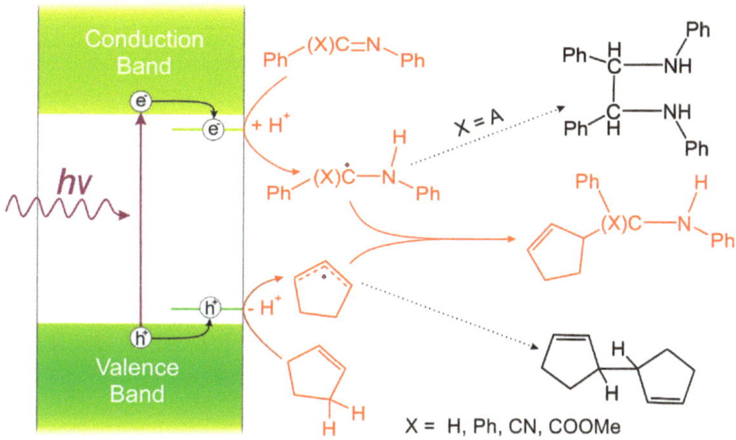

Figure 21 Photocatalytic addition of imines to cyclopentene (according to Refs. 91–93).

Ph–(X)C•–N(H)–Ph radical. Generated radicals can couple with the formation of a new C–C bond. Depending on the applied substrates (derivatives of *N*-benzylideneaniline and alkenes), the isolated yields of the main heterocoupling products reached even 80%. Other homocoupling products are also presented in Fig. 21.

The advantage of CdS application is clear taking into account the spectral range in which this material absorbs light (the bandgap energy of CdS is close to 2.4 eV but may vary depending on the material morphology). Unfortunately, a serious drawback of sulfides is their limited photostability.[94] This issue, however, can be diminished by a proper design of the photocatalyst and the reaction medium.[95]

4.3 Photocatalytic C–N Coupling

The C–N coupling reactions may proceed in a similar way. While the oxidative part of the process, taking place again at the surface of irradiated ZnS or CdS, consists of the allyl radical generation, the reductive part leads to the formation of *N*-centered radicals from 1,2-diazenes (Fig. 22).[96–98] The main C–N heterocoupling product is a derivative of allylhydrazine, produced with a yield up to 40% (calculated for the isolated products).

Formation of C–N bonds was achieved also in the synthesis of benzimidazoles from *o*-nitroaniline derivatives[99] or 2-nitrophenyl azides[100] and alcohols. Also, in these reactions, two organic substrates undergo

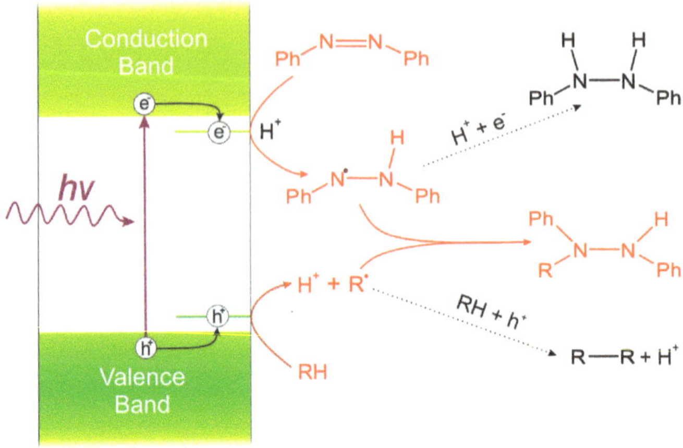

Figure 22 Photocatalytic addition of 1,2-diazenes to unsaturated ethers (e.g., 2,5-DHF) or olefins (e.g., cyclohexene; according to Ref. 84).

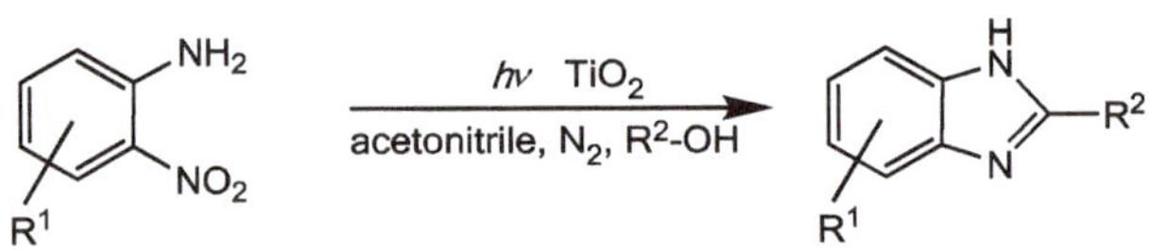

Figure 23 Photocatalytic production of benzimidazoles from *o*-nitroaniline derivatives.[99]

oxidation and reduction. The oxidative part involves $2h^+$ oxidation of alcohols to aldehydes, which react with *o*-nitroaniline to form the corresponding imine. This product undergoes a $4e^-$ reduction yielding benzimidazole derivatives (Fig. 23).

Alternatively to *o*-nitroaniline derivatives, 2-nitrophenyl azides can be used as substrates.[100] In this case, 2-nitrophenyl azide is first reduced to *o*-nitroaniline, and then the reaction proceeds according to the mechanism described previously. Interestingly, synthesis of benzimidazoles involves coupling of photocatalytically generated stable products of multielectron transfer reactions, and not photogenerated unstable radicals, as in the cases presented in Figs. 20–22. Because the synthesis of stable intermediates may be separated in space and time, the photocatalytic production of benzimidazoles does not fit perfectly into the typical scheme of photocatalysis type B, although it fits to the scheme presented in Fig. 1E.

5. IMPROVING THE EFFICIENCY OF PHOTOSYNTHETIC SYSTEMS

Photocatalytic selectivity and reaction efficiency depend on various factors, including structure, morphology, redox properties, surface chemistry, light absorption properties, lifetimes of photogenerated charges, and many others. Various modifications of photocatalysts, such as doping, deposition of metals or oxides, constructing semiconductor composites, or exposing selected crystal facets, may have a crucial influence on the performance of photocatalysts. In this section the influence of such modifications will be briefly reviewed.

5.1 Tailored Crystal Facets

DFT calculations of anatase {001} and {101} surfaces showed that the VB of {001} TiO_2 extends to higher energies compared to the valence band characteristic for the {101} surface. On the other hand, the bottom of the CB is situated at nearly the same energy for both {001} and {101} facets.[101] This difference in the energy levels leads to separation of charge carriers between various facets within the semiconductor crystal. Photogenerated electrons move selectively to the {101} facets, whereas photogenerated holes migrate to the {001} facets, at which reduction and oxidation occur, respectively.[102] In the case of rutile TiO_2, the {111} or {101} facets are preferred oxidation sites, whereas reduction prevails at the {110} facets.[103]

Chong et al.[104] studied glycerol conversion to hydroxyacetaldehyde (HAA) in aqueous solution on tailored TiO_2 crystals. The selectivity appeared to be facet dependent. At rutile with a high exposition of {110} facets, the selectivity of HAA production exceeded 90%. In the case of anatase with dominant {001} and {101} facets, the selectivity of HAA synthesis reached only 16% and 49%, respectively.

Cao et al.[105] studied oxidation of toluene to BzAD at anatase TiO_2 hollow spheres with exposed {001} facets. Anatase TiO_2 hollow spheres showed selectivity higher than 90% with respect to the photocatalytic oxidation of toluene to BzAD, which was a better result than that found for the commercial P25 photocatalyst. Interestingly, the formed BzA as an intermediate adsorbs at the surface of TiO_2 and forms a charge-transfer complex that absorbs visible light.[106] Its excitation leads to electron transfer to the conduction band and oxidation of the adsorbate to BzAD.

Yu et al.[107] studied reduction of CO_2 to CH_4 on anatase TiO_2 with different exposition of {001} and {101} facets. The ratio between the exposed {101} and {001} facets significantly influences the reduction of CO_2. Yu proposed that the optimal ratio between the exposed {101} and {001} facets was 45:55. In such material, electrons and holes were localized at {101} and {001} facets, respectively, which led to a hindered recombination of charge carriers and therefore the enhancement of redox reactions.

Oxidation of various alcohols on ZnO nanorods and commercially available ZnO photocatalysts was studied by Tang et al.[108] A good photoactivity and excellent selectivity toward the selective aerobic oxidation of alcohols to aldehydes under UV light irradiation was achieved for ZnO nanorods compared to commercial ZnO material. A comparison of conversion and selectivity observed for ZnO nanorods and commercial, not tailored, ZnO photocatalysts is shown in Table 7.

Yuan et al.[109] revealed that {040} and {110} facets of $BiVO_4$ play an important role in photocatalytic oxidation reactions. The researchers studied visible light–induced oxidation of *N*-*t*-butylbenzylamine on $BiVO_4$ crystals with different exposition of the {040} and {110} facets. The best conversion and selectivity were achieved for a {110}/{040} ratio of 1.17.

The presented examples demonstrate the importance of an efficient electron and hole separation. This can be achieved by application of tailored photocatalysts with exposed facets facilitating separation of reduction and oxidation reactions.

5.2 Doped Photocatalysts

Incorporation of small amounts of alien atoms into semiconductor crystal lattice leads to significant changes in the electronic properties of the photocatalyst. Dopants may affect the energy of conduction and valence band edges. They can create new electronic states within the bandgap of the semiconductor. Doping can also decrease the bandgap energy and extend the spectral range of photocatalytic activity, e.g., from UV to visible light (bathochromic shift).

Xiao et al.[110] studied the degradation of EDGs or EWGs in *p*-substituted phenols using Ag-doped TiO_2 as a photocatalyst. Ag^+ ions can capture efficiently the photogenerated electrons leading to an enhanced charge separation and plausibly improved superoxide generation.[111] The authors concluded that $O_2^{\bullet-}$ acts as a dominant oxidant responsible for decomposition of phenolic compounds, rather than $HO^{\bullet}$, $^1O_2$ or h^+. Phenols substituted with EDGs underwent more likely a nucleophilic attack by $O_2^{\bullet-}$, preferably at the *meta*-position (in terms of the carbon atom bound to the −OH group). EWGs substituted phenols underwent rather an

Table 7 Selective Oxidation of Alcohols in the Presence of ZnO Nanorods and Commercial ZnO Photocatalysts under UV Light Illumination ($\lambda = 365 \pm 15$ nm) for 8 h[108]

		ZnO Nanorods		**Commercial ZnO**	
Substrate	**Product**	**Conv. (%)**	**Sel. (%)**	**Conv. (%)**	**Sel. (%)**
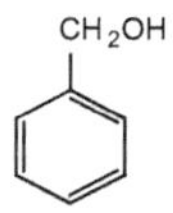	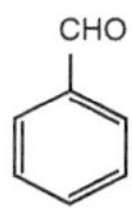	31	97	16	43
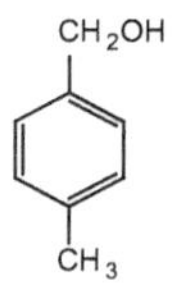	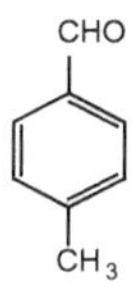	32	96	22	50
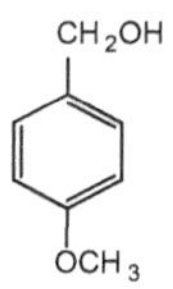	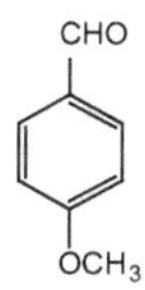	30	96	21	48
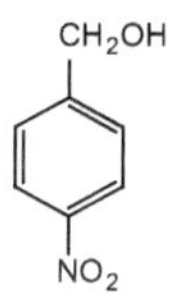	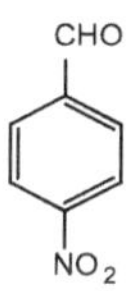	28	93	17	46
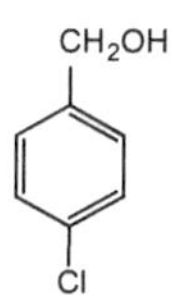	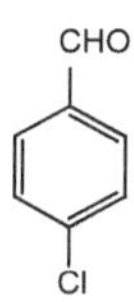	25	95	15	51
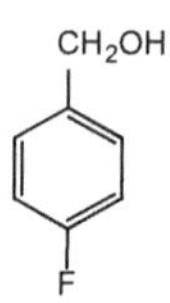	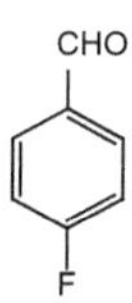	22	94	11	63
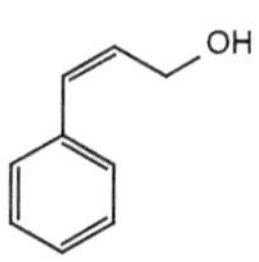	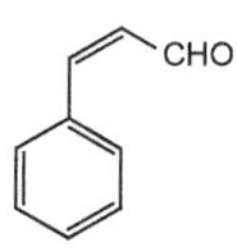	28	71	18	53
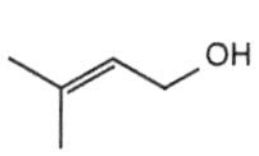	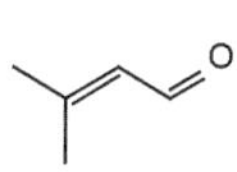	24	73	20	41

electrophilic attack by $O_2^{\bullet-}$. Moreover, the electrophilic and nucleophilic attacks by $O_2^{\bullet-}$ could simultaneously occur during degradation of *p*-chlorophenol.

Zhang et al.[112] studied photocatalytic oxidation of alcohols to aldehydes at N-doped TiO_2 under visible light illumination. A significantly enhanced conversion and selectivity of alcohol oxidation were observed on the N−TiO_2 photocatalyst compared to prepared or commercial nondoped TiO_2. For the $TiO_{1.966}N_{0.034}$ photocatalyst the best conversion of BzA to BzAD was achieved (19.6%; 2.5 times higher than for P25 TiO_2). The authors suggested that an increasing concentration of the N-dopant caused a decrease of the oxidation abilities of photogenerated holes. At the same time doping has a slight impact on the TiO_2 reducing properties. The CB potential of $TiO_{2-x}N_x$ was negative enough to enable reduction of oxygen (the conclusion was supported by EPR measurements), but the potential of N-states was too low to facilitate oxidation of H_2O to $HO^\bullet$ by photogenerated holes, Fig. 24. Under visible light irradiation, photogenerated $HOO^\bullet$ radicals react with oxidized alcohols to yield the corresponding aldehyde and hydrogen peroxide.

Tripathy et al.[113] studied photocatalytic oxidation of toluene on TiO_2 nanotubes and the influence of electronic properties of TiO_2 nanotubes (anatase, rutile, Ru-doped TiO_2) on the selectivity of various products (benzoic acid, BzAD, and BzA) formation. Benzoic acid was the main product formed on the anatase-based nanotubes (selectivity of *c.* 71%), while the selectivity of BzAD and BzA production was significantly lower (26% and 3%,

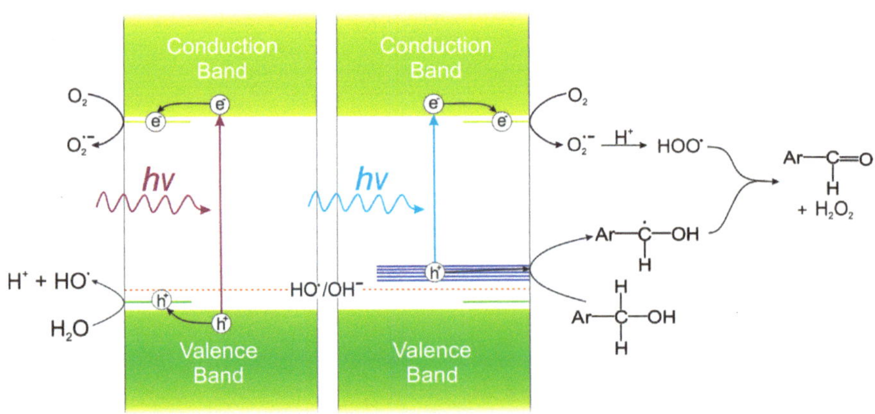

Figure 24 A plausible mechanism for photocatalytic oxidation of alcohols on visible light irradiation in the presence of the $TiO_{2-x}N_x$ photocatalyst.[112]

respectively). In the case of rutile-based nanotubes, a significant change in the product distribution was observed. BzAD was the main oxidation product (selectivity of 76%), while benzoic acid and BzA were formed with a selectivity of 14% and 10%, respectively. Even more pronounced changes were observed for Ru-modified rutile nanotubes (1 at%), which facilitated formation of BzAD (*c.* 89%) with a completely suppressed formation of benzoic acid. The authors postulated that the change in selectivity resulted from different abilities of anatase, rutile, and Ru-modified nanotubes to generate superoxide. Inhibition of $O_2^{\bullet-}$ generation prevents BzAD from further oxidation to benzoic acid. The doping with ruthenium results in the formation of $Ru^{3+/4+}$ electronic states, localized *c.* 0.4 eV below the conduction band of TiO_2. Electrons trapped there cannot reduce adsorbed O_2, and as a consequence, BzAD oxidation is inhibited.[114]

In the presented examples, doping improved charge separation, enhanced the overall reaction yield, tuned the absorption properties (bandgap engineering), and influenced the redox abilities of photogenerated holes ($N-TiO_2$) or electrons ($Ru-TiO_2$).

5.3 Cocatalysts

The use of cocatalysts may also increase the photocatalytic efficiency and selectivity of the photocatalytic synthesis. Cocatalysts can act in different ways: they can bind specifically the reactants to influence their adsorption at the surface, they can lower the activation energy of the reaction,[32] or they can act as electron/hole sinks improving the efficiency of multihole/electron processes.[115,116]

Metals, in particular noble metals, are often used as cocatalysts. Deposition of metals at a semiconductor surface results in the formation of a Schottky junction. As a consequence, an improved charge separation takes place. Moreover, deposition of metals on the semiconductor surface may lead to a localized surface plasmon resonance (LSPR) generation. This, in turn, can shift the spectral response of the photocatalyst to visible or even infrared light.

Ide et al.[117] investigated the selectivity of phenol synthesis on photocatalytic benzene oxidation on TiO_2 with deposited gold nanoparticles under air or CO_2 atmosphere. Gold nanoparticles show the LSPR absorption responsible for visible light–induced photocatalytic activity. Photocatalytic oxidation on 3%Au@P25 in the air led to phenol and many other products generation, including catechol, hydroxyquinone, and 1,2,3- and 1,2,4-trihydroxybenzene. However, when a CO_2 atmosphere was applied instead

Table 8 Photocatalytic Benzene to Phenol Oxidation in the Presence of Various Photocatalysts under Air or CO_2 Atmosphere. Irradiation with a Solar Simulator[117]

Catalyst	CO_2 Content/kPa (mg)	Conv. (%)	Yield (%)	Sel. (%)
P25	0[a]	34	7	21
	230 (120)	31	7	22
3%Au@P25	0[a]	13	8	62
	230 (120)	14	13	89
4%Au@A-6	0[a]	14	13	72
	230 (120)	13	12	95

[a]Under air at 100 kPa.

of air, the yield of phenol generation was significantly increased, while formation of other by-products was efficiently suppressed. A similar effect of CO_2 was observed when 4%Au@A-6 (A-6 - TiO_2 from Tynoc company) was studied. Yields and selectivity of benzene to phenol conversion are summarized in Table 8.

Shiraishi et al.[118] investigated the synthesis of benzimidazoles from *ortho*-arylenediamines and alcohols on TiO_2 decorated with platinum (Pt@TiO_2). Pt@TiO_2 enables an efficient benzimidazole production as a result of a one-pot photocatalytic/catalytic transformation on Pt@TiO_2, which involves a platinum-assisted photocatalytic oxidation on TiO_2 and a catalytic dehydrogenation on platinum particles. The authors postulated that the number and size of the platinum particles were crucial for a rapid and selective benzimidazole synthesis. The best conversion yield was observed for Pt(0.2)@TiO_2 [Pt(x)@TiO_2; x (wt%) = Pt/(TiO_2 + Pt)]. A lower selectivity was observed for photocatalysts with a higher platinum content. The diamine conversion and product yields on Pt(0.2)@TiO_2 are summarized in Table 9.

A photocatalytic reduction of styrene oxide to styrene on CeO_2, TiO_2, ZrO_2, Al_2O_3, and zeolite Y with deposited gold nanoparticles (Au-NPs) was reported by Ke et al.[119] Gold nanoparticles induced a visible light activity according to the LSPR effect. A comparison of performance of various photocatalysts in styrene oxide to styrene transformation is shown in Table 10. The highest conversion was observed for Au/Al_2O_3, but the highest selectivity was achieved for Au/CeO_2.

Tanaka et al.[120] studied oxidation of alcohols to aldehydes or ketones on Au@CeO_2 under irradiation with green light. An enhanced LSPR absorption was observed around 550 nm. The photocatalytic activity depended on the method of the cocatalyst deposition. Au/CeO_2 samples were prepared

Table 9 Synthesis of Benzimidazoles from *o*-Arylenediamines and Alcohols, According to Shiraishi et al.[118]

Irradiation Time (h)	Diamine	Alcohol	Catalyst	Diamine Conv. (%)	Product	Yield (%)
4	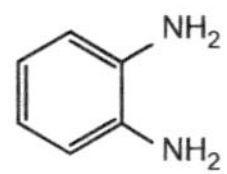	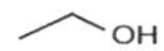	TiO_2	27	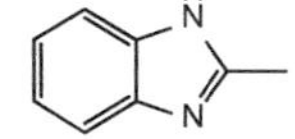	11
4			Pt/TiO_2	>99		93
4		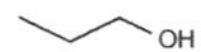	TiO_2	56	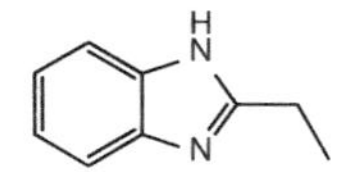	34
4			Pt/TiO_2	>99		>99
4			TiO_2	54	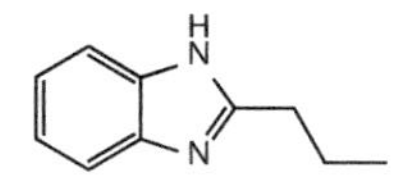	37
4			Pt/TiO_2	>99		>99
4		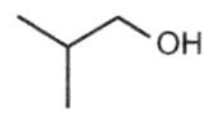	TiO_2	62	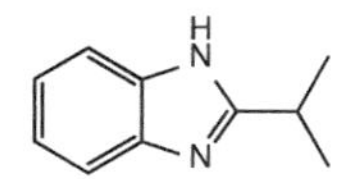	43
4			Pt/TiO_2	>99		95
12[a]		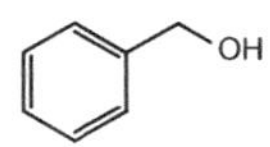	TiO_2	93	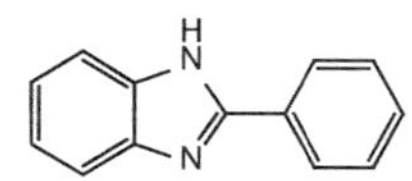	1.8
12[a]			Pt/TiO_2	>99		89
24[b]		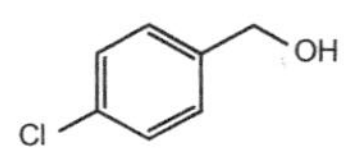	TiO_2	>99	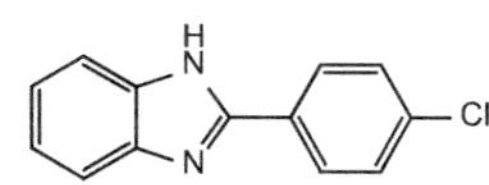	57
24[b]			Pt/TiO_2	>99		82
	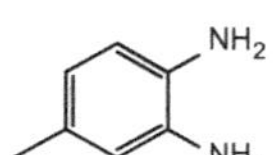		TiO_2	43	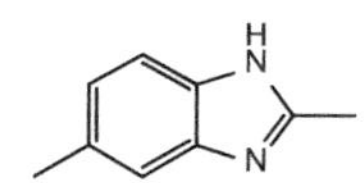	23
			Pt/TiO_2	>99		94

(Continued)

Table 9 Synthesis of Benzimidazoles from *o*-Arylenediamines and Alcohols, According to Shiraishi et al.[118]—cont'd

Irradiation Time (h)	Diamine	Alcohol	Catalyst	Diamine Conv. (%)	Product	Yield (%)
	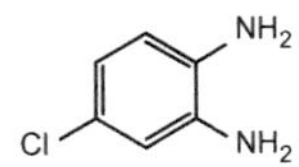		TiO_2	63	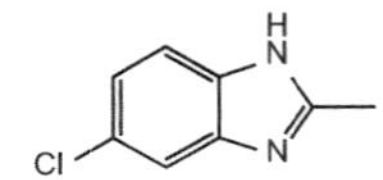	11
			Pt/TiO_2	>99		>99
	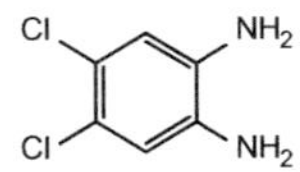		TiO_2	64	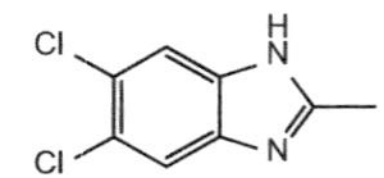	20
			Pt/TiO_2	>99		83

Reaction conditions: photocatalyst (10 mg), alcohol (10 mL), diamine (20 μmol), N_2 (1 atm), and $\lambda > 300$ nm.
[a]Photocatalyst (80 mg), diamine (10 μmol), alcohol (5 mL).
[b]Catalyst (5 mg), diamine (10 μmol), acetonitrile (5 mL).

Table 10 Reduction of Styrene Oxide to Styrene on Various Oxides with Deposited Au Nanoparticles (Reaction Time 16 h, 3 wt% Au-NPs)[119]

	Visible Light		No Light	
Catalyst	**Conv. (%)**	**Sel. (%)**	**Conv. (%)**	**Sel. (%)**
Au/CeO_2	19.7	88	4.4	72
Au/ZrO_2	17.4	72	7.8	75
Au/TiO_2	22.5	73	7.8	75
Au/Al_2O_3	23.9	60	5.8	51
Au/Y	4.5	62	1.6	55

applying a multistep (MS) or a single-step (SS) photodeposition procedure. MS-Au@CeO_2 samples exhibited higher activities than SS-Au@CeO_2 samples.

5.4 Composite Materials

Coupling of two semiconductors with different VB and CB energy levels, or a semiconductor with a nonsemiconducting support (e.g., oxide), may lead to an efficient charge separation, increased lifetimes of photogenerated charges, and an enhanced interfacial charge transfer between the photocatalyst and adsorbed reagents. These, in turn, improve the yields of photocatalytic reactions.

Liu et al.[121] studied oxidation of alcohols to aldehydes using 1D CdS/TiO_2 core-shell nanocomposites (CSNs) and CdS nanowires under visible light irradiation. In all tests, a higher conversion and yield were reached on 1D CdS/TiO_2 CSNs compared to CdS nanowires (Table 11). Authors concluded that photogenerated holes at the CdS core are blocked by TiO_2 shell, whereas electrons can be transferred to the TiO_2 shell. This enhances the lifetime of photogenerated charges. Molecular oxygen is reduced by electrons to superoxide radicals, taking part in alcohol oxidation. The role of holes in this system remains unclear since an efficient photocatalytic reaction requires efficient reduction and oxidation half reactions.

Epoxidation of olefins was studied on irradiated titanium(IV) oxide on a silica matrix.[122] Acetonitrile used as solvent, suppressed unwanted side reactions and resulted in the formation of alcohols or ketones. Reactions of several cyclic and linear olefins with O_2 in the presence of TiO_2 (anatase) or TiO_2-SiO_2 photocatalysts yielded the corresponding epoxides, as shown in Table 12. In general, TiO_2-SiO_2 offered a lower conversion yield, but higher selectivity of the epoxide formed.

Table 11 Photocatalytic Oxidation of Alcohols Into Corresponding Aldehydes Over CdS Nanowires and 1D CdS@TiO_2 Under Visible Light ($\lambda = 520 \pm 15$ nm; Irradiation for 8 h)[121]

		Conversion (%)		Yield (%)		Selectivity (%)	
Substrate	**Product**	**CdS**	**CdS@TiO_2**	**CdS**	**CdS@TiO_2**	**CdS**	**CdS@TiO_2**
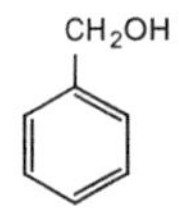	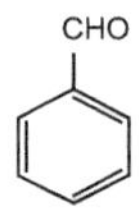	13	34	12	33	97	97
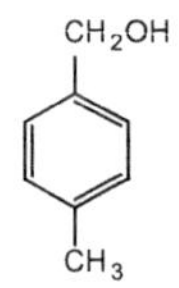	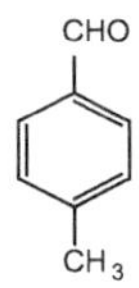	7	41	4	30	60	74
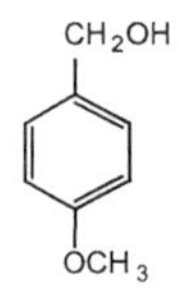	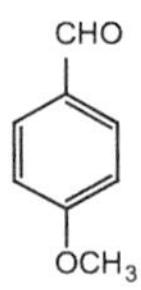	21	39	8	25	36	65
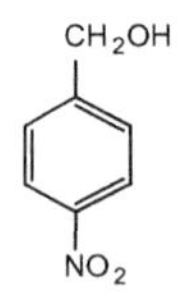	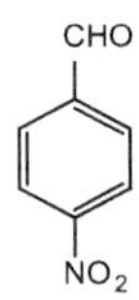	7	30	5	21	73	70
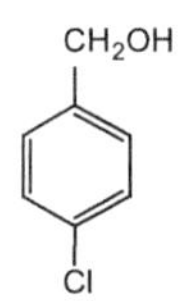	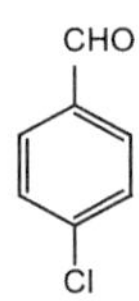	4	38	3	33	88	87
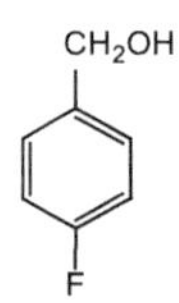	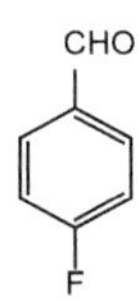	6	37	5	36	97	96

The mechanism of photocatalytic epoxidation of olefins on the TiO_2–SiO_2 catalyst is summarized in Fig. 25. Photoexcitation of the Ti oxide species (**I**) leads to an LMCT (ligand-to-metal charge transfer) excitation and electron transfer between the lattice oxygen ($O_L{}^{2-}$) and Ti^{4+} center, forming the excited $[Ti^{3+} - O_L{}^{-}]^*$ state (**II**). The reaction of **II** with O_2 produces two types of oxygen radicals, $O_3^{\bullet -}$ and $O_2^{\bullet -}$ (**III** and **IV**). The

Table 12 Results of Photocatalytic Epoxidation of Olefins at TiO_2-SiO_2 and TiO_2 (Anatase)[8]

Catalyst	System[a]	Substrate	Conv. (%)	Product	Sel. (%)
TiO_2-SiO_2	G/S	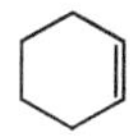	9	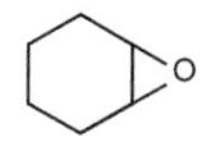	11
TiO_2-SiO_2	L/S		10		71
TiO_2	G/S		10		21
TiO_2	L/S		27		21
TiO_2-SiO_2	L/S	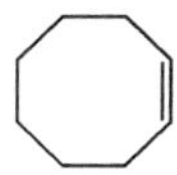	8	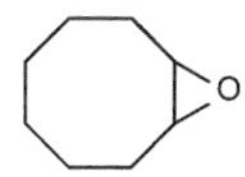	99
TiO_2	L/S		61		47
TiO_2-SiO_2	L/S	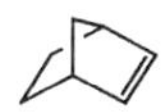	9	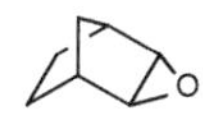	98
TiO_2	L/S		72		76
TiO_2-SiO_2	L/S	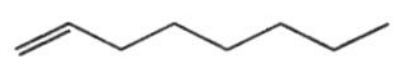	11	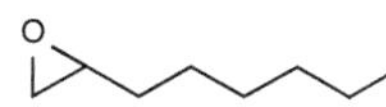	>99
TiO_2	L/S		26		25
TiO_2-SiO_2	L/S	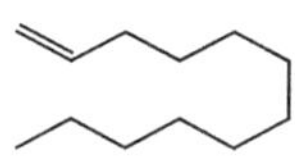	7	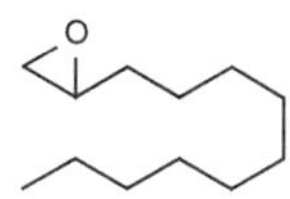	98
TiO_2	L/S		20		26
TiO_2-SiO_2	L/S	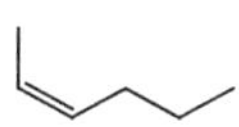	8	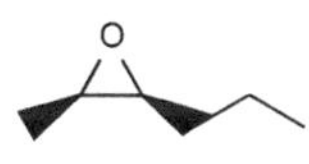	>99
TiO_2	L/S		32		33(9[b])
TiO_2-SiO_2	L/S	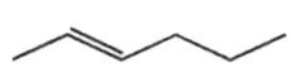	12	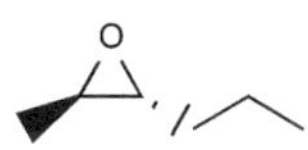	>99
TiO_2	L/S		73		12(36[c])

[a]L/S system: catalyst (10 mg), acetonitrile (10 mL), olefin (0.2 mmol), O_2 (1 atm), irradiation time (12 h). G/S system: catalyst (10 mg), olefin (0.2 mmol), O_2 (1 atm), irradiation time (12 h).
[b]Selectivity of trans epoxide.
[c]Selectivity of cis epoxide.

electrophilic $O_3^{\bullet-}$ formed at O_L^- (**III**) reacts directly with an olefin to produce the epoxide (route A). The vacant O_L^- sites on **II** and **IV** have also an electrophilic character, and therefore they behave as holes. In the absence of acetonitrile, a proton transfer from olefin to O_L^- takes place to produce an olefin radical (routes B, C), which reacts with $O_2^{\bullet-}$ to form side-products, such as cyclohexenol. In the presence of acetonitrile, proton transfer is suppressed because as a weak base, acetonitrile solvates the olefin and suppresses formation of the olefin radical. Therefore, route A prevails over routes B and C.

Dai et al.[123] studied the $CdS/g-C_3N_4$ composites for oxidation of alcohols and reduction of nitrobenzene to aniline under visible light irradiation. The $CdS/g-C_3N_4$ photocatalyst showed an enhanced

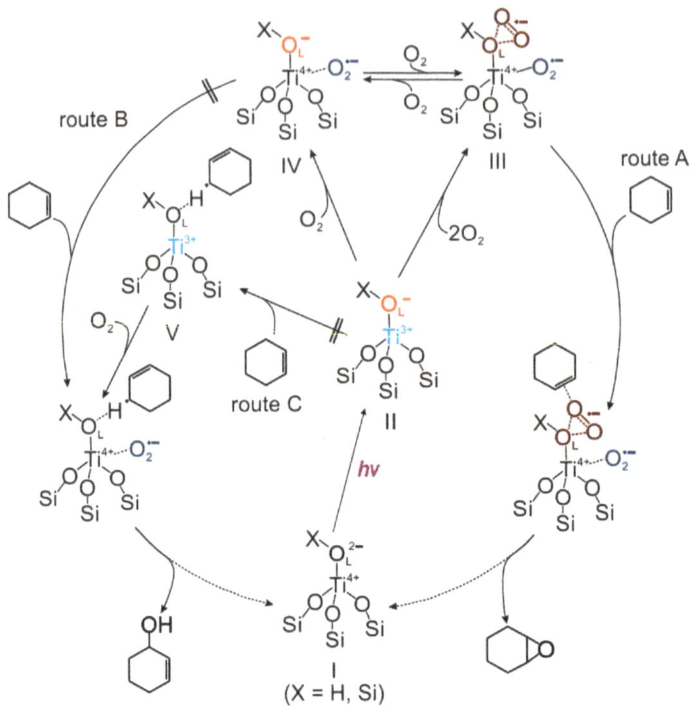

Figure 25 The mechanism of epoxidation of olefins on irradiated TiO_2—SiO_2 photocatalyst proposed by Shiraishi et al.[122]

photocatalytic activity compared to bare CdS and g-C_3N_4. In that composite system, photoexcited electrons from the CB of g-C_3N_4 can be transferred to the CB of CdS, whereas holes from the VB of CdS can be transferred to the VB of g-C_3N_4. As concluded by the authors, oxidation of aromatic alcohols to aromatic aldehydes under nitrogen atmosphere was achieved as a result of a direct oxidation with holes, whereas the reduction of nitrobenzene into aniline was the consequence of a direct nitrobenzene reduction with electrons (Fig. 26). After 4 h of illumination, the conversion of BzA and the yield of BzAD production reached the levels of 48.0% and 44.6%, respectively. At the same time, the conversion of nitrobenzene (49.2%) to aniline was characterized by a selectivity of 26.0%.

This example shows that composite materials may offer not only an improved charge separation but also an efficient space separation of reduction and oxidation reactions.

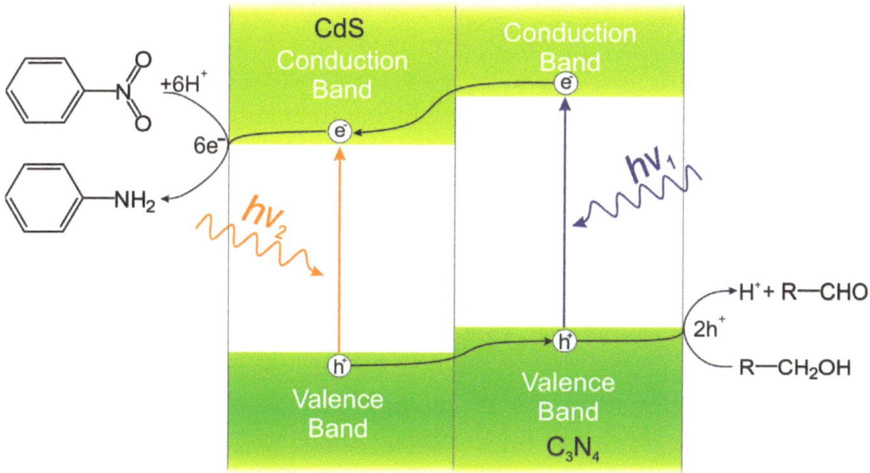

Figure 26 Proposed mechanism for the photocatalytic alcohol oxidation and the nitrobenzene reduction under N_2 atmosphere in the presence of CdS/g-C_3N_4 composite.[123]

6. CONCLUSIONS

In this work we tried to classify photocatalytic synthesis of organic compounds taking into account primary reduction and oxidation reactions, which play a pivotal role in the overall process. By understanding their nature, it should be easier to define properties for the optimal photocatalysts and reaction conditions. For instance, commonly applied titanium dioxide can be a good choice when oxidation of alcohols is considered or whenever generation of strongly oxidizing species, such as hydroxyl radicals, is required. However, such materials should not be expected to be optimal when strongly reductive conditions should be accomplished. For such reactions, in general, oxide photocatalysts will not be suitable. Sulfides and chalcogenides may offer here much milder oxidation conditions.

Redox properties of the photocatalyst, adsorption of reactants, and number of electrons participating in the reduction and oxidation reactions are examples of crucial parameters influencing the photocatalytic reactions. Redox properties of the photocatalysts can be determined using various methods, including the spectroelectrochemical approach developed recently in our laboratory.[73,124] Recognition of the density of electronic states may appear to be fundamental in understanding photoactivity and predicting applicability of the photocatalyst to synthesize chemicals. Based on such measurements, we were able to understand the differences in photocatalytic properties of rutile-TiO_2 and anatase-TiO_2.[72] Similar measurements revealed a high reducing ability of excited ZnS, at which generation of $CO_2^{\bullet-}$ radicals was confirmed.[88]

Noteworthy, the overall efficiency of any reaction is determined by the rate-limiting step. It means that independently of the goal of the designed synthesis, the efficiency of both reduction and oxidation half reactions should be taken into account. They are influenced, among others, by the number of electrons exchanged in the half reaction. Photocatalysis prefers one-electron, one-hole reactions over multi-charge processes. Therefore, activation of small molecules, often undergoing a one-electron exchange reaction, is an important step of many photocatalytic processes.

The presented mechanisms of the photocatalytic synthesis of chemicals perhaps do not cover all possible routes, although most of the photocatalytic reactions can be classified to one of the schemes in Fig. 1. Nevertheless, independent of the mechanism, photocatalysis may offer an important approach to sustainable synthesis of chemicals and fine chemicals and in future also to solar fuels.

ACKNOWLEDGMENTS

The work was supported by the Foundation for Polish Science (FNP) within the TEAM/2016-3/27 project.

REFERENCES

1. Kisch, H. *Semiconductor Photocatalysis: Principles and Applications;* John Wiley & Sons, 2014.
2. Wang, C-y.; Groenzin, H.; Shultz, M. J. *J. Am. Chem. Soc.* **2004,** *126,* 8094–8095.
3. Pillai, U. R.; Sahle–Demessie, E. *J. Catal.* **2002,** *211,* 434–444.
4. Jedsukontorn, T.; Meeyoo, V.; Saito, N.; Hunsom, M. *Chem. Eng. J.* **2015,** *281,* 252–264.
5. Hermes, N. A.; Corsetti, A. R.; Pacheco, A. S.; Lansarin, M. A. *J. Adv. Oxid. Technol.* **2015,** *18,* 315–321.
6. Park, H.; Choi, W. *Catal. Today* **2005,** *101,* 291–297.
7. Shiraishi, Y.; Saito, N.; Hirai, T. *J. Am. Chem. Soc.* **2005,** *127,* 12820–12822.
8. Shiraishi, Y.; Hirai, T. *J. Jpn. Petrol. Inst.* **2012,** *55,* 287–298.
9. Bui, T. D.; Kimura, A.; Ikeda, S.; Matsumura, M. *J. Am. Chem. Soc.* **2010,** *132,* 8453–8458.
10. Pang, X.; Chen, C.; Ji, H.; Che, Y.; Ma, W.; Zhao, J. *Molecules* **2014,** *19,* 16291–16311.
11. Bellardita, M.; Augugliaro, V.; Loddo, V.; Megna, B.; Palmisano, G.; Palmisano, L.; Puma, M. A. *Appl. Catal. A. Gen.* **2012,** *441,* 79–89.
12. Lv, K.; Guo, X.; Wu, X.; Li, Q.; Ho, W.; Li, M.; Ye, H.; Du, D. *App. Catal. B. Environ.* **2016,** *199,* 405–411.
13. Qi, K.; Zasada, F.; Piskorz, W.; Indyka, P.; Grybos, J.; Trochowski, M.; Buchalska, M.; Kobielusz, M.; Macyk, W.; Sojka, Z. *J. Phys. Chem. C* **2016,** *120,* 5442–5456.

14. Lv, K.; Lu, C. *Chem. Eng. Technol.* **2008,** *31,* 1272–1276.
15. Palmisano, G.; Addamo, M.; Augugliaro, V.; Caronna, T.; Di Paola, A.; López, E. G.; Loddo, V.; Marcì, G.; Palmisano, L.; Schiavello, M. *Catal. Today* **2007,** *122,* 118–127.
16. Huixian, S.; Zhang, T.; Hongliang, W.; Xiao, W.; Meng, H. *Chinese J. Catal* **2011,** *32,* 46–50.
17. Karam, F. F.; Kadhim, M. I.; Alkaim, A. F. *J. Chem. Sci.* **2015,** *13.*
18. Ohno, T.; Tokieda, K.; Higashida, S.; Matsumura, M. *Appl. Catal. A. Gen.* **2003,** *244,* 383–391.
19. Augugliaro, V.; Camera-Roda, G.; Loddo, V.; Palmisano, G.; Palmisano, L.; Parrino, F.; Puma, M. A. *App. Catal. B. Environ.* **2012,** *111,* 555–561.
20. Camera-Roda, G.; Augugliaro, V.; Cardillo, A.; Loddo, V.; Palmisano, G.; Palmisano, L. *Chem. Eng. J.* **2013,** *224,* 136–143.
21. Lee, J.; Choi, W.; Yoon, J. *Environ. Sci. Technol.* **2005,** *39,* 6800–6807.
22. Parrino, F.; Ramakrishnan, A.; Kisch, H. *Angew. Chem. Int. Ed.* **2008,** *47,* 7107–7109.
23. Zhang, P.; Wang, Y.; Li, H.; Antonietti, M. *Green Chem.* **2012,** *14,* 1904–1908.
24. Schoumacker, K.; Geantet, C.; Lacroix, M.; Puzenat, E.; Guillard, C.; Herrmann, J.-M. *J. Photochem. Photobiol. A* **2002,** *152,* 147–153.
25. Almquist, C. B.; Biswas, P. *Appl. Catal. A* **2001,** *214,* 259–271.
26. Boarini, P.; Carassiti, V.; Maldotti, A.; Amadelli, R. *Langmuir* **1998,** *14,* 2080–2085.
27. Du, P.; Moulijn, J. A.; Mul, G. *J. Catal.* **2006,** *238,* 342–352.
28. Shiraishi, Y.; Sugano, Y.; Ichikawa, S.; Hirai, T. *Catal. Sci. Technol.* **2012,** *2,* 400–405.
29. Shiraishi, Y.; Hirai, T. *J. Photochem. Photobiol. C* **2008,** *9,* 157–170.
30. Ni, X.; Ye, J.; Dong, C. *J. Photoch. Photobio A* **2006,** *181,* 19–27.
31. Dadashi-Silab, S.; Asiri, A. M.; Khan, S. B.; Alamry, K. A.; Yagci, Y. *J. Polym. Sci. A1* **2014,** *52,* 1500–1507.
32. Nakanishi, K.; Tanaka, A.; Hashimoto, K.; Kominami, H. *Phys. Chem. Chem. Phys.* **2017,** *19,* 20206–20212.
33. Nakanishi, K.; Yagi, R.; Imamura, K.; Tanaka, A.; Hashimoto, K.; Kominami, H. *Catal. Sci. Technol.* **2018,** *8,* 139–146.
34. Habisreutinger, S. N.; Schmidt-Mende, L.; Stolarczyk, J. K. *Angew. Chem. Int. Ed.* **2013,** *52,* 7372–7408.
35. Fujita, E. *Coord. Chem. Rev.* **1999,** *185,* 373–384.
36. Koppenol, W.; Rush, J. *J. Phys. Chem.* **1987,** *91,* 4429–4430.
37. Getoff, N. *Radiat. Phys. Chem.* **2003,** *67,* 617–621.
38. Baran, T.; Wojtyła, S.; Dibenedetto, A.; Aresta, M.; Macyk, W. *ChemSusChem* **2016,** *9,* 2933–2938.
39. Lingampalli, S.; Ayyub, M. M.; Rao, C. *ACS Omega* **2017,** *2,* 2740–2748.
40. Reli, M.; Kobielusz, M.; Matějová, L.; Daniš, S.; Macyk, W.; Obalová, L.; Kuśtrowski, P.; Rokicińska, A.; Kočí, K. *Appl. Surf. Sci.* **2017,** *391,* 282–287.
41. Ma, Y.; Wang, Z.; Xu, X.; Wang, J. *Chin. J. Catal.* **2017,** *38,* 1956–1969.
42. Nahar, S.; Zain, M.; Kadhum, A. A. H.; Hasan, H. A.; Hasan, M. R. *Materials* **2017,** *10,* 629.
43. Suzuki, T. M.; Takayama, T.; Sato, S.; Iwase, A.; Kudo, A.; Morikawa, T. *Appl. Catal. B* **2018,** *224,* 572–578.
44. Pan, Y.-X.; You, Y.; Xin, S.; Li, Y.; Fu, G.; Cui, Z.; Men, Y.-L.; Cao, F.-F.; Yu, S.-H.; Goodenough, J. B. *J. Am. Chem. Soc.* **2017,** *139,* 4123–4129.
45. Yanagida, S.; Ishimaru, Y.; Miyake, Y.; Shiragami, T.; Pac, C.; Hashimoto, K.; Sakata, T. *J. Phys. Chem.* **1989,** *93,* 2576–2582.
46. Joyce-Pruden, C.; Pross, J. K.; Li, Y. *J. Org. Chem.* **1992,** *57,* 5087–5091.
47. Matsushita, Y.; Kumada, S.; Wakabayashi, K.; Sakeda, K.; Ichimura, T. *Chem. Lett.* **2006,** *35,* 410–411.

48. Fukui, M.; Tanaka, A.; Hashimoto, K.; Kominami, H. *Chem. Commun.* **2017,** *53,* 4215–4218.
49. Fukui, M.; Kouda, H.; Tanaka, A.; Hashimoto, K.; Kominami, H. *ChemistrySelect* **2017,** *2,* 2293–2299.
50. Park, J. U.; Hong, M. J.; Go, P. G. H. *Bull. Korean Chem. Soc.* **2001,** *22,* 1213–1216.
51. Park, J.-W.; Kim, E.-K. *Bull. Korean Chem. Soc.* **2002,** *23,* 1229–1258.
52. Kohtani, S.; Yoshioka, E.; Saito, K.; Kudo, A.; Miyabe, H. *Catal. Commun.* **2010,** *11,* 1049–1053.
53. Kohtani, S.; Yoshioka, E.; Saito, K.; Kudo, A.; Miyabe, H. *J. Phys. Chem. C* **2012,** *116,* 17705–17713.
54. Kohtani, S.; Kamoi, Y.; Yoshioka, E.; Miyabe, H. *Catal. Sci. Technol.* **2014,** *4,* 1084–1091.
55. Blaser, H. U.; Malan, C.; Pugin, B.; Spindler, F.; Steiner, H.; Studer, M. *Adv. Synth. Catal.* **2003,** *345,* 103–151.
56. Corma, A.; Serna, P.; Concepción, P.; Calvino, J. J. *J. Am. Chem. Soc.* **2008,** *130,* 8748–8753.
57. Brezova, V.; Blažková, A.; Šurina, I.; Havlinova, B. *J. Photochem. Photobiol. A* **1997,** *107,* 233–237.
58. Flores, S. O.; Rios-Bernij, O.; Valenzuela, M. A.; Córdova, I.; Gómez, R.; Gutiérrez, R. *Top. Catal.* **2007,** *44,* 507–511.
59. Imamura, K.; Iwasaki, S-i.; Maeda, T.; Hashimoto, K.; Ohtani, B.; Kominami, H. *Phys. Chem. Chem. Phys.* **2011,** *13,* 5114–5119.
60. Wu, W.; Wen, L.; Shen, L.; Liang, R.; Yuan, R.; Wu, L. *Appl. Catal. B* **2013,** *130,* 163–167.
61. Ferry, J. L.; Glaze, W. H. *Langmuir* **1998,** *14,* 3551–3555.
62. Huang, H.; Zhou, J.; Liu, H.; Zhou, Y.; Feng, Y. *J. Hazard Mater.* **2010,** *178,* 994–998.
63. Brezová, V.; Tarábek, P.; Dvoranová, D.; Staško, A.; Biskupič, S. *J. Photochem. Photobiol. A* **2003,** *155,* 179–198.
64. Shiraishi, Y.; Togawa, Y.; Tsukamoto, D.; Tanaka, S.; Hirai, T. *ACS Catal.* **2012,** *2,* 2475–2481.
65. Wang, H.; Yan, J.; Chang, W.; Zhang, Z. *Catal. Commun.* **2009,** *10,* 989–994.
66. Pan, X.; Xu, Y.-J. *J. Phys. Chem. C* **2015,** *119,* 7184–7194.
67. Eskandari, P.; Kazemi, F.; Zand, Z. *J. Photochem. Photobiol. A* **2014,** *274,* 7–12.
68. Zhang, L.; He, X.; Xu, X.; Liu, C.; Duan, Y.; Hou, L.; Zhou, Q.; Ma, C.; Yang, X.; Liu, R. *Appl. Catal. B* **2017,** *203,* 1–8.
69. Van Thuan, D.; Khoa, N. T.; Kim, S.-W.; Chung, J. S.; Hur, S. H.; Kim, E. J.; Hahn, S. H.; Wang, M. *Catal. Let.* **2017,** *147,* 2440–2447.
70. Challagulla, S.; Tarafder, K.; Ganesan, R.; Roy, S. *Sci. Rep.* **2017,** *7,* 8783.
71. Dai, X.; Xie, M.; Meng, S.; Fu, X.; Chen, S. *Appl. Catal. B* **2014,** *158,* 382–390.
72. Buchalska, M.; Kobielusz, M.; Matuszek, A.; Pacia, M.; Wojtyła, S.; Macyk, W. *ACS Catal.* **2015,** *5,* 7424–7431.
73. Kobielusz, M.; Pilarczyk, K.; Świętek, E.; Szaciłowski, K.; Macyk, W. *Catal. Today* **2018,** *309,* 35–42.
74. Füldner, S.; Pohla, P.; Bartling, H.; Dankesreiter, S.; Stadler, R.; Gruber, M.; Pfitzner, A.; König, B. *Green Chem.* **2011,** *13,* 640–643.
75. Pal, B.; Torimoto, T.; Okazaki, K-i.; Ohtani, B. *Chem. Commun.* **2007,** 483–485.
76. Naya, S. i.; Niwa, T.; Kume, T.; Tada, H. *Angew. Chem. Int. Ed.* **2014,** *53,* 7305–7309.
77. Kiyonaga, T.; Fujii, M.; Akita, T.; Kobayashi, H.; Tada, H. *Phys. Chem. Chem. Phys.* **2008,** *10,* 6553–6561.
78. Tada, H.; Ishida, T.; Takao, A.; Ito, S. *Langmuir* **2004,** *20,* 7898–7900.

79. Dai, Y.; Li, C.; Shen, Y.; Lim, T.; Xu, J.; Li, Y.; Niemantsverdriet, H.; Besenbacher, F.; Lock, N.; Su, R. *Nature commun.* **2018,** *9,* 60.
80. Hörner, G.; Johne, P.; Künneth, R.; Twardzik, G.; Roth, H.; Clark, T.; Kisch, H. *Chem. Eur J.* **1999,** *5,* 208–217.
81. Bücheler, J.; Zeug, N.; Kisch, H. *Angew. Chem. Int. Ed.* **1982,** *21,* 783–784.
82. Zeug, N.; Bücheler, J.; Kisch, H. *J. Am. Chem. Soc.* **1985,** *107,* 1459–1465.
83. Künneth, R.; Twardzik, G.; Emig, G.; Kisch, H. *J. Photochem. Photobiol. A* **1993,** *76,* 209–215.
84. Kisch, H. *Acc. Chem. Res.* **2017,** *50,* 1002–1010.
85. Mitkina, T.; Stanglmair, C.; Setzer, W.; Gruber, M.; Kisch, H.; König, B. *Org. Biomol. Chem.* **2012,** *10,* 3556–3561.
86. Sun, X.; Jiang, D.; Zhang, L.; Wang, W. *Appl. Catal. B* **2018,** *220,* 553–560.
87. Su, F.; Mathew, S. C.; Möhlmann, L.; Antonietti, M.; Wang, X.; Blechert, S. *Angew. Chem. Int. Ed.* **2011,** *50,* 657–660.
88. Baran, T.; Dibenedetto, A.; Aresta, M.; Kruczała, K.; Macyk, W. *ChemPlusChem* **2014,** *79,* 708–715.
89. Aresta, M.; Dibenedetto, A.; Baran, T.; Wojtyła, S.; Macyk, W. *Faraday Discuss.* **2015,** *183,* 413–427.
90. Buchalska, M.; Kuncewicz, J.; Świętek, E.; Łabuz, P.; Baran, T.; Stochel, G.; Macyk, W. *Coord. Chem. Rev.* **2013,** *257,* 767–775.
91. Schindler, W.; Kisch, H. *J. Photochem. Photobiol. A* **1997,** *103,* 257–264.
92. Schindler, W.; Knoch, F.; Kisch, H. *Eur. J. Inorg. Chem.* **1996,** *129,* 925–932.
93. Gärtner, M.; Ballmann, J.; Damm, C.; Heinemann, F. W.; Kisch, H. *Photochem. Photobiol. Sci.* **2007,** *6,* 159–164.
94. Fermin, D. J.; Ponomarev, E. A.; Peter, L. M. *J. Electroanal. Chem.* **1999,** *473,* 192–203.
95. Tang, Y.; Hu, X.; Liu, C. *Phys. Chem. Chem. Phys.* **2014,** *16,* 25321–25329.
96. Künneth, R.; Feldmer, C.; Kisch, H. *Angewandte Chemie Inter. Ed.* **1992,** *31,* 1039–1040.
97. Herrmann, J.-M. *J. Photochem. Photobiol. A* **2010,** *216,* 85–93.
98. Reinheimer, A.; van Eldik, R.; Kisch, H. *J. Phys. Chem. B* **2000,** *104,* 1014–1024.
99. Annadhasan, M.; Selvam, K.; Swaminathan, M. *Synth. Commun.* **2012,** *42,* 1500–1508.
100. Selvam, K.; Krishnakumar, B.; Velmurugan, R.; Swaminathan, M. *Catal. Commun.* **2009,** *11,* 280–284.
101. Zheng, Z.; Huang, B.; Lu, J.; Qin, X.; Zhang, X.; Dai, Y. *Chem. Eur J.* **2011,** *17,* 15032–15038.
102. Ohno, T.; Sarukawa, K.; Matsumura, M. *New J. Chem.* **2002,** *26,* 1167–1170.
103. Aoyama, Y.; Oaki, Y.; Ise, R.; Imai, H. *Cryst. Eng. Comm* **2012,** *14,* 1405–1411.
104. Chong, R.; Li, J.; Zhou, X.; Ma, Y.; Yang, J.; Huang, L.; Han, H.; Zhang, F.; Li, C. *Chem. Commun.* **2014,** *50,* 165–167.
105. Cao, F.-L.; Wang, J.-G.; Lv, F.-J.; Zhang, D.-Q.; Huo, Y.-N.; Li, G.-S.; Li, H.-X.; Zhu, J. *Catal. Commun.* **2011,** *12,* 946–950.
106. Li, C.-J.; Xu, G.-R.; Zhang, B.; Gong, J. R. *App. Catal. B. Environ.* **2012,** *115,* 201–208.
107. Yu, J.; Low, J.; Xiao, W.; Zhou, P.; Jaroniec, M. *J. Am. Chem. Soc.* **2014,** *136,* 8839–8842.
108. Tang, Z.-R.; Yin, X.; Zhang, Y.; Xu, Y.-J. *RSC Adv.* **2013,** *3,* 5956–5965.
109. Yuan, B.; Chong, R.; Zhang, B.; Li, J.; Liu, Y.; Li, C. *Chem. Commun.* **2014,** *50,* 15593–15596.
110. Xiao, J.; Xie, Y.; Han, Q.; Cao, H.; Wang, Y.; Nawaz, F.; Duan, F. *J. Hazard Mater.* **2016,** *304,* 126–133.

111. Xiao, J.; Xie, Y.; Cao, H.; Nawaz, F.; Zhang, S.; Wang, Y. *J. Photochem. Photobiol. A* **2016,** *315,* 59–66.
112. Zhang, Z.; Luo, Z.; Yang, Z.; Zhang, S.; Zhang, Y.; Zhou, Y.; Wang, X.; Fu, X. *RSC Adv.* **2013,** *3,* 7215–7218.
113. Tripathy, J.; Lee, K.; Schmuki, P. *Angew. Chem. Int. Ed.* **2014,** *126,* 12813–12816.
114. Nishikawa, M.; Mitani, Y.; Nosaka, Y. *J. Phys. Chem. C* **2012,** *116,* 14900–14907.
115. Khan, M.; Al-Oufi, M.; Toseef, A.; Nadeem, M.; Idriss, H. *Catal. Lett.* **2018,** 1–10.
116. Yeh, T.-F.; Cihlář, J.; Chang, C.-Y.; Cheng, C.; Teng, H. *Mater. Today* **2013,** *16,* 78–84.
117. Ide, Y.; Nakamura, N.; Hattori, H.; Ogino, R.; Ogawa, M.; Sadakane, M.; Sano, T. *Chem. Commun.* **2011,** *47,* 11531–11533.
118. Shiraishi, Y.; Sugano, Y.; Tanaka, S.; Hirai, T. *Angew. Chem. Int. Ed.* **2010,** *122,* 1700–1704.
119. Ke, X.; Zhang, X.; Zhao, J.; Sarina, S.; Barry, J.; Zhu, H. *Green Chem.* **2013,** *15,* 236–244.
120. Tanaka, A.; Hashimoto, K.; Kominami, H. *J. Am. Chem. Soc.* **2012,** *134,* 14526–14533.
121. Liu, S.; Zhang, N.; Tang, Z.-R.; Xu, Y.-J. *ACS Appl. Mater. Inter.* **2012,** *4,* 6378–6385.
122. Shiraishi, Y.; Morishita, M.; Hirai, T. *Chem. Commun.* **2005,** 5977–5979.
123. Dai, X.; Xie, M.; Meng, S.; Fu, X.; Chen, S. *App. Catal. B. Environ.* **2014,** *158,* 382–390.
124. Świętek, E.; Pilarczyk, K.; Derdzińska, J.; Szaciłowski, K.; Macyk, W. *Phys. Chem. Chem. Phys.* **2013,** *15,* 14256–14261.

CHAPTER FOUR

Oxide Nanomaterials for Photoelectrochemical Hydrogen Energy Sources

Marta Radecka*,[1], Anna Kusior*, Anita Trenczek-Zajac* and Katarzyna Zakrzewska[§]

*Faculty of Materials Science and Ceramics, AGH University of Science and Technology, al. A. Mickiewicza 30, 30-059 Kraków, Poland

[§]Faculty of Computer Science, Electronics and Telecommunications, AGH University of Science and Technology, al. A. Mickiewicza 30, 30-059 Kraków, Poland

[1]Corresponding author: E-mail: radecka@agh.edu.pl

Contents

Abstract

Tremendous efforts aimed at the development of efficient, renewable, and ecologically safe hydrogen energy sources are still observed. The remarkable progress in the design and engineering of oxide nanomaterials to be applied as photoelectrodes in photoelectrochemical cells for water splitting is discussed in this chapter. This review focuses on fundamental contributions to the efficiency of solar–chemical conversion such as light harvesting, electron transport, and solid—solid junction. Our own results on 1D, 2D, and 3D TiO_2 along with those on heterostructures of TiO_2/CdS, TiO_2/SnO_2, TiO_2/MoS_2, and TiO_2/GO are presented on the background of a thorough literature overview.

Advances in Inorganic Chemistry, Volume 72
ISSN 0898-8838
https://doi.org/10.1016/bs.adioch.2018.05.001

1. INTRODUCTION

Solar hydrogen is a term that indicates hydrogen generated solely from the environmentally friendly sources such as the Sun radiation and water—the most abundant hydrogen supplier on the Earth. Photoelectrolysis of water, as an efficient means of formation of solar hydrogen, still constitutes the biggest challenge to the scientific community working on sustainable and renewable energy. The ingenious idea of water splitting into gaseous hydrogen and oxygen according to the scheme:

$$2h\nu + H_2O \rightarrow 1/2O_2(\text{gas}) + H_2(\text{gas}) \quad (1)$$

requiring photons of the energy $h\nu$ corresponding to the visible spectrum of light cannot be realized without an auxiliary medium that will absorb these photons.[1–5] From the point of view of thermodynamics, Reaction (1) occurs when the photon energy $h\nu$ exceeds 1.23 eV while water is transparent to light over this wavelength range ($\lambda < 1000$ nm). The search for suitable semiconducting materials playing the role of auxiliary light absorber has begun many years ago with the important work of Fujishima and Honda[1,2] demonstrating the photoelectrochemical cell (PEC) consisting of an n-type TiO_2 semiconductor acting as a photoanode immersed in a liquid electrolyte and a second Pt electrode. Reasonable efficiency of this first photoelectrochemical device was due to a certain pH bias imposed by separating the electrolyte surrounding the photoanode and Pt electrode by a permeable membrane.

Since the publication of this work, increasing efficiency of the solar-to-hydrogen conversion to at least 10% has been considered as the biggest challenge. It is worth noting at this point that another approach such as solar-to-electrical energy conversion with the use of the same semiconductor, namely TiO_2, was developed in parallel by Grätzel.[6,7] Photovoltaic (PV) cells of this type also suffer from low conversion efficiency. However, in spite of technical problems resulting from low efficiency, both solar-to-fuel and solar-to-electrical energy approaches still deserve particular attention.

In 1995, a special issue of *Accounts of Chemical Research* was devoted to the most important topics that could be treated as a Holy Grail in chemistry. One of the articles in the series,[8] written by Bard and Fox, described water splitting by sunlight to produce hydrogen. Even at that time, it was realized that this Holy Grail would be difficult to achieve. The authors of this paper defined their Holy Grail as "an efficient and long-lived system for splitting water to H_2 and O_2 with light in the terrestrial (AM1.5) solar spectrum at an intensity of one sun."

After 22 years, on the occasion of the 50th anniversary of the same journal, a "quest for perfection"—being a present-day interpretation of a Holy Grail—has been revisited.[9] It turns out that the main interest has shifted to other topics, but "the conversion and storage of solar energy in a sustainable fashion with minimal adverse impact on the planet" and "perfect catalysis" still remain central issues.[9] The examples of research performed in this field are given in Refs. 10 and 11 and involve solar fuels and surface chemistry.

During the past few decades, metal oxide semiconductors in a planar form were widely used as photoelectrodes in PECs.[12–15] Titanium dioxide (TiO_2) is still the most popular candidate for an n-type photoanode. Other metal oxides such a ZnO,[16–19] Fe_2O_3,[20–22] WO_3[23,24] are also considered while CuO and Cu_2O are proposed as p-type photocathodes.[25–28]

The most fundamental requirement for a semiconductor to be taken into account as a candidate for photoelectrode material in PECs is an appropriate match of its electronic band edges to the redox potential levels for water splitting, $E^o{}_{H^+/H_2}$ and $E^o{}_{O_2/H_2O}$, i.e., conduction band E_{CB} and valence band E_{VB} edges in vacuum level scale should fulfill the following conditions: $E^o{}_{H^+/H_2} < E_{CB}$ and $E^o{}_{O_2/H_2O} > E_{VB}$. As shown in Fig. 1, these conditions are met for all four generations of TiO_2 despite the fact that its bandgap is far too big to absorb efficiently within the whole solar spectrum. Bandgap engineering should concentrate particularly on the uprising of the valence band edge, which in undoped TiO_2 is quite below the O_2/H_2O potential level.[29–32]

The main drawbacks of the PEC devices based on bulk, planar photoelectrodes are[5,33–36]:

1. limited light absorption efficiency in the visible range due to restricted possibilities of light management (light harvesting) in such traditional structures;
2. fast recombination of photoexcited electrons and holes.

To overcome these limitations, four generations of TiO_2 (Fig. 1) were proposed and studied with the ultimate aim to use them in PECs.[37–45] All these generations are based on the idea of the modification of the electronic structure and especially reduction of the bandgap by doping.

The authors of this chapter have been involved in the photoelectrochemical studies of all four generations of TiO_2.[46–59] This research carried out over the last 15 years has led to the following conclusions.

The first generation includes pristine TiO_{2-x} with its different polymorphic forms, e.g., anatase and rutile, and a controlled departure from stoichiometry,

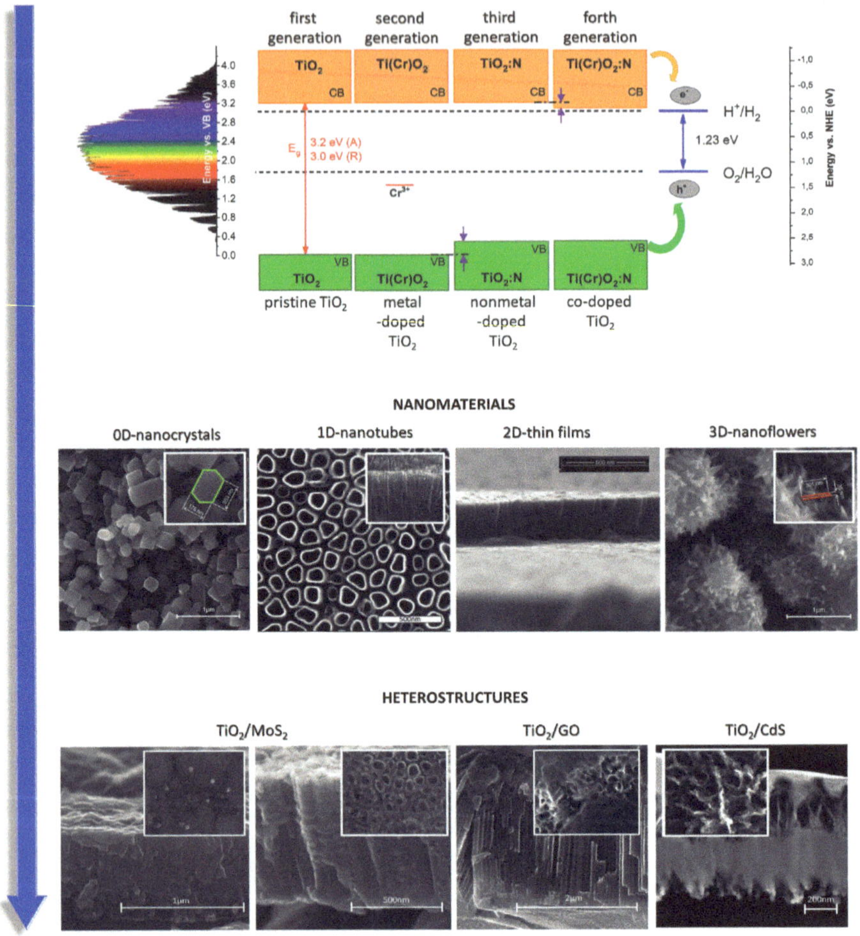

Figure 1 Historical evolution of materials for metal oxide semiconducting photoelectrodes in water splitting devices using photoelectrochemical cells. Four generations of planar TiO_2 photoelectrodes as contrasted with those in a form of nano- and heterostructures; band positions with respect to redox potential levels for water splitting. *A*, anatase; *CB*, conduction band; E_g, bandgap; *R*, rutile; *VB*, valence band.

x. Anatase represents a better match of the electronic structure than rutile; hence, it has been a more appropriate form of TiO_2 for the photoelectrochemical applications.[60–62] Moreover, a synergetic effect due to a coexistence of anatase and rutile within certain limits, i.e., 40%–80% of anatase, has been studied.[36,63–65]

As far as nonstoichiometric TiO_{2-x} is concerned, it was believed that as the bandgap could be a function of x, it might be possible to find such a composition of TiO_{2-x} that can result in bandgap reduction.[47,56,57] This idea was based on the fact that the bandgap energies of many suboxides of titanium, Ti_nO_{2n-1} (Magnéli phases, $n = 1/x$) with n below 9, are at least 10 times smaller than E_g of stoichiometric TiO_2.[57] However, this hypothesis was proven to be wrong since with an increase in x the bandgap first increases.[47] Similar results were obtained by Vratny and Micale.[66] Our studies indicated that the only advantage of using nonstoichiometric TiO_{2-x} is a decreased electrical resistivity of the photoanode.

The second generation of titanium dioxide involves substitutional cation (metal)—doped TiO_2. Donor and acceptor dopants such as Cr, V, and Nb have been tested with a moderate success.[47,67–75] Formation of an additional absorption feature over the visible range of the light spectrum has been reported[47] without any significant improvement in the photocurrent. The reason for this behavior can be traced back to the enhanced electron—hole recombination related to a buildup of the localized states within the bandgap of TiO_2.

The third generation of TiO_2 is based on incorporation of nonmetals such as carbon, nitrogen, and fluorine into the anionic sublattice of TiO_2.[42,76] In the case of nitrogen doping, it is possible to shift the valence band up toward the conduction band (Fig. 1), which causes narrowing of the bandgap. Our research on thin film photoanodes of TiO_{2-x}:N obtained under controllable conditions indicated that substitution of N in the anionic sublattice is greatly facilitated by departure from stoichiometry, x.[46,49,55] However, interstitial incorporation of nitrogen cannot be excluded; thus the improvement of the photoelectrochemical properties of the materials belonging to the third generation is hindered by the recombination of charge carriers at point defects.

The fourth generation was proposed to overcome the drawbacks of the second and third generation by assuming that simultaneous doping of both cationic and anionic sublattices of TiO_2 would solve the problem of pronounced recombination while at the same would shift the absorption spectrum toward the visible range. Our own contribution in this domain has been based on codoping of TiO_2 with N and Cr.[50,77] An increase in photocurrent has been observed for such thin film photoanodes.[78]

In contrast to the approach based on the electronic structure, modification of the microstructure and morphology has been proposed as an efficient solution for photoelectrodes in PECs. Nanomaterials (0D, 1D, 2D, 3D) and heterostructures of metal oxide semiconductors (Fig. 1) have proved[79] to be advantageous as compared with their bulk counterparts because of the:

1. increased efficiency due to larger surface area;
2. decreased electron–hole recombination loss due to the shorter distance for photogenerated holes to diffuse to the photoanode/electrolyte interface;
3. high quality of nanostructures with a low density of defects;
4. size-controlled variation of the nanomaterial electronic structure;
5. reduced energy loss due to light reflection because of naturally formed antireflecting layers.

However, the fact of having nanosized dimensions does not warrant that the material will perform successfully as a photoelectrode in PECs. As already pointed out,[80] the nanomaterials or nanostructured architectures have to be designed carefully to benefit fully from their special features while avoiding the detrimental effect of rapid recombination of charge photocarriers.

In this review, we intend to demonstrate the most important issues such as light harvesting, efficient charge transport, formation of solid–solid junctions, and the role they play in the successful design of nanostructured photoanodes for solar hydrogen generation. The content of this chapter is based on, but not limited to, our recent experience with 0D, 1D, 2D, and 3D metal oxide nanomaterials for photoelectrochemical applications.

2. LIGHT HARVESTING

There are numerous processes that decrease the efficiency of photovoltaic and photoelectrochemical devices. All of them are related to the complex mechanism of interaction with light, separation of photogenerated carriers, i.e., electrons and holes, and their transport to the reaction site usually located at the surface, as well as rate-determining interfacial transformation at the electrode surface in the case of hydrogen generation by water splitting.

As the first step consists of absorption of sunlight by the semiconductor of a given bandgap energy E_g, the performance of the device strongly depends on the maximum utilization of the whole spectrum, especially its visible range. The striking example of the low efficiency of about 2%, resulting from a substantial mismatch between the bandgap of about 3 eV (UV range) and the longer wavelengths provided by solar radiation (visible and infrared), can be seen in the case of TiO_2, which are otherwise excellent photocatalysts.[81] Because of its wide bandgap, TiO_2 can capture only 1% of sunlight energy.[80]

Light harvesting belongs to a more general class of approaches to the problem of increasing the efficiency of all processes that rely on light absorption as a first step of light-induced phenomena as shown in Fig. 2. Photon or light management is a general term that describes how incoming light can be utilized efficiently to contribute fully to the process.

Biological entities, including green plants, bacteria, archaea, etc. have developed almost perfect light-harvesting strategies that allow for natural photosynthesis as a powerful source of energy and fuel. In natural photosynthesis, the solar energy is converted to the chemical energy as a result of light absorption by antenna protein–pigment complexes containing chlorophyll, followed by multistep photoinduced energy transfer from the absorption site to the reaction center.[82–84] In a system of weakly coupled pigments, the energy transfer taking place within the antenna complexes is well accounted for by the Förster theory, i.e., Förster resonance energy transfer (FRET).[82,84] The nonradiative energy transfer between two isolated chromophore molecules, donor (D) and acceptor (A), depends on the distance between them and when it is large enough can be described by the dipole–dipole approximation of the Coulomb interaction.[84] The overlap between the emission spectrum of the donor and the absorbance spectrum of the acceptor is an essential requirement for efficient energy transfer. The FRET model is no longer valid when the interaction between D and A becomes stronger.[84]

Strategies based on bio-inspired artificial photoprocesses in human-engineered systems, such as solar cells and photoelectrochemical as well as photocatalytic reactors, aim at applying the light-harvesting paradigms by mimicking Nature.[85] The systems in which artificial light harvesting is realized comprise dendrimers, organogels, porphyrin arrays/assemblies, biopolymer assemblies, and organic–inorganic hybrid materials.[86–90] Moreover, periodically or randomly textured inorganic nanomaterials,[91,92] semiconducting quantum dots (QDs),[93,94] and plasmonic metal nanoparticles[95,96] are usually considered, for which the idea is shown in Fig. 2.

Periodic or random patterning aims at increasing the path of light in the material, thus enhancing the possibility of absorption to occur. This strategy usually involving photolithography[97,98] inevitably leads to surface roughening that helps to avoid optical losses due to the specular reflection of light. In this way, the periodically or randomly textured surface affects the scattering of light and acts as an antireflective (AR) coating. Many examples

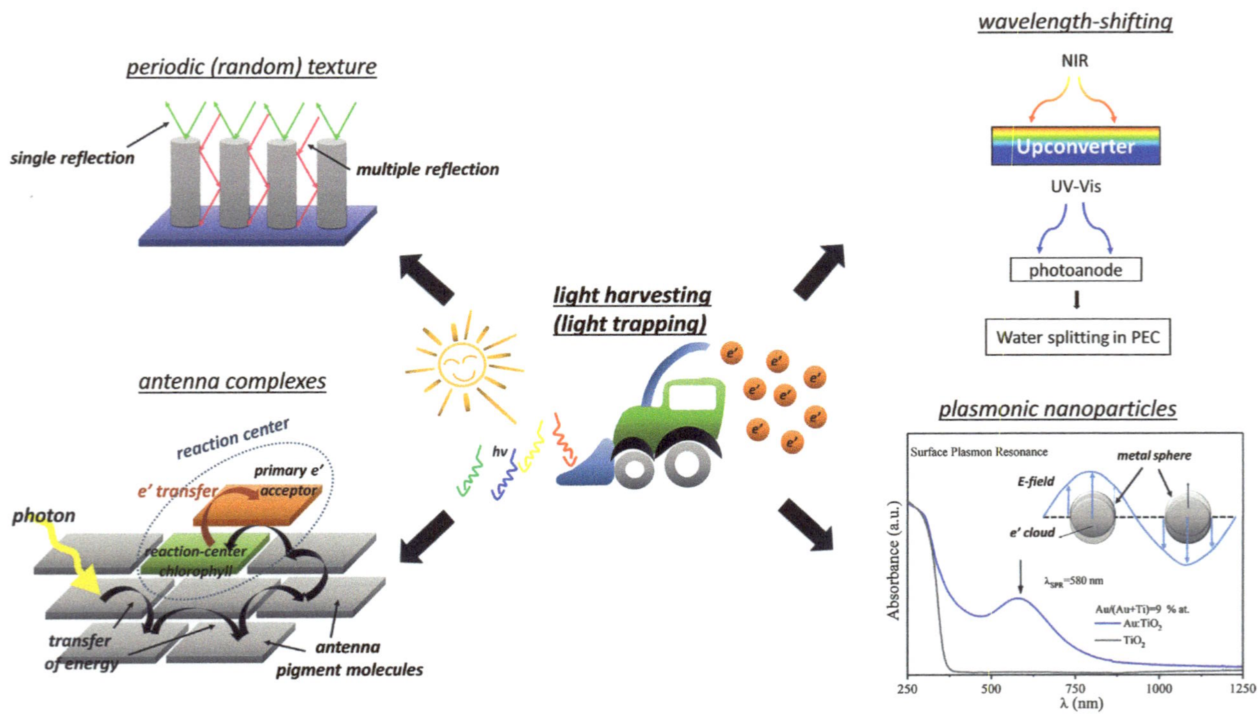

Figure 2 Light-harvesting schemes in various systems: periodically or randomly textured materials, antenna complexes in biological matter, up-conversion in photoelectrochemical processes, plasmonic nanoparticles. *NIR*, near infrared; *PEC*, photoelectrochemical cell. *Data of plasmon effect taken from Zakrzewska, K.; Radecka, M.; Kruk, A.; Osuch, W. Solid State Ionics **2003**, 157 (1–4), 349–356.*

of this type of light-trapping scenario have been demonstrated[99–103]; however, it is worth noting that 1D nanomaterials in the form of vertical nanowires and nanotubes demonstrate inherently AR behavior.[104–108]

Photonic crystals represent another important example of the periodic texturing.[109] A novel, lithography-free method for fabricating subwavelength-scale hierarchically nanotextured surfaces by direct molding of TiO_2 nanoparticles has been recently proposed.[110] Many authors claim[111–114] that improvement in the solar-to-fuel efficiency is due to this process. However, a critical review by Eftekhari et al.[80] raises doubt about the significance of enhanced absorption of light in the nanostructured materials. These authors claim that the surface roughness is not a critical parameter for the photocurrent. Patterning of the electrode surface helps to decrease the path the hole has to travel to reach the surface, thus affecting only charge transfer.

Semiconductor QDs are luminescent nanomaterials that generate light after photoexcitation while their emission wavelength can be tuned by adjusting the size and shape of QDs due to quantum confinement effects.[82,115–117] Significant attention has been paid to developing a new generation of light-harvesting materials based on QDs that can control the charge carrier dynamics and transfer in PV and photoelectrochemical devices.

Noble metal nanoparticles (Au, Ag, Pt) embedded in a semiconducting photoelectrode may be applied as light-harvesting elements due to surface plasmon absorption (surface plasmon resonance, SPR) or by increasing the photon path length within the semiconductor surface region.[118–122] Plasmonic photocatalysis involves dispersion of noble metal nanoparticles in the bulk or at the surface of a photocatalytic semiconductor to create an additional absorption feature usually within the visible range of the light spectrum. The localized SPR can be treated as the resonant photoinduced collective oscillation of valence electrons. The SPR results in an absorption peak at the frequency of photons that corresponds to the natural frequency of surface electrons oscillating in the electric field of positive nuclei. The local electric field is affected by the dielectric properties of the medium surrounding a metal nanoparticle, its size and shape. By changing the composition, shape, and size of plasmonic nanoparticles, as well as the dielectric constant of the nonconducting matrix, the resonant wavelength can be adapted to match any part of the entire solar spectrum. In particular, the absorption peak shifts to longer wavelengths if the surrounding medium has a higher dielectric constant or the size of the nanoparticle increases.

Larger metal nanoparticles increase the average path length of unabsorbed photons by scattering, which results in an increase in the absorption of light in the region close to the photoelectrode surface. The authors of this review studied SPR in TiO_2 thin films with noble metal inclusions, so-called nanocermets. The application of these films to hydrogen generation has been demonstrated.[56,58]

Most of these concepts came to life while attempting to develop efficient PV devices. These efforts aimed at modification of the structure of a solar cell based on matching its spectroscopic sensitivity to the existing sunlight spectrum.[123] The reverse approach to fit the sunlight spectrum to the sensitivity of the device via a wavelength-shift operation, i.e., down- or up-conversion, has been recently proposed.[124–128] This approach represents a real light management strategy that will allow to avoid detrimental consequences of the bandgap engineering such as increased recombination rate of photoinduced charge carriers due to, e.g., defects created by doping.[47]

In conventional solar cells, thermalization of charge carriers, which are generated by the absorption of high-energy photons, is one of the major mechanisms responsible for energy loss.[124] Down-conversion that transforms a high-energy photon into one or more lower-energy photons that can be matched to the spectral sensitivity peak of the Si cells has been shown to improve the efficiency of solar cells.

On the other hand, in the case of PECs for hydrogen generation with wide bandgap photoelectrodes, the up-conversion process is required. As it is necessary to harvest efficiently the visible range of the solar spectrum, whereas photons of too low energy cannot be absorbed in a wide bandgap semiconductor, transmission of the subbandgap photons is one of the major energy loss mechanisms.

Down-conversion can be based on quantum cutting or Stokes-shifted emission.[127] Quantum cutting is a direct conversion of the energy of one absorbed photon into two (or more) emitted low-energy photons. It can be realized[128] by the inclusion of layers, e.g., SiO_2, containing Ln^{3+} ions codoped with Yb^{3+} at the front of the solar cell. Stokes-shifted luminescence is based on absorption of short-wavelength light (300–500 nm) and emission at the wavelengths where the quantum efficiency of PV devices is high. Lanthanide-doped inorganic phosphors, colloidal QDs, organolanthanide complexes, organic dyes are examples of Stokes-shifting materials.[127]

On the contrary, photon up-conversion represents anti-Stokes optical processes that convert two or more low-energy pump photons, even from incoherent light sources, to higher-energy photons, preferably in the UV range of the light spectrum, by sequential absorption or ion–ion energy transfer.[129,130] The up-conversion process can be realized by rare-earth–doped nanocrystals or triplet–triplet annihilation (TTA) and is facilitated by selective excitation of strongly absorbing sensitizer chromophores that internally convert to the long-lived lowest energy triplet excited state with high quantum efficiency. The examples of application of up-conversion to water splitting[131,132] and even to the detection of cancer cells are known.[133]

3. EFFICIENT CHARGE TRANSFER

Besides strong absorption of photons within the wide wavelength range of the solar spectrum, the remedy of which can be found in improved light harvesting,[125,131,132] an efficient charge transfer is another, very important requirement for semiconductor photoelectrodes in water splitting devices.[134] Fast charge transfer depends strongly on the form of the material, with 1D nanorods and nanotubes being considered the best suited for this purpose. Nanotubes offer a natural way for fast channeling of charge carriers as shown in Fig. 3. The geometrical distance to be covered between the site (usually in the bulk) of electron–hole creation and the surface where the

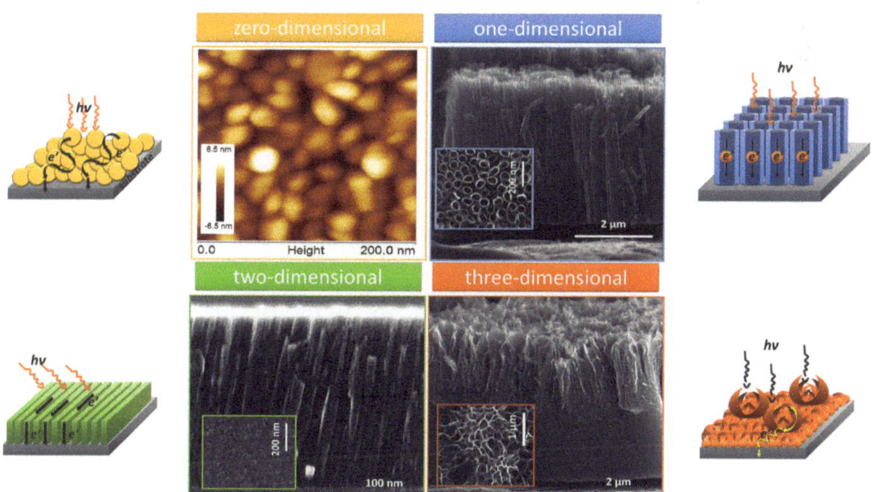

Figure 3 Illustration of the ways the photogenerated electron transfer takes place depending on the dimensionality of nanostructures; hv, photon energy.

photocatalytic process takes place is responsible for efficient charge transfer; the electron–hole recombination rate is equally important, as it determines the lifetimes of photoinduced charges.

Charge separation after photogeneration being a consequence of the absorption event is absolutely necessary to avoid both bulk and surface recombination (Fig. 4). Efficient electron and hole separation in space can be realized when a heterojunction between grains is formed. The electric field created within the region of a junction prevents charge carriers from recombination. In the case of a single photoanode material, a highly negative flat band potential is a good prognostics for efficient charge carrier separation and improved performance in PECs.

The important parameters for characterization of a PEC are the negative flat band potential, photocurrent density, and the shape of the current versus potential characteristics under illumination. The photocurrent depends on the competition between hole transfer across the interface and electron–hole recombination in the bulk and at the surface of the semiconductor electrode. A high photocurrent will be observed if the electron and hole are separated effectively before the system relaxes to the ground state. This requires efficient kinetics for electron transfer across the solid/solution interface. The extent of collection of minority carriers (holes in the case of n-type semiconductor) from the region beyond the depletion layer is dictated by the diffusion process.

4. TIO_2 NANOSTRUCTURES OF DIFFERENT DIMENSIONALITY—RESULTS OF AUTHORS' INDIVIDUAL RESEARCH

In this section we present results of research performed by our group on the growth, characterization, and photoelectrochemical performance of TiO_2 nanostructures of different dimensionalities, 1D, 2D, and 3D, in PECs.

Fig. 5 demonstrates some representative results concerning TiO_2 nanotubes grown by Ti-foil anodization in different electrolytes. It was found[51,135–138] that the morphological quality of nanotubes, their length, diameter, and mode of growth, depends on many factors among which the most important are the type and composition of the electrolyte. Here, we present the effect of ethylene glycol (EG) or glycerol (GL) water solutions in polar organic electrolytes containing 1 wt.% NH_4F on the growth mode of nanotubes. The geometry of titania nanotube arrays grown vertically from a substrate and having large surface area appears to be ideal

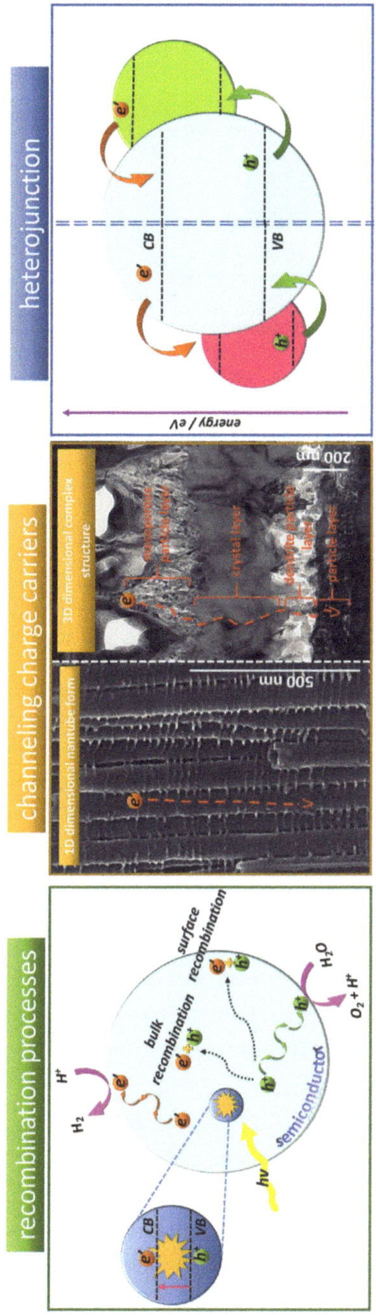

Figure 4 Recombination processes and possible pathways leading to separation of photoinduced charge carriers.

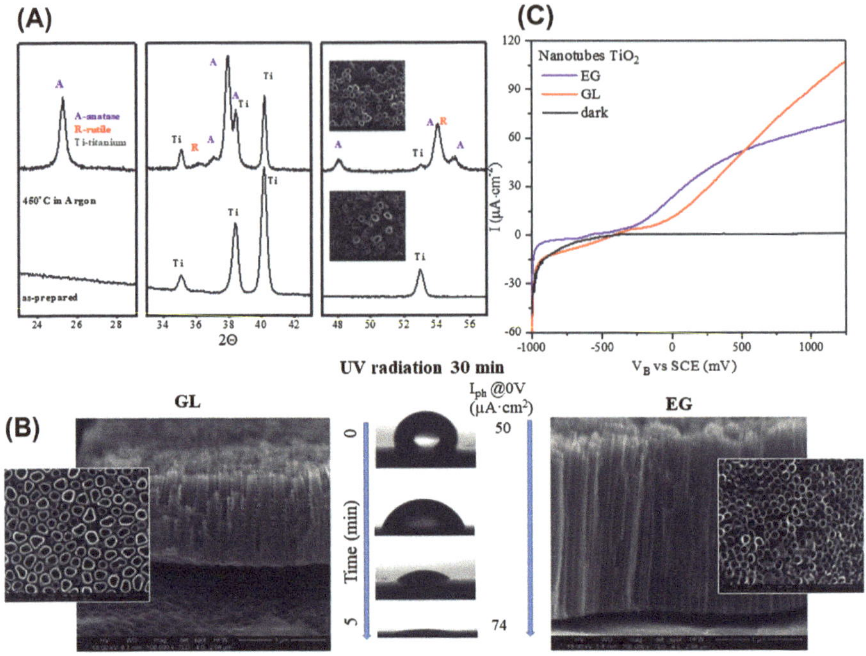

Figure 5 Representative results obtained by the authors of this review for 1D nanostructured TiO_2 grown by anodization of Ti-foil in different liquid electrolytes. (A) XRD patterns and SEM images for as-prepared and annealed nanotubes; (B) SEM images of nanotubes grown in liquid electrolyte containing glycerol (GL) or ethylene glycol (EG) (top-view and cross-section); in between: time development of shape of water droplet after initial UV irradiation interposed with an increase in the photocurrent; (C) photocurrent density as function of the applied voltage for the photoanodes shown in (B). *Based on Kusior, A.; Wnuk, A.; Trenczek-Zajac, A.; Zakrzewska, K.; Radecka, M.* Int. J. Hydrog. Energy ***2015***, 40 *(14), 4936–4944; Radecka, M.; Wnuk, A.; Trenczek-Zajac, A.; Schneider, K.; Zakrzewska, K.* Int. J. Hydrog. Energy ***2015***, 40 *(1), 841–851.*

for water photolysis, facilitates penetration of electrolyte, and provides more efficient charge transfer. The structure of nanotubes prevents the randomization of the grain network and increases contact points for a good electrical connection.

Fig. 5A shows a comparison of X-ray diffraction (XRD) patterns and SEM top-view images for anodized and heat-treated nanotubes. The prepared samples are clearly amorphous, and only Ti reflexes from the foil substrate can be seen. SEM images revealed that the surface morphology of prepared TiO_2 samples was not well defined, with nanotubes that were grown in different shapes. To make them suitable for application in PECs, the amorphous TiO_2 should be transformed into crystalline anatase or rutile, or a mixture of anatase and rutile phases. Annealing in an argon atmosphere at 450°C indeed transformed nanostructures into crystalline phase. Two polymorphic forms of TiO_2, i.e., rutile and anatase, were observed after annealing as indicated in Fig. 5A, while the anatase fraction strongly dominated over that of rutile. In this case, the nanotubes were well developed, with sharp contours and clearly defined shape as seen in SEM images.

Fig. 5B shows cross-sectional SEM images of the nanotubes obtained after anodization in two different electrolytes. The insets include the top-view images of the structures. It can clearly be seen that the nanotubes grown under these conditions are open at the top and closed at the bottom. Anodization in both investigated electrolytes leads to the formation of self-organized nanotubes, but the resulting structures are different. For the electrolytes based on EG, the tubes have very smooth walls and a smaller diameter (55–65 nm). For electrolytes based on EG, tube length is over 2 mm after 1 h of anodization, while for GL it is about 1.2 mm after 2 h. Experiments with various viscous electrolytes demonstrated that in the case of electrolyte containing GL, the tube length increased during the first 1 h and finally saturated at 1.1 mm. Moreover, in this case the sidewalls of the tubes are rough with bigger diameters of about 100–115 nm and an average distance of 30–50 nm between the nanotubes.

In between the SEM images in Fig. 5B, one can observe the time development of shape of water droplet after an initial 30 min of UV irradiation. The increase in the photocurrent with time after irradiation can be correlated with the progressing hydrophilicity of the TiO_2 nanotubes.

Wettability involves interactions between a liquid and a solid in contact, and it is expressed by the value of the measured contact angle (CA). Higher wettability results in a smaller CA. If the water droplet CA on a surface is smaller than 90°, i.e., when water has a tendency to spread over the surface, the surface is considered hydrophilic. By contrast, at higher CA one can consider a surface being hydrophobic. A considerable decrease in water CA can be beneficial, providing better access to the nanotubes for the electrolyte, thus improving the PEC performance. In the present work, water CAs for most samples were found to be in the range from 90° to 120°. As we can see in Fig. 5B, the CA of H_2O droplets gradually decreased after UV irradiation (c.1.0 mW/cm^2), and the nanotubes consequently changed to a superhydrophilic state. The correlation between the photocurrent and wettability can be clearly seen. The highest photocurrent occurred when the water CA decreased to 33° as shown in Ref. 138. The empirical relationship between CA and PEC efficiency was studied by S.A. Vanalakar et al.[139] and W. Yang.[140] The authors suggested the CA measurement as an efficient tool for the determination of the quality of photoelectrodes in PECs.

A typical photocurrent versus voltage response $I_{ph}(V)$ to white light as compared with dark current is given in Fig. 5C. The flat band potential, V_{FB}, was determined as the potential of the photocurrent onset. It was found that the photoelectrochemical behavior of TiO_2 is highly sensitive to the morphology and crystalline structure of the nanotubes. When using the electrolyte with GL, TiO_2 nanotubes are characterized by higher photocurrent in comparison with the electrolyte based on EG. The V_{FB} was negative and close to −0.60 V for samples grown in EG electrolyte and −0.50 V for those grown in GL electrolyte. It can be seen that for the nanotubes deposited using the EG electrolyte, high anodic photocurrent densities were achieved and the slope of the anodic photocurrent curve for these photoanodes was steep, which indicated fast electronic transfer through the photoanode−electrolyte interface. By contrast, the anodic photocurrents recorded for the nanotubes obtained in the GL electrolyte were relatively higher than in the case of the EG electrolyte, but they did not reach a saturation level.

Fig. 6 shows some representative results concerning nonstoichiometric TiO_y thin films deposited by dc-pulsed magnetron sputtering from a metallic Ti target in the reactive $Ar + O_2$ atmosphere as an example of a 2D nanostructure. The deposition rate was controlled by plasma emission spectroscopy, the optical method that allowed to keep a predefined departure from stoichiometric composition, 2-y in the synthesized films.[57,141,142] The parameter y was determined on the basis of RBS and Nuclear Reactions Analysis (NRA) investigations.[143]

In the case of thin films, the XRD patterns are usually measured in glancing incidence geometry (GID). The analysis of GID patterns shown in Fig. 6A revealed that the films were composed of two TiO_2 polymorphic forms: anatase and rutile, accompanied by an amorphous phase.[57,141,142] The relative intensities of A(101) anatase and R(110) rutile diffraction lines varied with y. For the highest y, when the target was oxidized, because of the low rate of deposition, the films showed the highest degree of crystallization. Under these conditions both anatase and rutile phases occurred, with anatase as the dominating phase. With decreasing y, the sputtering was performed from the target, with less oxidized surface, and a systematic evolution of the microstructure was observed, i.e., the rutile content increased, while the anatase phase gradually disappeared. Finally, films with the highest

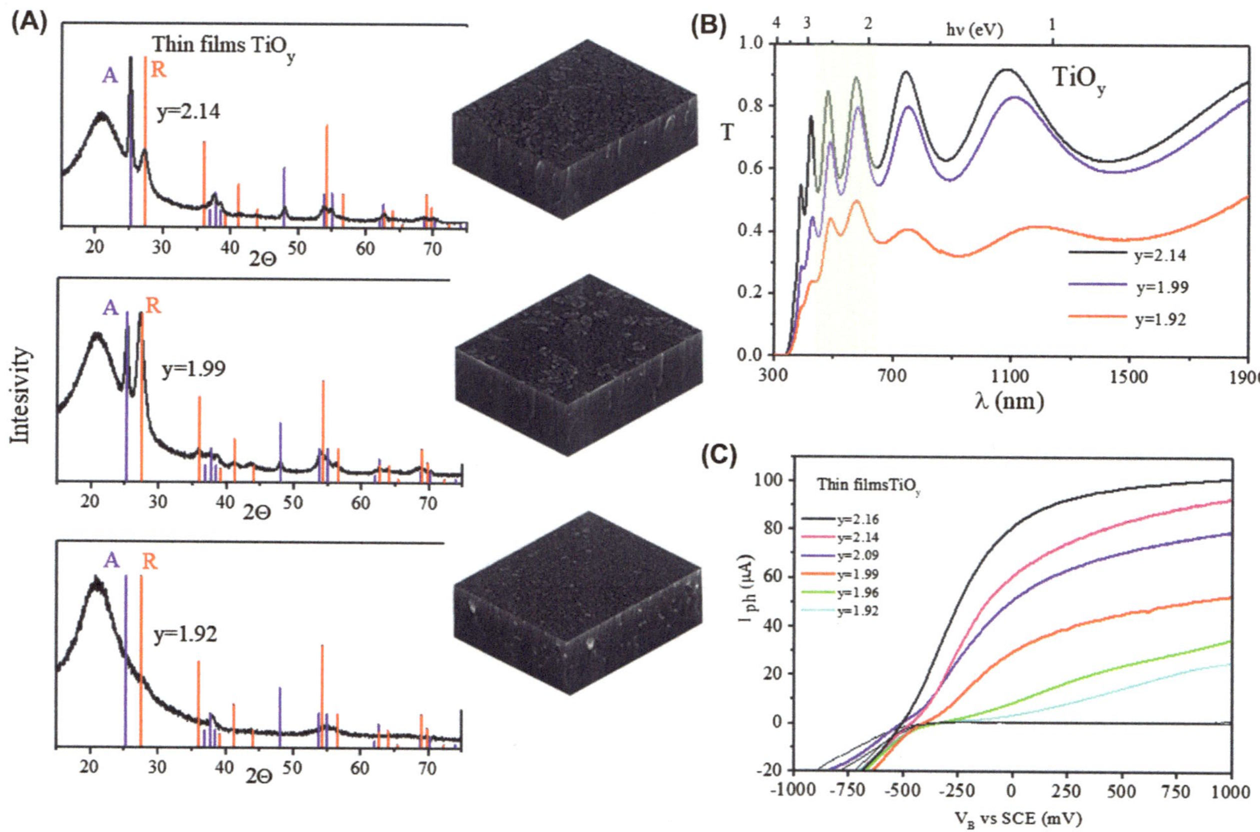

Figure 6 Representative results obtained by the authors of this review for 2D thin films of TiO_y with well-controlled departure from stoichiometry, 2-y, deposited by plasma emission–controlled magnetron sputtering. (A) XRD patterns and SEM images of thin TiO_y with different y; (B) spectral dependence of the transmission coefficient T; (C) photocurrent density as function of the applied voltage for TiO_y thin film photoanodes with different y. *Based on Brudnik, A.; Gorzkowska-Sobaś, A.; Pamuła, E.; Radecka, M.; Zakrzewska, K.* J. Power Sources ***2007***, 173 *(2), 774–780; Kollbek, K.; Sikora, M.; Kapusta, C.; Szlachetko, J.; Brudnik, A.; Kusior, E.; Zakrzewska, K.; Radecka, M.* Radiat. Phys. Chem. ***2013***, 93, *40–46; Zakrzewska, K.* Adv. Mater. Sci. Eng. ***2012***, 2012, *1–13.*

departure from stoichiometry were obtained at the highest sputtering rate from an almost metallic target. These films were amorphous with finely dispersed rutile grains. Insets in Fig. 6A present the SEM images (both top-surface and cross-section) of nonstoichiometric TiO_y thin films for which the pronounced changes in the film morphology with increasing departure from stoichiometry could be seen. In general, these films could be characterized by a columnar growth perpendicular to the substrate. Two kinds of nanosized features could be distinguished in SEM top-surface image of films with the biggest y: one of 10 nm for smaller features and the other of about 20–30 nm for larger protrusions. Some of the columns were wider while some of them were broken due to the interrupted growth. Columns were ended with caps, which could be assumed to form larger features in the top-surface image. On increasing departure from a stoichiometric composition, the columnar growth was less visible. This result is consistent with XRD analysis, which showed that this sample had lost its crystallinity and became, to a large extent, amorphous.

The optical properties have been found equally sensitive to changes in the oxidation state of the sputtered target and the resulting departure from stoichiometry as shown in Fig. 6B. The films with the lowest departure from stoichiometry 2-y, obtained by sputtering when the target was oxidized, were optically uniform and were characterized by quite high transmittance over the visible range of the light spectrum. The optical absorption edge of these films was relatively sharp and corresponded to the absorption edge of stoichiometric TiO_2. For films with smaller y, the average transmittance decreased over the visible range. An additional absorption band was formed over the visible range of the light spectrum at about 1–2 eV and could be accounted for by additional electronic states formed within the forbidden bandgap of TiO_y that should be traced back directly to the departure from the oxygen stoichiometry. It should be pointed out that the bandgap for TiO_2 rutile is of about 3.0 and 3.2 eV for anatase. Thus, the increase in the optical bandgap with the increasing y could be related to the growing anatase content in the films.

The changes in y and the microstructure of the films were found to be followed closely by the photoelectrochemical properties as shown in Fig. 6C. Dynamic photocurrent response of the photoanode based on TiO_y thin films was shown to be due to the photogeneration of charge carriers and the relaxation mechanism.[57]

Fig. 6C presents the I_{ph}–V curves for the TiO_y films with different y. It can be seen that for the films with the highest y, high anodic photocurrent densities were achieved and the slope of the anodic photocurrent curve for these films was steep, giving evidence for the fast electronic transfer through the photoanode–electrolyte interface. By contrast, the anodic photocurrents recorded for the films with small y were relatively low and did not reach a saturation level.

Fig. 7 illustrates some representative results concerning flowerlike TiO_2 obtained by a rapid chemical oxidation process at moderate temperature[48,51,52] as an example of a 3D nanostructure. The preparation procedure consisted of the following steps. First the titanium foils were cleaned in

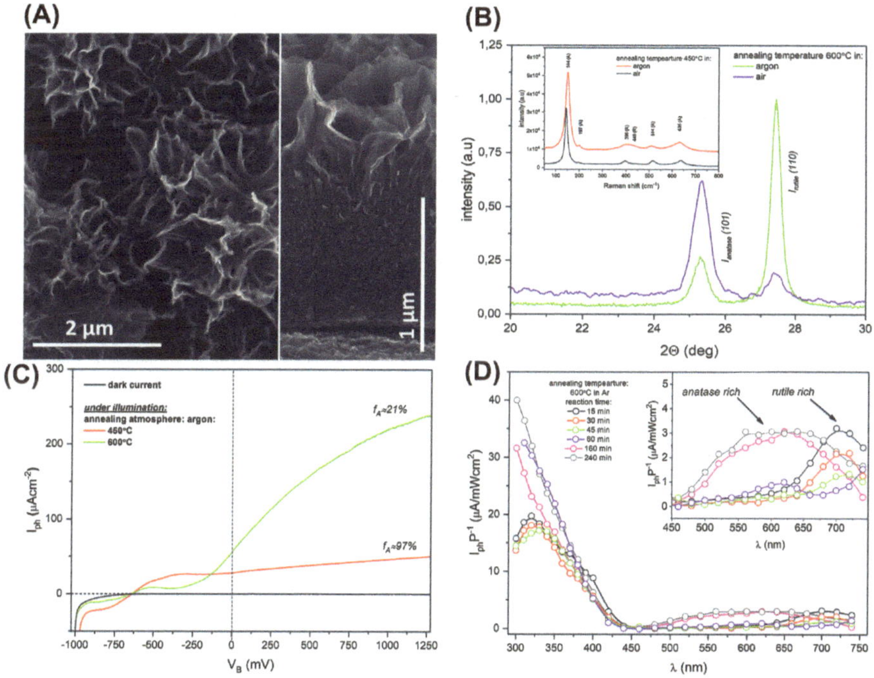

Figure 7 Representative results obtained by the authors of this review for 3D flowerlike nanostructures of TiO_2 obtained by chemical oxidation. (A) SEM top-view and cross-sectional images; (B) XRD patterns of the samples annealed in argon or air at 600°C—inset includes Raman spectra of 3D TiO_2 annealed at 450°C; (C) photocurrent density as function of the applied voltage for the same samples as in (B); (D) spectral dependence of normalized photocurrent for the samples annealed in Ar at 600°C for different reaction time intervals. *Based on Trenczek-Zajac, A.; Kusior, A.; Lacz, A.; Radecka, M.; Zakrzewska, K.* Mater. Res. Bull. ***2014***, 60, *28–37; Kusior, A.; Wnuk, A.; Trenczek-Zajac, A.; Zakrzewska, K.; Radecka, M.* Int. J. Hydrog. Energy ***2015***, 40 *(14), 4936–4944; Trenczek-Zajac, A.; Kusior, A.; Radecka, M.* Int. J. Hydrog. Energy ***2016***, 41 *(18), 7548–7562.*

acetone and ethanol. Then they were etched in 35% hydrochloric acid for 2 min, washed in distilled water, and dried in argon (100°C) for 30 min. After etching, the samples were oxidized in 30% hydrogen peroxide at 80°C for different periods (15–240 min). To improve their crystallinity, the obtained nanostructures were annealed at 450°C and 600°C in argon and air.[48,51,52]

Porous sample morphology shown in Fig. 7A indicates a large number of nanoflowers. The cross section in SEM analysis revealed that flowerlike structures consisted of three parts: particle layer, spongelike form, and the nanoflower itself.

Independently of the type of gas atmosphere (Ar or air), calcination at 450°C resulted in crystallization of nanoflowers as seen in Fig. 7B. The inset in the same figure includes Raman spectra of the samples annealed at 450°C in argon and air. A strong peak at around 144 cm^{-1} was observed, which could be assigned to the E_g mode of anatase (Ti–O stretching vibrations). The bands at 197, 396, 514, and 635 cm^{-1} were characteristic for anatase. The remaining band at 448 cm^{-1} could be attributed to the thermodynamically stable rutile phase. A temperature increase to 600°C affects the phase composition of the obtained nanostructured layers. XRD diffraction patterns of TiO_2 flowerlike nanostructures annealed at 600°C in argon and in air that are shown in Fig. 7B confirm that when the nanoflowers were annealed in argon, rutile was the predominant form of TiO_2. Annealing in air resulted in an increased anatase-to-rutile content.

Current density under white light as a function of the applied voltage is presented in Fig. 7C for nanostructures of TiO_2 annealed in Ar atmosphere at 450 and 600°C. The flat band potential (V_{FB}) derived from the onset of the anodic photocurrent was close to −650 mV.

It can be seen that for the photoanodes annealed at 600°C, high anodic photocurrent densities were achieved, but the plateau region could not be reached. By contrast, the anodic photocurrents recorded for flowerlike nanostructures annealed at 450°C were relatively low, but the saturation level was attained, which was associated with a fast electronic transfer through the photoanode–electrolyte interface. For TiO_2 nanostructures annealed for 3 h in Ar at 450°C, the weight fraction of anatase was approximately 97% and 21% at 600°C.

Anatase is believed to have better properties over rutile in terms of photocatalytic and photoelectrochemical applications. On the other hand, the synergistic effect of anatase–rutile is well known. According to the literature,[4] the optimum phase composition of anatase–rutile mixtures varies in different reports; however, high photoactivity is observed only for multiphase composites prepared in situ. Some authors[144] attribute the synergetic effect to the electronic band alignment of anatase and rutile. It was postulated that the coexistence of anatase and rutile phases enhanced charge carrier separation.

A typical spectral dependence of the normalized photocurrent density, $I_{ph}P^{-1}$ (where P is the power density of the light source), versus wavelength, λ, in the PEC with TiO_2 flowerlike nanostructures is presented in Fig. 7D. The photoresponse is very high over the UV range due to the fundamental absorption edge of TiO_2. An additional, broad peak in the photocurrent appears over the visible range of the light spectrum between 500 and 650 nm. The position of the maximum photocurrent in the visible range depends on the anatase—rutile content.

Fig. 8 presents a recapitulation of photoelectrochemical parameters obtained for photoanodes discussed in the context of Figs. 5—7. Based on the calculation of the maximum power, the fill factor (FF), maximum photocurrent I_{max}, and maximum potential V_{max} were assigned and marked in Fig. 8A. The values of I_{max} and V_{max} depend on the shape of the current versus voltage characteristics under illumination.

A comparison of the flat band potential, V_{FB}, photocurrent density, and FF values, for TiO_2 photoanode materials with different dimensionality, i.e., 1D nanotubes, 2D thin films, and 3D nanoflowers, is given in Fig. 8B—D, respectively. All the values of V_{FB} are highly negative, which is advantageous for water splitting. The photocurrent density can reach values as high as 80—90 mA/cm^2. The FF is related to the shape of the I(V) characteristics.

In all the cases a large scatter of data can be seen, and it is very difficult to find the decisive factors that determine these parameters. Undoubtedly, the microstructure details of the photoanode material are the most important, whereas in the case of thin films they depend on the oxygen-to-titanium ratio y.

5. SOLID—SOLID JUNCTION

Solid—solid junction can be composed by combining:

- a semiconductor SC with a metal Me to form a Schottky junction (SC/Me);
- two or more different semiconductors to form SC/SC heterojunction;
- semiconductors of the same chemical composition but of different polymorphic forms, i.e., a multiphase heterojunction;

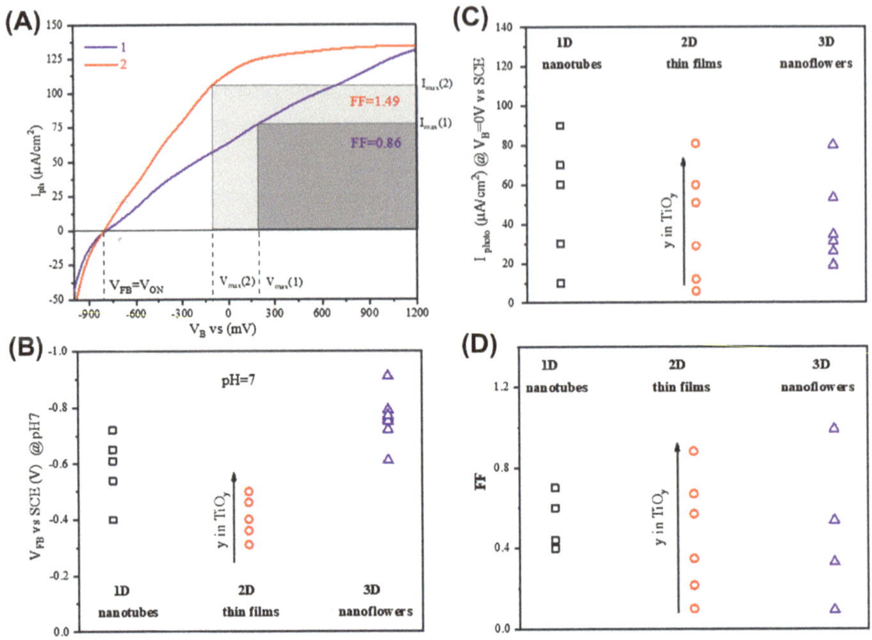

Figure 8 Comparison of parameters that have been derived from the basic photoelectrochemical cell characteristics, i.e., the photocurrent density I_{ph} versus applied voltage V_B for 1D, 2D, and 3D nanostructures of TiO_2 studied by the authors of this work. (A) Graphical explanation of how the fill factor (FF) and flat band potential V_{FB} are extracted from I_{ph} versus V_B; (B) typical flat band potential V_{FB} values; (C) typical photocurrent density I_{ph} values; (D) typical values of the FF. *Data from Trenczek-Zajac, A.; Kusior, A.; Lacz, A.; Radecka, M.; Zakrzewska, K.* Mater. Res. Bull. ***2014***, 60, *28–37; Kusior, A.; Wnuk, A.; Trenczek-Zajac, A.; Zakrzewska, K.; Radecka, M.* Int. J. Hydrog. Energy ***2015***, 40 *(14), 4936–4944; Trenczek-Zajac, A.; Kusior, A.; Radecka, M.* Int. J. Hydrog. Energy ***2016***, 41 *(18), 7548–7562; Brudnik, A.; Gorzkowska-Sobaś, A.; Pamuła, E.; Radecka, M.; Zakrzewska, K.* J. Power Sources ***2007***, 173 *(2), 774–780; Radecka, M.; Wnuk, A.; Trenczek-Zajac, A.; Schneider, K.; Zakrzewska, K.* Int. J. Hydrog. Energy ***2015***, 40 *(1), 841–851; Kollbek, K.; Sikora, M.; Kapusta, C.; Szlachetko, J.; Brudnik, A.; Kusior, E.; Zakrzewska, K.; Radecka, M.* Radiat. Phys. Chem. ***2013***, 93, *40–46; Zakrzewska, K.* Adv. Mater. Sci. Eng. ***2012***, 2012, *1–13.*

- semiconductors of the same chemical composition but of different type of conductivity (n-p), i.e., a homojunction.

Photoelectrodes based on solid–solid junctions can provide new functionalities:

- extend the light absorption spectral range;
- enhance charge separation and suppress charge recombination;
- protect the narrow-bandgap semiconductor from photocorrosion.

Fig. 9 demonstrates possible schemes of formation of solid–solid junctions as regarded from the point of view microstructure.

In the case of the metal oxide layer deposited onto a substrate, we can imagine either simple sensitization by covering the metal oxide with active species or first structuring of the metal oxide followed by its sensitization as shown in Fig. 9. In the case of simple sensitization (left-hand side of Fig. 9), one can consider either total (layer-by-layer deposition of $Ti/TiO_2/Ti$) or partial coverage (MoS_2 islands on continuous TiO_2 layer). Partial coverage may take a form of isolated particles, the example of which are CdS dots deposited onto a continuous TiO_2 layer. Structuring/sensitization (right-hand side of Fig. 9) is based on deposition of nanoparticles (Ag, CdS) or nanoislands (MoS_2) onto a previously nanostructured semiconductor, e.g., TiO_2 in a form of nanotubes or nanoflowers.

Formation of solid–solid junctions is believed to improve charge carrier separation leading to an increase in the recombination time. Therefore, mostly it is the charge transfer that is affected this way. However, one cannot exclude enhanced light harvesting, especially when structuring and sensitization is involved.

Fig. 10 presents the most important concepts in the application of solid–solid junction to water splitting from the point of view of the electronic structure. To consider a possible use of a semiconductor in PECs, the diagrams similar to that shown in Fig. 10A have to be consulted.[5,145,146] In Fig. 10A, the band edge positions E_{VB} and E_{CB} with respect to the redox potential levels E_{H_2/H^+} and E_{O_2/H_2O} are compared for the chosen metal oxide and sulfide semiconductors in a liquid electrolyte of $pH = 0$. Band edge positions can be corrected for any pH value by adding a pH-dependent term of $2.303RT/F \times pH$, but this does not affect the band alignment. As one can see, TiO_2 (anatase, rutile), CdS, graphene oxide (GO), and layered MoS_2 fulfill the thermodynamic criteria for simultaneous water splitting ($E_{H_2/H^+} < E_{CB}$ and $E_{O_2/H_2O} > E_{VB}$ at vacuum scale).

Formation of junctions between different materials has a strong influence on the alignment of band edges as illustrated in Fig. 10B. Taking into account the band diagram only, three types of heterostructures can be distinguished.[4] Type I corresponds to a situation under which the conduction band minimum of one semiconductor E_{CB1} is below E_{CB2} of the other one, and its valence band maximum E_{VB1} is above E_{VB2}. For type I heterostructures, both an electron and a hole photogenerated in the second semiconductor of a wider bandgap transfer to the first one, thus no spatial charge separation occurs. When both the conduction band minimum E_{CB1} and

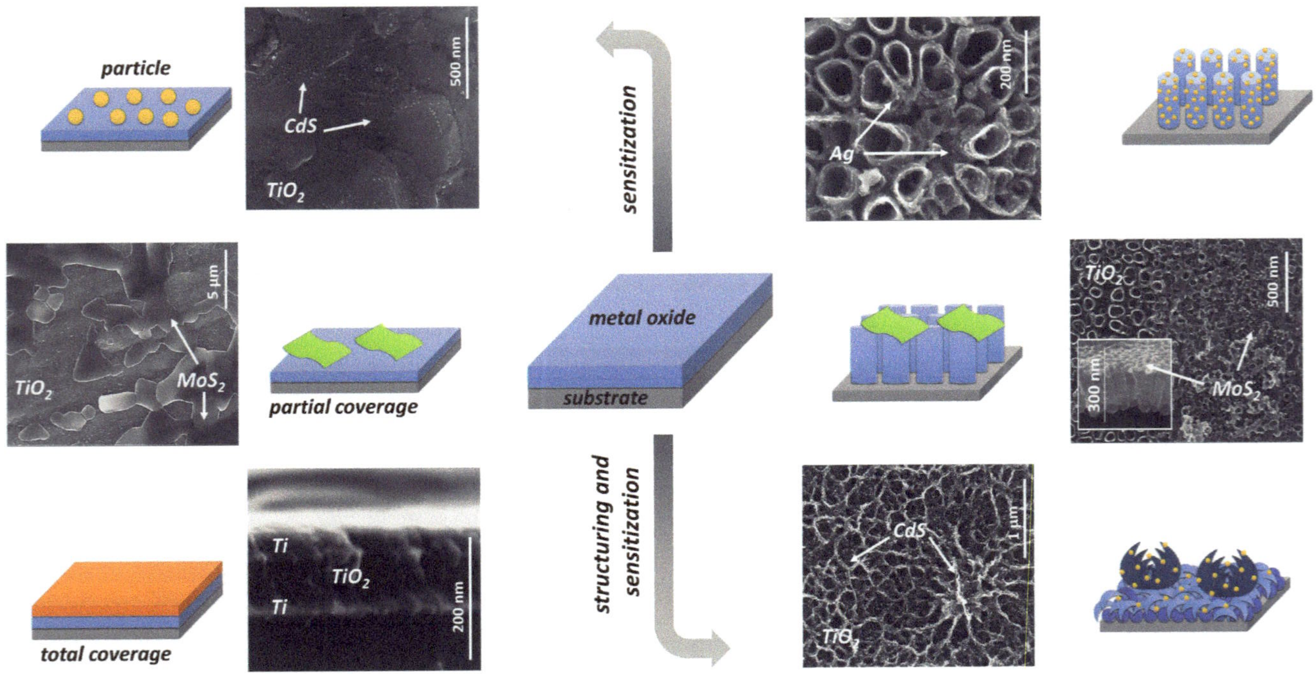

Figure 9 Microstructural modifications of solid–solid junction based on simple sensitization (left-hand side) of metal oxide and structuring of the metal oxide followed by sensitization (right-hand side). Sensitization involves total, partial, or particle coverage with different oxides, selenides, or metals.

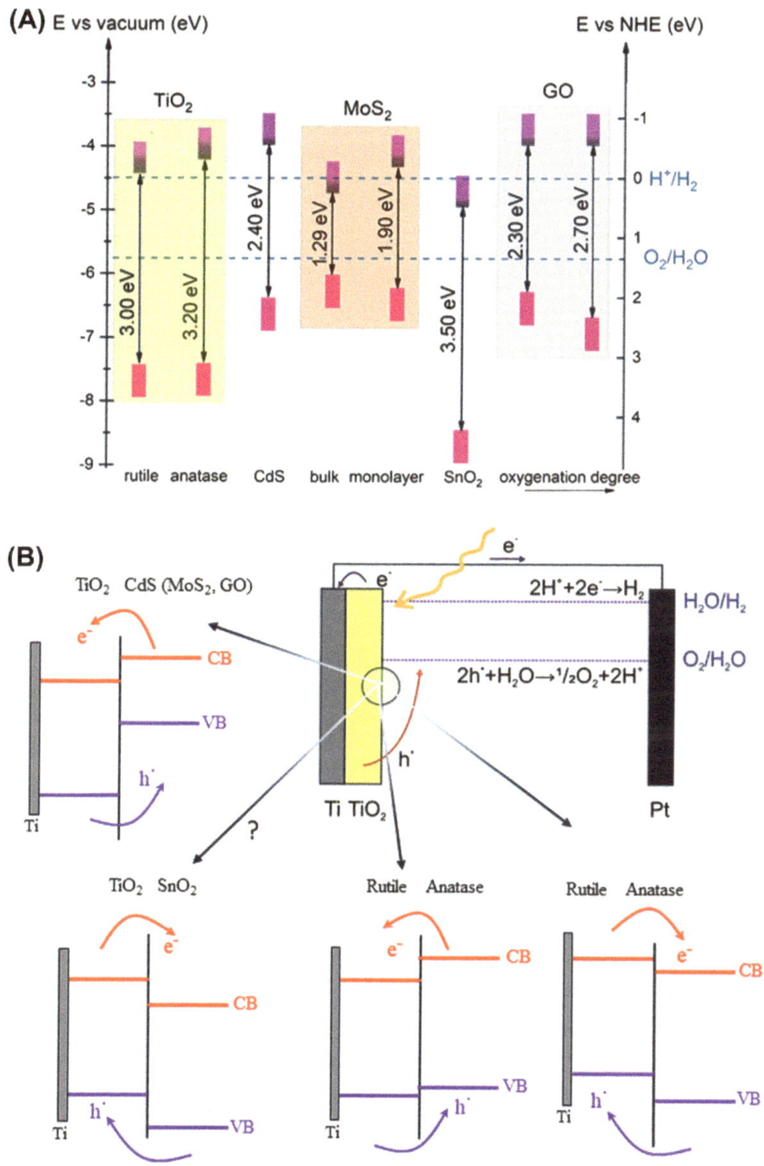

Figure 10 Electronic structure of chosen photoanode materials: TiO_2 (anatase, rutile), CdS, MoS_2, SnO_2, graphene oxide (GO) (A), and their solid–solid junctions as well as a detailed illustration of electron and hole transfer in the photoelectrochemical cell for water splitting on illumination of the both constituents (B); all band edges are given at 25°C and pH = 0 with respect to the absolute vacuum scale. *Data of Fig. 10A taken from Li, J.; Wu, N.* Catal. Sci. Technol. ***2015***, 5 *(3), 1360–1384; Baik, S.S.; Im, S.; Choi, H.J.* Sci. Rep. ***2017***, 7 *(November 2016), 1–8; Yeh, T.F.; Cihlář, J.; Chang, C.Y.; Cheng, C.; Teng, H. Mater.* Today ***2013***, 16 *(3), 78–84.*

valence band maximum E_{VB1} of one semiconductor are below those of the second one (type II), electrons and holes are separated in space by moving in different directions. Type III, i.e., the so-called broken-gap is similar to type II as far as charge transfer is concerned, but with much bigger energetic separation between band edges of the two semiconductors. Configuration of type II is the only one suitable for hydrogen evolution in the PEC because of the proper position of the corresponding band edges, providing the required direction of flow of charge carriers. In the photoelectrolysis, the reduction and oxidization reactions occur separately at different electrodes as shown in Fig. 10B. In a PEC with an n-type semiconductor photoanode, the holes oxidize water at the anode surface, while hydrogen is generated at the cathode. Photoelectrolysis involves the transport of holes toward the electrolyte and electrons via an external circuit from the anode to the cathode. Therefore, the sequence of layers in a solid—solid junction is important.

The example of a type II heterojunction is shown in Fig. 10B for TiO_2/CdS, which is known to be the most thoroughly studied multicomponent photocatalyst.[4,5,48,52] On visible light irradiation, electron—hole pairs are generated in CdS—low bandgap material of only E_g (CdS) = 2.4 eV. Therefore, the photogenerated holes are left where created, due to the higher VB position of CdS with respect to that of TiO_2. The electrons are transferred to TiO_2 and thus become spatially separated from CdS, reducing the recombination probability. In the case of the type II heterojunction between SnO_2 and TiO_2, both being wide-bandgap semiconductors of E_g (SnO_2) = 3.7 eV, E_g (TiO_2) = 3.0—3.2 eV, the charge carriers are photogenerated in both semiconductors on UV light irradiation. However, due to their relative band positions (see Fig. 10B), the charge carriers are separated and the electrons are transferred to the CB of SnO_2. Moreover, all photogenerated holes are accumulated in the VB on TiO_2. However, the only possible configuration in PECs is that when electrons from SnO_2 are transferred efficiently to an external circuit via the supporting conducting contact. Thus, a system Ti/TiO_2/SnO_2 shown in Fig. 10B may not work correctly. The same argument applies to anatase/rutile TiO_2 phase junctions, which can operate as photoanodes only when one assumes that CB of anatase is above that of rutile.[62—65] Then, the electrons photoexcited to the CB of the anatase drop to the CB of rutile and are correctly transferred to the Ti contact layer. An opposite configuration of band edges, i.e., when the CB of anatase is lower than that of rutile was reported by some research groups.[4,64,79,134,147]

Graphenelike layered structures, such as GO, reduced graphene oxide (RGO), the metal dichalcogenides MoS_2, WS_2, $MoSe_2$, WSe_2, black phosphorus, and hexagonal boron nitride, are currently one of the most sought-after class of materials. Sun et al.[148] argued that the electron transfer rate between TiO_2 and a graphene layer depends on the dimensionality of the semiconductor. It was suggested[5,148] that 2D–2D nanocomposites are characterized by much more efficient charge separation.

Although the advantages of the incorporation of RGO flakes in the TiO_2 matrix are well known, unfortunately, a detailed understanding of the role of graphene and its derivatives is still very controversial and scarce in the literature. Some studies[53,149,150] suggest that the RGO can facilitate electron injection and assist in electron transport, whereas other reports[54,149] claim that RGO can act as a sensitizer in semiconductor oxide films.

Looking at the band positions of single constituents (Fig. 10A), it can be concluded that TiO_2/MoS_2 represents a heterojunction of type I. As a consequence, the charge carriers photogenerated in TiO_2 should transfer to the narrow-bandgap semiconductor (MoS_2). In this case, neither an improvement in charge carrier separation nor the enhancement of photocatalytic activity could be expected. However, several research groups observed an improved photoactivity for TiO_2/MoS_2. This effect could be accounted for by assuming the transfer of the electrons photogenerated in TiO_2 to the MoS_2 layer that acted as a preferential path—a "highway" for the transport of electrons to the surface.

On the other hand, a type II heterojunction has been reported in the literature[151,152] for TiO_2/MoS_2. In fact, positions of the conduction bands E_{CB} in the TiO_2/MoS_2 heterostructure are close and a possibility of the synergetic effect[151] cannot be excluded.

It has been demonstrated[153] that a semiconducting RGO can act directly as a photocatalyst for H_2 generation. As it is well known, the CB edge of RGO is mainly formed by the antibonding $\pi^\star$ orbital, corresponding to a potential of -0.52 V versus NHE, $pH = 0$, while the VB maximum of RGO is mainly composed by the O 2p orbitals. Therefore, the bandgap of RGO could be easily engineered by upshifting its VB maximum as a result of reduction of a degree of oxidation. Therefore, it is not surprising that the semimetallic and semiconducting graphene could be fabricated to form various kinds of graphene-based heterojunctions suitable for photoelectrochemistry and photocatalysis.

6. NANOHETEROSTRUCTURES OF TIO_2 WITH SNO_2, CDS, MOS_2, AND GO—RESULTS OF AUTHORS' INDIVIDUAL RESEARCH

In this section we present results of research performed by our group on the growth, characterization, and photoelectrochemical performance of nanostructured TiO_2 sensitized with nanoparticles of CdS and SnO_2 as well as with islands of 2D-layered materials such as GO and MoS_2. The idea behind this presentation is to illustrate the role solid–solid junctions play in the PEC.

Nanostructured TiO_2 substrates were prepared as described previously (Section 4). Sensitization by nanoparticles was performed by a SILAR method, i.e., successive ionic layer adsorption and reaction, by immersion in a liquid.[154,155] The main mechanism of SILAR is a subsequent deposition of cations and anions from separate precursor solutions.[154]

Fig. 11 shows some representative results concerning 3D nanoflowers and 1D nanotubes of TiO_2 covered partially with CdS and SnO_2 nanoparticles.

SEM images of CdS nanoparticles deposited from 0.01 M water-based precursors on flowerlike TiO_2 nanostructures of 0.7 μm thickness are presented in Fig. 11A while those concerning SnO_2 synthesized in solution of $SnCl_4 \cdot 5H_2O$ and NaOH at room temperature, in Fig. 11B. Analysis of the images allowed us to conclude that synthesis carried out in water-based solutions led to evenly distributed 2–4 nm CdS nanoparticles.[48,52] Morphology of this type turned out to be appropriate for application of photoanodes in PECs as shown in Fig. 11C.

The photocurrent–voltage characteristics of flowerlike nanostructures of TiO_2 of different thickness 0.7 and 1.4 μm indicated negligible current measured at positive potentials V_B without illumination (dark current). However, illumination of the photoanode with white light resulted in a remarkable increase in the photocurrent. Thinner TiO_2 nanoflowers exhibited much better photocurrent response than thicker ones. This positive effect was also observed after sensitization with CdS. It could be accounted for by the fact that in the case of photoanodes based on films, the optimum thickness should be around $L_D + W$, where L_D represents the diffusion length and defines the region within which electron–hole pairs generation is fully effective and W is the depletion layer width. In our previous papers[50,51,59] we have shown that the width of depletion layer W reaches the values between 30 and 100 nm. The diffusion length L_D

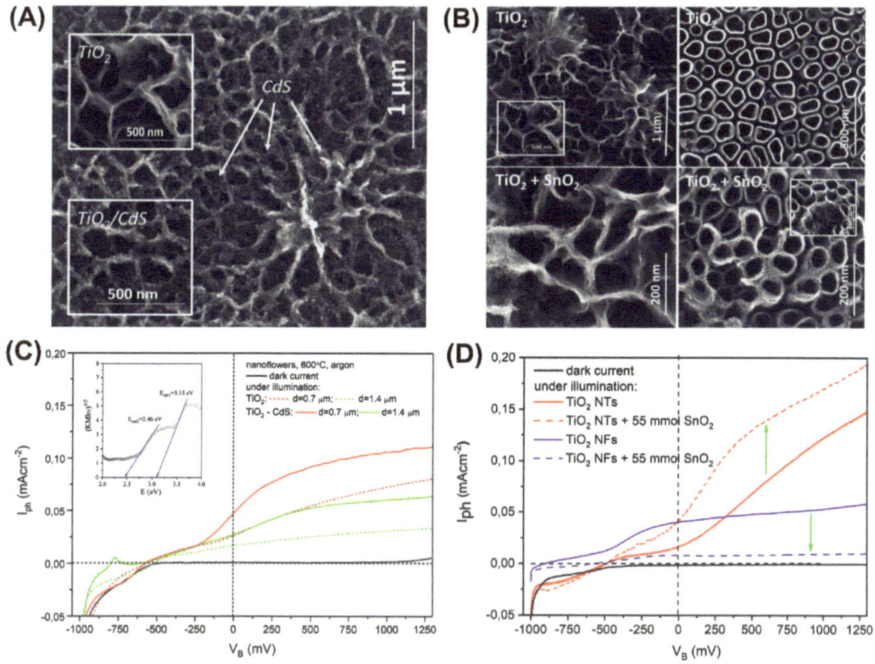

Figure 11 Representative results of the authors for nanostructured 3D and 1D materials sensitized with nanoparticles of CdS or SnO_2. (A) SEM images of TiO_2 nanoflowers with and without CdS; (B) SEM images of TiO_2 nanoflowers and nanowires with and without SnO_2; (C) photocurrent density I_{ph} versus bias voltage V_B for TiO_2 nanoflowers with and without CdS; (D) photocurrent density I_{ph} versus bias voltage V_B for TiO_2 nanoflowers and nanotubes with and without SnO_2; optical bandgap E_{opt} determined from Kubelka–Munk method is given in the inset of Fig. 11C. *Data of Figs. 11A and C taken from Trenczek-Zajac, A.; Kusior, A.; Lacz, A.; Radecka, M.; Zakrzewska, K.* Mater. Res. Bull. ***2014***, 60, *28–37; Trenczek-Zajac, A.; Kusior, A.; Radecka, M.* Int. J. Hydrog. Energy ***2016***, 41 *(18), 7548–7562; Figs. 11B and D contain unpublished data.*

equals 70 nm for anatase and 10 nm for rutile TiO_2.[5] For thicknesses above the critical value of 100–200 nm, the nanostructure will play a crucial role.

For TiO_2 nanoflowers modified with CdS nanoparticles, higher anodic photocurrent densities and more rectangular-like shape of the current–voltage curve under illumination were observed with respect to those without CdS. These features usually imply a fast electronic transfer through the photoanode–electrolyte interface. CdS has a longer charge carrier diffusion length in comparison with titanium dioxide.[5] In addition, CdS nanoparticles are isolated from each other and form heterojunction with TiO_2. Deposition of 0D CdS onto 3D TiO_2 can effectively reduce the charge recombination. The effect of cadmium sulfide deposition on the optical properties of TiO_2/CdS was analyzed from the spectral dependence of reflectance, and the optical bandgap energy (E_{opt}) was calculated from the Kubelka–Munk function $(KM\cdot h\nu)^2$ versus photon energy E as presented in the inset of Fig. 11C. The bandgap energy of TiO_2 is 3.32 eV while that of CdS is 2.35 eV.

The influence of the SnO_2 nanoparticles on the photoelectrochemical properties of the photoanode TiO_2 in the form of 1D (nanotubes) and 3D (nanoflowers) is shown in Fig. 11D. Depending on the dimensionality, 1D or 3D, of TiO_2 substrate, tin dioxide nanoparticles improve or reduce the photocurrent. This effect can be related to the morphology presented in Fig. 11B. SnO_2 nanoparticles deposit preferentially on the walls of TiO_2 nanotubes, thus creating a different type of heterostructure than that in the case of SnO_2 covering 3D nanoflowers of TiO_2.

GO and molybdenum sulfide can also be used as sensitizing agents in the case of nanostructured substrates. A novel method for MoS_2 electrodeposition that is environmentally friendly, as it does not require toxic reagents and solvents, has been proposed.[54,156] Titanium dioxide nanotubes 1D TiO_2 and 2D TiO_2 layers were used as scaffolds for supporting 2D materials in the form of GO and MoS_2 islands as shown in the SEM images in Fig. 12A and B, respectively. Partial coverage is clearly seen in both cases.

The results of impedance spectroscopic measurements are presented as Bode and Nyquist plots in Fig. 12C. The complex admittance plots in a form of circles are typically approximated by the simplest equivalent circuit composed of a resistor and a CPE (constant phase element) in series. However, in the case of GO deposited onto 1D TiO_2, a more complicated circuit had to be taken into account.[54] The impedance spectra of photoanodes are affected by deposition of MoS_2 and GO, which means that the electrical properties change.

Comparative results of the photoelectrochemical measurements performed in a PEC with the photoanode composed of MoS_2 and GO-modified TiO_2 are presented in Fig. 12D. Two parameters, the flat band potential and photocurrent normalized to that of unmodified TiO_2, $I_{modified}/I_{TiO_2}$, are considered. As can be seen, both 1D and 2D unmodified TiO_2 anodes were characterized by negative flat band potentials of similar values of 462 mV for 1D TiO_2, and 442 mV for 2D TiO_2.

The flat band potential was not affected by GO deposition, but MoS_2 had a positive impact on its value shifting it down to -518 mV. In the case of sensitization with either GO and MoS_2, the photocurrent of the modified electrodes was greater than that observed for the unmodified

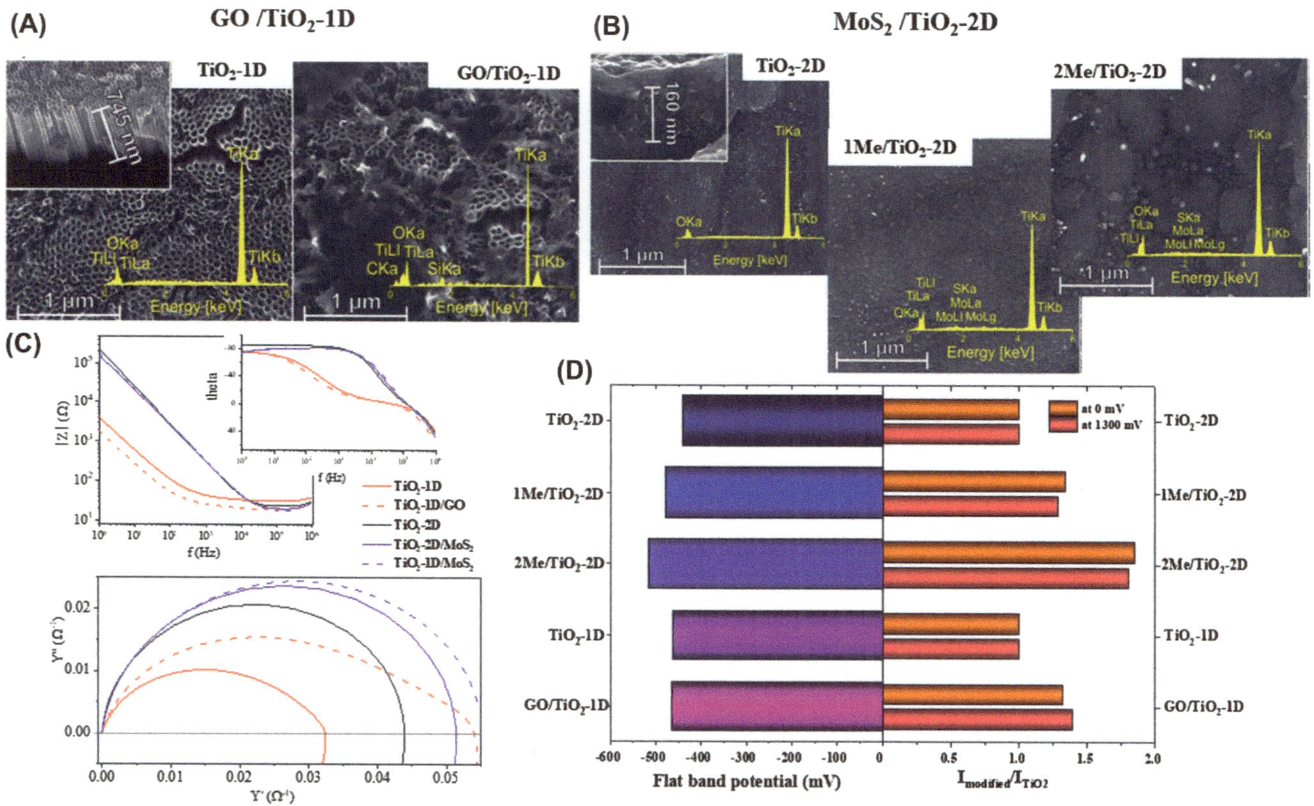

Figure 12 Representative results of the authors for nanostructured 1D and 2D materials sensitized with islands of graphene oxide (GO) and MoS_2. (A) SEM images of TiO_2 nanotubes with and without GO; (B) SEM images of TiO_2 2D layers with and without MoS_2; (C) impedance Bode and Nyquist plots of samples shown in Fig. 12A and B; (D) comparison of parameters (flat band potential and normalized photocurrent) of photoanodes based on materials shown in Fig. 12A and B. *Adopted from Trenczek-Zajac, A.; Banas, J.; Radecka, M.* RSC Adv. ***2016****, 6 (105), 102886–102898.*

ones. Faster transport of electrons was proposed.[54] Combined with the lowest value of the flat band potential, this resulted in the highest photocurrent response.

Moreover, TiO_2 with a second phase in the form of islands facilitates the penetration of the electrolyte into the electrode and enhances photoactivity.

7. CONCLUSIONS AND PERSPECTIVES

This review has addressed some fundamental issues concerning application of oxide nanomaterials as photoelectrodes in PECs for solar hydrogen generation. Processes and mechanisms, such as light harvesting, charge transfer, and engineering of solid–solid junctions, have been identified as most important for efficient conversion of solar-to-chemical energy. Representative results of the authors' own studies on the photoanode nanomaterials have been presented in the context of the leading studies performed by other scientific groups.

Fig. 13 gives a summary of all issues discussed in this chapter with an indication of the authors' own research in this field.

The main conclusion that can be drawn from this recapitulative diagram is the fundamental difficulty to extract and isolate only one contribution to the problem of efficiency of PECs. For example, certain light-harvesting schemes lead to a detrimental effect in terms of charge carrier transfer. Formation of solid–solid junctions, especially those based on nanostructuring

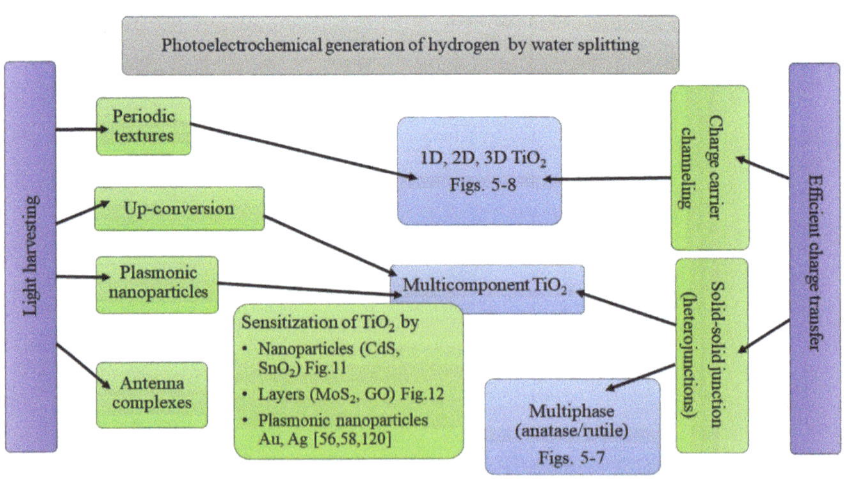

Figure 13 Schematic diagram summarizing the main issues discussed in this review.

and sensitization that help in separation of photoinduced charge carries, cannot be treated without taking into account the improved light harvesting. Careful coordination of all these effects based on their mutual enhancement is necessary to realize finally the dream of a Holy Grail in photochemistry—solar hydrogen produced from efficient (more than 10%) water splitting.

At this point it is appropriate to highlight some issues that seem to constitute a perspective for solar hydrogen sources. In our opinion they include the following:

- Bio-inspired artificial photoprocesses based on inorganic antenna complexes for extremely efficient charge transfer
- Materials for efficient photon up-conversion as a relatively new light-harvesting scheme
- Materials of novel architectures comprising multijunctions that could provide both efficient light harvesting and charge transfer

Research focusing on these materials and processes is already under way as indicated by the most recent publications.[85,131,157–162] Mimicking nature seems to be very attractive especially as far as the artificial photosynthesis is concerned. Biological systems not only demonstrate high ability to harvest light but also provide fast transport channels for the photogenerated charge carriers to the reaction site at which the process itself takes place. Intelligent behavior of the living cells, manifesting itself in a mechanism of protection against damage caused by too strong illumination, as well as adaptation to conditions of insufficient illumination, seems to be extremely difficult to reproduce in the artificial devices based on inorganic matter. However, despite these limitations, inorganic antenna complexes for efficient charge transfer have been presented[85] as the example of optically chiral double vertical fibers of TiO_2 shows. Enhanced light harvesting has been successfully demonstrated with a matrix of QDs tuned to different ranges of the light spectrum,[163] which is a realization of the excellent idea, encountered in living organisms, to cover the widest possible spectrum by using a variety of elements with a selective absorption features.

A relatively new light-harvesting scheme represents an active manipulation of light in the up-conversion process, particularly attractive in the case of application of wide-bandgap semiconductors such as TiO_2 as photoanodes. Near-infrared—driven processes of photoelectrochemical water splitting are advantageous as they allow to benefit from the best properties of wide-bandgap metal oxides with the simultaneous increase in the solar energy conversion efficiency as the infrared range constitutes 50% of the Sun spectrum. Two types of up-converting systems have been proposed[131,162]:

the lanthanide compounds Ln, e.g., YF_3: Yb^{3+}, Tm^{3+} in TiO_2 and carbon quantum dots (CQDs) loaded on $CdSe/TiO_2$ and $SrTiO_3$. However, because of numerous sources of losses accompanying up-conversion process, there is still a challenge to reach a reasonable level of efficiency of water splitting with the photoelectrodes modified this way that would allow them to compete with direct processes induced by UV or visible illumination in pristine metal oxide photoelectrodes.

Novel architectures comprising multijunctions represent another strategy that would ultimately not only improve light harvesting and facilitate charge carrier transport but also affect the spatial separation of photoelectron and holes. Hierarchical growth of heterostructures with quite promising performance as electrodes in PECs has been recently reported.[164] Branched nanoforest and nanoporous structures, as well as Janus and multicompartment "patchy" particles, are the best examples of this trend.

Taking all that into account, one may hope that in near future new progress toward achievement of the highly efficient generation of solar hydrogen will be announced.

ACKNOWLEDGMENTS

Origin research cited in this report was financed by numerous NCN grants as acknowledged in Refs. 48, 51, 52, 54, 138, and 142.
The work on the purposes of this review was performed within the framework of the project no. 2016/21/B/ST8/00457 of the National Science Centre, Poland.

REFERENCES

1. Fujishima, A.; Honda, K. *Nature* **1972,** *238,* 37–38.
2. Fujishima, A.; Honda, K.; Kikuchi, S. *J. Soc. Chem. Ind. Japan* **1969,** *72* (1), 108–113.
3. Hisatomi, T.; Kubota, J.; Domen, K. *Chem. Soc. Rev.* **2014,** *43* (22), 7520–7535.
4. Marschall, R. *Adv. Funct. Mater.* **2014,** *24* (17), 2421–2440.
5. Li, J.; Wu, N. *Catal. Sci. Technol.* **2015,** *5* (3), 1360–1384.
6. Goate; APP; O'Regan, B.; Grätzel, M. *Lett. to Nat.* **1991,** *353,* 737–740.
7. Grätzel, M. *Nature* **2001,** *414* (6861), 338–344.
8. Bard, A. J.; Fox, A. M. *Acc. Chem. Res.* **1995,** *28,* 141–145.
9. Burrows, C. J. *Acc. Chem. Res.* **2017,** *50* (3), 445.
10. Kamat, P. V. *Acc. Chem. Res.* **2017,** *50* (3), 527–531.
11. Nocera, D. G. *Acc. Chem. Res.* **2017,** *50* (3), 616–619.
12. Chen, L.; et al. *J. Phys. Chem. C* **2013,** *117* (42), 21635–21642.
13. Hodes, G.; Cahen, D.; Manassen, J.; David, M. *J. Electrochem. Soc.* **1980,** *127* (10), 2252–2254.
14. Sun, K.; Saadi, F. H.; Lichterman, M. F.; Hale, W. G.; Wang, H.; Zhou, X. *Proc. Natl. Acad. Sci. U. S. A.* **2015,** *112* (12), 3612–3617.
15. Rocheleau, R. E.; Miller, E. L.; Misra, A. **1998,** *624* (97), 3–10.
16. Rao, T. N. *J. Electrochem. Soc.* **1997,** *144* (1), 179–185.
17. Nakabayashi, S.; Kira, A. *J. Phys. Chem. Chem.* **1989,** *93,* 5543–5548.

18. Schoenmakers, G. H.; Vanmaekelbergh, D.; Kelly, J. J. *J. Phys. Chem.* **1996,** *100* (8), 3215–3220.
19. Maeda, Y.; Fujishima, A.; Honda, K. *Bull. Chem. Soc. Jpn.* **1982,** *55,* 3373–3376.
20. Sivula, K.; Le Formal, F.; Grätzel, M. *ChemSusChem* **2011,** *4* (4), 432–449.
21. Wang, J.; et al. *ACS Appl. Mater. Inter.* **2017,** *9* (1), 381–390.
22. Candea, R. *Electrochim. Acta* **1981,** *26* (12), 1803–1808.
23. Kalanur, S. S.; Yoo, I. H.; Seo, H. *Electrochim. Acta* **2017,** *254,* 348–357.
24. Hilliard, S.; et al. *Sustain. Energy Fuels* **2017,** *1* (1), 145–153.
25. Luo, J.; Steier, L.; Son, M. K.; Schreier, M.; Mayer, M. T.; Grätzel, M. *Nano Lett.* **2016,** *16* (3), 1848–1857.
26. Khan, K. A.; Leung, Y. K.; Kos, J. F. *Renew. Energy* **1997,** *11* (3), 293–298.
27. Takeuchi, M.; Weichman, F.; Morosawa, K.; Kawakami, M.; Nagasaka, H. *Appl. Surf. Sci.* **1988,** *33* (34), 972–979.
28. Fernando, C. A. N.; Wetthasinghe, S. K. I. *Sol. Energy Mater. Sol. Cells* **2000,** *63* (3), 299–308.
29. Shaban, Y. A.; Khan, S. U. M. *Int. J. Hydrogen Energy* **2008,** *33* (4), 1118–1126.
30. Frites, M.; Khan, S. U. M. V. *Electrochem. Commun.* **2009,** *11* (11), 2257–2260.
31. Yin, W. J.; Tang, H.; Wei, S. H.; Al-Jassim, M. M.; Turner, J.; Yan, Y. *Phys. Rev. B Condens. Matter Mater. Phys.* **2010,** *82* (4), 1–6.
32. Chen, H.-Z.; Zhang, Y.-Y.; Gong, X.; Xiang, H. *J. Phys. Chem. C* **2014,** *118* (5), 2333–2337.
33. Nguyen, O.; Krins, N.; Laberty-Robert, C. H. *Mater. Sci. Semicond. Process.* **2018,** *73* (February 2017), 2–12.
34. Katoh, R.; Murai, M.; Furube *Chem. Phys. Lett.* **2008,** *461* (4–6), 238–241.
35. Ozawa, K.; et al. *J. Phys. Chem. Lett.* **2014,** *2* (1), 1953–1957.
36. Sang, L.; Zhao, Y.; Burda, C. *Chem. Rev.* **2014,** *114* (19), 9283–9318.
37. Antonucci, V.; Giordano, N.; Bart, J. C. *Int. J. Hydrog. Energy* **1982,** 7 (10), 769–774.
38. Zhen, C.; Wang, L.; Liu, L.; Liu, G.; Lu, G. Q.; Cheng, H.-M. *Chem. Commun.* **2013,** *49* (55), 6191–6193.
39. Salvador, P. *Sol. Energy Mater.* **1982,** *6,* 241–250.
40. Chandiran, A. K.; et al. *J. Phys. Chem. C* **2010,** *114* (37), 15849–15856.
41. Wang, H.; Lewis, J. P. *J. Phys. Condens. Matter* **2006,** *18* (2), 421–434.
42. Khan, S. U. M.; Al-Shahry, M.; Ingler, W. B. *Science* **2002,** *297,* 2243–2245.
43. Tanyi, A. R.; et al. *Electrochim. Acta* **2015,** *178,* 240–248.
44. Chen, D.; Jiang, Z.; Geng, J.; Wang, Q.; Yang, D. *Ind. Eng. Chem. Res.* **2007,** *46* (9), 2741–2746.
45. Liu, Q.; Ding, D.; Ning, C.; Wang, X. *Int. J. Hydrogen Energy* **2015,** *40* (5), 2107–2114.
46. Brudnik, A.; Bućko, M.; Radecka, M.; Trenczek-Zajac, A.; Zakrzewska, K. *Vacuum* **2008,** *82* (10), 936–941.
47. Radecka, M.; Rekas, M.; Trenczek-Zajac, A.; Zakrzewska, K. *J. Power Sources* **2008,** *181* (1), 46–55.
48. Trenczek-Zajac, A.; Kusior, A.; Lacz, A.; Radecka, M.; Zakrzewska, K. *Mater. Res. Bull.* **2014,** *60,* 28–37.
49. Trenczek-Zajac, A.; et al. *J. Nanosci. Nanotechnol.* **2012,** *12* (6), 4703–4709.
50. Zakrzewska, K.; et al. *Int. J. Hydrogen Energy* **2015,** *40* (1), 815–824.
51. Kusior, A.; Wnuk, A.; Trenczek-Zajac, A.; Zakrzewska, K.; Radecka, M. *Int. J. Hydrogen Energy* **2015,** *40* (14), 4936–4944.
52. Trenczek-Zajac, A.; Kusior, A.; Radecka, M. *Int. J. Hydrogen Energy* **2016,** *41* (18), 7548–7562.
53. Kołacz, K.; Gajewska, M.; Komornicki, S.; Radecka, M. *Int. J. Hydrogen Energy* **2016,** *41* (18), 7538–7547.

54. Trenczek-Zajac, A.; Banas, J.; Radecka, M. *RSC Adv.* **2016,** *6* (105), 102886–102898.
55. Kollbek, K.; Sikora, M.; Kapusta, C.; Trenczek-Zajac, A.; Radecka, M.; Zakrzewska, K. *Open Chem.* **2015,** *13* (1), 857–868.
56. Gorzkowska-Sobas, A.; Kusior, E.; Radecka, M.; Zakrzewska, K. *Surf. Sci.* **2006,** *600* (18), 3964–3970.
57. Brudnik, A.; Gorzkowska-Sobaś, A.; Pamuła, E.; Radecka, M.; Zakrzewska, K. *J. Power Sources* **2007,** *173* (2), 774–780.
58. Zakrzewska, K.; Radecka, M.; Kruk, A.; Osuch, W. *Solid State Ionics* **2003,** *157* (1–4), 349–356.
59. Radecka, M.; Zakrzewska, K.; Wierzbicka, M.; Gorzkowska, A.; Komornicki, S. *Solid State Ionics* **2003,** *157,* 379–386.
60. Luttrell, T.; Halpegamage, S.; Tao, J.; Kramer, A.; Sutter, E.; Batzill, M. *Sci. Rep.* **2015,** *4,* 1–8.
61. Fox, M. A.; Dulay, M. T. *Chem. Rev.* **1997,** *93,* 341–357.
62. Sclafani, A.; Herrmann, J. M. C. *J. Phys. Chem.* **1996,** *100* (32), 13655–13661.
63. Bickley, R. I.; Gonzalez-Carreno, T.; Lees, J. S.; Palmisano, L.; Tilley, R. J. D. *J. Solid State Chem.* **1991,** *92* (1), 178–190.
64. Hurum, D. C.; Agrios, A. G.; Gray, K. A.; Rajh, T.; Thurnauer, M. C. *J. Phys. Chem. B* **2003,** *107* (19), 4545–4549.
65. Ma, Y.; Wang, X. L.; Jia, Y. S.; Chen, X. B.; Han, H. X.; Li, C. *Chem. Rev.* **2014,** *114* (19), 9987–10043.
66. Vratny, F.; Micale, F. *Trans. Faraday Soc.* **1963,** *59,* 2739–2749.
67. Maruska, H. P.; Ghosh, A. K. *Sol. Energy Mater.* **1979,** *1* (3–4), 237–247.
68. Hermann, J.-M.; Disider, J.; Pichat, P. *Chem. Phys. Lett.* **1984,** *108* (6), 618–622.
69. Claverie, J.; Verniolle, J.; Campet, G.; Doumers, J.-P.; Hagenmuller, P. *Mater. Lett.* **1981,** *16,* 1019–1025.
70. Maruska, H. P.; Gosh, A. *Sol. Energy Mater.* **1979,** *1,* 411–429.
71. Metikoš-Huković, M.; Ceraj-Cerić, M. *Mater. Res. Bull.* **1988,** *23* (11), 1535–1544.
72. Monnier, A. *J. Electrochem. Soc.* **1980,** *127* (7), 1576–1579.
73. Wang, Y.; Cheng, H.; Hao, Y.; Ma, J.; Li, W.; Cai, S. *Thin Solid Films* **1999,** *349* (1–2), 120–125.
74. Wilke, K.; Breuer, H. D. *J. Photochem. Photobiol. Chem.* **1999,** *121* (1), 49–53.
75. Zhao, G.; Kozuka, H.; Lin, H.; Yoko, T. *Thin Solid Films* **1999,** *339* (1–2), 123–128.
76. Fang, W. Q.; et al. *Chem. Eur. J.* **2014,** *20* (36), 11439–11444.
77. Kollbek, K.; et al. *Appl. Surf. Sci.* **2016,** *380,* 73–82.
78. Zhu, W.; et al. *Phys. Rev. Lett.* **2009,** *103* (22), 1–4.
79. Li, Y.; Zhang, J. Z. *Laser Photonics Rev.* **2010,** *4* (4), 517–528.
80. Eftekhari, A.; Babu, V. J.; Ramakrishna, S. *Int. J. Hydrogen Energy* **2017,** *42* (16), 11078–11109.
81. Concina, I.; Ibupoto, Z. H.; Vomiero, A. *Adv. Energy Mater.* **2017,** *1700706,* 1700706–1700735.
82. Kundu, S.; Patra, A. *Chem. Rev.* **2017,** *117* (2), 712–757.
83. Croce, R.; Van Amerongen, H. *Nat. Chem. Biol.* **2014,** *10* (7), 492–501.
84. Ostrumov, E.; Jumper, C.; Scholes, G. Photoelectrochemical Water Splitting: Materials, Processes and Architectures. In *RSC Energy & Environmental Series No. 9: Cambridge, UK;* Lewerenz, H.-J., Laurence, P., Eds.; 2017; pp 389–405.
85. Liu, S.; et al. *Nat. Commun.* **2012,** *3,* 1215–1216.
86. Zeng, Y.; Li, Y. Y.; Chen, J.; Yang, G.; Li, Y. Dendrimers: A Mimic Natural Light-Harvesting System. *Chem. Asian J.* **2010,** *5* (5), 992–1005.
87. Choi, M. S.; Aida, T.; Yamazaki, T.; Yamazaki, I. *Angew. Chem. Int. Ed.* **2001,** *40* (17), 3194–3198.

88. Ajayaghosh, A.; Praveen, V. K.; Vijayakumar, C.; George, S. J. *Angew. Chem. Int. Ed.* **2007,** *46* (33), 6260–6265.
89. Kumar, C. V.; Duff, M. R. *J. Am. Chem. Soc.* **2009,** *131* (44), 16024–16026.
90. Takeda, H.; et al. *Chem. Commun.* **2009,** *40,* 6032–6034.
91. Battaglia, C.; et al. *H. Nano* **2012,** *6* (3), 2790–2797.
92. Jiang, J.; Gu, F.; Shao, W.; Li, C. *Ind. Eng. Chem. Res.* **2012,** *51* (7), 2838–2845.
93. Yang, H.; Bin; Miao, J.; Hung, S. F.; Huo, F.; Chen, H. M.; Liu, B. *ACS Nano* **2014,** *8* (10), 10403–10413.
94. Zhu, G.; Pan, L.; Xu, T.; Sun, Z. *ACS Appl. Mater. Interfaces* **2011,** *3,* 3146–3151.
95. Nbelayim, P.; Kawamura, G.; Kian Tan, W.; Muto, H.; Matsuda, A. *Sci. Rep.* **2017,** 7 (1), 1–12.
96. Abd-Ellah, M.; Moghimi, N.; Zhang, L.; Thomas, J. P.; McGillivray, D.; Srivastava, S.; Leung, K. T. *Nanoscale* **2016,** *8* (3), 1658–1664.
97. Asim, N.; Ahmadi, S.; Alghoul, M. A.; Hammadi, F. Y.; Saeedfar, K.; Sopian, K. *Int. J. Photoenergy* **2014,** *2014,* 1–19.
98. Ji, L.; et al. *Nat. Nanotechnol.* **2015,** *10* (1), 84–90.
99. Huang, S.; et al. *ACS Appl. Mater. Interfaces* **2014,** *6,* 12111–12118.
100. Shi, Y.; Wang, X.; Liu, W.; Yang, T.; Yang, F. *J. Opt.* **2014,** *16* (7), 075706–075713.
101. Tang, C.; Cheng, N.; Pu, Z.; Xing, W.; Sun, X. *Angew. Chem. Int. Ed.* **2015,** *54* (32), 9351–9355.
102. Wei, Y.; Ke, L.; Kong, J.; Liu, H.; Jiao, Z.; Lu, X.; Du, H.; Sun, X. W. *Nanotechnology* **2012,** *23* (23), 235401–235409.
103. Wang, N.; Wang, D.; Li, M.; Shi, J.; Li, C. *Nanoscale* **2014,** *6* (4), 2061–2066.
104. Rao, P. M.; Cho, I. S.; Zheng, X. *Proc. Combust. Inst.* **2013,** *34* (2), 2187–2195.
105. Zhang, Z.; Hossain, M. F.; Takahashi, T. *Int. J. Hydrog. Energy* **2010,** *35* (16), 8528–8535.
106. Dang, H.; Singh, V. P.; Guduru, S.; Rajaputra, S.; Chen, Z. D. *Nano Res.* **2015,** *8* (10), 3186–3196.
107. Barreca, D.; Fornasiero, P.; Gasparotto, A.; Gombac, V.; Maccato, C.; Montini, T.; Tondello, E. *ChemSusChem* **2009,** *2* (3), 230–233.
108. Momeni, M. M.; Ghayeb, Y.; Davarzadeh, M. *J. Electroanal. Chem.* **2015,** *739,* 149–155.
109. Cai, J.; et al. *Inside Chem.* **2017,** *2* (6), 877–892.
110. Ram, S. K.; et al. *J. Phys. D Appl. Phys.* **2015,** *48* (36), 365101–365111.
111. Tilley, S. D.; Cornuz, M.; Sivula, K.; Grätzel, M. *Angew. Chem. Int. Ed.* **2010,** *49* (36), 6405–6408.
112. Li, J.; et al. *Energy Environ. Sci.* **2014,** 7 (11), 3651–3658.
113. Dotan, H.; et al. *Nat. Mater.* **2013,** *12* (2), 158–164.
114. Tsui, K.; Zhang, Y.; Yang, S.; Fan, Z. *Nano Lett.* **2014,** *14,* 2123–2129.
115. Dimova-Malinovska, D. *J. Phys. Conf. Ser.* **2016,** *682* (1), 012004–012014.
116. Wang, X.; Xiao, Y.; Zeng, D.; Xie, C. *CrystEngComm* **2015,** *17* (5), 1151–1158.
117. Kim, K.; Kim, M. J.; Kim, S. I.; Jang, J. H. *Sci. Rep.* **2013,** *3,* 1–8.
118. Zakrzewska, K. *Thin Solid Films* **2004,** *451–452,* 93–97.
119. Hirano, K.; Suzuki, E.; Ishikawa, A.; Moroi, T.; Shiroishi, H.; Kaneko, M. *J. Photochem. Photobiol. Chem.* **2000,** *136* (3), 157–161.
120. Yoon, J. W.; Sasaki, T.; Koshizaki, N. *Thin Solid Films* **2005,** *483* (1–2), 276–282.
121. Maeda, K. Photocatalysis, Fundamentals and Prospectives. In *RSC Energy & Environmental No. 14: Cambridge, UK;* Schneider, J., Bahnemann, D., Ye, J., Puma, G., Dionysiou, D., Eds.; 2016; pp 295–317.
122. Awazu, K.; et al. *J. Am. Chem. Soc.* **2008,** *130* (5), 1676–1680.
123. Cho, S.; et al. *Apl. Mater.* **2014,** *10703,* 1–5.

124. Trupke, T.; Green, M. A.; Würfel, P. Improving Solar Cell Efficiencies by down-Conversion of High-Energy Photons. *J. Appl. Phys.* **2002,** *92* (3), 1668–1674.
125. Stanley, C.; Mojiri, A.; Rosengarten, G. *Nanophotonics* **2016,** *5* (1), 161–179.
126. Abrams, Z. R.; Niv, A.; Zhang, X. *J. Appl. Phys.* **2011,** *109* (11), 1149051–1149059.
127. Chen, G.; Seo, J.; Yang, C.; Prasad, P. N. *Chem. Soc. Rev.* **2013,** *42* (21), 8304.
128. Dimova-Malinovska, D. *J. Phys. Conf. Ser.* **2016,** *682* (1), 012004–012013.
129. Akizuki, N.; Aota, S.; Mouri, S.; Matsuda, K.; Miyauchi, Y. *Nat. Commun.* **2015,** *6,* 1–6.
130. Asahi, S.; Teranishi, H.; Kusaki, K.; Kaizu, T.; Kita, T. *Nat. Commun.* **2017,** *8,* 1–9.
131. Nguyen, T.; et al. *J. Nanosci. Nanotechnol.* **2017,** *17* (October), 7647–7650.
132. Khnayzer, R. S.; Blumhoff, J.; Harrington, J. A.; Haefele, A.; Deng, F.; Castellano, F. N. *Chem. Commun.* **2012,** *48* (2), 209–211.
133. Wang, K.; Zhang, R.; Sun, N.; Li, X.; Wang, J.; Cao, Y.; Pei, R. *ACS Appl. Mater. Interfaces* **2016,** *8* (39), 25834–25839.
134. Li, J.; Wu, N. *Catal. Sci. Technol.* **2015,** *5* (3), 1360–1384.
135. Roy, P.; Berger, S.; Schmuki, P. *Angew. Chem. Int. Ed.* **2011,** *50* (13), 2904–2939.
136. Regonini, D.; Bowen, C. R.; Jaroenworaluck, A.; Stevens, R. *Mater. Sci. Eng. R Rep.* **2013,** *74* (12), 377–406.
137. Indira, K.; Mudali, U. K.; Nishimura, T.; Rajendran, N. *J. Bio- Tribo-Corrosion* **2015,** *1* (4), 1–28.
138. Radecka, M.; Wnuk, A.; Trenczek-Zajac, A.; Schneider, K.; Zakrzewska, K. *Int. J. Hydrog. Energy* **2015,** *40* (1), 841–851.
139. Vanalakar, S. A.; Mali, S. S.; Pawar, R. C.; Tarwal, N. L.; Moholkar, A. V.; Kim, J. H.; Patil, P. S. *J. Appl. Phys.* **2012,** *112* (4), 0443021–0443027.
140. Yang, W.; Peng, Q.; Chen, R.; Wen, Y.; Shan, B. *Phys. Procedia* **2013,** *48,* 220–227.
141. Zakrzewska, K.; Brudnik, A.; Radecka, M.; Posadowski, W. *Thin Solid Films* **1999,** *343–344,* 152–155.
142. Kollbek, K.; Sikora, M.; Kapusta, C.; Szlachetko, J.; Brudnik, A.; Kusior, E.; Zakrzewska, K.; Radecka, M. *Radiat. Phys. Chem.* **2013,** *93,* 40–46.
143. Zakrzewska, K. *Adv. Mater. Sci. Eng.* **2012,** *2012,* 1–13.
144. Scanlon, D. O.; et al. *Nat. Mater.* **2013,** *12* (9), 798–801.
145. Baik, S. S.; Im, S.; Choi, H. J. *Sci. Rep.* **November 2016,** *2017* (7), 1–8.
146. Yeh, T. F.; Cihlář, J.; Chang, C. Y.; Cheng, C.; Teng, H. *Mater. Today* **2013,** *16* (3), 78–84.
147. Buchalska, M.; Kobielusz, M.; Matuszek, A.; Pacia, M.; Wojtyła, S.; Macyk, W. *ACS Catal.* **2015,** *5* (12), 7424–7431.
148. Sun, J.; Zhang, H.; Guo, L.; Zhao, L. *ACS Appl. Mater. Interfaces* **2013,** *5,* 13035–13041.
149. Li, X.; Shen, R.; Ma, S.; Chen, X.; Xie, J. *Appl. Surf. Sci.* **2018,** *430,* 53–107.
150. Yeh, T.-F.; Syu, J.-M.; Cheng, C.; Chang, T.-H.; Teng, H. *Adv. Funct. Mater.* **2010,** *20* (14), 2255–2262.
151. Ren, X.; Qi, X.; Shen, Y.; Xiao, S.; Xu, G.; Zhang, Z.; Huang, Z.; Zhong, J. *J. Phys. D Appl. Phys.* **2016,** *49* (31), 315304–315313.
152. Nguyen, H.; et al. *J. Nanomater.* **2017,** *2017,* 1–6.
153. Xie, G.; Zhang, K.; Guo, B.; Liu, Q.; Fang, L.; Gong, J. R. *Adv. Mater.* **2013,** *25* (28), 3820–3839.
154. Pawar, S. M.; Pawar, B. S.; Kim, J. H.; Joo, O. S.; Lokhande, C. D. *Curr. Appl. Phys.* **2011,** *11* (2), 117–161.
155. Rühle, S.; Shalom, M.; Zaban, A. *ChemPhysChem* **2010,** *11* (11), 2290–2304.
156. Trenczek-Zajac, A.; Banas, J.; Radecka, M. *Int. J. Hydrog. Energy* **2018,** *43,* 6824–6837.

157. Li, X.; Yu, J.; Low, J.; Fang, Y.; Xiao, J.; Chen, X. *J. Mater. Chem. A* **2015,** *3* (6), 2485–2534.
158. Dutta, S. K.; Mehetor, S. K.; Pradhan, N. *J. Phys. Chem. Lett.* **2015,** *6* (6), 936–944.
159. Fang, M.; Dong, G.; Wei, R.; Ho, J. C. *Adv. Energy Mater.* **2017,** *1700559,* 1–25.
160. Lee, D.; Grigoropoulos, C. P. Hierarchical Nanostructures for Energy Devices. In *RSC Nanoscience & Nanotechnology No. 55: Cambridge, UK;* Nuzzo, R., Rocha, J., Liu, X., Eds.; 2015; pp 174–203.
161. Chen, K.; et al. *Nanoscale* **2018,** *10* (8), 4130–4137.
162. Fan, W.; Bai, H.; Shi, W. *CrystEngComm* **2014,** *16* (15), 3059–3067.
163. Mackowski, S. *J. Phys. Condens. Matter* **2010,** *22* (19), 1–17.
164. Cobo, S.; et al. *Nat. Mater.* **2012,** *11* (9), 802–807.

CHAPTER FIVE

Charge-Transporting Materials for Perovskite Solar Cells

Sadia Ameen*, M. Shaheer Akhtar§, Hyung-Shik Shin* and Mohammad Khaja Nazeeruddin¶,1

*Energy Materials & Surface Science Laboratory, Solar Energy Research Center, School of Chemical Engineering, Chonbuk National University, Jeonju, Republic of Korea
§New & Renewable Energy Material Development Center (NewREC), Chonbuk National University, Jeonbuk, Republic of Korea
¶Group for Molecular Engineering of Functional Materials (GMF), Institute of Chemical Science and Engineering, Faculty of Basic Science, Ecole Polytechnique Federale de Lausanne (EPFL), Sion, Switzerland
[1]Corresponding author: E-mail: mdkhaja.nazeeruddin@epfl.ch

Contents

Abstract

The power conversion efficiency of perovskite solar cells (PSCs) has been certified as ~22.1%, approaching the best single crystalline silicon solar cells. The improvement in the performance of PSCs could be achieved through the testing of novel materials in the device. This review briefly discusses the systematic introduction about several inorganic and organic electron-transporting materials (ETMs) and hole-transporting materials (HTMs) for efficient PSCs. The transport mechanism of electrons and holes in different ETMs/HTMs is also discussed on the basis of energy band diagrams with respect to the perovskite absorber. Moreover, the introduction of appropriate interfacial materials, hybrid ETMs, and doping is discussed to optimize the interfacial electronic properties between the perovskite layer and the charge-collecting electrode.

Advances in Inorganic Chemistry, Volume 72
ISSN 0898-8838
https://doi.org/10.1016/bs.adioch.2018.05.009

1. INTRODUCTION

Solar energy is abundant low-cost renewable energy, environmentally friendly, and contributes largely to the photovoltaic (PV) research. The solar cell as the photovoltaic device utilizes the solar energy and converts the sunlight into electricity through photoelectric effects. In general, the solar cells are classified into three generations, i.e., the first generation, which is wafer based; the second generation is based on thin films; and the third generation applies to organic photovoltaic devices. Till now, silicon solar cells are dominating the photovoltaic industries with the reported power conversion efficiency (PCE) of ~25%–26%. However, the emergence of the perovskite solar cell (PSC) as a new class of the third generation of solar cells is competent to silicon solar cells with the certified PCE of ~22.1%.[1–4] The photovoltaic performances of PSCs are mostly attributed to the intrinsic properties of organohalide perovskite such as a suitable and tunable band gap,[5] a high optical absorption coefficient,[6] high carrier mobility,[7] long carrier diffusion length,[8] and the capability for ambipolar carrier transport.[9] The PSCs are comprised of the absorber layer ($CH_3NH_3PbX_3$), which is sandwiched between the electron-transporting layer (ETL) and the hole-transporting layer (HTL). On photoexcitation, the perovskite absorber injects an electron and a hole into the n- and the p-type charge-transporting materials, respectively, generating free carriers. The electron travels through the mesoporous film and then through the external circuit to reach the cathode. In parallel, the oxidized perovskite is restored back to the ground state by the reduced component of the hole-transporting material (see Fig. 1).[10,11,18] Thus the hole extracted into the hole-transporting material diffuses toward the counter electrode and recombines with an electron, providing a current. The thickness of the perovskite pigment dictates the amount of current generated (J_{SC}), and the open circuit potential (V_{OC}) is determined by the difference between the Fermi energy level of titania and the highest occupied molecular orbital (HOMO) energy level of hole-transporting materials.[12] The PCE is a product of short circuit current, open circuit potential, and fill factor (FF) divided by the intensity of incident light as shown in the formula: $\eta = i_{ph} V_{oc} ff/I_s$.

Fig. 1 shows charge transfer (CT) processes when a light is absorbed by the perovskite. Electrons are separated (1) from the holes and injected (2) into the conduction band (CB) of the electron-transporting material before migrating to the anode. Meanwhile, the holes generated in the perovskite

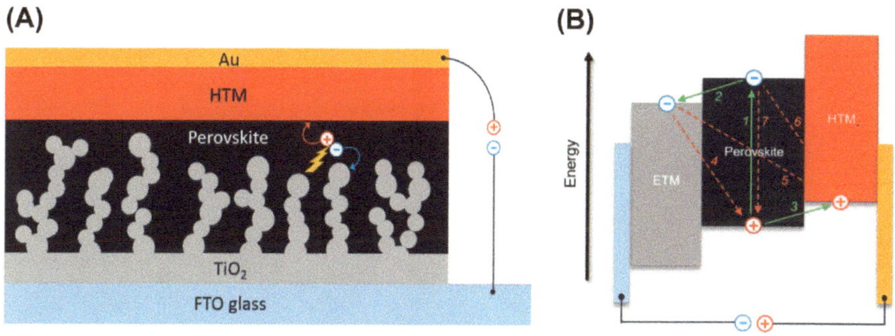

Figure 1 (A) Structure of typical mesoscopic n-i-p device configuration in a PSC. (B) Schematic illustration of energy levels and charge-transfer processes in PSCs. *ETM*, electron-transporting material; *HTM*, hole-transporting material; *PSC*, perovskite solar cell. *Reproduced from Burschka, J.; Pellet, N.; Moon, S. J.; Humphry-Baker, R.; Gao, P.; Nazeeruddin, M. K.; Grätzel, M.* Nature, ***2013***, 499, *316 with permission of Nature Communications.*

are transferred (3) to the HOMO level of the hole-transporting material, before injection of the hole into the cathode. Electron and hole injections occur quite efficient due to the high diffusion lengths of charge carriers. Undesirable CT processes such as back CT recombination of the charge carriers at three ETM/perovskite/HTM interfaces also occur (4–6), including nonradiative recombination (7). The kinetics of these processes is largely expected to control the overall photovoltaic conversion efficiency.

A perovskite absorber is based on a hybrid crystalline material ABX_3, where A is an organic cation (typically $R{-}NH_3^+$), B is a metal cation (typically Pb^{+2} or Sn^{+2}), and X is a halide anion, forming an octahedral in the unit cell that possesses a three-dimensional (3D) crystal structure (see Fig. 2A).[2] The hybrid perovskite materials have significant advantages for optoelectronic applications due to their combined electrical and optical properties in terms of high absorption coefficient,[13] ambipolar charge transport properties,[7] long carrier diffusion lengths,[8,14] and extremely low exciton binding energy.[15] In addition, the ability to tune the perovskite band gap by simple chemical substitution (i.e., by manipulation of the "A" cation and "X" halide substitution),[5,16,17] together with their facility to be solution-processable from inexpensive precursors and simple fabrication methods, makes them compete with the existing silicon or Cu(InGa)Se_2 (CIGS) photovoltaic technologies. A typical PSC exploits a sandwich n-i-p configuration,[10] where the perovskite is deposited on top of a mesoporous titanium dioxide (TiO_2) layer, which behaves as an ETM, and covered by an HTM, crucial to facilitate the extraction of holes from

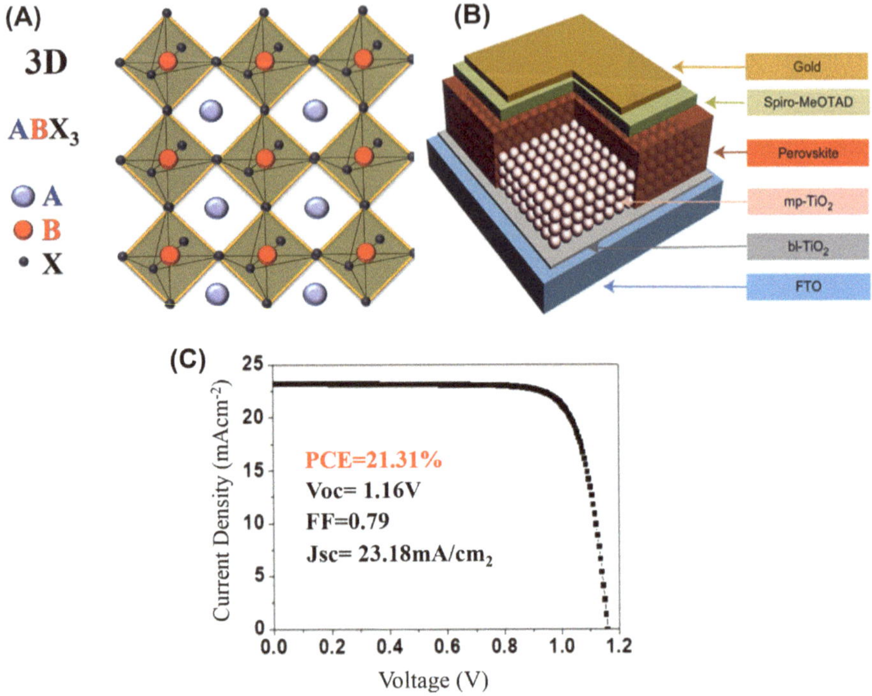

Figure 2 (A) Perovskite crystal structure of hybrid organic–inorganic metal trihalide ABX_3, where A is the organic cation, B is the metal Pb^{2+} cation, and X is the halide anion. (B) Schematic of perovskite solar cell structure and (C) Current density-voltage (IV) data of $(FAPbI_3)_{0.85}(MAPbBr_3)_{0.15}$ perovskite with PTAA as hole-transporting layer. *Reproduced from Ref. K. T. Cho, S. Paek, G. Grancini, C. Roldán-Carmona, P. Gao, Y. Lee, M. K. Nazeeruddin, Energy Environ. Sci.,* ***10****, 2017, 621 with permission of Nature Publications.*

perovskite to the back contact (see Fig. 2B). Other sandwiched structure has also been developed such as n-i-p planar[3] or inverted (p-i-n) architecture.[18] Both ETM and HTM and their interfaces with the perovskite are of paramount importance in governing charge collection and extraction to the contacts, and thus the overall device performances. By judiciously manipulating a perovskite absorber with nonstoichiometric precursors, several groups have obtained the PCE (over ~21%) of PSCs (Fig. 2C) and intense electroluminescence yield (0.5%) with a cell configuration of FTO/TiO/perovskite/Spiro-OMeTAD (or Spiro-MeOTAD)/Au (fluorine-doped tin oxide [FTO]) and an optimized composition $[(FAPbI_3)_{0.85}(MAPbBr_3)_{0.15}]$ in the precursor solutions.[19] Further replacing the methylammonium with formamidinium cation and cesium, one could realize a combination of various ratios of methylammonium, formamidinium, and cesium cations, which are expected to enhance short circuit current and stability.[20] It has become clear that the performance of the PSC very much depends on the control of morphology, crystallinity, and crystal size, which can be optimized by solvent engineering of deposition methods and compositional engineering of the precursor solutions.

PSCs have achieved unprecedented performance, but the most critical issues are related to the commercialization and the stability of PSCs. The commercial requirements could be compensated through a selective choice of the designing of device structures, modifications of ETMs, and the development of different HTMs. The selection and designing of ETMs/HTMs are critical because they prevent interfacial recombination, which directly affects the open circuit voltage (V_{OC}) and FF and the PCE of PSCs.[21] In this review, the fundamentals of selecting ETMs of different metal oxides, hybrid materials, and dopants are discussed. In addition, modification of the interface between the ETL and the perovskite layer as well as the interface between the ETL and the metal electrode through the incorporation of the interfacial layer is discussed to improve the charge separation, charge injection, charge collection, and charge transport properties. These parameters are interlinked to achieve the overall performance. Herein, the wide ranges of HTMs such as organic (small molecules (SM)) and inorganic HTMs are also elaborated to explain their role toward the charge extraction at the HTM–perovskite interface and also in determining the PCE and overall performance of PSCs.

2. ELECTRON TRANSPORT MATERIALS IN PSCs

The ETLs are important in the extraction and transportation of photogenerated electrons in PSCs. ETLs serve the main purpose of transporting the photo-generated electrons from the absorber to transparent conducting oxides (TCOs) and simultaneously blocks the holes. The unwanted recombination of electrons and holes usually happens during the charge transport process and wastes the charges, such as electrons ($\bar{e}$) in the ETL recombined with holes (h^+) in the perovskite layer, $\bar{e}$ in the perovskite layer recombined with h^+ in the HTL, and even $\bar{e}$ recombined with h^+ in HTLs. Thus the enhancement of CT to inhibit charge recombination is an effective way to improve the performance of PSCs. For this, the ETL plays a significant role, which is also called the blocking layer in PSCs. In addition, a satisfactory ETL contributes to lowering the dark current and series resistance and also promotes the crystallization of the perovskite layers.[22] Thus a careful selection of ETLs with high electron mobility is significant to achieve the high photovoltaic performance.

2.1 Inorganic Electron-Transporting Materials

To date, the majority of research into PSCs utilizes compact titanium oxide (TiO_2) and mesoporous titanium oxide (TiO_2) layer, which is directly deposited on TCO substrates. 2,2′,7,7′-Tetrakis(N,N-di-p-methoxyphenylamine)-9,9′-spirobifluorene (Spiro-OMeTAD) as an HTL deposited on top of the perovskite layer in an n-i-p device architecture.[23] The main reasons to use TiO_2 (anatase) are due to its favorable CB edge, −4.3 to −4.0 eV, in relation to several perovskites with a wide band gap of 3.2 eV as well as good hole-blocking ability due to its deep valence band (VB) position. The synthesis of TiO_2 with reduced oxygen vacancies and defects is critical to improving the charge extraction and collection in PSCs. T. Park et al. spin coated the planar TiO_2 with ionic salt copper iodide (CuI) as an ETL. Herein, the incorporation of p-type CuI on the TiO_2 surface modified the TiO_2 band alignment, which resulted to barrier-free contacts and enhanced V_{OC}. The polarity of the CuI-modified TiO_2 surface pulled electrons to the interface between the perovskite and the TiO_2 and thus contributed to electron extraction and reduced the recombination. The CuI-modified TiO_2 ETL exhibited a PCE of ~19.0% with high reproducibility and negligible J—V hysteresis.[24] In other report, a mesoporous TiO_2 ETL with a TiO_2-blocking layer was selected for the fabrication of the PSC of architecture Ag/PANI-NPs/$CH_3NH_3PbI_3$/mp-anatase-TiO_2/bl-TiO_2/FTO. The reasonable PCE of ~6.29% with a high short circuit current (J_{SC}) of ~17.97 mA/cm^2 and V_{OC} of ~0.877 V were achieved.[25] R. Jose reported a PSC module by employing solution-grown TiO_2 nanorods (NRs) as an ETL. The fabrication of an intermediate thin TiO_2 layer between FTO and NRs significantly contributed to higher PCE of ~10.5% and higher shelf-life durability.[26] M. K. Nazeeruddin et al. deposited aligned TiO_2 nanobundles (TiO_2 NBs) on titanium (Ti)-coated FTO substrates as an efficient ETL. The fabricated PSC with the TiO_2 NB ETL with a newly designed HTM, 4-(3,5-bis(trifluoromethyl) phenyl)-7-(5′-hexyl-[2,2′-bithiophen]-5-yl)benzo[c][1,2,5]thiadiazole(CF-BTz-ThR, Structure-5, HTM-43) achieved a PCE of ~15.4% with a J_{SC} of ~22.42 mA/cm^2 and a V_{OC} of ~1.02 V. Herewith, the surface view (Fig. 3A,B) displays the visible NBs with an average diameter of ~50 nm, and the porous nature of the NBs is depicted in the inset of Fig. 3B.[27] P. Qin et al. reported the high efficiency of solid-state PSCs based on highly

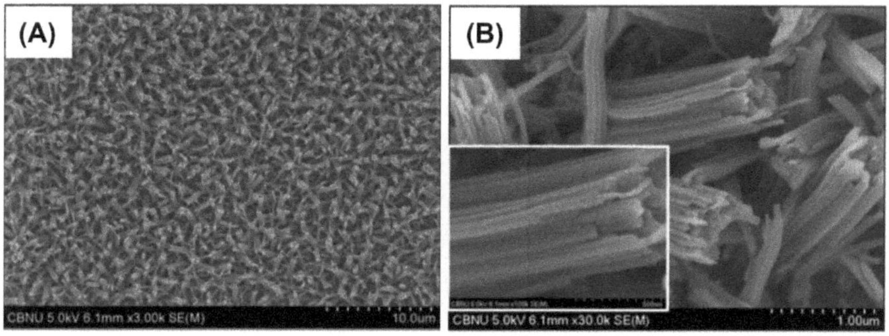

Figure 3 (A) Low and high magnification (B) modes of aligned TiO_2 NBs/FTO thin film. Inset shows selected high-magnification image. *Reproduced from Ameen, S.; Nazim, M.; Akhtar, M. S.; Nazeeruddin, M. K.; Shin, H. S.* Nanoscale ***2017****, 9, 17544–17550 with permission of Royal Society of Chemistry.*

transparent anatase TiO_2 nanotube array as an ETL. The submicrometer TiO_2 nanotube arrays infiltrated with $CH_3NH_3PbI_3$ and Spiro-MeOTAD as an HTM resulted in an overall PCE of ~14.8% under AM 1.5G illumination with a striking J_{SC} of ~22.6 mA/cm^2, a V_{OC} of ~1007 mV, and an FF of ~0.64. For comparison, the nanoparticle (NP)-based device was also fabricated under the same conditions that displayed low J_{SC} and V_{OC}, and a relatively lower PCE of ~12.8%.[28] Novel 3D TiO_2 nanowire (NW) architectures were synthesized via surface reaction–limited pulsed chemical vapor deposition (SPCVD) technique by Y. Yu et al. for the fabrication of PSCs. Compared to other 1D nanostructure arrays, the synthesized 3D TiO_2 NW architecture loaded a large amount of perovskite loading, which contributed to high light-harvesting efficiency and promising electron-transport property. The circuit current increased, while V_{OC} and FF decreased with the TiO_2 length. The PCE of ~9.0% was achieved with ~600-nm long 3D TiO_2 NW structures.[29] Two-dimensional (2D) TiO_2 nanosheet array films were perpendicularly grown on FTO glass substrates through a simple hydrothermal method and sensitized with CdS quantum dots (QDs). The photoelectrode of CdS-sensitized TiO_2 nanosheet arrays exhibited a photocurrent intensity of ~3.24 mA/cm^2 at a potential of 0 V versus Ag/AgCl, which was much higher than bare TiO_2 nanosheet arrays (~0.48 mA/cm^2).[30] Using anatase TiO_2 film with 3D network nanoporous structure, H. Tao et al. fabricated the PSC with the optimized PCE of ~9.19%. For PSC fabrication, a thin Ti film was deposited on FTO substrate by magnetron sputtering, and then a layer of network nanoporous TiO_2 nanotubes film was formed.[31] X. He et al. synthesized anatase TiO_2

single-crystalline NRs with a length of ~30 ± 10 nm by a one-pot solvothermal approach. A pinhole-free compact layer of TiO_2 NRs was spin coated on TCOs for the fabrication of PSCs that exhibited the PCE of ~17.58%.[32]

Although TiO_2 is a preferred ETL for PSCs, it requires high-temperature sintering that increases the cost and limits the applications for flexible plastic substrate.[33–36] On the other hand, zinc oxide (ZnO) metal oxide could be a promising alternative to TiO_2 HTMs. The electron mobility of ZnO nanostructures is relatively higher than that of TiO_2, and ZnO could be processed at the room temperature.[37–39] For example, X. Li et al. introduced an ionic liquid of 1-ethyl-3-methylimidazolium hexafluorophosphate ([EMIM]PF_6) as an interlayer between the ZnO and perovskite layers. The PSC achieved a small J–V hysteresis with a PCE of ~13.50%.[40] Liu et al. fabricated efficient flexible PSCs by using solution-processed ZnO thin film at room temperature as an ETL.[41] In continuation, Ameen et al. spin coated ZnO QDs over the surface of graphene-deposited ITO-PET thin-film flexible substrates, and the surface modifications were studied through AFM analysis, as shown in Fig. 4. The changes in the contact angle measurements were accredited to the application of APJeT plasma, which introduced the polar group moieties.[42] Yang et al. reported efficient solution-processed ZnO-based PSCs with a PCE of up to ~16.1%.[43] Song et al. boosted the PCE of ZnO-based PSCs to ~18.9% by introducing a triple-cation perovskite absorber.[44] S. Yeon utilized the self-assembled monolayer (SAM) of ZnO ETLs with sequential deposited perovskite active layers. The insertion of SAMs improved the quality of PbI_2 layers and the perovskite layers. Leveraged by SAM modification, ZnO-based PSCs achieved an unprecedentedly high PCE of ~18.82% with a V_{OC} of ~1.13 V, a J_{SC} of ~21.72 mA/cm^2, and an FF of ~0.76.[45] B. Song et al. adopted the sol-gel (SG) and hydrolysis–condensation (HC) techniques to study their effects on the performance of PSCs. The HC-ZnO films displayed higher conductivity and achieved a higher device performance, with a PCE of ~12.9%, whereas SG-ZnO–based PSCs displayed a low PCE of ~10.9% PCE.[46] A PSC based on ZnO NRs was prepared, and its photovoltaic performance was investigated by N. G. Park et al. $CH_3NH_3PbI_3$-infiltrated ZnO NRs showed a PCE of ~11.13% with a J_{SC} of ~20.08 mA/cm^2, a V_{OC} of ~991 mV, and an FF of ~0.56. The external quantum efficiency (EQE) of ~80% in almost the entire wavelength range from ~400 to ~750 nm was observed, and the integrated current density of ~20.03 mA/cm^2 was calculated from EQE data, which was in good agreement with the observed

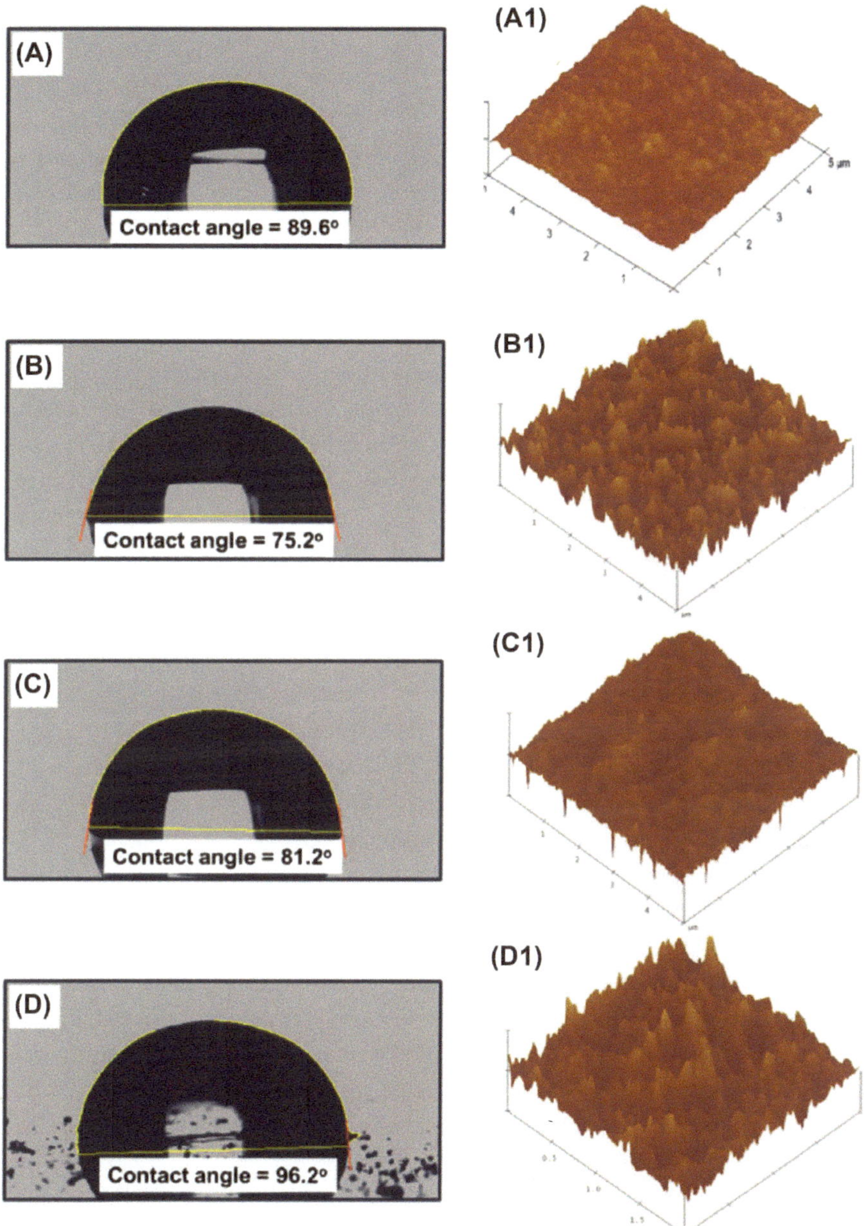

Figure 4 Contact angles and the corresponding 3D AFM images of (a, a1) ITO-PET, (b, b1) O_2 plasma-treated ITO-PET, (c, c1) ITO-PET/Gr, and (d, d1) ITO-PET/Gr/ZnO-QDs(AP-JeT) thin films. *Reproduced from Ameen, S.; Akhtar, M. S.; Seo, H. K.; Nazeeruddin, M. K.; Shin, H. S.* J. Phys. Chem. C ***2015**,* 119, *10379 with permission of American Chemical Society.*

J_{SC}.[47] P. Ruankham et al. reported full coverage of the perovskite layer onto ZnO NR substrates with less pinholes that generated a longer charge carrier lifetime, which resulted to a PCE of ~3.05%.[48] Liu et al. designed all-inorganic PSCs with novel structure (FTO/NiO_x/$CsPbI_2Br$/ZnO@C_{60}/Ag) in which the ZnO@C_{60} bilayer as an ETL demonstrated high carrier extraction efficiency and low leakage loss. The fabricated all-inorganic $CsPbI_2Br$ PSC yielded a PCE as high as ~13.3% with a V_{OC} of ~1.14 V, a J_{SC} of ~15.2 mA/cm^2, and an FF of ~0.77. Importantly, the obtained all-inorganic PSCs without encapsulation exhibited only ~20% PCE loss with thermal treatment at 85°C for 360 h, which largely outperformed the organic species–containing PSCs. Fig. 5A displays the cross-sectional scanning electron microscopy (SEM) images of the ZnO@C_{60} bilayer–based PSC, exhibiting that all of the thin films are well layered. From the EDS line scanning of the cross section (Fig. 5B), the morphology and elemental distribution of the C_{60} layer and ZnO layer are confirmed, where the ZnO@C_{60} bilayer ETL displays a uniform layer-by-layer morphology. The schematic view of electron extraction based on a single ZnO NP layer, single C60 layer, and ZnO@C_{60} bilayer is shown in Fig. 5C. Fig. 5D shows the time-resolved photoluminescence (TRPL) spectra of the $CsPbI_2Br$ perovskite thin films based on different ETLs. The average decay time (τ_{ave}) of a glass/perovskite sample (~1.45 ns) is longer than that of glass/perovskite/ETL samples, which is associated with the charge extraction process by ETLs. Likewise, the corresponding value of τ_{ave} of ZnO@C_{60}-, ZnO-, and C_{60}-based devices are ~0.73, ~1.43, and ~1.04 ns, respectively, which represents the ability of carrier extraction for these three types of ETLs. Clearly, photoluminescence (PL) quenching in the C_{60}- and ZnO@C_{60}-based devices are faster than the single ZnO–based device. Moreover, the ZnO@C_{60} bilayer exhibited even higher electron extraction efficiency than the single C_{60} layer. Aside from the reduced trap-state density, it was likely attributed to the good gradient energy band matching between the perovskite layer and ZnO@C_{60} bilayer, which favored the photogenerated electron to be extracted from the perovskite layer to the metal electrode effectively[49a,b]

Tin oxide (SnO_2) is another emerging ETL for PSCs. SnO_2 displays wider band gap,[49c] higher electron mobility (two orders of magnitude higher than TiO_2),[50] and efficient electron injection rate from perovskite into SnO_2. Hagfeldt et al. applied atomic layers deposited SnO_2 ETL which yielded almost hysteresis-free PCEs of over ~18%.[51] Murugadoss et al. deposited the SnO_2 layer using spin-coated $SnCl_2$ solution. A planar solar

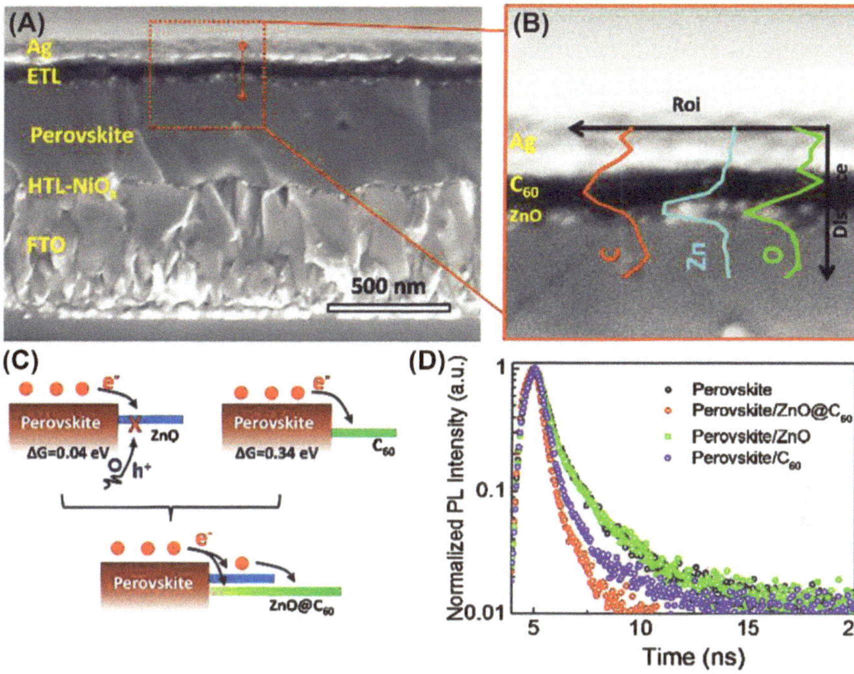

Figure 5 (A) Cross-sectional SEM image of the ZnO@C60 bilayer–based PSC. (B) The corresponding EDS line scanning of the selected area marked within a *red rectangle*. (C) Schematic view of different ETL with band gap alignment. (D) TRPL spectra of CsPbI2Br thin films based on different ETLs. *ETL*, electron-transporting layer; *HTL*, hole-transporting layer; *PSC*, perovskite solar cell; *SEM*, scanning electron microscopy; *TRPL*, time-resolved photoluminescence. *Reproduced from Do Sung, S.; Kang, M. S.; Choi, I. T.; Kim, H. M.; Kim, H.; Hong, M.; Kim, H. K.; Lee, W. I.* Chem. Commun. ***2014***, 50, *14161; Zhu, J. Y.; Niu, K.; Li, M.; Lin, M. Q.; Li, J. H., Wang, Z. K.;* Org. Elec. ***2018***, 54, 9; *Cui, E. J.; Meng, F.; Zhang, H.; Cao, K.; Yuan, H.; Cheng, Y.; Huang, F.; Wang, M.* ACS Appl. Mater. Interfaces ***2014***, 6, *22862 with permission of American Chemical Society.*

cell fabricated with perovskite-infiltrated SnO_2 exhibited a PCE of ~8.38% with a J_{SC} of ~18.99 mA/cm^2, a V_{OC} of ~0.96 mV, and an FF of ~45%.[52] T. Ma et al. applied the SnO_2 compact ETL for PSCs and compared their performance with PSCs based on a TiO_2 thin layer. SnO_2 ETL–based PSCs showed a higher J_{SC} of ~30.99 mA/cm^2 and lower V_{OC} (~0.86 V), an FF of ~0.39, and a PCE of ~10.30% that was higher than the conventional TiO_2 ETL–based PSCs.[53] Zhang et al. reported the planar PSC based on spin-coated SnO_2 solution that achieved almost hysteresis-free PCEs of ~19.9%.[54] An effective in situ template self-etching strategy was adopted by Gao et al. to grow a novel hollow domed SnO_2

nanotubes array with ZnO NRs as sacrifice templates.[55] Using such SnO_2 nanotube as an ETL for PSCs demonstrated a high photocurrent density of ~15.9 mA/cm^2 at a based-voltage of ~760 mV that exhibited a stable PCE of ~12.1% under the simulated AM 1.5G one sun illumination.[55] In the attempt to decrease fabrication cost of PSCs, several other metal oxides such as tungsten trioxide (WO_3), $ZnSnO_4$, strontium titanate ($SrTiO_3$), and aluminum oxide (Al_2O_3) are also explored as ETLs. Gheno et al. claimed the fabrication of cheap, expendable, and efficient PSCs by using the WO_3 ETL with drastically improved device stability and the PCE of ~9.5%.[56] WO_3 NRs and nanosheet arrays were directly grown on the FTO glass substrates and applied as ETLs in PSCs. For this work, the electrospraying method was used to fabricate uniform and dense layers of WO_3 to initiate the growth, followed by a hydrothermal step to synthesize the NRs and nanosheet arrays by adjusting the solution composition in an efficient manner. WO_3 nanostructures were further treated with aqueous solution of $TiCl_4$ to reduce the charge recombination at the perovskite/ WO_3 interface, which displayed a high PCE of ~11.24% for devices based on WO_3 nanosheet arrays.[57] Lee et al. synthesized lithium silicotungstate (Li_4SiW12O40, termed Li-ST) by a solution process at low temperature and applied Li-ST as an ETL for planar PSCs. Herein, by employing 60-nm-thick Li-ST layer, the fabricated PSC displayed the PCE of ~14.26% with a J_{SC} of ~22.16 mA/cm^2, a V_{OC} of ~0.993 mV, and an FF of ~64.81%. The obtained PCE was significantly higher than that of the PSC employing a TiO_2 layer processed at the same temperature (PCE = ~12.27%).[58] Room temperature–sputtered niobium oxide (Nb_2O_5) was applied as an ETL by Ma et al. The amorphous Nb_2O_5 ETL without any heat treatment delivered a best PCE of ~17.1% for planar PSCs. Interestingly, the crystalline Nb_2O_5 that was annealed at high temperature of 500°C resulted a very similar PCE of ~17.2%.[59] Sundaram et al. applied an ETL of barium tin oxide ($BaSnO_3$) of different morphologies for PSCs. The fabricated cells were tested in ambient condition using low-cost graphitic carbon as a counter electrode. The PCE of ~2.91% was recorded for rod-based active layer compared to particle (~1.93%).[60] Wu et al. applied efficient mesoporous ternary oxide Zn_2SnO_4 as an ETL for highly stable and hysteresis-free PSCs. Importantly, mesoporous Zn_2SnO_4 significantly accelerated the crystallization of the perovskite layers, and with simple one-step spin coating, a uniform bilayer structure of perovskite was obtained. The fabricated device exhibited a PCE of ~13.34%, which was even higher than that achieved with the commonly used TiO_2

under the similar experimental conditions (~9.1%).[61] Wu et al. explored $SrTiO_3$ (STO) for the first time as an ETL for the fabrication of PSCs. The steady-state PL quenching and transient absorption experiments revealed efficient photoelectron transfer from $CH_3NH_3PbI_{3-x}Cl_x$ to STO. The application of STO formed a uniform perovskite layer with large grains and complete surface coverage, which resulted to a high shunt resistance and improved performance. PSCs with meso-STO exhibited a V_{OC} of ~1.01 V, which was ~25% higher than the value of ~0.81 V achieved in the control device with the conventional meso-TiO_2. In addition, an increase of ~17% in the FF was credited to the change in the thickness of the meso-STO layer.[62]

2.2 Organic Electron-Transporting Materials

The organic charge-transporting materials have attracted much attention due to their adjustable energy level, facile synthesis, good solubility, and low processing cost. Fullerene (C_{60}) and its derivatives 1′,1″,4′,4″-tetrahydro-di[1,4]methanonaphthaleno[1,2:2′,3′,56,60:2″,3″][5,6]fullerene-C60 (ICBA), phenyl-C61-butyric acid methyl ester ($PC_{61}BM$), and [6,6]-phenyl-C71-butyric acid methyl ester ($PC_{71}BM$)) are the most widely used n-type organic materials for ETLs in PSCs[63,64] as they are processed at low temperature, eliminate photocurrent hysteresis, and display improved charge injection/separation efficiency from the perovskite absorber to fullerene.[65] Jeng et al. reported $PC_{71}BM$ PSCs with enhanced V_{OC},[66] and perovskite/fullerene planar heterojunction hybrid PSCs were fabricated by Sun et al.[67] The influence of organic ETLs was studied for metal oxide–free PSCs.[68] J. Huang et al. applied fluorinated perylene diimide (FPDI) as a novel organic ETL in planar heterojunction PSCs. Dense and uniform perovskite film with high substrate coverage was obtained when the FPDI film was treated by chloroform for half an hour, and the FF of the PSC increased from ~30.44% to ~55.20%. The PCE of the best device reached to ~7.93%, which was comparable with that (~8.25%) of the conventional ZnO ETL–based PSC device prepared by the same method.[69] Huang et al.[70] reported the elimination of photocurrent hysteresis by fullerene passivation in planar heterojunction PSCs. Nie et al. fabricated a planar PSC $PC_{61}BM$ as the ETL with a PCE of ~18%.[71] Two fullerene C_{60} derivatives—2-(3-ethoxycarbonylpropyloxy)-(5,8)-dihydronaphthyl-(6,7)-[6,6]-C60 (EDNC) and 2-benzyloxy-(5,8)-dihydronaphthyl-(6,7)-[6,6]-C60 (BDNC)—were introduced into p-i-n–type planar heterojunction PSCs as ETLs by S. M. Dai et al. to investigate electrochemical, charge-transporting, and film-forming properties of fullerene ETLs, and the resulted photovoltaic performance were compared with $PC_{61}BM$.[72] L.S. Liao et al. reported an excellent ETL that simultaneously enhanced the stability and efficiency of n-i-p planar PSCs. Large π-conjugated graphdiyne (GD) was introduced into cross-linkable fullerene [6,6]-phenyl-C61-butyric styryl dendron ester (PCBSD) to improve the film orientation. The orientated C-PCBSD:GD film was favorable for the growth and crystallization of the subsequent perovskite films and provided the merits of superior electron mobility, efficient charge extraction, and energy-level tailoring. In addition, the thermally annealed C-PCBSD:GD film provided an adhesive film network with sufficient solvent resistance. Consequently, the PSC delivered a PCE of ~20.19% with obviously improved cell stability.[73] Inverted p-i-n PSCs were fabricated with phenyl-C61-butyric acid methyl ester (PCBM) and PCBM/C_{60} ETLs. The PSCs with the PCBM ETL resulted in a large photocurrent hysteresis when the perovskite/PCBM layer was not thermally treated. In contrast, the PSCs with PCBM/C_{60} layer exhibited no hysteresis effect even though thermal treatment was not undertaken. This result indicated that the PCBM/C60 layer suppressed the defect sites at the interface layers and facilitated the charge transports to the electrodes, thereby minimizing the hysteresis behavior.[74] S. Chang et al. applied a new fullerene derivative (isobenzofulvene—C_{60}—epoxide [IBF—E_p]) that served as an ETL for PSCs, both in the normal and inverted device configurations. The fabricated PSC provided higher photovoltaic device performance as compared to PCBM (6.9% vs. 2.5% in the normal and 9.0% vs. 5.3% in the inverted device configuration). Moreover, IBF—E_p devices showed superior tolerance to high humidity (~90%).[75] In other work, a mesoporous graphene/polymer composite was used as an ETL in PSCs, and a high efficiency of ~13.8% under AM 1.5G solar illumination was obtained. The network structure of highly conducting graphene and granular-like polyaniline contained well-defined pores that favored fast electronic channels. Due to the high heat transmission coefficient and low isoelectric point of mesoporous graphene-based ETLs, the PSC device displayed good chemical and thermal stability.[115] Other ETLs such as PEHT, PCBM, and PEHT:PCBM are also explored for the performance of PSCs.[63,76–79]

2.3 Doped Electron-Transporting Materials

The doping is an effective technique that plays a significant role in altering CB position of TiO_2 through a downward shift to increase electron injection from the perovskite film into the TiO_2 or through an upward shift to increase the V_{OC} of PSCs.[80] Moreover, the surface morphology is also affected through doping, or the crystallinity of perovskite might change and thereby influence the performance of PSCs.[81] The n- or p-doping of TiO_2 with different metal ions is known to modify the electron injection/transport ability and to improve the photovoltaic performances. So far, a wide range of elements have been investigated for doping TiO_2 nanomaterials such as lanthanum,[82] niobium,[83] lithium,[84,85] indium,[86] cadmium[87], yttrium,[88] tungsten,[89] neodymium,[90] and aluminum.[91] In the thrust of improving the electronic properties and to reduce the trap-state density of TiO_2, Y. Xiao et al. studied the impact of doping of TiO_2 NPs with different loadings of tungsten (W) on the photovoltaic properties of HTL-free PSCs. The doping with W (less than 1000 ppm) improved the PCEs of PSCs by promoting the electron conductivity in the TiO_2 layer, which facilitated electron transfer and collection. With the incorporation of W, the average PCE of PSCs was increased from ~9.1% for the undoped samples to ~10.53% for the 1000-ppm W-doped samples that mainly originated from the increase of J_{SC} and FF. The champion cell exhibited an impressive PCE of ~12.06% with 1000-ppm W-doped TiO_2 films.[92] M. C. Wu et al. focused the improvement of photovoltaic properties by the doping of TiO_2 with zinc (Zn-doped TiO_2) as the ETL for PSCs. The Zn doping with less than ~5.0 mol% enhanced the absorption behavior, electrical conductivity, and charge separation efficiency. In contrast, when Zn doping was increased to ~7.0 mol%, the charge separation properties were decreased. Conclusively, the Zn doping increased the J_{SC} from ~18.5 to ~22.3 mA/cm^2, and the PCE was significantly improved from ~11.3% to 14.0%.[93] High-quality, fully covered aluminum (Al)–doped ZnO (AZO) thin film of thickness ~20 nm was sputtered on ITO substrate by a sputtering method. Compared to the cell based on undoped ZnO, PSCs using AZO as an ETL showed better stability, high PCE of ~17.6%, comparable J_{SC}, higher V_{OC} (~1.07 V), and FF. The improvement in the photovoltaic performance was credited to high conductivity of AZO and better band matching with perovskite.[94] The investigations of zirconium (Zr) incorporation into the TiO_2 ETL were performed by D. S. Ginger et al. Compared with Zr-free TiO_2, PSCs employing Zr–TiO_2 electrodes with pyridine treatment gave rise to an increase in overall PCE and a decrease in hysteresis.[95] S. F. Liu et al. performed niobium (Nb) doping to TiO_2 by a facile one-pot chemical bath process and applied this Nb-doped TiO_2 as a high-quality ETL for planar PSCs. Compared with pure TiO_2, the Nb-doped TiO_2 was more efficient for photogenerated electron injection and extraction, showed higher conductivity, higher mobility, and lower trap-state density. A PSC with 1% Nb-doped TiO_2 yielded a PCE of more than ~19%, with the credibility of restoring about ~90% of its initial efficiency after storing for 1200 h in air.[96] The yttrium-doped TiO_2 compact layer was used for the fabrication of PSCs by Zhou et al. to improve the conductivity and carrier extraction ability of the ETLs that enhanced the PCE to ~19.3%.[112] Tin (Sn)-doped TiO_2 films were successfully deposited by low-temperature solution method and applied as the ETL by G. Shao et al. Compared with pristine TiO_2, the Sn-doped TiO_2 was more efficient for photo-generated electron extraction and transport as well as showed lower trap-state density and higher conductivity. Herewith, higher J_{SC} was obtained due to the enhanced carrier transport velocity and improved charge collection efficiency, which contributed to a higher FF and better energy transfer efficiency. The optimized PSC using the Sn-doped TiO_2 ETL achieved a PCE of ~17.2%, which was almost ~29.3% higher than the undoped TiO_2.[97] Lithium (Li) is reported to passivate the defects in the TiO_2 film, especially the defects on the surface that increases the conductivity of TiO_2. Li-doped TiO_2 is beneficial to the performance of the PSCs, and especially the hysteresis of the current–voltage (I–V) curves.[85,98] S. Li et al. claimed Fe^{3+} doping improved the conductivity of TiO_2 compact layers that boosted the photovoltaic performance of PSCs. Compared with the pristine TiO_2 ETL, the optimized doping (1 mol%) of Fe^{3+} resulted in the improvement of PCE from ~16.02% to ~18.60%.[99] W. Lee et al. demonstrated a solution-processed hybrid cathode interfacial layer composed of polyethyleneimine ethoxylate and a lithium quinolate complex that improved the photovoltaic performance of the planar heterojunction PSCs. The hybrid cathode modifier effectively lowered the work function of ITO cathode, which resulted to efficient electron transport and collection at ITO. Consequently, the PSCs with hybrid ETLs generated a maximum PCE of ~15.21%, which was ~25% better than ETLs without hybrid composition.[100]

Carbon materials,[101,102] such as single-walled carbon nanotubes (SWNTs),[103,104] graphene,[105,106] and graphene oxide,[107] are emerging dopants for ETLs and HTMs in PSCs. Z. Yang et al. successfully reported the doping of graphene quantum dots (GQDs) into PCBM and its application as the ETL for PSCs, which enhanced the conductivity of PCBM together with the enhanced charge extraction and reduced the trap-state density of perovskite films. A low doping ratio of ~0.5 wt% boosted the V_{OC}, J_{SC}, FF, and PCE of ~17.56%.[108] G. Kakavelakis et al. explored reduced graphene oxide (rGO) and PCBM as ETLs for planar inverted PSCs, which exhibited a hysteresis-free PCE of ~14.5%, compared with ~12.9% for the pristine PCBM—based device. After a comprehensive spectroscopic characterization of both devices, it was clear that the introduction of rGO in the PCBM ETL resulted in a significant increase of the ETL conductivity, together with reduced series resistance and surface roughness. As a result, a significant PL quenching of such perovskite/ETL was observed, which confirmed the increase in measured J_{SC}.[109] C. Kuang et al. reported for the first time the doping of GD, a novel 2D carbon material, into the PCBM layer for the fabrication of inverted PSC of configuration ITO/PEDOT:PSS/$CH_3NH_3PbI_{3-x}Cl_x$/PCBM:GD/C_{60}/Al to improve the electron transport. The optimized PCE of ~14.8% was achieved, and an average PCE of PCBM:GD-based devices was observed with ~28.7% enhancement (~13.9% vs. ~10.8%) compared with that of pristine PCBM-based PSCs. Both the space charge—limited current and PL quenching measurements ascribed to the better coverage on the perovskite layer improved electrical conductivity, strong electron mobility, and efficient charge extraction. The small hysteresis and stable power output also favored the performance of PCBM:GD-based devices. The enhanced device performance indicated the improvement of film conductivity and interfacial coverage based on GD doping, which delivered a high PCE of the devices.[110]

2.4 Hybrid Electron-Transporting Materials

The careful surface engineering is highly necessary in pursuit of better performance by the selection of ETLs with smooth and compact surface, appropriate CB, superior electron mobility, and conductivity to attenuate excessive charge accumulation and guarantee effective electron injection. Recently, many bilayer-structured ETLs consisting of metal oxides and organic materials, such as TiO_2/PCBM,[81] graphene/TiO_2,[106] polyethyleneimine ethoxylate/Y—TiO_2,[111] and SnO_2/C60, are reported successfully in PSCs with superior device performances.[112] Solution-processed carbon quantum dots (CQDs)/TiO_2 composite ETLs with negligible absorption in the visible spectral range for PSCs achieved an unprecedented efficiency of ~19% under standard illumination test conditions. It was found that a CQDs/TiO_2 combination increased both the V_{OC} and J_{SC} as compared with pristine TiO_2.[128] The effects of the $BaTiO_3/TiO_2$ mesoporous double layer (MDL) and its application as an ETL for PSCs enhanced the photovoltaic performance. The PCE was enhanced from ~9.89% for the TiO_2

mesoporous single layer to ~12.4% for the $BaTiO_3/TiO_2$ MDL. The larger $CH_3NH_3PbI_3$ crystals resulted in the better light absorption and improved J_{SC}. Moreover, V_{OC} was also improved by suppressing a charge recombination, which was attributable to the fewer $CH_3NH_3PbI_3$ crystal boundaries and band structure of the $BaTiO_3/TiO_2$ MDL. The results suggested that using $BaTiO_3$ as the second layer of the TiO_2 mesoporous layer was a promising way to boost the performance of PSCs.[113] S. W. Tong et al. demonstrated mesoporous graphene/polymer composite–based scaffold as a promising ETL for high performance and thermally stable PSCs. A mesoporous graphene/polymer composite as the ETL achieved a high efficiency of ~13.8% under AM 1.5G solar illumination. Due to the high heat transmission coefficient and low isoelectric point of the mesoporous graphene-based ETL, the PSC device exhibited good chemical and thermal stability.[114] Heterogeneous TiO_2 and Al_2O_3 composites were employed as a mesoporous scaffold in formamidinium lead trihalide ($FAPbI_{3-x}Cl_x$)–based PSCs to modify the surface properties of a mesoporous layer. It was found that the quality and morphology of the perovskite film were strongly affected by the TiO_2/Al_2O_3 ratio in the mesoporous film. The PCE of PSCs was improved to ~11.0% for a PSC with a 7:3 TiO_2/Al_2O_3 composite.[115] Large π-conjugated GD was introduced into cross-linkable fullerene PCBSD by M. Lia et al. to improve the film orientation. Raman spectroscopy and 2D grazing incidence X-ray diffraction (GIXRD) measurements revealed that a strong π–π stacking interaction occurred between GD and cross-linkable PCBSD (C-PCBSD), generated a face-on stacked composite film that delivered a PCE of ~20.19%.[116] A. Apostolopoulou et al. reported PSCs with an ETL of a composite mesoporous TiO_2–In_2O_3 thin film. The selection of In_2O_3 was on the basis of its slightly higher CB placement compared with TiO_2, which facilitated the charge transport of the photogenerated electrons from the CB edge of the perovskite to the TiO_2 by reducing the recombination rate. The PSC of configuration FTO/c-TiO_2/mp-TiO_2/(In_2O_3)/$CH_3NH_3I_{3-x}Cl_x$/P3HT/Au achieved the optimum PCE of ~12.9% with improved J_{SC} and V_{OC}.[117] W. Li et al. utilized the compact layer of ZnO/Zn_2SnO_4 (ZSO) composite as a promising alternative to TiO_2. ZnO/ZSO composite possessed higher carrier mobility and suitable band gap structure, and therefore, the PCE of PSCs based on the ZnO/ZSO composite compact layer reached ~12.03% at AM 1.5G solar light of 100 mW/cm^2, which was ~18% higher than that of PSC based on the TiO_2 compact layer.[118] Amorphous tungsten oxide (WOx) was successfully employed as the ETL in PSCs.

However, WOx device suffered from inherent charge recombination, resulted in low V_{OC} and FF. Thus, Eze et al. utilized WOx/fullerene C_{60} to enhance the PCE of PSCs to ~16.07% with a V_{OC} of ~0.93 V, a J_{SC} of ~22.15 mA/cm^2, and an FF of ~0.78, which were higher than those of PSCs based on the pristine WOx ETL.[119] M. Che et al. reported a novel scaffold layer composed of TiO_2−ZrO_2 composite for PSCs. Compared with pure TiO_2 NPs, the relatively larger ZrO_2 NPs increased the film roughness and enhanced the light-scattering effect in TiO_2−ZrO_2 composite films. The device exhibited an outstanding PCE of ~11.41%.[120] M. A. Mclachlan et al. investigated the ETM bilayer to form all low-temperature processed, thermally stable, efficient PSCs with negligible hysteresis. The components of the bilayer created a symbiosis that resulted in improved devices. An SG-derived ZnO layer facilitated improved energy level alignment and enhanced charge carrier extraction and PCBM layer to reduce hysteresis and enhanced perovskite thermal stability.[121] $PC_{60}BM/SnO_2$ bilayer ETLs were successfully prepared via a solution process for the fabrication of PSCs. Y. Chen et al. reported the introduction of $PC_{60}BM$ into the SnO_2 layer that dramatically enhanced the performance of planar PSCs. The results showed that the existence of $PC_{60}BM$ enhanced the crystalline quality and grain boundary integration in the perovskite films. The J−V analysis, absorption spectra, PL spectra, and TRPL spectra showed enhanced light harvesting of perovskite films and carrier extraction from perovskite films to $PC_{60}BM/SnO_2$ bilayer.[122] P. S. Chandrasekhar et al. fabricated the PSCs using the ETL of graphene/ZnO nanocomposite (G/ZnO NC). The effects of varying the graphene concentration in the G/ZnO NC films from 0 to 1 wt% on the photovoltaic performance of PSCs were studied. It was reported that ~0.75 wt% graphene concentration in the G/ZnO NC films exhibited an optimum PSC performance with J_{SC} and PCEs ranged from ~15.54 to ~19.97 mA/cm^2 and ~7.01−~10.34%, respectively, as compared with pristine ZnO. The enhancement in photovoltaic performance was attributed to the superior growth of the perovskite thin film and enhanced electron transport/extraction on using the graphene network.[123] Solar cells fabricated with HTL-free architecture under ambient air conditions with ZnO as the ETL achieved an efficiency of ~3.02%. Herewith, the incorporation of graphene oxide (GO) in ZnO matrix was critical to improve the performance. The maximum efficiency of ~4.52% was achieved for the device fabricated with ZnO/GO composite. The reported efficiency was not good as compared to other reported literature, but the device construction/architecture showed the potential of reducing the material and processing cost by avoiding expensive HTLs and avoiding processing under high-vacuum environments.[124] X. Zhao et al. incorporated chemically modified graphene into nanocrystal SnO_2 for highly efficient planar PSCs. The modification of SnO_2 with highly conductive 2D naphthalene diimide graphene increased the surface hydrophobicity and formed van der Waals interaction between the surfactant and the organic–inorganic hybrid lead halide perovskite compounds. The PCE of ~20.2% was achieved with an improved FF of ~82%, which was mainly attributed to the augmented charge extraction and transport.[125] H. J. Snaith et al. reported a low-cost, solution-based deposition procedure by utilizing nanocomposites of graphene and TiO_2 NPs as ETLs in meso-superstructured PSCs. The work demonstrated that graphene/metal oxide nanocomposites showed the potential to contribute significantly toward the development of low-cost PSCs. Graphene nanoflakes provided superior charge collection in the nanocomposites, enabled the entire device fabrication at temperature not higher than 150°C. The PSC showed remarkable photovoltaic performance with a PCE up to ~15.6%.[126] A solution-processed CQDs/TiO_2 was used as a novel CQDs/TiO_2 ETL in conjunction with a planar n-i-p heterojunction, and an unprecedented efficiency of ~19% was achieved under standard illumination test conditions. It was found that a CQDs/TiO_2 combination increased both the V_{OC} and J_{SC} as compared to pristine TiO_2.[127]

2.5 Interfacial Layers in Pervoskite Solar Cells

The selection of under layer for perovskite films is a good choice to have a beneficial influence on the interface matching and electrical stabilization. The application of insulating layer could effectively decline the degradation rate of perovskite. The optimal insulating layer has a capability to suppress the charge carrier recombination process and improves the PCE of PSCs. H. Sia et al. adopted the promising approach through the interface modulation by inserting the Al_2O_3 insulating layer at the ZnO/perovskite interface, which exhibited the PCE of ~15.55%.[128] The charge recombination and transport are the key issues in achieving improvements in the device performance. It is well-known that the surface passivation might contribute to better charge carrier separation and hinders recombination when the electrons excited from the perovskite absorber layer and injected into the CB of TiO_2.[129,130] As a consequence, the reduced recombination contributes to higher V_{OC} and PCE values.[131] The recent attempts have demonstrated that the deposited insulating layers on the surface of semiconductors could efficiently increase the physical separation of injected electron–hole couple and thereby retard the recombination reactions.[132] Such conformal coating films are analogous to a core/shell structure, where buffer shell layers are grown on ETLs to provide insulating barriers for impeding electron reverse transport through the film.[133] The passivation of TiO_2 via a thin layer with wide band gap energy, such as MgO and Al_2O_3, is expected to prevent a direct contact of the HTM with TiO_2. Lee et al. reported improved FF and J_{SC} from the passivation effect of the Al_2O_3 layer for the mp-TiO_2–based cells. The charge separation and charge transport in the interface were managed separately in PSCs.[134] Recently, Beloqui et al. reported Al_2O_3 conformal coating on the mesoporous TiO_2 ETL, which facilitated both the retardation of interfacial recombination dynamics and improved the device performance.[135] W. D. Hu et al. demonstrated the copper iodide (CuI)/poly (3,4-ethylenedioxythiophene)-poly(styrenesulfonate) (PEDOT: PSS) double layers as the anode buffer layer for efficient hole transport in PSCs. An obvious enhancement of V_{OC} and PCE was observed, and a PCE of ~14.3% was achieved. In case of single PEDOT:PSS–based device, a low PCE of ~12.9% was observed. Herewith, the inorganic CuI buffer layer significantly enhanced the hole extraction from the perovskite film, and the average exciton lifetime was reduced to ~2.7 ns. Moreover, the devices with double-HTL structure displayed better long-time stability. The PCE remained ~88% of the initial value after 720-h storage.[136] Another work on Al_2O_3-passivated TiO_2 was reported by L.W. Yin et al. In this work, Al_2O_3 layer–passivated TiO_2 NRs array was applied as ETL for formamidine lead iodine ($FAPbI_3$)–based PSCs. The modification of TiO_2 NRs by the Al_2O_3 layer efficiently enhanced the charge transport and retarded the charge recombination at the interface between the ETL and $FAPbI_3$ layer based on the evaporation-fabricated $FAPbI_3$ absorption layer and Al_2O_3–TiO_2 ETL, which displayed excellent photoelectronic performance, exhibited a high PCE of ~14% with negligible hysteresis and excellent reproducibility and stability. Herein, Al_2O_3 passivation induced an obvious fluorescence-quenching phenomenon and reduced electron–hole recombination rate. The greatly improved performances were attributed to the factors similar to that of the Al_2O_3 buffer layer, which enhanced $TiO_2/FAPbI_3$ interface contacts and alleviated the electron-trapping effect resulted from the surface defects at the TiO_2/perovskite interface. In addition, the evaporation synthetic route was helpful to get extremely homogeneous $FAPbI_3$ film without any pinholes and compact interface.[137] E. M. J. Johansson et al. showed the feasibility of using a high–band gap semiconductor such as Al_2O_3 as the underlayer in PSCs and opened up pathways to use atomic layer deposited (ALD) Al_2O_3 underlayers for flexible PSCs. An optimal Al_2O_3 layer effectively blocked electron recombination at the perovskite/FTO interface and sufficiently transported electrons through tunneling. PSCs fabricated with the Al_2O_3 layer demonstrated a high PCE of ~16.2%.[138] H. J. Snaith et al. deposited HTMs within an insulating mesoporous buffer layer, which comprised Al_2O_3 NPs and prevented the metal electrode migration while allowing for precise control of the HTM thickness. This resulted to an improvement in the FF and prevented degradation of the device after 350 h of operation.[139]

The ordered hierarchical TiO_2 nanostructures were scattered on compact TiO_2 layer to form a quasi-mesoscopic scaffold of scattered distribution, which offered enough growth space for perovskite grains and promoted the ordered growth of perovskite grains. The fabricated PSCs showed PCE of ~16.56% that was much higher than the PCE (~11.38%) of PSCs with conventional small TiO_2 NPs as scaffold and the PCE (~6.07%) of planar PSCs with a compact TiO_2 layer. The PCEs of PSCs with heretical TiO_2 and TiO_2 NPs remained ~47% and ~22% of the initial PCE values aged for 30 days in air, indicated better stability and moisture resistance of heretical TiO_2 scaffold based PSCs.[140] SiO_2 NPs were applied as an alternative to mesoporous TiO_2 or Al_2O_3 scaffolds by D. Zou et al. It was revealed that SiO_2-based PSCs showed much higher PCE of ~16.2% with higher V_{OC} and FF. Moreover, the hysteresis for SiO_2-based PSC was less severe than for Al_2O_3-based PSCs.[141] In other work, a new ETL was synthesized by removing the conventional TiO_2 ETL. The CT was investigated by EIS measurements, which indicated an additional kinetic process between FTO and perovskite and reduced the interfacial charge accumulation severely. In addition, a new ETL was proved to be more advantageous in avoiding interfacial degradation and improving stability of the device. The new ETL-based PSC showed less hysteresis effect and exhibited an outstanding PCE of ~18.16%.[142] For the first time, H. Li et al. reported the surface passivation of perovskite layers with chemically modified graphene oxides, which behaved as an efficient interlayer to reduce the interfacial recombination and enhanced the hole extraction. Herein, the passivation affected mainly the interaction between functional group (4-fluorophenyl) and under-coordinated Pb ions. The resulted PSCs achieved the high PCE of ~18.75% with an enhanced V_{OC} of ~1.11 V. Ultrafast spectroscopy, photovoltage/photocurrent transient decay, and electronic impedance spectroscopy characterizations revealed the effective passivation effects and the energy loss mechanism.[143]

3. HOLE-TRANSPORTING MATERIALS IN PSCs

An HTM is an indispensable component in achieving high-efficiency PSCs and play vital roles such as blocking electron transfer to the anode, extracting photo-generated holes from the perovskite and transporting these charges to the back-contact metal electrode, and preventing direct contact between the perovskite layer and the metal electrode to improve the device's stability. HTMs improve the hole-transfer efficiency[144] and also influence V_{OC} by determining the splitting of the quasi-Fermi energy levels of the perovskite.[145–147] The HTMs in efficient PSCs could be categorized into inorganic and organic materials. E. Nouri et al. highlighted tetratriphenylamine-substituted Zn phthalocyanine (TPA-ZnPc, Structure-1, HTM-1) as a new HTM and also studied the importance of a buffer layer between the perovskite layer and the HTL. The efficiency of the PSC was ~9.0% for HTM-1 but increased to ~13.65% in the presence of Al_2O_3 as a buffer layer.[148] The organic HTMs usually contain small molecular compounds,[149–152] such as Spiro-OMeTAD, and conducting polymers,[153–156] including poly(3-hexylthiophene-2,5-diyl) (P3HT). Insertion of suitable HTMs between the perovskite layer and metal electrode promotes the separation of electrons and holes in the functional layer interface and thus reduces the charge recombination and improves the performance of PSCs. For example, Chen et al. synthesized three small-molecule HTMs based on SAF core named as (Structure-2, CW3, HTM-2), (CW4, HTM-3), and (CW5, HTM-4). The results revealed that the steric interaction of HTMs had an effect on morphological coverage. With proper alkyl chains, (HTM-3) possessed preferable coverage on top of perovskite layer, which resulted to a high photocurrent and a better PCE of ~16.56%.[157] Inspired from the crucial role of HTMs in improving the device performance and the stability of PSCs, this section aims to provide in-depth analysis of various organic and inorganic HTMs employed for the fabrication of PSCs.

3.1 Organic Hole-Transporting Materials

To date, Spiro-OMeTAD is the most widely used HTM, but it suffers from low conductivity in its pristine form, and therefore, dopants/additives, such as cobalt complexes, 4-*tert*-butylpyridine (TBP), and lithium bis(trifluoromethyl sulfonyl)imide (LiTFSI), are required for improving the conductivity.[10,18,131,158,159] Moreover, the stability of the Spiro-OMeTAD HTM is

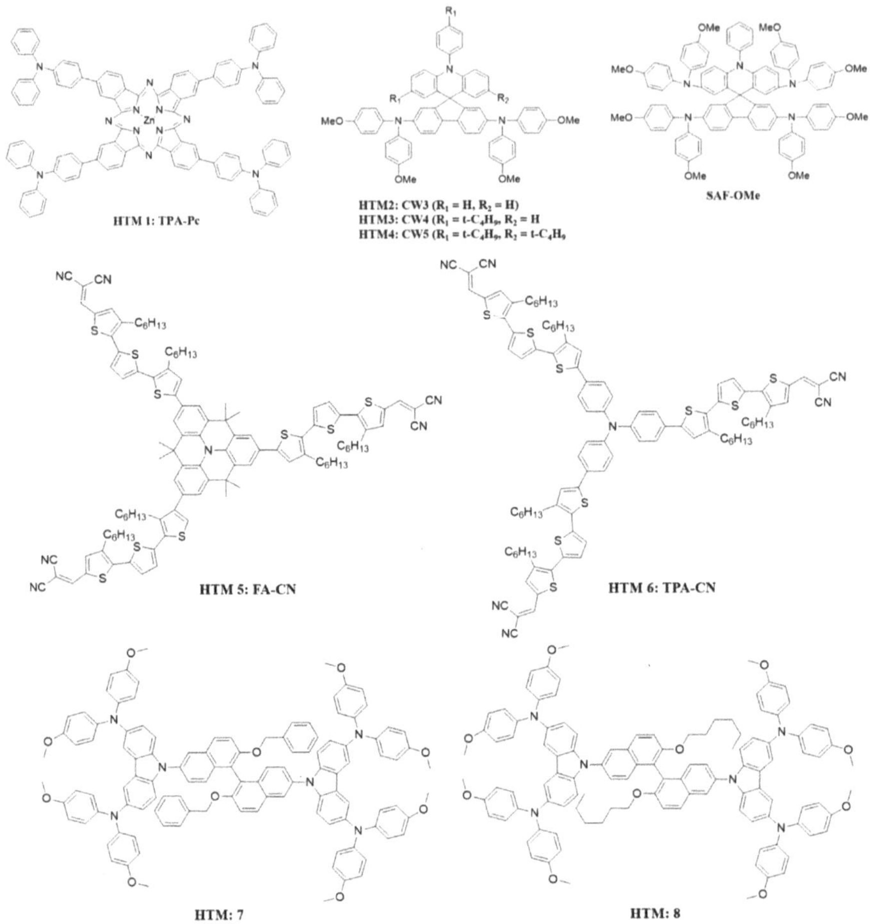

Structure-1

still limited due to dopants, and thus, dopant-free HTMs are paramount to improve the stability of the PSCs.[160–164] M. K. Nazeeruddin et al. reported molecularly engineered star-shaped D−π−A molecules by incorporating a rigid quinolizino acridine (FA-CN, Structure-1, HTM-5) and a flexible triphenylamine (TPA-CN, Structure-1, HTM-6) as the central unit and donor part, alkyl-substituted terthiophene as a conjugated bridge, and malononitrile as an electron acceptor part. The prepared central donor unit was connected to the conjugated linkers by Pd-catalyzed Suzuki coupling reaction. The corresponding aldehyde derivative was obtained by Vilsmeier–Haack reaction using $POCl_3$/dimethylformamide. Finally, the Knoevenagel condensation of the aldehyde derivative with malononitrile in the presence

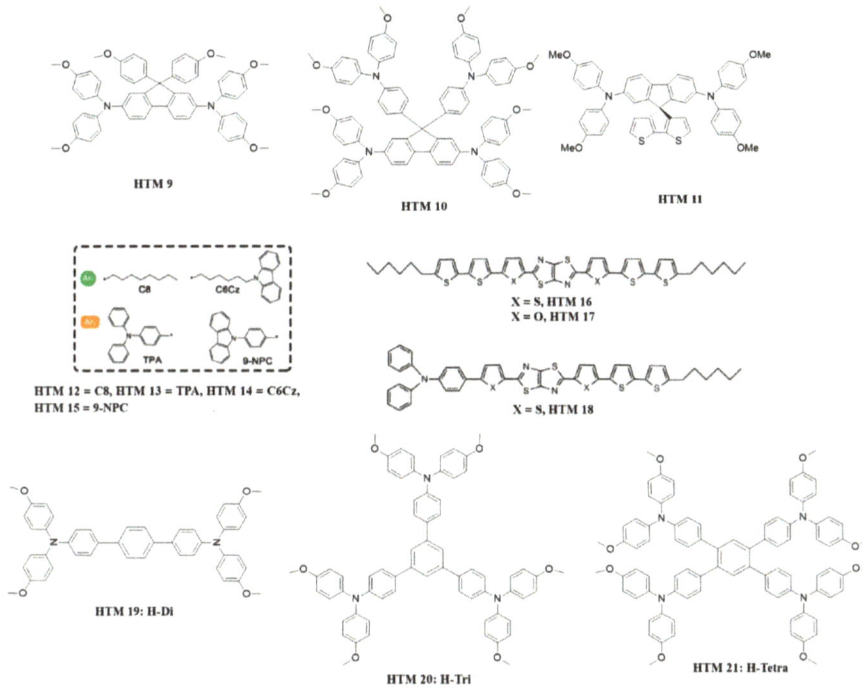

Structure-2

of triethylamine led to the final molecules HTM-5 and HTM-6. The stability of the unsealed devices based on dopant-free HTM-5 and HTM-6 was significantly enhanced compared to the doped Spiro-OMeTAD.[165] The ultraviolet—visible (UV—vis) absorption and emission spectra of HTM-5 and HTM-6 molecules in dichloromethane (CH_2Cl_2) solution are shown in Fig. 6A. HTM-5 displays absorption bands at ~418 and ~527 nm, whereas the absorption bands of HTM-6 are blue shifted with less intensity, which might due to the less delocalized electrons. The UV—vis spectra of the perovskite $(FAPbI_3)0.85(MAPbBr_3)0.15$ films, with and without the new HTMs, were measured and compared with Spiro-OMeTAD data, as shown in Fig. 6B. From Fig. 6B, in the presence of the new HTM, TiO_2/perovskite/HTM films show a significant enhancement in the absorption between ~400 and ~560 nm, which is attributed to colored HTM-5 and HTM-6 molecules. In the case of TiO_2/perovskite/Spiro-OMeTAD film, the slightly improved absorption is due to the oxidized Spiro-OMeTAD that has absorption in the visible region.[166] From Fig. 6C, the HOMO energy levels are estimated as −5.30 and −5.38 eV

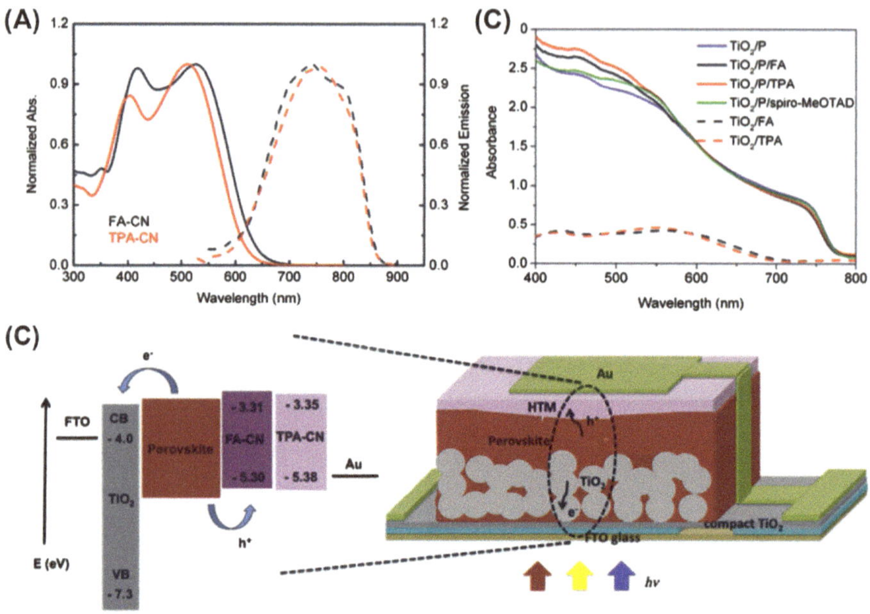

Figure 6 (A) UV—vis absorption (*solid line*) and emission (*dashed line*) spectra of FA-CN (*black line*) and TPA-CN (*red line*) in CH_2Cl_2 solution. (B) The absorption spectra of the perovskite films with different HTMs. (C) Device architecture and energy-level diagram. *Reproduced from Burschka, J.; Dualeh, A.; Kessier, F.; Baranoff, E.; Cevey, N-L.; Yi, C.; Nazeeruddin, M. K.; Grätzel, M.* J. Am. Chem. Soc. ***2011***, 133, *18042 with permission of American Chemical Society.*

for HTM-5 and HTM-6, respectively, indicating that the hole injection from the perovskite to the HTM is feasible. Herein, the HTM-6 molecule with the triphenylamine-based core is found to be lower in energy compared with HTM-5, which might due to weak electron-donating effect.

A novel HTM (Q221, Structure-1, HTM-7, 9,9′-(2,2′-bis(benzyloxy)-[1,1′-binaphthalene]-6,6′-diyl)bis(N3,N3,N6,N6-tetrakis(4-methoxyphenyl)-9H-carbazole-3,6-diamine)) was synthesized by introducing benzyl groups onto 1,10-bi-2-naphthol central core as edge chains and bis(4-methoxy-phenyl)amine-substituted 9H-carbazole as donor groups. A reference molecule (Q222, Structure-1, HTM-8, 9,9′-(2,2′-bis(hexyloxy)-[1,1′-binaphthalene]-6,6′-diyl)bis(N3,N3,N6,N6-tetrakis(4-methoxyphenyl)-9H-carbazole-3,6-diamine)) was also synthesize with hexyl edge chains. The introduction of edge chains influenced the molecular orbital energy levels. HTM-7—based PSCs with carbon counter electrodes exhibited the highest PCE of ~10.37% at a low doping level of Li-TFSI/TBP (15 mM/100 mM), whereas HTM-8–based PSCs exhibited ~8.87%. On comparison, Li-TFSI/TBP (15 mM/100 mM)–doped HTM-7–based PSCs showed much better photovoltaic parameters than doped with (30 mM/200 mM) Li-TFSI/TBP.[167] Boschloo et al.[168] introduced two novel low-cost fluorene-based molecules, i.e., (Structure-2, HTM-9 and HTM-10), for the fabrication of PSCs. The PSCs delivered a good solubility and charge-transport ability with considerably high PCEs of ~17.2% and ~18.0% for HTM-9 and HTM-10, respectively. Nazeeruddin et al. engineered a novel small dissymmetric fluorene–dithiophene (FDT, Structure-2, HTM-11). The X-ray crystallography and molecular docking simulations confirmed the interactions between the HTM-11 and the perovskite layers. Herewith, not only the methoxy groups of HTM-11 interacted with the perovskite's methyl ammonium sites but also the dithiophene moiety in HTM-11 took part in a thiophene–iodine interaction at the perovskite surface, which promoted the effective transfer of the photo-generated holes into the HTM. HTM-11–based PSCs yielded the PCEs of ~20.2% that was comparatively better than Spiro-OMeTAD–based PSCs[169] L. Bai et al. designed a series of diarylfluorene-based molecules, i.e., the triphenylamine (TPA, Structure-2, HTM-13) or 9-phenylcarbazole (9-NPC, Structure-2, HTM-15) was taken as the arm and the octyl (C8, Structure-2, HTM-12) or the 9-hexyl carbazole (C6Cz, Structure-2, HTM-14) was taken as the side chains, namely C8-TPA (HTM-12–HTM-13), C8-9-NPC (HTM-12–HTM-15), C6Cz-TPA (HTM-14–HTM-13), and C6Cz-9-NPC (HTM-14–HTM-15), and applied as HTMs without any dopants for p-i-n type PSCs. Compared to the reference device fabricated with PEDOT: PSS as the HTMs, these novel HTMs favored the deposition of perovskite film with larger grain size and less grain boundaries. The hole extraction at the HTMs/perovskite film interface and the hole transport in HTMs were also promising for diarylfluorene-based HTMs. The high PCE of ~13.85% was achieved; however, a low PCE of ~11.90% accomplished by using PEDOT:PSS as an HTM. The flexible devices were also fabricated under the same condition by using poly(ethylene naphthalene-2,6-dicarboxylate) (PEN) as substrate and a maximum PCE ~9% was achieved. From Fig. 7A, the absorption is enhanced from ~400 to ~550 nm in the visible region range when the perovskite films are casted onto the novel HTMs, which favors the enhancement in light harvesting. To detect the hole-transport capability of HTMs in perovskite devices, steady-state PL spectra of the perovskite/HTMs bilayer films are measured with the excitation wavelength at ~500 nm, as shown in Fig. 7(b). The samples show a dramatic quenching

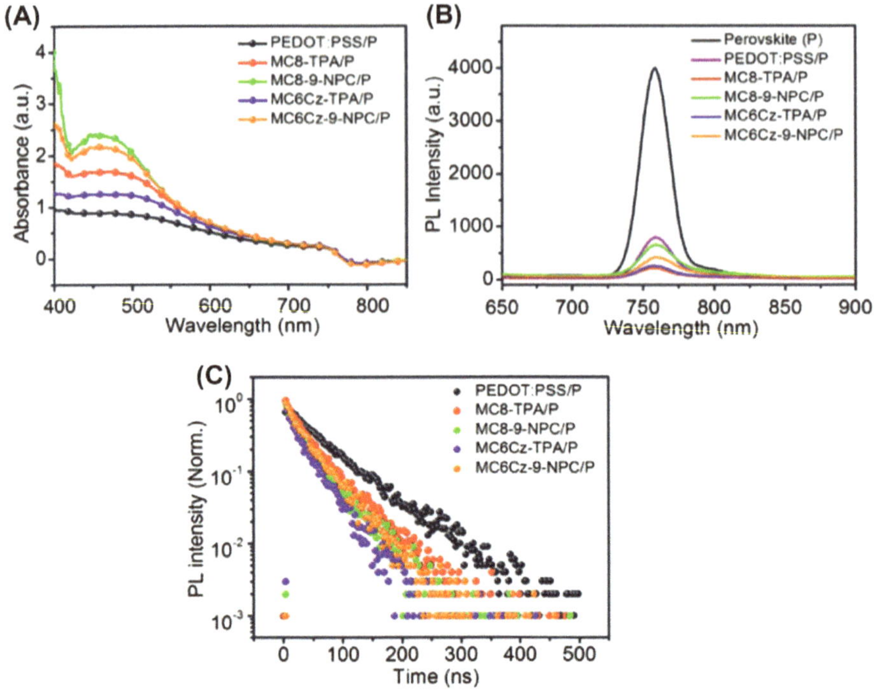

Figure 7 Optical characterization. (A) UV–vis absorption spectra of perovskite films deposited on different HTMs. (B) Steady-state photoluminescence (PL) spectra and (C) time-resolved PL decay transients taken at the peak emission wavelength (760 nm) of perovskite film based on different hole-transporting materials (HTMs). The excitation wavelength of PL spectra is 500 nm. *Reproduced from Bai, L.; Wang, Z.; Han, Y.; Zuo, Z.; Liu, B.; Yu, M.; Zhang, H.; Lin, J.; Xia, Y.; Yin, C.; Xie, L.; Chen, Y.; Lin, Z.; Wang, J.; Huang, W.* Nano Energy ***2018***, 46, *241 with permission of Elsevier.*

at ~760 nm when these designed HTMs are applied. The obtained results indicate that the hole–electron separation at the interface between the perovskite layer and HTLs is effective, which is ascribed to the good film morphology and a comparable hole transport mobility. Furthermore, TRPL decay transients of perovskite films based on different HTMs are conducted, as shown in Fig. 7(c). The average PL decay time (τ) of ~64.31 ns was observed for perovskite on PEDOT:PSS, ~50 ns for perovskite on (HTM-12–HTM-13), ~35.37 ns for perovskite on (HTM-12–HTM-15), ~36.74 ns for perovskite on (HTM-14–HTM-13), and ~44.23 ns for perovskite on (HTM-14–HTM-15). The results of steady-state PL spectra and TRPL decay measurements confirmed fast and effective hole extraction and charge dissociation for perovskite layer based on the novel

HTLs.[170] J. Niu et al. reported a superior HTM by modifying the commonly used PEDOT:PSS HTM by GO. The hole mobility was increased from 5.55×10^{-5} to 1.57×10^{-4} cm^2/V s, which lead to efficient hole extraction and low current leakage and thus resulted to ~20% higher PCE as compared to the control device without the GO doping. Compared with the PSC using the pristine PEDOT:PSS as an HTM, the device based on GO-doped PEDOT:PSS showed much higher efficiency, due to the reduction in the charge recombination, and enhanced charge separation and collection. Herein, the decreased current leakage and increased V_{OC} were attributed to the PCE improvement.[171]

H. S. Shin et al. adopted a solution processable, p-type thiazolo[5,4-d] thiazole–based small organic molecules as HTMs in PSCs. The HOMO energy levels of HTMs were tuned to match with perovskite by incorporating different spacer functional group such as thiophene (RTzR, Structure-2, HTM-16), furan (RFTzR, Structure-2, HTM-17), and triphenylamine (TPTzR, Structure-2, HTM-18). The high PCE of ~10.60% was observed for the PSC fabricated with HTM-17, whereas the other fabricated PSCs with HTM-16 and HTM-18 displayed inferior PCEs of ~4.37% and ~8.63%, respectively. Noticeably, HTM-17 presented a high FF with slightly low V_{OC} of ~0.955 V, which might be related to its low series resistance. The designed HTM-17 molecule acted promisingly for providing a high efficiency in PSC application. Among the synthesized HTMs, HTM-17 displayed higher oxidation potential compared with the other two situated at the onset values +0.32 eV. The HOMO of HTM-17 was calculated as −5.38 eV, which was slightly positive than that of $CH_3NH_3PbI_3$ (−5.44 eV). Fig. 8A describes the schematic representation, and the energy level representation of the devices is shown in Fig. 8B.[172] X. Liu et al. designed three simple arylamine molecules, i.e., (H-Di, Structure-2, HTM-19), (H-Tri, Structure-2, HTM-20), and (H-Tetra, Structure-2, HTM-21), with a benzene core through simple molecular engineering and applied as efficient HTMs. It was noted that the molecule with three arms (HTM-20) showed a lower HOMO level and a better film morphology on the perovskite layer than the molecules with two or four arms ((HTM-19, HTM-21). For PSCs, the HTM-20–based device showed better performance compared with the HTM-19– or HTM-21–based PSCs. The PL and impedance spectroscopy demonstrated that HTM-20 improved the hole–electron separation efficiency and decreased the charge recombination. Moreover, the HTM-20–based PSC showed a comparable performance and a much less material cost than the conventional

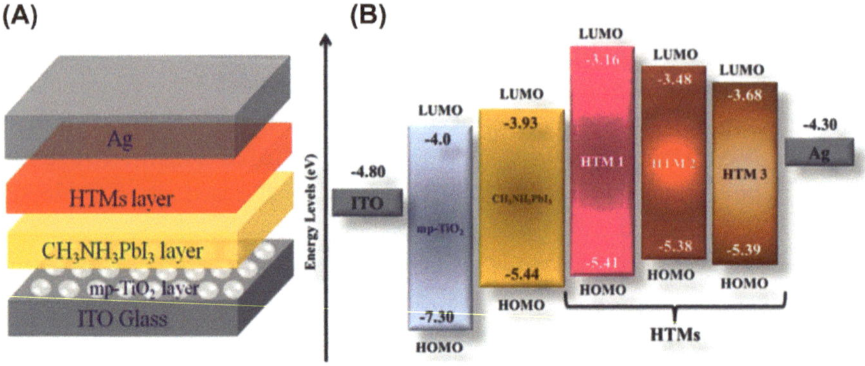

Figure 8 (A) Schematic and (B) energy level diagram of the fabricated PSCs. *Reproduced from Nazim, M.; Ameen, S.; Akhtar, M. S.; Nazeeruddin, M. K.; Shin, H. S.* Sol. Energy Mater. Sol. Cell. ***2018***, 180, *334 with permission of Elsevier.*

Spiro-OMeTAD.[173] M. K. Nazeeruddin et al. synthesized four center symmetrical star-shaped HTMs, i.e., 5,10,15-trihexyl-3,8,13-trimethoxy-10,15-dihydro-5H-diindolo[3,2-a:3′,2′-c]carbazole (KR122, Structure-3, HTM-22); 5,10,15-trihexyl-3,8,13-tris(4-methoxyphenyl)-10,15-dihydro-5H-diindolo[3,2-a:3′,2′-c]-carbazole (KR131, Structure-3, HTM-23); 4,4′,

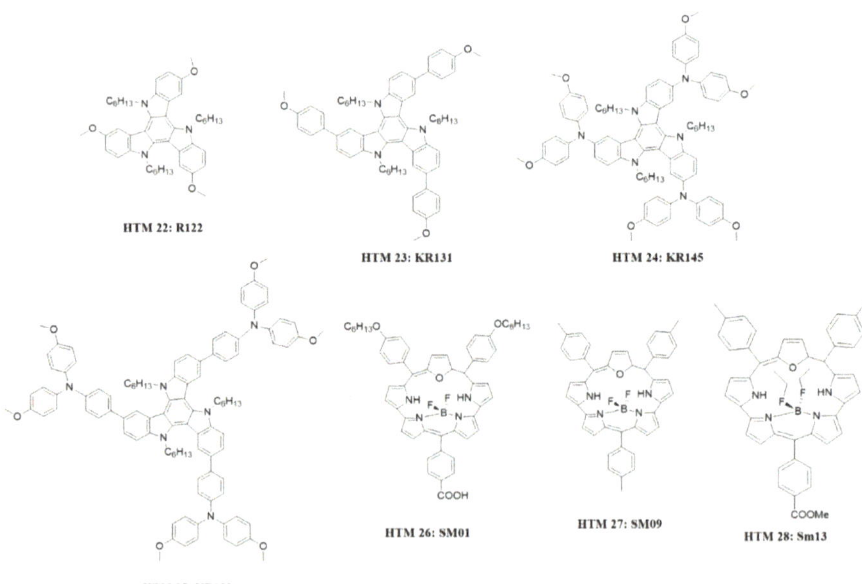

Structure-3

4″-(5,10,15-trihexyl-10,15-dihydro-5Hdiindolo[3,2-a:3′,2′-c]carbazole-3, 8,13-triyl)tris(N,N-bis(4-methoxyphenyl)aniline) (KR133, Structure-3, HTM-25); and 5,10,15-trihexyl-N3,N3,N8,N8,N13,N13-hexakis(4-methoxyphenyl)-10,15-dihydro-5H-diindolo[3,2-a:3′,2′-c]carbazole-3,8, 13-triamine (KR145, Structure-3, HTM-24), which comprised planar triazatruxene core and electron-rich methoxy-engineered side arms for $(FAPbI_3)_{0.85}$ $(MAPbBr_3)_{0.15}$—based PSCs. These HTMs were of low cost synthesized from relatively cheap starting materials by adopting facile preparation procedure, without using expensive and complicated purification techniques. The synthesized HTMs showed suitable HOMO with respect to the VB level of the perovskite, and TRPL indicated that hole injection from the VB of perovskite into the HOMO of triazatruxene-based HTMs was relatively more efficient as compared with that of well-studied Spiro-OMeTAD. Remarkable PCE over ~18% was achieved using (HTM-23) with the perovskite absorber.[174] S. B. Mane et al. reported boryl oxasmaragdyrin—based molecules, i.e., (SM01, Structure-3, HTM-26), (SM09, Structure-3, HTM-27), and (SM13, Structure-3, HTM-28), and applied as efficient HTMs for PSCs. These oxasmaragdyrins demonstrated complementary absorption spectra in the low-energy region, good redox reversibility, good thermal stability, suitable energy levels with $CH_3NH_3PbI_3$, and high hole mobility. A remarkable PCE of ~16.5% and $V_{OC} = 1.09$ V, $J_{SC} = 20.9$ mA/cm^2, and FF = 72% were achieved with (HTM-27) which was close to Spiro-OMeTAD with the PCE of ~18.2%, $V_{OC} = 1.07$ V, $J_{SC} =$ ~22.9 mA/cm^2, and FF = ~74%. In contrast, a poor photovoltaic performance of PSCs using (HTM-26) was observed due to the interactions of terminal carboxylic acid functional group with $CH_3NH_3PbI_3$. As depicted in Fig. 9A, two reversible oxidation waves and one reversible reduction wave are observed for (HTM-26) and (HTM-27), whereas two reversible oxidation and two reversible reduction waves are observed for (HTM-28). The energy level diagram (Fig. 9B with HOMO and LUMO energy levels convert directly from the first oxidation potential and the first reduction potential, respectively. As seen from the energy level diagram, the HOMO energy levels of small molecules are more positive than that of perovskite methylammonium lead iodide ($MAPbI_3$) but more negative than the HOMO energy level of Spiro-OMeTAD, suggesting that these boryl oxasmaragdyrins are effective HTMs for $MAPbI_3$-based PSCs. Noticeably, the HOMO and LUMO potentials of (HTM-28) are shifted to more positive values after the substitution of electron-donating ethoxy groups on the boron center.[175] R. Grisorio et al. designed phenothiazine-based molecules, (PTZ1,

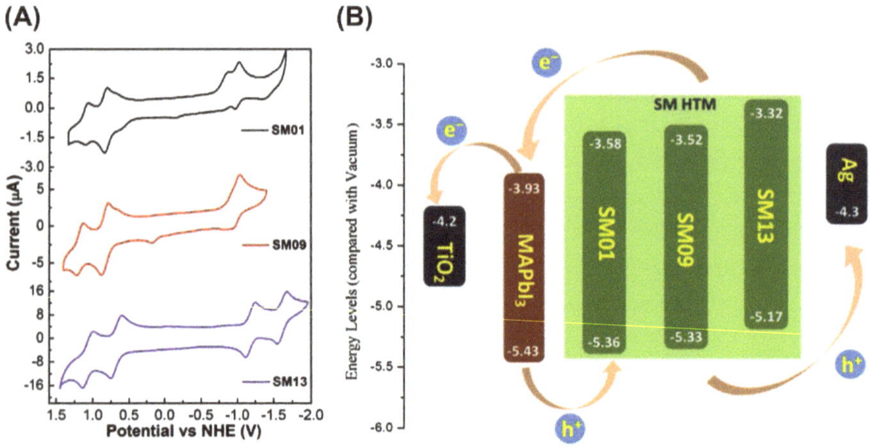

Figure 9 (A) CV spectra for SM molecules and (B) energy level diagram for SM HTMs. *HTM*, hole-transporting material. *Reproduced from Mane, S. B.; Sutanto, A. A.; Cheng, C. F.; Xie, M. Y.; Chen, C.; Leonardus, M.; Yeh, S. C.; Beyene, B. B.; Diau, E. W. G.; Chen, C. T.; Hung, C. H.* ACS Appl. Mater. Interfaces ***2017****, 9, 31950 with permission of American Chemical Society.*

Structure-4, HTM-29) and (PTZ2, Structure-4, HTM-30), through straight forward Buchwald–Hartwig and Suzuki–Miyaura cross-couplings, respectively, by binding the suitable donor groups (diarylamine or triarylamine) to a phenothiazine core. Phenothiazine-based structures were applied for the first time as HTMs in solution-processed lead trihalide perovskite–based PSCs. A dramatic effect exerted by the presence of phenylene spacers was observed on the relevant photovoltaic performances. The PCEs measured under AM 1.5G sun increased from ~2.1% (HTM-29) to a remarkable ~17.6% (HTM-30) which was competent to the state-of-the-art Spiro-OMeTAD (~17.7%). The results supported that phenothiazine-based compounds are promising as cost-effective HTMs in PSCs.[176] J. Lee reported a green-solvent-processable, polymeric HTM comprising benzothiadiazole (BT) and benzo[1,2-b:4,5:b′]dithiophene (BDT) (asy-PBTBDT, Structure-4, HTM-31). This new HTM was well matched with perovskites in terms of energy levels and attained a high μ_h [1.13×10^{-3} $cm^2/(V\ s)$] even without the use of dopants. Using HTM-31, a PCE of ~18.3% (~91% retention after 30 days without encapsulation under 50%–75% relative humidity) without dopants and a PCE of ~20.0% with dopants bis(trifluoromethanesulfonyl)imide were achieved. This work was a first report for a green-solvent-processable hole-transporting polymer, which exhibited the highest efficiencies reported so far for n-i-p devices with

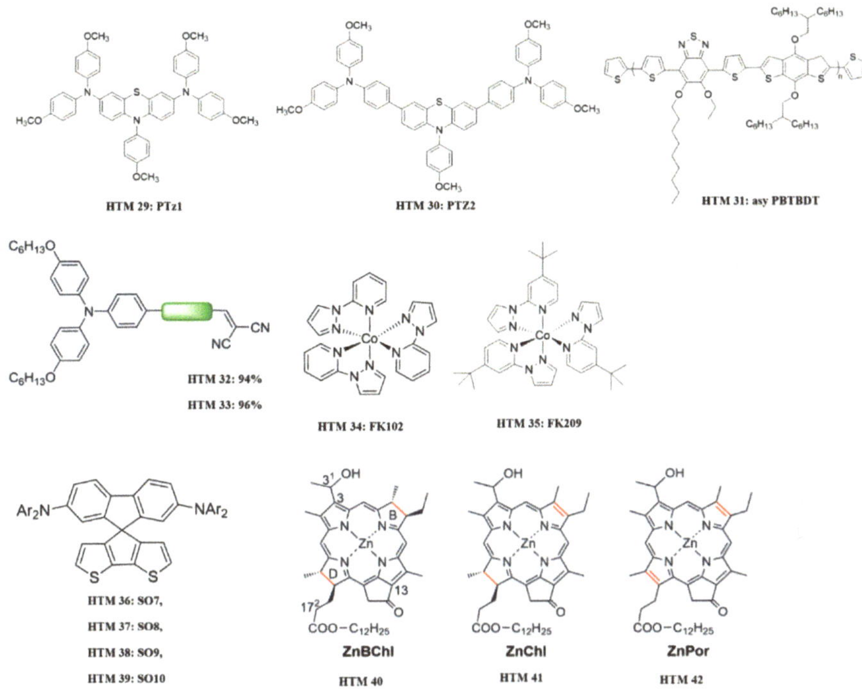

Structure-4

and without the dopants.[177] S. M. Zakeeruddin et al.[178] developed two new donor—acceptor (D—A)-substituted S,N-heteroacene—based molecules and investigated them as HTMs for PSCs. The optical and electrochemical characterization showed that the energy levels of both HTMs were suitable for their use in PSCs. Consequently, PCEs of ~17.7% and ~16.1% are achieved from PSCs based on (Structure-4, HTM-1, HTM-32) and (Structure-4, HTM-2, HTM-33), respectively. The optoelectronic properties in terms of series resistance, conductivity, and charge carrier recombination were further examined to unfold the potential of these new HTMs. TRPL spectroscopy confirmed that the hole injection from the VB of perovskite into HTMs followed the trend, which was in accordance with the position of the HOMO. The champion device based on (HTM-32) exhibits a higher PCE of ~17.7% with a J_{SC} of ~22.5 mA/cm^2, V_{OC} of ~1043 mV, and an FF of ~72.4% in comparison with a PCE of ~16.1% (J_{SC} = ~21.7 mA/cm^2, V_{OC} = ~1015 mV, and an FF = ~69.9%) for (HTM-33)—based PSCs. Interestingly, the V_{OC} values slightly decrease for (HTM-33), which is ascribed to the relatively deeper

HOMO level of (HTM-32) than (HTM-33) and is consistent with the electrochemical measurements. The PSC fabricated under similar conditions using Spiro-MeOTAD as the HTM displayed a maximum PCE of ~19.8% (V_{OC} = ~1100 mV, J_{SC} = ~23.0 mA/cm^2, FF = ~75.4%). Furthermore, the device parameters are statistically collected from a batch of 10 identical devices which results to an average PCEs of ~16.8 ± 0.61 and ~15.5 ± 0.56 for (HTM-32) and (HTM-33), respectively, as shown in Fig. 10. H. Xi et al. performed chemical p-doping in affecting the charge-transport properties of HTMs and the photovoltaic performance of PSCs. Two types of representative dopants, including lithium bis(trifluoromethane)sulfonimide (LiTFSI)

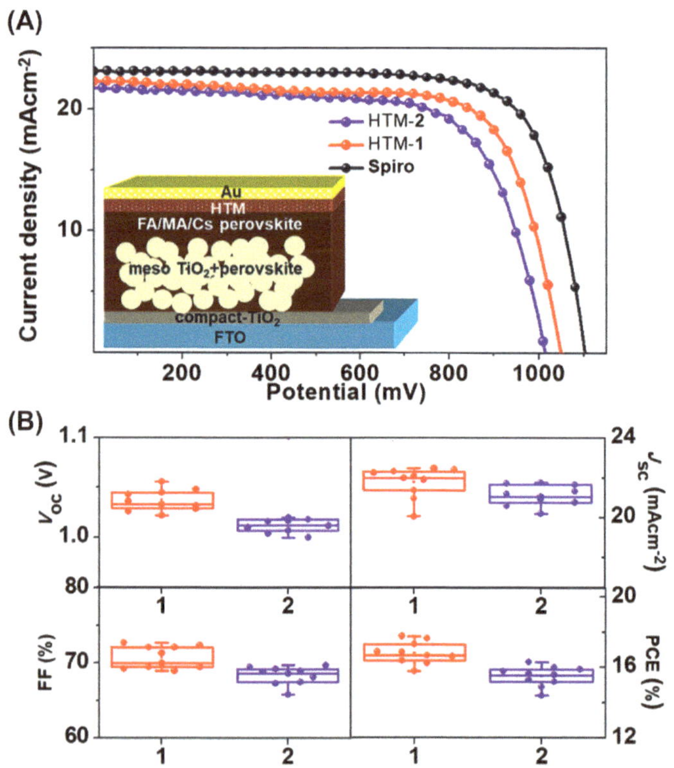

Figure 10 (A) Current density—voltage (J—V) curves (backward scan) for the triple-cation perovskite devices based on Spiro-MeOTAD, HTM-1, and HTM-2 recorded at a scan rate of 0.05 V/s under the irradiation of simulated AM 1.5G (inset shows the device architecture) and (B) statistics (J—V metrics) for 10 different solar cells based on HTM-1 and HTM-2. *Reproduced from Xi, H.; Tang, S.; Ma, X.; Chang, J.; Chen, D.; Lin, Z.; Zhong, P.; Wang, H.; Zhang, C.* ACS Omega ***2017****, 2, 326 with permission of American Chemical Society.*

and two Co(III) complexes tris[2-(1H-pyrazol-1-yl)-4-tert-butylpyridine] cobalt(III)tris[bis(trifluoromethylsulfonyl)imide] (Structure-4, FK209, HTM-35) and tris-[2-(1H-pyrazol-1-yl)pyridine]cobalt(III)tris[bis(trifluoromethylsulfonyl)imide] (Structure-4, FK102, (HTM-34), were employed as the p-type dopant to dope Spiro-OMeTAD HTM. Both the dopants facilitated the generation of oxidized Spiro-OMeTAD radical cations and improved the hole mobility. Co-doping of (HTM-35) and LiTFSI were necessary to achieve an optimal doping property and best device performance with a PCE of ~17.8% compared with that of the HTM-35–doped device (~13.5%) and the LiTFSI-doped device (~15%). UV–vis absorption, space charge–limited current measurements, and steady-state and TRPL measurements confirmed that with the co-doping of the two kinds of p-dopants in a proper ratio, the doped Spiro-OMeTAD exhibited high charge carrier mobility and charge carrier transfer/collection capability.[179] M. K. Nazeeruddin et al. synthesized four different fluorene–dithiophene derivative–based HTMs (Structure-4, SO7–SO10, HTM 36–39) via a facile route and were successfully used for the fabrication of formamidinium lead bromide PSCs. Under AM 1.5G illumination, the mesoscopic PSC with HTM-36 displayed an outstanding V_{OC} of ~1.5 V. The photovoltaic performance is on par with the device using the state-of-the-art Spiro-OMeTAD as HTM, which delivered a V_{OC} of ~1.47 V and a maximum PCE of ~6.9%, as depicted in Fig. 11. Herewith, the deeper HUMO levels for newly synthesized fluorene–dithiophene derivatives eventually made the synthesized molecules as promising HTMs for PSCs. The PCE of ~7.1% (V_{OC} = ~1.50 V, J_{SC} = ~6.9 mA/cm^2, and FF = ~0.69) is achieved by employing HTM-36 for the device based on FTO/compact TiO_2/meso-TiO_2/$FAPbBr_3$/HTM/Au architecture. All the new HTMs display strikingly high V_{OC} values of ~1.45 V with the exception of HTM-37 that shows slightly lower V_{OC} of ~1.43 V. The photovoltaic performance of PSCs employing four new HTMs was comparable with the device using the state-of-the-art Spiro-OMeTAD. The higher V_{OC} of the device relative to Spiro-OMeTAD counterpart was rationalized by considering the deeper HOMO level of HTM-36.[180] M. Li et al. designed three zinc-coordinated chlorophyll derivatives with bacteriochlorin, chlorin, and porphyrin macrocycles, namely, (ZnBChl, Structure-4, HTM-40), (ZnChl, Structure-4, HTM-41), and (ZnPor, Structure-4, HTM-42), respectively, and employed as HTMs in PSCs. The difference in the π backbones of these HTMs caused differences in their photophysical properties and thus different hole extraction abilities, as revealed by steady-state PL spectra. The PCEs of

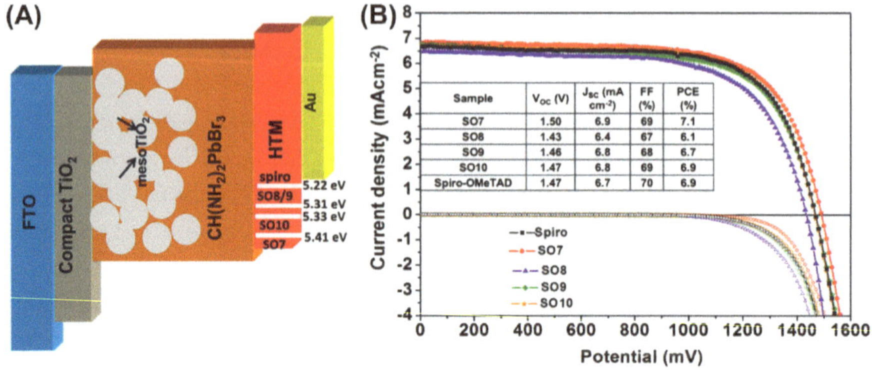

Figure 11 (A) Device architecture based on FTO/compact TiO_2/meso-TiO_2/$FAPbBr_3$/HTM/Au and (B) photovoltaic performance of $FAPbBr_3$ perovskite devices with Spiro-OMeTAD and SO7–SO10 as HTMs recorded at a scan rate of 0.1 V/s under simulated AM 1.5G solar light of 100 mW/cm^2 photon flux. Open symbols represent dark current. (Inset: summarized photovoltaic parameters derived from current–voltage curves). *Reproduced from Arora, N.; Orlandi, S.; Dar, M. I.; Aghazada, S.; Jacopin, G.; Cavazzini, M.; Mosconi, E.; Gratia, P.; Angelis, F. D.; Pozzi, G.; Graetzel, M.; Nazeeruddin, M. K.* ACS Energy Lett. ***2016***, 1, *107 with permission of American Chemical Society.*

PSCs with a typical mesoporous structure, FTO/compact TiO_2/mesoporous TiO_2/$CH_3NH_3PbI_3$/HTM/Ag, were ~8.26%, ~11.88%, and ~0.68% for HTM-40, HTM-41, and HTM-42, respectively. A low PCE of HTM-42–based PSCs was partially attributed to the small energy gap of HOMO levels between HTM-42 and $CH_3NH_3PbI_3$ perovskite. Thereafter, the energy gap was slightly increased by shifting the HOMO level of $CH_3NH_3PbI_3$ perovskite downward by incorporating formamidinium and bromide ions into the crystal lattice of $CH_3NH_3PbI_3$, which improved the PCE of HTM-42–based PSCs to ~4.04%, indicated better energy alignment between ZnPor and the modified perovskite.[181] M. K. Nazeeruddin et al. constructed a series of the HTMs with strategically placed aliphatic substituents to investigate the relationship between the chemical structure of the HTMs and the photovoltaic performance. PSCs employing the HTMs demonstrated PCE values in the range of ~9%–~16.8%, which highlighted the importance of the optimal molecular structure. Due to the ease of synthesis and moieties employed, it offered a wide range of possible structural modifications. This class of molecules had a great potential for structural optimization in order to realize simple and efficient small molecule–based HTMs for PSCs.[182] T. Gatti et al. studied the covalent functionalization of rGO flakes with different organic functional groups to increase the stability and homogeneity of their dispersion within a P3HT HTM. After preparation and characterization of a number of functionalized rGO@P3HT blends, the performance was tested for two containing the highest percentage of dispersed rGO as HTMs in PSCs and compared their performance with that of pristine P3HT and of the standard Spiro-OMeTAD HTM. The obtained results revealed the big influence of the morphology adopted by the single rGO flakes contained in the composite HTM in driving the final device performance and allowed to distinguish one of these blends as a promising material for the fabrication of highly reproducible PSCs.[183] S. Ameen et al. highlighted the application of newly designed HTM, i.e., benzothiadiazole:4-(3,5-bis(trifluoromethyl)phenyl)-7-(5′-hexyl-[2,2′-bithiophen]-5-yl)benzo[c][1,2,5] thiadiazole (CF-BTz-ThR, Structure-5, HTM-43) and synthesized aligned TiO_2 NBs as ETLs for

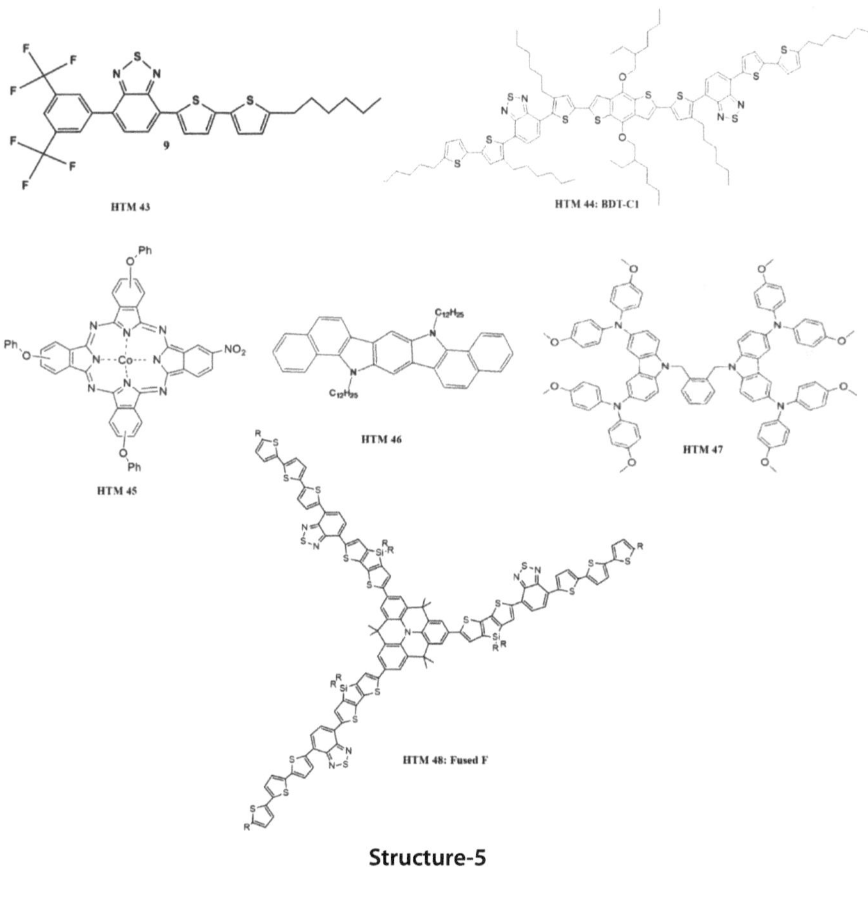

Structure-5

PSCs. The aligned TiO_2 NBs were grown on titanium (Ti)-coated FTO substrates using a facile hydrothermal method. The newly designed HTM-43 with suitable HOMO favored the effective hole injection from the perovskite layer deposited on aligned TiO_2 NBs thin film. The fabricated PSCs demonstrate a PCE of ~15.4% with a J_{SC} of ~22.42 mA/cm^2 and V_{OC} of ~1.02 V. The efficiency data showed the importance of proper molecular engineering while highlighting the advantages of dopant-free HTMs in PSCs. The I−V characteristics of the fabricated Au/HTM-43/$CH_3NH_3PbI_3$/aligned TiO_2 NBs/FTO PSCs are studied, as shown in Fig. 12A. The effect of the aligned TiO_2 NBs structure on the performance of PSCs is compared with that of PSC based on TiO_2 NPs. The Au/HTM-43/$CH_3NH_3PbI_3$/aligned TiO_2 NBs/FTO PSC displays a J_{SC} ~22.44 mA/cm^2 and a V_{OC} ~1.02 mV, resulted in a PCE of ~15.4%. The PCE of Au/HTM-43/$CH_3NH_3PbI_3$/TiO_2 NPs/FTO device is ~13.6% with a J_{SC} of ~20.73 mA/cm^2 and V_{OC} of ~1.02 V. The performance of the reference Au/Spiro-MeOTAD/$CH_3NH_3PbI_3$/TiO_2 NPs/FTO device (PCE ~10.4%) is inferior to the HTM-43−based devices. The superior J_{SC} of Au/HTM-43/$CH_3NH_3PbI_3$/aligned TiO_2 NBs/FTO PSCs was attributed to the morphology of the vertically arranged TiO_2 NBs. This morphology likely enhanced the photocurrent by improving the charge extraction/collection. It was also probable that the aligned TiO_2 NBs thin film facilitated electron transport, resulting to the observed J_{SC}

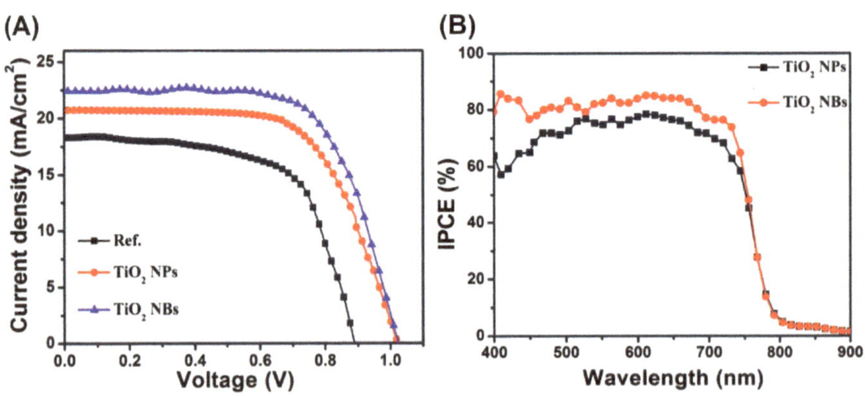

Figure 12 (A) J−V curves and (B) incident photon-to-electron conversion efficiency (IPCE) curves of the fabricated Au/CF-BTz-ThR/$CH_3NH_3PbI_3$/TiO_2 NPs/FTO and Au/CF-BTz-ThR/$CH_3NH_3PbI_3$/aligned TiO_2 NBs/FTO perovskite solar cells. *Reproduced from Ameen, S.; Nazim, M.; Akhtar, M. S.; Nazeeruddin, M. K.; Shin, H. S.* Nanoscale ***2017***, 9, *17544−17550 with permission of Royal Society of Chemistry.*

and FF. Compared with the TiO_2 NPs film, the aligned TiO_2 NBs were able to reduce the number of grain boundaries on the film surface thereby reduced the corresponding series resistance. As a result, the J_{SC} and FF of the PSC based on the aligned TiO_2 NBs ETL were superior to the traditional TiO_2 NPs ETL—based PSCs. Moreover, a high J_{SC} for the aligned TiO_2 NBs—based PSC is substantiated by the incident photon-to-electron conversion efficiency (IPCE) spectrum (Fig. 12B).[27] C. Chen et al. designed and synthesized an oval molecular material, (BDT-C1, Structure-5, HTM-44), in which the core unit, benzodithiophene (BDT), was functionalized by thiophene (TP) and benzo-[c][1,2,5]thiadiazole (BTZ) derivatives to generate extended π—conjugation. HTM-44—based PSCs showed high hole mobility and high conductivity in its pristine form in combination with appropriate energy-level alignment with respect to $CH_3NH_3PbI_3$ and $PC_{70}BM$ and thus was promising in PSCs as dopant-free HTMs and in organic solar cells (OSCs) as donor materials. The champion PSCs based on HTM-44 showed an average PCE of ~13.4% (scan forward: ~13.9%; scan backward: PCE = ~12.9%, scan rate: 10 mV/s). Although the average efficiency obtained was slightly lower than that of reference devices based on Spiro-OMeTAD (~13.7%), the HTM-44—based devices displayed better stability. HTM-44 as a donor material in OSCs also showed good performance in combination with $PC_{70}BM$ as acceptor material, and an efficiency of ~6.1% was obtained.[184] J. J. Guo et al. synthesized a novel asymmetric 2,9,16-triphenoxy-23-nitrophthalocyaninatocobalt ($CoPcNO_2$-OPh, Structure-5, HTM-45) and employed as a promising HTM in solid-state PSCs. The three phenoxy groups introduced into phthalocyanine complex improved the solubility and inhibited the aggregation effectively. This material showed strong absorption and high thermal stability. The optimized devices achieved a high V_{OC} of ~1.04 V with a maximum PCE of ~8.24% under AM 1.5G standard conditions.[185] I. Liman et al. introduced indolocarbazole-based small-molecule derivative, 8,16-didodecyl-8,16-dihydrobenzo[a]benzo[6,7]-indolo[2,3-h]carbazole (C12-carbazole, Structure-5, HTM-46) as an environmentally stable, cost-effective, and highly efficient HTM. In contrast to the PCE of ~9.62% exhibited by the Spiro-OMeTAD—based PSC, the HTM-46—based device under the same experimental conditions demonstrated an enhanced PCE of ~11.26%. The improved photovoltaic performance of the HTM-46—based device was attributed to the lowering of carrier recombination by better hole extraction ability of HTM-46, which demonstrated remarkably higher hole mobility compared with Spiro-OMeTAD.[186] M. K. Nazeeruddin et al.

developed new hole-transporting twin molecule (V886, Structure-5, HTM-47) based on methoxy diphenyl amine–substituted carbazole and applied for PSCs, which displayed a PCE of ~16.91%. The investigated HTM was synthesized in two steps from commercially available and relatively inexpensive starting reagents.[187] Qin et al. reported the synthesis, characterization, and PV application of a novel conjugated quinolizino acridine–based small-molecular HTM, termed as (Fused-F, Structure-5, HTM-48) for the application of PSCs. The synthesized HTM exhibited a suitable HOMO level and hole mobility to effectively extract holes from perovskite, which facilitated the corresponding PSC to obtain a PCE of ~12.8%.[188] In another report, two HTMs based on thiophene core, (2,3,4,5-tetra[4,4′-bis(methoxyphenyl) aminophen-4″-yl]-thiophene (H111, Structure-6, HTM-49) and 4,4′,5,5′-tetra[4,4′-bis(methoxyphenyl) aminophen-4″-yl]-2,2′-bithiophene (H112, Structure-6, HTM-50), were developed.[189] These HTMs (HTM-49 and HTM-50) showed superior air stability and deeper HOMO levels of ~5.31 and ~5.29 eV, respectively. The improved PCEs of ~14.7% and ~14.9% were respectively obtained for HTM-49– and HTM-50–based PSCs. S. Pauloa et al. synthesized and performed optical and electrochemical characterization of CQDs. The cyclic voltammetry measurements revealed that the HOMO and LUMO energy values of the CQDs were appropriate to ensure both efficient hole-transfer and electron-blocking capability from the perovskite to CQDs. Preliminary photovoltaic devices based on CQDs as HTMs exhibited a PCE of ~3%. This poor PCE might be responsible for the relatively low performance of PSCs.[186] L. Calióa et al. implemented 5,10,15-tribenzyl-5H-diindolo[3,2-a:3′,2′-c]-carbazole (TBDI, Structure-6, HTM-51) as a new HTM for p-i-n PSCs. An efficient and competitive PCE of ~15% with negligible hysteresis was obtained, which supports the potential of the dopant-free triazatruxene HTM for PSCs.[190] In other work, three HTMs based on carbazole structures, namely (SGT-405, Structure-6, HTM-53) and (SGT-407, Structure-6, HTM-54) that belong to the three-arm type structure and (SGT-404, Structure-6, HTM-52) for two-arm type structure, were applied for PSCs.[191] Among these synthesized HTMs, HTM-53 showed the ability to form a crystalline structure when coated as a film and exhibited the highest conductivity (~4.2×10^{-8} S/cm) with a remarkable PCE of ~14.79% and V_{OC} of ~1.02 V. This performance was comparable with that of the Spiro-OMeTAD–based device of a PCE of ~15.23% and a V_{OC} of ~1.03 V. Y. Li inspired by two molecules—4,40-bis(N-carbazolyl)-1,10-biphenyl (CBP) and

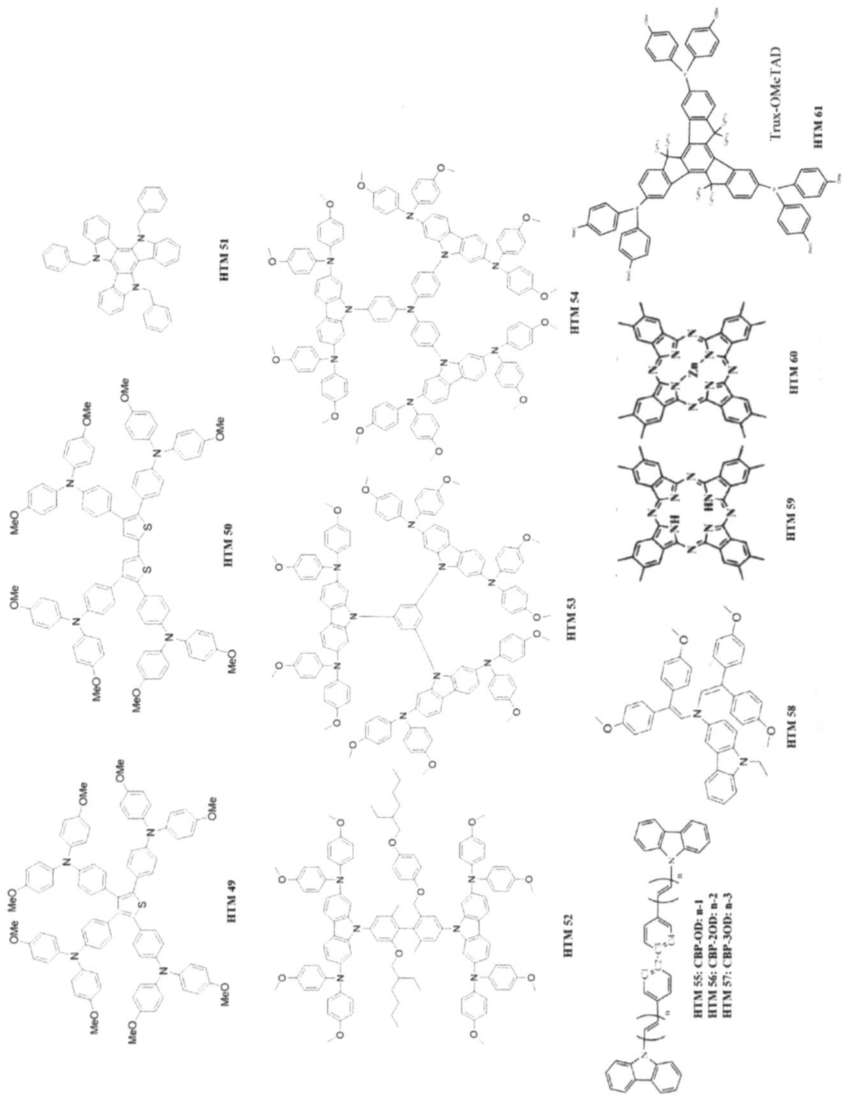
HTM 49
HTM 50
HTM 51
HTM 52
HTM 53
HTM 54
HTM 55: CBP-OD: n-1
HTM 56: CBP-2OD: n-2
HTM 57: CBP-3OD: n-3
HTM 58
HTM 59
HTM 60
HTM 61
Trux-OMeTAD

Structure-6

2TPA-2-DP—and synthesized three new organic molecules—(CBP-OD, Structure-6, HTM-55), (CBP-2OD, Structure-6, HTM-56), and (CBP-3OD, Structure-6, HTM-57)—for PSCs. The HOMO energy level of designed three molecules was lower than that of 2TPA-2-DP but higher than that of CBP. HTM-55 showed the largest hole mobility because of the larger transfer integral, which was even larger than that of CBP indicating a superior potential to be an HTM candidate. On the contrary, hole mobilities of HTM-56 and HTM-57 were similar to or less than that of CBP. In addition, all designed molecules displayed the much larger hole mobility than 2TPA-2-DP.[192] M. K. Nazeeruddin et al. designed a simple carbazole-based conjugated enamine (V950, Structure-6, HTM-58) and incorporated into a PSC, which displayed high PCE close to ~18%. HTM-58 was synthesized via an extremely simple route (one step, no expensive catalysts, no column chromatography, or sublimation purification) from commercially available and relatively inexpensive starting reagents, resulted in more than one order of magnitude lower cost of the final product compared with the commercial Spiro-OMeTAD.[193] Two octamethyl-substituted phthalocyanines (HMe2Pc, Structure-6, HTM-59) and zinc phthalocyanine (ZnMe2Pc, Structure-6, HTM-60) as dopant-free HTMs were demonstrated for planar PSCs. HTM-59 illustrated a very similar HOMO energy level and a slightly smaller band gap compared with that of HTM-60, which resulted in very close V_{OC} values based on different phthalocyanine HTMs. By varying the central metals, the HTM-59 film exhibited higher carrier mobility, better film coverage, and stronger surface hydrophobicity than the HTM-60 film, which was consistent with higher J_{SC} and FF values and better long-term film stability for HTM-59–based devices. When HTMs were vacuum deposited, the HTM-59–based devices showed the highest PCE of ~15.59%, which was slightly higher than the ~14.88% PCE of HTM-60–based devices.[194a] Li, Jen, and Chen et al.[194b] designed a planar and fully conjugated HTM, (Trux-OMeTAD, Structure-6, HTM-61) using truxene as the core. The planar truxene core enabled a face-on packing resulted in an excellent hole mobility of $2.3 \times 10^{-3} cm^2/Vs$, about two orders of magnitude higher than Spiro-OMeTAD. Six aliphatic chains in the core improved the hydrophobicity of molecules to induce a large contact angle of 90.4 degrees, which protected the underlying perovskite film from moisture. Due to the high hole mobility, HTM-61 served as a dopant-free HTM, and the resulted inverted PSC showed an impressive PCE of ~18.6%.[195] J. Y. Zhu et al.[196a] synthesized a solution-processed HTM for planar PSCs by incorporating CrO_3 aqueous solution into PEDOT:PSS aqueous dispersion. PEDOT:PSS–CrO_3 composite films displayed a high conductivity and interface modification and also acted as a good underlayer for the growth of the crystal perovskite films. The perovskite film deposited on the PEDOT:PSS–CrO_3 underlayer demonstrated high quality with large-scale domains and good film uniformity, which resulted in obvious improvement of the absorption in almost whole visible-light range. The resulting PSC device showed a PCE of ~16.9%.[196b] S. S. Mali[196c] modified the commercial available carbon ink with methylammonium lead iodide (MAI) and used as an HTM for ambipolar PSCs. The fabricated optimized PSC of glass/FTO/mp-TiO_2/$MAPbI_{3-x}Cl_x$/carbon-MAI/carbon configuration exhibited a PCE of ~13.87%, with V_{OC} ~0.997 V, J_{SC} = ~21.41 mA/cm^2, and FF of ~0.65. Furthermore, the air stability was tested at room temperature in open atmosphere. The results revealed that, although carbon-based devices showed lower PCE (~13.87%) compared with Spiro-MeOTAD HTM (~15%), the fabricated PSCs could retain >90% after water exposure and >20 times in ambient air for more than 160 days. Furthermore, the large area of the device (>1 cm^2) showed a PCE of ~13.04% with J_{SC} = ~21.47 mA/cm^2, V_{OC} = ~0.996 V, and FF = ~0.61.[197] A. L. Palmaa et al. prepared rGO by an easy and efficient reduction method and applied as an HTM in PSCs. The rGO-based device showed a PCE of ~6.6%, which outperformed the one achieved with the Spiro-OMeTAD (PCE = ~6.5%), after the shelf-life test. The light soaking test confirmed the better durability of rGO-based PSCs with respect to the Spiro-OMeTAD. Moreover, by V_{OC} decay measurements, rGO not only acted as the HTM but also played a role of an antirecombination layer that reduced the charge carriers recombination pathways and increased their lifetime.[198] D. Selvakumar et al. synthesized heteroatom (boron)-doped reduced graphene oxide (B-rGO) papers by the simple self-assembly process of graphene oxide through the deoxygenation in argon atmosphere at 400°C. The phase purity and crystallinity of the B-rGO paper were studied by X-ray diffraction analysis, and the morphology was analyzed by SEM analysis. From the Raman spectra, it was observed that the incorporation of boron into the rGO paper exhibited an improved conductivity and mobility. The current density–voltage (J–V) characteristics revealed that the B-rGO paper could be effectively used as an HTM for PSCs. The largest PCE of ~8.96% was achieved for the FTO/TiO_2/$CH_3NH_3PbI_3$/B-rGO/FTO–based PSC with an active area of 1 cm^2.[199] D. Zheng et al. introduced the unprecedentedly zero-dimensional dopant of carbon nano-onions (CNOs) and the functionalized oxidized CNOs (ox-CNOs) to modify the PEDOT:PSS HTM. Besides the merits of high conductivity and suitable energy level, the CNOs- and ox-CNOs—modified PEDOT:PSS HTMs provided a superior perovskite crystalline film with large-scale grains and orderly grain boundaries, which resulted an enhancement of PCE from ~11.07% to ~15.26%. Moreover, by suppressing the corrosion effect of PEDOT:PSS on ITO electrode, a dramatic improvement in the device stability was noticed.[200]

3.2 Inorganic HTMs

The use of inorganic p-type materials as hole-transport media offers the double advantage to reduce the cost of the cell and enhances its resistance to degradation. Inorganic semiconductors are good candidates in PSCs owing to their high hole mobility, easily processed via solution from corresponding precursors, and low production costs.[201] Few inorganic materials such as copper iodide (CuI),[202] copper thiocyanate (CuSCN),[203,204] nickel oxide,[205,206] carbon,[207,208] vanadium oxide (V_2O_5),[209,210] MoOx and WOx,[211] Cu_2ZnSnS_4,[212] $CuCrO_2$,[213] $CuAlO_2$,[214] $CuGaO_2$,[226] and CuS[230] have been tested for the fabrication of PSCs. The quest for good inorganic HTMs must be guided by the need of high carrier mobility and defect-free interface with the absorbing layer to minimize carrier recombination.

M. I. Hossain et al. used wxAMPS and SCAPS software to compute the key characteristics of PSCs with various HTM layers including Spiro-OMeTAD, NiO, CuI, CuSCN, and Cu_2O. The computations were carried out by considering defect-free perovskite and HTM layers. The results showed that PSCs containing Cu_2O as the HTM out perform all other devices with an organic or inorganic HTM with an exceeding PCE of ~24%. These results indicated that there was a possibility to further increase the performance of PSCs and to reduce their cost by replacing the expensive and moisture sensitive Spiro-OMeTAD by copper oxide.[215] S. Kumar Muduli et al. demonstrated for the first time liquid-exfoliated few layers of 2D black phosphorus (BP) nanosheets as an HTM for PSCs. The photoelectron spectroscopy measurements confirmed that the low-laying VB level of BP nanosheets (−5.2 eV) was favorable for hole injection from $CH_3NH_3PbI_3$ ($MAPbI_3$). The results showed ~25% improvement in PCE of ~16.4% for BP nanosheets + Spiro-OMeTAD as an HTM as compared with Spiro-OMeTAD (PCE = ~13.1%). When BP nanosheets were exclusively utilized as an HTM, a PCE of ~7.88% was noted. The PL quenching of $MAPbI_3$ and impedance measurements further confirmed the charge extraction ability of BP nanosheets.[216] Y. Wei et al. demonstrated the high performance of inverted planar PSCs based on the novel inorganic HTM of silver (Ag)-doped NiO_x (Ag:NiO_x). The density-functional theory calculation revealed that Ag preferred to occupy the substitutional Ni site (AgNi) and behaved as an acceptor in NiO lattice. Compared with the pristine NiO_x films, appropriate Ag doping increased the optical transparency, work function, electrical conductivity, and hole mobility of NiO_x films. Moreover, the $CH_3NH_3PbI_3$ perovskite films grown on Ag:NiO_x exhibited better crystallinity, higher coverage, and smoother surface with densely packed larger grains than those grown on the pristine NiO_x film. Consequently, the Ag:NiO_x HTL boosted the PCE of PSCs from ~13.46% (for the pristine NiO_x–based device) to ~16.86% (for the 2 at.% Ag:NiO_x-based device). Furthermore, the environmental stability of PSCs based on Ag:NiO_x HTMs was dramatically improved compared with devices based on organic HTMs and pristine NiO_x HTMs.[217] In another report, Z. Hu et al. fabricated high-performance inverted planar PSCs using SG-processed yttrium (Y)-doped NiO thin films as HTMs. Y-doped NiO thin films doped with different Y concentrations were successfully prepared through a simple SG process. The Y-doping significantly improved the electrical conductivity of NiO thin film, and the photovoltaic performance of Y-doped NiO HTL-based PSC devices out performed that of the pristine NiO HTL–based device. Notably, the PSC using a 5%Y–NiO HTL exhibited the champion performance with V_{OC} of ~1.00 V, a J_{SC} of ~23.82 mA/cm^2, an FF of ~68%, and a PCE of ~16.31%, which resulted in a ~27.62% enhancement in PCE as comparison with the NiO device. The enhanced performance of the Y-doped NiO device could be attributed to the improved hole mobility, the high-quality compact active layer morphology, the more efficient charge extraction from the perovskite absorber, and the lower recombination probability of charge carriers.[218] NiO is an important HTM for preparing inverted planar PSCs. Apart from some complicated preparation methodologies such as doping routes, pulsed laser deposition, atomic layer deposition, and high-temperature spray pyrolysis, the simpler solution-processed method was found promising for high-quality NiO HTMs for efficient inverted planar PSCs. J. Tang et al. used the sophisticated solvothermal method to synthesize highly crystalline NiO NCs. The oleylamine ligands were used to control the nucleation and growth of NiO NCs and achieved good colloidal stability in toluene, which was prepared for the high-quality NiO HTM. The related measurements and analysis revealed that the as-prepared NiO HTM showed faster hole extraction, more effectively suppressed recombination and fewer surface trap states than the typical poly(3,4-ethylenedioxythiophene)/poly(styrenesulfonic acid) HTM and finally contributed to superior photovoltaic performance of the corresponding device.[219] K. C. Icli et al. prepared NiO nanoparticulate dispersion in isopropanol without using surfactants or additives and deposited as the HTM on top of the perovskite layer at room temperature in a mesoscopic n-i-p cell for polymeric component—free PSC applications. This strategy lead to deposition of NiO NPs on the perovskite layer without damaging the underneath perovskite layer and inhibited the CT between individual NPs, which was confirmed by SEM, PL quenching, and J—V measurements. UV photoelectron spectroscopy analysis showed excellent matching of the band alignment of NiO layers with that of perovskite. A PCE of ~10.89% was achieved after optimizing the nickel oxide layer thickness and compared with a hole conductor—free device. The J—V measurements and electrochemical impedance spectroscopy revealed that the NiO layer possessed an excellent electron-blocking ability and reduced the recombination rate, which also in turn stabilized the power output and hysteresis of the cells. This strategy was believed to be applicable for other metal oxides employed in charge selective layers of perovskite and OSCs.[220] In other work, NiO as the HTL was widely explored and deposited for the fabrication of PSCs. X. Yin et al. employed a simple solution-processed NiO_x film as the HTM in PSCs. When the thickness of the perovskite layer increased from ~270 to ~380 nm, the light absorption and photogenerated carrier density were enhanced and the transporting distance of electrons and holes would also increase at the same time, which resulted in a large CT resistance and a long hole-extracted process in the device. The best PCE of ~15.71% was obtained with a J_{SC} of ~20.51 mA/cm^2, a V_{OC} of ~988 mV, and an FF of ~77.51% with almost no hysteresis. A stable efficiency of ~15.10% was caught at the maximum power point. The device configuration and energy level diagram of the fabricated PSCs with an inverted planar structure are presented in Fig. 13A,B. A typical surface and cross-sectional SEM image of the device fabricated with $MAPbI_3$ precursor solution is demonstrated in Fig. 13C,D, showing that the thicknesses of NiO_x, $MAPbI_3$, PCBM, and silver are approximately ~49.4, ~334.2, ~68.5, and ~93.2 nm, respectively.[221] Cui et al.[222a] have sputtered NiO films for the fabrication of inverted PSCs. Very recently, M. B. Islam et al. fabricated highly stable, low-temperature processed planar lead halide perovskite ($MAPbI_{3-x}Cl_x$) solar cells with NiO_x interfaces. The PSCs maintained over ~85% of the initial efficiency for more than 670 h, at the

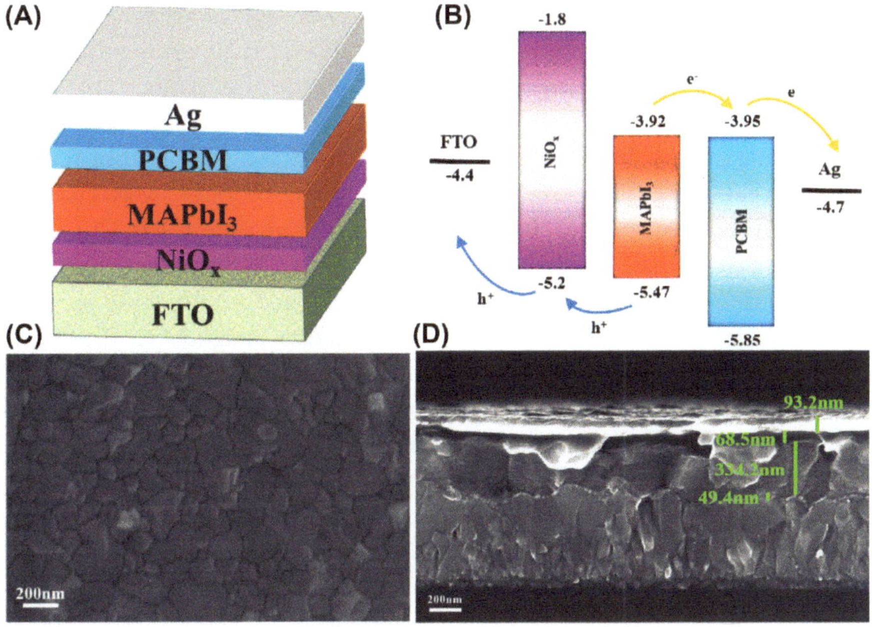

Figure 13 (A) The schematic of NiO_x-based perovskite solar cell, (B) the energy band level of solar cell, (C) the SEM image of perovskite layer, and (D) the cross-sectional SEM image of the device. *SEM*, scanning electron microscopy *Reproduced from Yin, X.; Yao, Z.; Luo, Q.; Dai, X.; Zhou, Y.; Zhang, Y.; Zhou, Y.; Luo, S.; Li, J.; Wang, N.; Lin, H.* ACS Appl. Mater. Interfaces ***2017****, 9, 2439 with permission of American Chemical Society.*

maximum power point tracking (MPPT) under 1 sun illumination (no UV-light filtering) at 30°C, and over 73% of the initial efficiency for more than 1000 h, at the accelerating aging test (85°C) under the same MPPT condition. Herein, the key factor for the prolonged lifetime of the devices was the sputter-deposited polycrystalline NiO_x HTL, which obtained the highest PCE of ~15.2% at the optimized NiO_x condition.[222b−d] H. Lee et al. fabricated inverted PSCs incorporating Radio Frequency (RF) sputtered NiO thin films as an HTM and window materials. The PSC fabricated with the NiO thin films showed the highest performance owing to excellent transmittance and wider bandgap along with moderate conductivity. With further optimization, the PSCs exhibited a J_{SC} of ~20 mA/cm^2 and a PCE of ~12.4%.[223] The sputtered NiO_x-derived PSCs displayed the limitations of low FF and J_{SC}, and thus to overcome this limitation, magnesium doping of NiO was performed. The fabricated PSCs showed a high PCE of ~18.5%, along with negligible hysteresis, improved ambient stability, and high reproducibility. In addition, good uniformity was also demonstrated over an area of 100 cm^2. This simple and well-established approach constituted a reliable and scale method paving the way for the commercialization of PSCs.[220a] The effect of the NiO_x thickness on the photovoltaic properties in planar type PSCs of configuration ITO/NiO_x/perovskite/PCBM/Ag was investigated by M. Yanagida et al. For this work, the flat thin film of NiO_x on ITO was deposited and the thickness of NiO_x was changed from ~20 to 120 nm by changing the sputtering time.[224] J. Cui et al. reported that planar PSCs showed significant PCE improvement from ~4.88% to ~6.13% by introducing a homogeneous and uniform NiO blocking interlayer fabricated with the reactive magnetron sputtering method. The sputtered NiO layer exhibited enhanced crystallization, high transmittance, and uniform surface morphology. The PCE of the sputtered NiO—based PSC was further promoted to ~9.83% when a homogeneous and dense perovskite layer was formed with solvent engineering technology. This work showed an impressive V_{OC} of ~1.10 V which was about ~33% higher than that of devices using the conventional spray pyrolysis of NiO onto TCO.[222a] The solution-processed inorganic $CuGaO_2$ was applied as an HTM in the PSCs with n-i-p configuration. The optimized device exhibited an impressive PCE of ~18.51%, which was better than that of the devices using organic Spiro-OMeTAD as the HTM. More importantly, it was found that the employment of $CuGaO_2$ as the HTM improved the long-term stability of PSCs.[225] In other work, a cost-effective spin-coated CuI was used as an HTM in PSCs. The CuI-based PSC recorded the PCE of ~2.22% (max) on the day of production and PCE of ~6.52% (max) after the 20 days of production. Moreover, the CuI-based PSC had a smaller hysteresis than the Spiro-based PSC, suggesting that CuI was highly promising alternative hole conductor for PSCs.[226] Q. Wu et al. applied Cu_2ZnSnS_4 (CZTS) NPs as a novel inorganic HTM for organolead halide PSCs for the first time and achieved a PCE of ~12.75%. The size of CZTS NPs was optimized, and the effects of the CZTS HTM on the optical absorption, crystallinity, morphology of the perovskite film, and the interface between the perovskite layer and the Au electrode were investigated and compared with the case of the Spiro-MeOTAD HTM.[214] A solution-processed CuO_x film was successfully integrated as the HTM for inverted planar heterojunction PSCs as depicted in Fig. 14A. The CuO_x layer was fabricated by simply spin-coating a copper acetylacetonate ($Cu(acac)_2$) chloroform solution onto ITO glass with high transparency in the visible range. The compact and pinhole-free perovskite film with large grain domains was grown on the CuO_x film. The PSC with the structure of

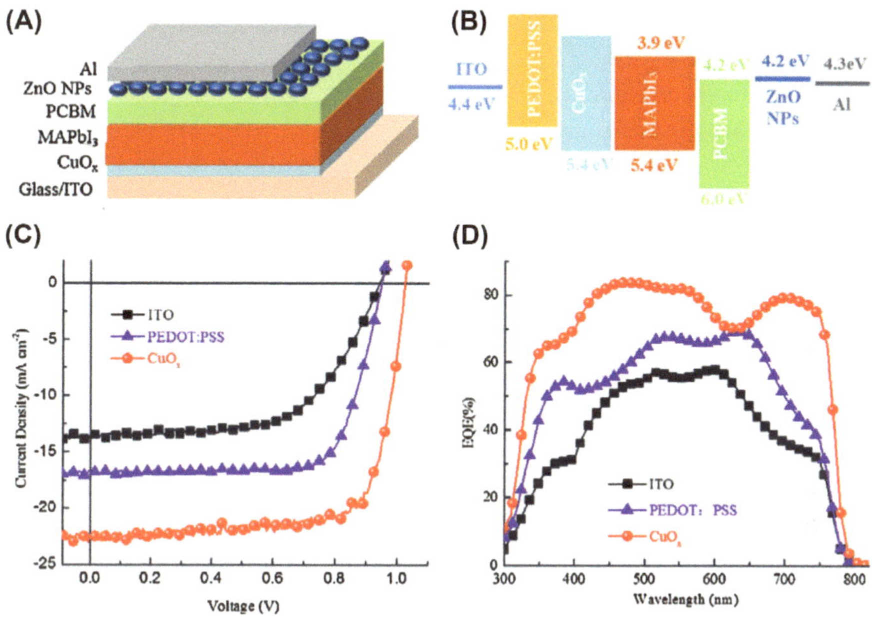

Figure 14 (A) Schematic of the device structure and (B) the energy-level diagram of the corresponding device. (C) J—V curves and (D) external quantum efficiency (EQE) spectra of the PSCs without or with different HTLs. *Reproduced from Yu, Z. K.; Fu, W. F.; Liu, W. Q.; Zhang, Z. Q.; Liu, Y. J.; Yan, J. L.; Ye, T.; Yang, W. T.; Li, H. Y.; Chen, H. Z.* Chinese Chem. Lett. ***2017***, 28, *13; Huang, C.; Fu, W.; Li, C. Z., et al.* J. Am. Chem. Soc. ***2016***, 138, *2528; Mali, S. S.; Kim, H.; Kim, H. H.; Park, G. R.; Shim, S. E., Hong, C. K.* Mat. Today Chem. ***2017***, 4, *53 with permission of Elsevier.*

ITO/CuO_x/$MAPbI_3$/$PC_{61}BM$/ZnO/Al is fabricated and displays a best PCE of ~17.43% under standard AM 1.5G simulated solar irradiation with a V_{OC} of ~1.03 V, a J_{SC} of ~22.42 mA/cm^2, and an FF of ~0.76, which is significantly higher and more stable than the device fabricated with PEDOT:PSS (~11.98%) as the HTM under the same experimental conditions. The enhanced performance was attributed to the efficient hole extraction through the CuO_x layer as well as the high-quality $CH_3NH_3PbI_3$ films grown on the CuO_x. The results indicated that low-cost and solution-processed CuO_x film was a promising HTM for the high performance of PSCs with better stability. The energy level diagram of the device with different HTMs is shown in Fig. 14B. The devices with CuO_x as an HTL were optimized by adjusting the concentration of the CuO_x precursor. As expected, PSCs without an HTM show poor performance with low V_{OC}, J_{SC}, and FF. After incorporating a layer of PEDOT:PSS as the HTM, the J_{SC} increases from ~13.73 to

~16.85 mA/cm^2, FF increases from 0.60 to 0.75, V_{OC} increases slightly from ~0.94 to ~0.95 V, and the PCE significantly improved to ~11.98% from ~7.72% (Fig. 14C). The low V_{OC} and J_{SC} resulted from the low work function and the inability to block electrons efficiently of PEDOT:PSS, as well as the small crystalline grains of $MAPbI_3$ film on PEDOT:PSS. When CuO_x is used as the alternative, the best PCE of ~17.43% that substantially surpassed the performance of the PEDOT: PSS-based device, with a significantly enhanced J_{SC} of ~22.42 mA/cm^2 and V_{OC} of ~1.03 V, was obtained. FF slightly increased to ~0.76 compared with the one based on PEDOT:PSS. The enhanced J_{SC} was interpreted as a consequence of the better charge extraction ability of CuO_x and larger grain domains and reduced grain boundaries of the $MAPbI_3$ films, whereas the higher V_{OC} is resulted from the higher work function of CuO_x that rendered the ohmic contact at the $CuO_x/MAPbI_3$ interface and thus reduced V_{OC} loss. The EQE spectra of PSCs without and with different HTMs are shown in Fig. 14D. CuO_x as an HTM significantly improves the photocurrent in the region from ~350 to ~800 nm, which agrees well with the enhanced J_{SC}.[49b]

M. K. Nazeeruddin et al. mixed perovskite in the form of $(FAPbI_3)_{0.85}$ $(MAPbBr_3)_{0.15}$ in combination with Copper(I) thiocyanate (CuNCS) as p-type hole conductor that leads to over ~16% PCE under full sun illumination and yielded a remarkable monochromatic IPCE of ~85%. The devices displayed a J_{SC} of ~21.8 mA/cm^2, a V_{OC} of 1100 mV, an FF of 0.69, and a PCE of ~16.6%. Under similar conditions, the device without CuSCN showed a PCE of ~9.5%, with a significant decrease in the J_{SC} (from ~21.8 to ~15.64 mA/cm^2) and V_{OC} (from 1100 to 900 mV). The high J_{SC} with CuSCN was mainly due to the effective CT between perovskite and CuSCN, followed by the fast hole transport through CuSCN to the Au. Fig. 15A shows the UV–vis spectra of the perovskite and perovskite/CuSCN samples. The absorption of the sample is all very similar, reflecting the similar perovskite thickness (around 700 nm) for all the samples fabricated. Fig. 15B shows the PL emission of the perovskite layer. The perovskite PL emission peak is centered at around ~795 nm. When CuSCN layer is on top, a dramatic quenching of the PL signal is noticed. With respect to the pristine perovskite, a quenching of about ~95% is noticed in the presence of the HTM. This shows that with CuSCN, the quenching of the PL signal reflects an efficient hole transfer to the CuSCN. The fast quenching provides evidence that hole transfer happens at the perovskite/CuSCN interface.[227] S. Ito et al. also used the

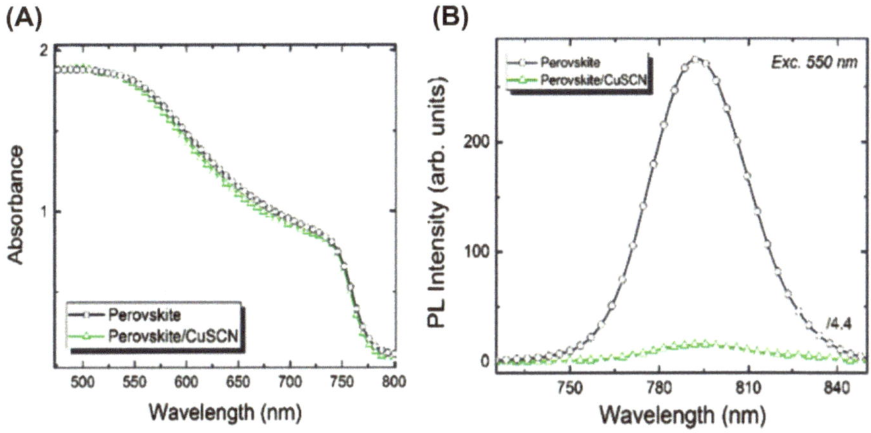

Figure 15 (A) UV–vis absorption spectra of the perovskite and perovskite/CuSCN layer samples on glass substrates. (B) CW-PL spectra of the samples, excitation at 550 nm. *Reproduced from Madhavan, V. E.; Zimmermann, I.; Carmona, C. R.; Grancini, G.; Buffiere, M., Belaidi, A.; Nazeeruddin, M. K.* ACS Energy Lett. ***2016***, 1, *1112 with permission of American Chemical Society.*

sequential fabrication technique to fabricate planar structure PSCs using CuSCN as the HTM. The I–V curves of FTO/planar TiO_2/(with/without porous TiO_2)/$CH_3NH_3PbI_3$/CuSCN/Au PSCs using the MAI-dripping or MAI-soaking methods were measured. A smooth less-pinhole $CH_3NH_3PbI_3$ perovskite layer is fabricated to obtain photocurrent densities as high as ~17 mA/cm^2. However, the FFs of PSCs are less than porous TiO_2 layers due to the partial short circuit in the PSCs, as shown in Fig. 16.[228] H. Rao et al. applied p-type inorganic hole-selective contact material, CuS nanoparticles (CuS NPs), in an inverted planar heterojunction (PHJ) PSC. The ITO modified with CuS NPs successfully tuned the surface work function from ~4.9 to ~5.1 eV but did not affect the surface roughness and transmittance, which could effectively reduce the interfacial carrier injection barrier and facilitate high hole extraction efficiency between the perovskite and ITO layers. After optimization, the maximum PCE of ~16% with low J–V hysteresis and excellent stability were achieved. Herewith, the low-cost solution-processed and stable CuS NPs was an alternative to interfacial modification material for industrial production in PSCs.[229] Kamat et al. employed CuI as the HTM, which was deposited by drop casting, and obtained a PCE of ~6.0%.[230] In 2014, Christians et al. reported the deposition of the CuI HTL by automated drop casting (modified doctor blade method), which resulted in a PCE of ~6.0%.[231] The similar HTL

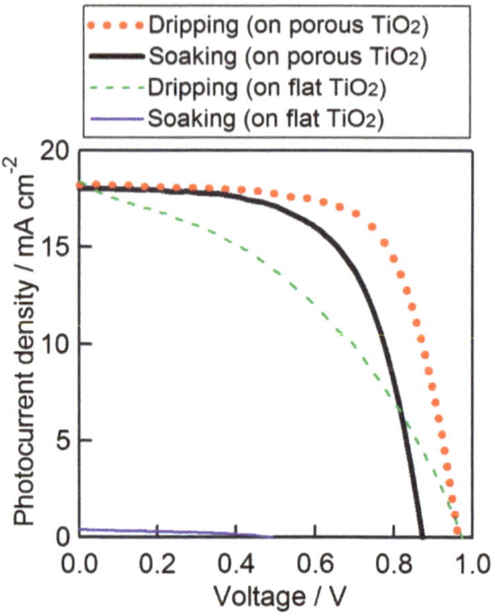

Figure 16 Photocurrent density–voltage curves of $CH_3NH_3PbI_3$ perovskite solar cells by the MAI-soaking and MAI-dripping methods on with/without porous TiO_2 layers. *Reproduced from Ito, S.; Tanaka, S.; Nishino, H.* J. Phys. Chem. Lett. ***2015****, 6, 881 with permission of American Chemical Society.*

was explored again in 2015 by Sepalage et al. Here also the CuI HTL was deposited on TCO by manual doctor blade method and a PCE of ~7.5% was achieved.[232] Again in 2015, Huangfu et al. deposited CuI thick film as the HTL and reported a PCE of ~5.8%.[233] Herein, the CuI thick film was deposited by convenient airbrush process, which exhibited a high conductivity and wide band gap. Organolead halide PSCs with different thickness of mesoporous layers and CuI HTL were fabricated and the performances of the cells were mainly controlled by the thickness of TiO_2 mesoporous layers. Under optimized conditions, a PCE of ~5.8% was achieved with a J_{SC} of ~22.3 mA/cm^2, a V_{OC} of ~614 mV, and an FF of ~42%. However, the V_{OC} remained low in comparison with the state-of-the-art perovskite-based solar cells, which might be due to the high recombination in CuI devices.[233] P. Nazari et al. also showed interest in CuI and adopted the simultaneous deposition method for preparing thin CuI film on TCO substrate. The interface engineering was conducted to grow CuI as an HTM at the interface of Cu contact and perovskite layer to achieve the reproducibility and durability of PSCs. The prepared devices showed ~9.24% and ~8.3% PCEs, respectively, for 0.1 and 1 cm^2 active areas.[234] Solution-processed $CuIn_{0.1}Ga_{0.9}(S_{0.9}Se_{0.1})_2$ (CIGSSe) nanocrystals were applied as a novel inorganic HTM for the fabrication of PSCs. A PCE of ~9.15% was obtained for CIGSSe-based devices with improved stability, compared with devices using Spiro-OMeTAD as the HTM.[235] D. Wang reported the novel interface-engineered HTL incorporating vanadium pentoxide (V_2O_5) and PEDOT:PSS (PVO) bilayer for efficient charge transport in the devices. The PSCs incorporating the PVO bilayer HTL exhibited ~20% higher PCE than conventional PEDOT-based devices. The PSCs incorporating the PVO bilayer HTL demonstrated superior electronic properties as evaluated using impedance spectroscopy measurements. The recombination resistance (Rrec) of the bilayer-based devices was ~57% higher than that of the reference cells. In addition to high charge selectivity, the bilayer PSCs exhibited low interfacial capacitance originating from electrode polarization and almost zero hysteresis. Furthermore, the bilayer-based devices were more stable than PEDOT-only devices and retained ~95% of their initial PCE even after 18 days of testing.[236]

4. CONCLUSIONS

In this review, we have investigated the progress of HTMs and ETMs and discussed their effects on PCEs of PSCs. The HTMs and ETMs play an indispensable role in hole transportation and charge extraction of PSCs. Driven by the need of cost-effective and high-efficient PSCs, more meaningful attempts are made in the designing and modifications of HTMs/ETMs to open new opportunities for highly efficient PSCs. The review summarizes several points to improve the performance of PSCs. First, the interface compatibility of ETMs and HTMs, a good energy level alignment between the HTMs/ETMs and perovskite layer, and novel methods for depositing thin layers of HTMs and ETMs are discussed because these factors are related to the minimization of recombination losses and essential for achieving high-performance PSCs. Second, certain structural configurations employing the introduction of carbon materials (graphene oxide, rGO, etc.) could improve the charge extractions of corresponding ETMs. Third, to improve the performance of PSCs, the doping process is ensuring to obtain sufficient conductivity and to shift the Fermi level toward the HOMO level. In summary, optimization of the ETMs and HTMs is an efficient method to enhance the efficiency improvement and stability as well as improve the overall performance of PSCs.

REFERENCES

1. Kojima, A.; Teshima, K.; Shirai, Y.; Miyasaka, T. *J. Am. Chem. Soc.* **2009,** *131,* 6050.
2. Cho, K. T.; Paek, S.; Grancini, G.; Roldán-Carmona, C.; Gao, P.; Lee, Y.; Nazeeruddin, M. K. *Energy Environ. Sci.* **2017,** *10,* 621.
3. Jeon, N. J.; Noh, J. H.; Kim, Y. C.; Yang, W. S.; Ryu, S.; Seok, S. I. *Nat. Mater.* **2014,** *13,* 897.
4. Green, M. A.; Ho-Baillie, A.; Snaith, H. J. *Nat. Photon.* **2014,** *8,* 506.
5. Noh, J. H.; Im, S. H.; Heo, J. H.; Mandal, T. N.; Seok, S. I. *Nano Lett.* **2013,** *13,* 1764.
6. De Wolf, S.; Holovsky, J.; Moon, S. J.; Loper, P.; Niesen, B.; Ledinsky, M.; Haug, F. J.; Yum, J. H.; Ballif, C. *J. Phys. Chem. Lett.* **2014,** *5,* 1035–1039.
7. Stranks, S. D.; Eperon, G. E.; Grancini, G.; Menelaou, C.; Alcocer, M. J. P.; Leijtens, T.; Herz, L. M.; Petrozza, A.; Snaith, H. J. *Science* **2013,** *342,* 341.
8. Dong, Q.; Fang, Y.; Shao, Y.; Mulligan, P.; Qiu, J.; Cao, L.; Huang, J. *Science* **2015,** *347,* 967.
9. Snaith, H. J. *J. Phys. Chem. Lett.* **2013,** *4,* 3623.
10. Burschka, J.; Pellet, N.; Moon, S. J.; Humphry-Baker, R.; Gao, P.; Nazeeruddin, M. K.; Grätzel, M. *Nature* **2013,** *499,* 316.
11. D'Innocenzo, V.; Srimath Kandada, A. R.; De Bastiani, M.; Gandini, M.; Petrozza, A. *J. Am. Chem. Soc.* **2014,** *136,* 17730.
12. Jeon, N. J.; Noh, J. H.; Yang, W. S.; Kim, Y. C.; Ryu, S.; Seo, J.; Il Seok, S. *Nature* **2015,** *517,* 476.
13. Paufler, P. *Acta Crystallogr. Sect. A Found. Crystallogr.* **2012,** *68,* 523.
14. Xing, G.; Mathews, N.; Sun, S.; Lim, S. S.; Lam, Y. M.; Grätzel, M.; Mhaisalkar, S.; Sum, T. C. *Science* **2013,** *342,* 344–347.
15. Miyata, A.; Mitioglu, A.; Plochocka, P.; Portugall, O.; Wang, J. T.-W.; Stranks, S. D.; Snaith, H. J.; Nicholas, R. J. *ArXiv Prepr.* **2015**. ArXiv150407025.
16. Pellet, N.; Gao, P.; Gregori, G.; Yang, T.-Y.; Nazeeruddin, M. K.; Maier, J.; Grätzel, M. *Angew. Chem. Int. Ed.* **2014,** *53,* 3151.
17. Lee, M. M.; Teuscher, J.; Miyasaka, T.; Murakami, T. N.; Snaith, H. J. *Science* **2012,** *338,* 643.
18. Liu, M.; Johnston, M. B.; Snaith, H. J. *Nature* **2013,** *501,* 395.
19. Bi, D.; Gao, P.; Scopelliti, R.; Oveisi, E.; Luo, J.; Grätzel, M.; Hagfeldt, A.; Nazeeruddin, M. K. *Adv. Mater.* **2016,** *28,* 2910.
20. Saliba, M.; Matsu, T.; Seo, J.-Y.; Domanski, K.; Correa-Baena, J.-P.; Nazeeruddin, M. K.; Zakeeruddin, S. M.; Tress, W.; Abate, A.; Hagfeldt, A.; Grätzel, M. *Energy Environ. Sci.* **2016,** *9,* 1989.
21. Barbé, J.; Tietze, M. L.; Neophytou, M.; Murali, B.; Alarousu, E.; Labban, A. E.; Abulikemu, M.; Yue, W.; Mohammed, O. F.; McCulloch, I.; Amassian, A.; Gobbo, S. D. *ACS Appl. Mater. Interfaces* **2017,** *9,* 11828.
22. Ke, W.; Fang, G.; Wang, J.; Qin, P.; Tao, H.; Lei, H.; Liu, Q.; Dai, X.; Zhao, X. *ACS Appl. Mater. Interfaces* **2014,** *6,* 15959.
23. Ameen, S.; Rub, M. A.; Kosa, S. A.; Alamry, K. A.; Akhtar, M. S.; Shin, H. S.; Seo, H. K.; Asiri, A. M.; Nazeeruddin, M. K. *ChemSusChem* **2016,** *9,* 10.
24. Byranvand, M. M.; Kim, T.; Song, S.; Kang, G.; Ryu, S. U.; Park, T. *Adv. Energy Mater* **2018,** *8,* 1702235.
25. Ameen, S.; Akhtar, M. S.; Seo, H. K.; Shin, H. S. *Langmuir* **2014,** *30,* 12786.
26. Fakharuddin, A.; Giacomo, F. D.; Palma, A. L.; Matteocci, F.; Ahmed, I.; Razza, S.; Epifanio, A. D.; Licoccia, S.; Ismail, J.; Carlo, A. D.; Brown, T. M.; Jose, R. *ACS Nano* **2015,** *9,* 8420.
27. Ameen, S.; Nazim, M.; Akhtar, M. S.; Nazeeruddin, M. K.; Shin, H. S. *Nanoscale* **2017,** *9,* 17544–17550.

28. Qin, P.; Paulose, M.; Dar, M. I.; Moehl, T.; Rora, N. A.; Gao, P.; Varghese, O. K.; Gräetzel, M.; Nazeeruddin, M. K. *Small* **2015,** *11,* 5533.
29. Yu, Y.; Li, J.; Geng, D.; Wang, J.; Zhang, L.; Andrew, T. L.; Arnold, M. S.; Wang, X. *ACS Nano* **2015,** *9,* 564.
30. Yao, H.; Fu, W.; Yang, H.; Ma, J.; Sun, M.; Chen, Y.; Zhang, W.; Wu, D.; Lv, P.; Li, M. *Electrochim. Acta* **2014,** *125,* 258.
31. Tao, H.; Ke, W.; Wang, J.; Liu, Q.; Wan, J.; Yang, G.; uojia Fang, G. *J. Power Sour.* **2015,** *290,* 144.
32. He, X.; Wu, J.; Tu, Y.; Xie, Y.; Dong, J.; Jia, J.; Wei, Y.; Lan, Z. *J. Power Sour.* **2016,** *332,* 366.
33. Sun, S.; Salim, T.; Mathews, N.; Duchamp, M.; Boothroyd, C.; Xing, G.; Sum, T. C.; Lam, Y. M. *Energy Environ. Sci.* **2014,** *7,* 399.
34. Wojciechowski, K.; Saliba, M.; Leijtens, T.; Abate, A.; Snaith, H. J. *Energy Environ. Sci.* **2014,** *7,* 1142.
35. Wang, P.; Zhang, J.; Chen, R.; Zeng, Z.; Huang, X.; Wang, L.; Xu, J.; Hu, Z.; Zhu, Y. *Electrochim. Acta* **2017,** *227,* 180.
36. Liao, W.; Zhao, D.; Yu, Y.; Grice, C. R.; Wang, C.; Cimaroli, A. J.; Schulz, P.; Meng, W.; Zhu, K.; Xiong, R. G. *Adv. Mater.* **2016,** *28,* 9333.
37. Yang, G.; Tao, H.; Qin, P. L.; Ke, W. J.; Fang, G. J. *J. Mater. Chem. A* **2016,** *4,* 3970.
38. Liang, Z. Q.; Zhang, Q. F.; Jiang, L.; Cao, G. Z. *Energy Environ. Sci.* **2015,** *8,* 3442.
39. Liu, H. F.; Huang, Z. R.; Wei, S. Y.; Zheng, L. L.; Xiao, L. X.; Gong, Q. H. *Nanoscale* **2016,** *8,* 6209.
40. Zhang, W.; Ren, Z.; Guo, Y.; He, X.; Li, X. *Electrochim. Acta* **2018,** *268,* 539.
41. Liu, D.; Kelly, T. L. *Nat. Photonics* **2014,** *8,* 133.
42. Ameen, S.; Akhtar, M. S.; Seo, H. K.; Nazeeruddin, M. K.; Shin, H. S. *J. Phys. Chem. C* **2015,** *119,* 10379.
43. You, J.; Meng, L.; Song, T. B.; Guo, T. F.; Yang, Y.; Chang, W. H.; Hong, Z.; Chen, H.; Zhou, H.; Chen, Q.; Liu, Y.; De Marco, N.; Yang, Y. *Nat. Nanotechnol.* **2016,** *11,* 75.
44. Song, J.; Liu, L.; Wang, X.-F.; Chen, G.; Tian, W.; Miyasaka, T. *J. Mater. Chem. A* **2017,** *5,* 13439.
45. Azmi, R.; Hadmojo, W. T.; Sinaga, S.; Lee, C. L.; Yoon, S. C.; Jung, I. H.; Jang, S. Y. *Adv. Energy Mater.* **2018,** *8,* 1701683.
46. Zhao, H.; Zhang, K. C.; Wang, Z. W.; Huang, P.; Zhu, K.; Li, Z. D.; Li, D. H.; Yuan, L. G.; Zhou, Y.; Song, B. *ACS Appl. Mater. Interfaces* **2017,** *9,* 26234.
47. Son, D. Y.; Im, J. H.; Kim, H. S.; Park, N. G. *J. Phys. Chem. C* **2014,** *118,* 16567.
48. Ruankhama, P.; Wongratanaphisan, D.; Gardchareon, A.; Phadungdhitidhad, S.; Choopun, S.; Sagawad, T. *Appl. Surf. Sci.* **2017,** *410,* 393.
49. (a) Liu, C.; Li, W.; Zhang, C.; Ma, Y.; Fan, J.; Mai, Y. *J. Am. Chem. Soc.* **2018,** *140,* 3825. (b) Yu, Z. K.; Fu, W. F.; Liu, W. Q.; Zhang, Z. Q.; Liu, Y. J.; Yan, J. L.; Ye, T.; Yang, W. T.; Li, H. Y.; Chen, H. Z. *Chinese Chem. Lett.* **2017,** *28,* 13. (c) Tiwana, P.; Docampo, P.; Johnston, M. B.; Snaith, H. J.; Herz, L. M. *ACS Nano* **2011,** *5,* 5158.
50. Wali, Q.; Fakharuddin, A.; Jose, R. *J. Power Sources* **2015,** *293,* 1039.
51. Correa Baena, J. P.; Steier, L.; Tress, W.; Saliba, M.; Neutzner, S.; Matsui, T.; Giordano, F.; Jacobsson, T. J.; Kandada, A. R. S.; Zakeeruddin, S. M.; Petrozza, A.; Abate, A.; Nazeeruddin, M. K.; Graetzel, M.; Hagfeldt, A. *Energy Environ. Sci.* **2015,** *8,* 2928.
52. Murugadoss, G.; Kanda, H.; Tanaka, S.; Nishino, H.; Ito, S.; Imahori, H.; Umeyama, T. *J. Power Sources* **2016,** *307,* 891.
53. Dong, Q.; Shi, Y.; Wang, K.; Li, Y.; Wang, S.; Zhang, H.; Xing, Y.; Du, Y.; Bai, X.; Ma, T. *J. Phys. Chem. C* **2015,** *119,* 10212.

54. Jiang, Q.; Zhang, L.; Wang, H.; Yang, X.; Meng, J.; Liu, H.; Yin, Z.; Wu, J.; Zhang, X.; You, J. *Nat. Energy* **2016,** *1,* 1.
55. Gao, C.; Yuan, S.; Cao, B.; Yu, J. *Chem. Eng. J.* **2017,** *325,* 378.
56. Gheno, A.; Pham, T. T. T.; Bin, C. D.; Bouclé, J.; Ratier, B.; Vedraine, S. *Sol. Ener. Mater. Sol. Cells* **2017,** *161,* 347.
57. Mahmood, K.; Swain, B. S.; Kirmania, A. R.; Amassian, A. *J. Mater. Chem. A* **2015,** *3,* 9051.
58. Choi, Y. H.; Kim, H. B.; Yang, I. S.; Sung, S. D.; Choi, Y. S.; Kim, J.; Lee, W. I. *ACS Appl. Mater. Interfaces* **2017,** *9,* 25257.
59. Ling, X.; Yuan, J.; Liu, D.; Wang, Y.; Zhang, Y.; Chen, S.; Wu, H.; Jin, F.; Wu, F.; Shi, G.; Tang, X.; Zheng, J.; Liu, S.; Liu, Z.; Ma, W. *ACS Appl. Mater. Interfaces* **2017,** *9,* 23181.
60. Roy, A.; Selvaraj, P.; Sujatha Devi, P.; Sundaram, S. *Mater. Lett.* **2018,** *219,* 166.
61. Bera, A.; Sheikh, A. D.; Haque, M. A.; Bose, R.; Alarousu, E.; Mohammed, O. F.; Wu, T. *ACS Appl. Mater. Interfaces* **2015,** *7,* 28404.
62. Bera, A.; Wu, K.; Sheikh, A.; Alarousu, E.; Mohammed, O. F.; Wu, T. *J. Phys. Chem. C* **2014,** *118,* 28494.
63. Chiang, H.; Tseng, Z. L.; Wu, C. G. *J. Mater. Chem.* **2014,** *2,* 15897.
64. Heo, J. H.; Han, H. J.; Kim, D.; Ahn, T. K.; Im, S. H. *Energy Environ. Sci.* **2015,** *8,* 1602.
65. Paek, N.; Cho, H.; Choi, H.; Jeong, J. S.; Lim, J. Y.; Hwang, J. K.; Lee, J. K. *J. Phys. Chem. C* **2014,** *118,* 25899.
66. Y Jeng, J.; Chiang, Y. F.; Lee, M. H.; Peng, S. R.; Guo, T. F.; Chen, P.; Wen, T. C. *J. Mater. Chem. A* **2013,** *2,* 15897.
67. Sun, S.; Salim, T.; Mathews, N.; Duchamp, M.; Boothroyd, C.; Xing, G.; Sum, T. C.; Lam, Y. M. *Adv. Mater* **2014,** *25,* 3727.
68. Yella, O.; Lee, A.; Minguez, Y. H.; Espallargas, G.; Graetzel, M.; Nazeeruddin, M. K.; Bolink, H. J. *Adv. Energy Mater* **2014,** *4,* 1400345.
69. Huang, J.; Gua, Z.; Zuo, L.; Ye, T.; Chen, H. *Sol. Energy* **2016,** *133,* 331.
70. Shao, Y.; Xiao, Z.; Bi, C.; Yuan, Y.; Huang, J. *Nat. Commun.* **2014,** *5,* 5784.
71. Nie, W.; Tsai, H.; Asadpour, R.; Blancon, J. C.; Neukirch, A. J.; Gupta, G.; Crochet, J. J.; Chhowalla, M.; Tretiak, S.; Alam, M. A.; Wang, H. L.; Mohite, A. D. *Science* **2015,** *347,* 522.
72. Dai, S. M.; Deng, L. L.; Zhang, M. L.; Chen, W. Y.; Zhu, P.; Wang, X.; Li, C.; Tan, Z.; Xie, S. Y.; Huang, R. B.; Zheng, L. S. *Inorg. Chim. Acta.* **2017,** *468,* 146.
73. Lia, M.; Wang, Z. K.; Kang, T.; Yang, Y.; Gao, X.; Hsu, C. S.; Li, Y.; Liao, L. S. *Nano Energy* **2018,** *43,* 47.
74. Namkoong, G.; Mamun, A. A.; Ava, T. T. *Org. Electron.* **2018,** *56,* 163.
75. Chang, S.; Han, G. D.; Weis, J. G.; Park, H.; Hentz, O.; Zhao, Z.; Swager, T. M.; Gradec, S. *ACS Appl. Mater. Interfaces* **2016,** *8,* 8511.
76. Wang, Q.; Shao, Y.; Dong, Q.; Xiao, Z.; Yuan, Y.; Huang, J. *Energy Environ. Sci.* **2014,** *7,* 2359.
77. Malinkiewicz, O.; Yella, A.; Lee, Y. H.; Espallargas, G. M.; Graetzel, M.; Nazeeruddin, M. K.; Bolink, H. J. *Nat. Photonics* **2014,** *8,* 128.
78. You, J.; Hong, Z.; Yang, Y.; Chen, Q.; Cai, M.; Song, T. B.; Chen, C. C.; Lu, S.; Liu, Y.; Zhou, H.; Yang, Y. *ACS Nano* **2014,** *8,* 1674.
79. Seo, J.; Park, S.; Kim, Y. C.; Jeon, N. J.; Noh, J. H.; Yoon, S. C.; Seok, S. I. *Energy Environ. Sci.* **2014,** *7,* 2642.
80. Yang, D.; Yang, R.; Zhang, J.; Yang, Z.; Liu, S.; Li, C. *Energy Environ. Sci.* **2015,** *8* (11), 3208.
81. Tao, C.; Neutzner, S.; Colella, L.; Marras, S.; Kandada, A. R. S.; Gandini, M.; De Bastiani, M.; Pace, G.; Manna, L.; Caironi, M. *Energy Environ. Sci.* **2015,** *8,* 2365.

82. Gao, X. X.; Ge, Q. Q.; Xue, D. J.; Ding, J.; Ma, J. Y.; Chen, Y. X.; Zhang, B.; Feng, Y.; Wan, L. J.; Hu, J. S. *Nanoscale* **2016,** *8,* 16881.
83. Chen, B. X.; Rao, H. S.; Li, W. G.; Xu, Y. F.; Chen, H. Y.; Kuang, D. B.; Su, C. Y. *J. Mater. Chem. A* **2016,** *4,* 5647.
84. Arora, N.; Dar, M. I.; Abdi-Jalebi, M.; Giordano, F.; Pellet, N.; Jacopin, G.; Friend, R. H.; Zakeeruddin, S. M.; Grätzel, M. *Nano Lett.* **2016,** *16,* 7155.
85. Giordano, F.; Abate, A.; Correa Baena, J. P.; Saliba, M.; Matsui, T.; Im, S. H.; Zakeeruddin, S. M.; Nazeeruddin, M. K.; Hagfeldt, A.; Graetzel, M. *Nat. Commun.* **2016,** *7,* 10377.
86. Peng, J.; Duong, T.; Zhou, X.; Shen, H.; Wu, Y.; Mulmudi, H. K.; Wan, Y.; Zhong, D.; Li, J.; Tsuzuki, T.; et al. *Adv. Energy Mater.* **2017,** *7,* 1601768.
87. Li, Y.; Guo, Y.; Li, Y.; Zhou, X. *Electrochim. Acta* **2016,** *200,* 29.
88. Qin, P.; Domanski, A. L.; Chandiran, A. K.; Berger, R.; Butt, H. J.; Dar, M. I.; Moehl, T.; Tetreault, N.; Gao, P.; Ahmad, S.; Nazeeruddin, M. K.; Grätzel, M. *Nanoscale* **2014,** *6,* 1508.
89. Xiao, Y.; Cheng, N.; Kondamareddy, K. K.; Wang, C.; Liu, P.; Guo, S.; Zhao, X. Z. *J. Power Sour.* **2017,** *342,* 489.
90. Kim, D.; Han, G.; Seong, W.; Lee, J.; Kim, B.; Park, N.; Hong, K.; Lee, S.; Jung, H. *ChemSusChem* **2015,** *8,* 2392.
91. Pathak, S. K.; Abate, A.; Ruckdeschel, P.; Roose, B.; Gödel, K. C.; Vaynzof, Y.; Santhala, A.; Watanabe, S. I.; Hollman, D. J.; Noel, N.; Sepe, A.; Wiesner, U.; Friend, R.; Snaith, H. J.; Steiner, U. *Adv. Funct. Mater.* **2014,** *24,* 6046.
92. Xiao, Y.; Cheng, N.; Kondamareddy, K. K.; Wang, C.; Liu, P.; Guo, S.; Zhao, X. Z. *J. Power Sour.* **2017,** *342,* 489.
93. Wu, M. C.; Chan, S. H.; Jao, M. H.; Su, W. F. *Sol. Ener. Mater. Sol. Cells* **2016,** *157,* 447.
94. Tseng, Z. L.; Chiang, C. H.; Chang, S. H.; G Wu, C. *Nano Energy* **2016,** *28,* 311.
95. Nagaoka, H.; Ma, F.; de Quilettes, D. W.; Vorpahl, S. M.; Glaz, M. S.; Colbert, A. E.; Ziffer, M. E.; Ginger, D. S. *J. Phys. Chem. Lett.* **2015,** *6,* 669.
96. Yin, G.; Ma, J.; Jiang, H.; Li, J.; Yang, D.; Gao, F.; Zeng, J.; Liu, Z.; Liu, S. F. *ACS Appl. Mater. Interfaces* **2017,** *9,* 10752.
97. Cai, Q.; Zhang, Y.; Liang, C.; Li, P.; Gu, H.; Liu, X.; Wang, J.; Shentu, Z.; Fan, J.; Shao, G. *Electrochim. Acta* **2018,** *261,* 227.
98. Heo, J. H.; You, M. S.; Chang, M. H.; Yin, W.; Ahn, T. K.; Lee, S. J.; Sung, S. J.; Kim, D. H.; Im, S. H. *Nano Energy* **2015,** *15,* 530.
99. Gu, X.; Wang, Y.; Zhang, T.; Liu, D.; Zhang, R.; Zhang, P.; Wu, J.; David Chen, Z.; Li, S. *J. Mater. Chem. C* **2017,** *5,* 10754.
100. Lee, W.; Choi, J.; Jung, J. W. *Org. Electr.* **2017,** *51,* 404.
101. Qian, X.; Liu, H.; Huang, C.; Chen, S.; Zhang, L.; Li, Y.; Wang, J.; Li, Y. *Sci. Rep.* **2015,** *5,* 7756.
102. Qian, X. M.; Liu, H. B.; Li, Y. L. *Chin. Sci. Bull.* **2013,** *58,* 2686.
103. Habisreutinger, S. N.; Leijtens, T.; Eperon, G. E.; Stranks, S. D.; Nicholas, R. J.; Snaith, H. J. *J. Phys. Chem. Lett.* **2014,** *5,* 4207.
104. Habisreutinger, S. N.; Leijtens, T.; Eperon, G. E.; Stranks, S. D.; Nicholas, R. J.; Snaith, H. J. *Nano Lett.* **2014,** *14,* 5561.
105. Yang, N.; Liu, Y.; Wen, H.; Tang, Z.; Zhao, H.; Li, Y.; Wang, D. *ACS Nano* **2013,** *7,* 1504.
106. Wang, J. T. W.; Ball, J. M.; Barea, E. M.; Abate, A.; Webber, J. A.; Huang, J.; Saliba, M.; Mora-Sero, I.; Bisquert, J.; Snaith, H. J. *Nano Lett.* **2014,** *14,* 724.
107. Feng, S. L.; Yang, Y. G.; Li, M.; Wang, J. M.; Cheng, Z. D.; Li, J. H.; Ji, G. W.; Yin, G. Z.; Song, F.; Wang, Z. K.; Li, J. Y.; Gao, X. Y. *ACS Appl. Mater. Interfaces* **2016,** *8,* 14503.

108. Yang, Z.; Xie, J.; Arivazhagan, V.; Xiao, K.; Qiang, Y.; Huang, K.; Hu, M.; Cui, C.; Yu, X.; Yang, D. *Nano Energy* **2017,** *40,* 345.
109. Kakavelakis, G.; Maksudov, T.; Konios, D.; Paradisanos, I.; Kioseoglou, G.; Stratakis, E.; Kymakis, E. *Adv. Energy Mater.* **2017,** *7,* 1602120.
110. Kuang, C.; Tang, G.; Jiu, T.; Yang, H.; Liu, H.; Li, B.; Luo, W.; Li, X.; Zhang, W.; Lu, F.; Fang, J.; Li, Y. *Nano Lett.* **2015,** *15,* 2756.
111. Zhou, H. P.; Chen, Q.; Li, G.; Luo, S.; Song, T. B.; Duan, H. S.; Hong, Z.; You, J. B.; Liu, Y. S.; Yang, Y. *Science* **2014,** *345,* 542.
112. Wang, C. L.; Zhao, D. W.; Grice, C. R.; Liao, W. Q.; Yu, Y.; Cimaroli, A.; Shrestha, N.; Roland, P. J.; Chen, J.; Yu, Z. H.; Liu, P.; Cheng, N.; Ellingson, R. J.; Zhao, X. Z.; Yan, Y. F. *J. Mater. Chem. A* **2016,** *4,* 12080.
113. Okamoto, Y.; Suzuki, Y. *J. Phys. Chem. C* **2016,** *120,* 13995.
114. Tong, S. W.; Balapanuru, J.; Fu, D.; Loh, K. P. *ACS Appl. Mater. Interfaces* **2016,** *8,* 29496.
115. Numata, Y.; Sanehira, Y.; Miyasaka, T. *ACS Appl. Mater. Interfaces* **2016,** *8,* 4608.
116. Li, M.; Wang, Z. K.; Kang, T.; Yang, Y.; Gao, X.; Hsu, C. S.; Li, Y.; Liao, L. S. *Nano Energy* **2018,** *43,* 47.
117. Apostolopouloua, A.; Sygkridoua, Andreas, D.; Kalarakisd, A. N.; Stathatos, E. *Sol. Energy Mater. Sol. Cells* **2017,** *166,* 100.
118. Li, W.; Jiang, Q.; Yang, J.; Luo, Y.; Li, X.; Hou, Y.; Zhou, S. *Sol. Energy Mater. Sol. Cells* **2017,** *159,* 143.
119. Eze, V. O.; Seike, Y.; Mori, T. *Org. Electron.* **2017,** *46,* 253.
120. Che, M.; Zhu, L.; Zhao, Y. L.; Yao, D. S.; Gu, X. Q.; Song, J.; Qiang, Y. H. *Mater. Sci. Semi. Proc.* **2016,** *56,* 29.
121. Zhang, J.; Hong Tan, C.; Du, T.; Morbidoni, M.; Lin, C. T.; Xu, S.; Durrant, J. R.; McLachlan, M. A. *Sci. Bull.* **2018,** *63,* 343.
122. Chen, Y.; Xu, C.; Xiong, J.; Zhang, Z.; Zhang, X.; Yang, J.; Xue, X.; Yang, D.; Zhang, J. *Org. Electr.* **2018,** *58,* 294.
123. Chandrasekhar, P. S.; Komarala, V. K. *RSC Adv.* **2017,** *7,* 28610.
124. Ahmed, M. I.; Hussain, Z.; Mujahid, M.; Khan, A. N.; Javaid, S. S.; Habib, A. *AIP Adv.* **2016,** *6,* 065303.
125. Zhao, X.; Tao, L.; Li, H.; Huang, W.; Sun, P.; Liu, J.; Liu, S.; Sun, Q.; Cui, Z.; Sun, L.; Shen, Y.; Yang, Y.; Wang, M. *Nano Lett.* **2018,** *18,* 2442.
126. Wang, J. T. W.; Ball, J. M.; Barea, E. M.; Abate, A.; Webber, J. A. A.; Huang, J.; Saliba, M.; Sero, I. M.; Bisquert, J.; Snaith, H. J.; Nicholas, R. J. *Nano Lett.* **2014,** *14,* 724.
127. Li, H.; Shi, W.; Huang, W.; Yao, E. P.; Han, J.; Chen, Z.; Liu, S.; Shen, Y.; Wang, M.; Yang, Y. *Nano Lett.* **2017,** *17,* 2328.
128. Sia, H.; Liaoa, Q.; Zhang, Z.; Lia, Y.; Yang, X.; Zhang, G.; Kang, Z.; Zhang, Y. *Nano Energy* **2016,** *22,* 223.
129. Li, J. F.; Zhang, Z. L.; Gao, H. P.; Zhang, Y.; Mao, Y. L. *Inside Chem. A* **2015,** *3,* 19476.
130. Mali, S. S.; Shim, C. S.; Park, H. K.; Heo, J.; Patil, P. S.; Hong, C. K. *Chem. Mater.* **2015,** *27,* 1541.
131. Bi, D.; Tress, W.; Dar, M. I.; Gao, P.; Luo, J.; Renevier, C.; Schenk, K.; Abate, A.; Giordano, F.; Baena, J. P. C.; Decoppet, J. D.; Zakeeruddin, S. M.; Nazeeruddin, M. K.; Gratzel, M.; Hagfeldt, A. *Sci. Adv.* **2016,** *2,* 1501170.
132. Niu, G.; Li, W.; Meng, F.; Wang, L.; Dong, H.; Qiu, Y. *J. Mater. Chem. A* **2014,** *2,* 705.
133. Palomares, E.; Clifford, J. N.; Haque, S. A.; Lutz, T.; Durrant, J. R. *J. Am. Chem. Soc.* **2003,** *125,* 475.
134. Lee, Y. H.; Luo, J. S. H.; Son, M. K.; Gao, P.; Cho, K. T.; Seo, J.; Zakeeruddin, S. M.; Grätzel, M.; Nazeeruddin, M. K. *Adv. Mater.* **2016,** *28,* 3966.

135. M-Beloqui, M.; Lanzetta, L.; Palomares, E. *Chem. Mater.* **2016,** *28,* 207.
136. Hu, W. D.; Agnese, C. D.; Wang, X. F.; Chen, G.; Zhen Li, M.; Song, J. X.; Wei, Y. J.; Miyasaka, T. *J. Phot. Photobio. A: Inside Chem.* **2018,** *357,* 36.
137. Zhang, Y. N.; Lia, B.; Zhang, L. Y.; Yin, L. W. *Sol. Ener. Mater. Sol. Cells* **2017,** *170,* 187.
138. Zhang, J.; Hultqvist, A.; Zhang, T.; Jiang, L.; Ruan, C.; Yang, L.; Cheng, Y.; Edoff, M.; Johansson, E. M. J. *ChemSusChem* **2017,** *10,* 3810.
139. Guarnera, S.; Abate, A.; Zhang, W.; Foster, J. M.; Richardson, G.; Petrozza, A.; Snaith, H. J. *J. Phys. Chem. Lett.* **2015,** *6,* 432.
140. Hou, X.; Pan, L.; Huan, S.; Yang, W. O.; Chen, X. *Electrochim. Acta* **2017,** *236,* 351.
141. Yu, X.; Chen, S.; Yan, K.; Cai, X.; Hu, H.; Peng, M.; Chen, B.; Dong, B.; Gao, X.; Zou, D. *J. Power Sour.* **2016,** *325,* 534.
142. Dong, M.; Wang, Q.; Zhang, F.; Chen, S.; Zhang, J.; Bian, T.; Ma, L.; Wang, Y.; Shi *Nano Energy* **2017,** *38,* 358.
143. Li, H.; Tao, L.; Huang, F.; Sun, Q.; Zhao, X.; Han, J.; Shen, Y.; Wang, M. *ACS Appl. Mater. Interfaces* **2017,** *9,* 38967.
144. Zhu, L.; Xiao, J.; Shi, J.; Wang, J.; Lv, S.; Xu, Y.; Luo, Y.; Xiao, Y.; Wang, S.; Meng, Q.; Li, X.; Li, D. *Nano Res.* **2015,** *8,* 1116.
145. Chen, S.; Hou, Y.; Chen, H.; Richter, M.; Guo, F.; Kahmann, S.; Tang, X.; Stubhan, T.; Zhang, H.; Li, N.; Gasparini, N.; Quiroz, C. O. R.; Khanzada, L. S.; Matt, G. J.; Osvet, A.; Brabec, C. J. *Adv. Energy Mater.* **2016,** *6,* 6, 1600132.
146. Elumalai, N. K.; Uddin, A. *Energy Environ. Sci.* **2016,** *9,* 391.
147. Shao, Y.; Yuan, Y.; Huang, J. *Nat. Energy* **2016,** *1,* 15001.
148. Nouri, E.; Krishna, J. V. S.; Kumar, C. V.; Dracopoulos, V.; Giribabu, L.; Mohammadi, M. R.; Lianos, P. *Electrochim. Acta* **2016,** *222,* 875.
149. Zhang, T.; Yang, M.; Zhao, Y.; Zhu, K. *Nano Lett.* **2015,** *15,* 3959.
150. Chen, H.; Bryant, D.; Troughton, J.; Kirkus, M.; Neophytou, M.; Miao, X.; Durrant, J. R.; McCulloch, I. *Chem. Mater.* **2016,** *28,* 2515.
151. Liu, X.; Zhang, F.; Liu, X.; Sun, M.; Wang, S.; Li, D.; Meng, Q.; Li, X. *J. Energy Chem.* **2016,** *25,* 702.
152. Wang, H.; Sheikh, A. D.; Feng, Q.; Li, F.; Chen, Y.; Yu, W.; Alarousu, E.; Ma, C.; Haque, M. A.; Shi, D.; Wang, Z.-S.; Mohammed, O. F.; Bakr, O. M.; Wu, T. *ACS Photonics* **2015,** *2,* 849.
153. Guo, Y.; Liu, C.; Inoue, K.; Harano, K.; Tanaka, H.; Nakamura, E. *J. Mater. Chem. A* **2014,** *2,* 13827.
154. Qin, P.; Tetreault, N.; Dar, M. I.; Gao, P.; McCall, K. L.; Rutter, S. R.; Ogier, S. D.; Forrest, N. D.; Bissett, J. S.; Simms, M. J.; Page, A. J.; Fisher, R.; Graetzel, M.; Nazeeruddin, M. K. *Adv. Energy Mater.* **2015,** *5,* 1400980.
155. Chen, W.; Bao, X.; Zhu, Q.; Zhu, D.; Qiu, M.; Sun, M.; Yang, R. *J. Mater. Chem. C* **2015,** *3,* 10070.
156. Yang, W. S.; Noh, J. H.; Jeon, N. J.; Kim, Y. C.; Ryu, S.; Seo, J.; Seok, S. I. *Science* **2015,** *348,* 1234.
157. Li, M. H.; Hsu, C. W.; Shen, P. S.; Cheng, H. M.; Chi, Y.; Chen, P.; Guo, T.-F. *Chem. Commun.* **2015,** *51,* 15518.
158. Zhou, H.; Chen, Q.; Li, G.; Luo, S.; Song, T. B.; Duan, H. S.; Hong, Z.; You, J.; Liu, Y.; Yang, Y. *Science* **2014,** *345,* 542.
159. Li, X.; Bi, D.; Yi, C.; Decoppet, J.-D.; Luo, J.; Zakeeruddin, S. M.; Hagfeldt, A.; Grätzel, M. *Science* **2016,** *353,* 58.
160. Zhang, F.; Liu, X.; Yi, C.; Bi, D.; Luo, J.; Wang, S.; Li, X.; Xiao, Y.; Zakeeruddin, S. M.; Grätzel, M. *ChemSusChem* **2016,** *1,* 9.
161. Zhao, X.; Zhang, F.; Yi, C.; Bi, D.; Bi, X.; Wei, P.; Luo, J.; Liu, X.; Wang, S.; Li, X.; Zakeeruddin, S. M.; Gratzel, M. *J. Mater. Chem. A* **2016,** *4,* 16330.

162. Jeon, S.; Thakur, U. K.; Lee, D.; Wenping, Y.; Kim, D.; Lee, S.; Ahn, T. K.; Park, H. J.; Kim, B.-G. *Org. Electron.* **2016,** *37,* 134.
163. Huang, C.; Fu, W.; Li, C.-Z.; Zhang, Z.; Qiu, W.; Shi, M.; Heremans, P.; Jen, A. K.-Y.; Chen, H.; Am, J. *Chem. Soc.* **2016,** *138,* 2528.
164. Cheng, M.; Aitola, K.; Chen, C.; Zhang, F.; Liu, P.; Sveinbjörnsson, K.; Hua, Y.; Kloo, L.; Boschloo, G.; Sun, L. *Nano Energy* **2016,** *30,* 387.
165. Paek, S.; Qin, P.; Lee, Y.; Cho, K. T.; Gao, P.; Grancini, G.; Oveisi, E.; Gratia, P.; Rakstys, K.; Al-Muhtaseb, S. A.; Ludwig, C.; Ko, J.; Nazeeruddin, M. K. *Adv. Mater.* **2017,** *29,* 1606555.
166. Burschka, J.; Dualeh, A.; Kessier, F.; Baranoff, E.; Cevey, N.-L.; Yi, C.; Nazeeruddin, M. K.; Grätzel, M. *J. Am. Chem. Soc.* **2011,** *133,* 18042.
167. Zong, X.; Qiao, W.; Chen, Y.; Sun, Z.; Liang, M.; Xue, S. *Tetrahedron* **2017,** *73,* 3398.
168. Hua, Y.; Zhang, J.; Xu, B.; Liu, P.; Cheng, M.; Kloo, L.; Johansson, E. M. J.; Sveinbjörnsson, K.; Aitola, K.; Boschloo, G.; Sun, L. *Nano Energy* **2016,** *26,* 108.
169. Saliba, M.; Orlandi, S.; Matsui, T.; Aghazada, S.; Cavazzini, M.; Correa-Baena, J. P.; Gao, P.; Scopelliti, R.; Mosconi, E.; Dahmen, K. H.; De Angelis, F.; Abate, A.; Hagfeldt, A.; Pozzi, G.; Graetzel, M.; Nazeeruddin, M. K. *Nat. Energy* **2016,** *1,* 15017.
170. Bai, L.; Wang, Z.; Han, Y.; Zuo, Z.; Liu, B.; Yu, M.; Zhang, H.; Lin, J.; Xia, Y.; Yin, C.; Xie, L.; Chen, Y.; Lin, Z.; Wang, J.; Huang, W. *Nano Energy* **2018,** *46,* 241.
171. Niu, J.; Yang, D.; Ren, X.; Yang, Z.; Liu, Y.; Zhu, X.; Zhao, W.; Liu, S. *Org. Electron.* **2017,** *48,* 165.
172. Nazim, M.; Ameen, S.; Akhtar, M. S.; Nazeeruddin, M. K.; Shin, H. S. *Sol. Energy Mater. Sol. Cell.* **2018,** *180,* 334.
173. Liu, X.; Kong, F.; Jin, S.; Chen, W.; Yu, T.; Hayat, T.; Alsaedi, A.; Wang, H.; Tan, Z.; Chen, J.; Dai, S. *ACS Appl. Mater. Interfaces* **2017,** *9,* 27657.
174. Rakstys, K.; Abate, A.; Dar, M. I.; Gao, P.; Jankauskas, V.; Jacopin, G.; Kamarauskas, E.; Kazim, S.; Ahmad, S.; Grätzel, M.; Nazeeruddin, M. K. *J. Am. Chem. Soc.* **2015,** *137,* 16172.
175. Mane, S. B.; Sutanto, A. A.; Cheng, C. F.; Xie, M. Y.; Chen, C.; Leonardus, M.; Yeh, S. C.; Beyene, B. B.; Diau, E. W. G.; Chen, C. T.; Hung, C. H. *ACS Appl. Mater. Interfaces* **2017,** *9,* 31950.
176. Grisorio, R.; Roose, B.; Colella, S.; Listorti, A.; Suranna, G. P.; Abate, A. *ACS Energy Lett.* **2017,** *2,* 1029.
177. Lee, J.; Byranvand, M. M.; Kang, G.; Son, S. Y.; Song, S.; Kim, G. W.; Park, T. *J. Am. Chem. Soc.* **2017,** *139,* 12175.
178. Arora, N.; Wetzel, C.; Dar, M. I.; Mishra, A.; Yadav, P.; Steck, C.; Zakeeruddin, S. M.; Bäuerle, P.; Grätze, M. *ACS Appl. Mater. Interfaces* **2017,** *9,* 44423.
179. Xi, H.; Tang, S.; Ma, X.; Chang, J.; Chen, D.; Lin, Z.; Zhong, P.; Wang, H.; Zhang, C. *ACS Omega* **2017,** *2,* 326.
180. Arora, N.; Orlandi, S.; Dar, M. I.; Aghazada, S.; Jacopin, G.; Cavazzini, M.; Mosconi, E.; Gratia, P.; Angelis, F. D.; Pozzi, G.; Graetzel, M.; Nazeeruddin, M. K. *ACS Energy Lett.* **2016,** *1,* 107.
181. Li, M.; Li, N.; Hu, W.; Chen, G.; Sasaki, S.; Sakai, K.; Ikeuchi, T.; Miyasaka, T.; Tamiaki, H.; Wang, X. F. *ACS Appl. Energy Mater.* **2018,** *1,* 9.
182. R. Tiazkis, S. Paek, M. Daskeviciene, T. Malinauskas, M. Saliba, J. Nekrasovas, V. Jankauskas, S. Ahmad, V. Getautis, M.K. Nazeeruddin, Sci. Rep. 7, 150 DOI:10.1038/s41598-017-00271-z.
183. T. Gatti, F. Lamberti, P. Topolovsek, M.A. Aguye, R. Sorrentino, L. Perino, M. Salerno, L. Girardi, C. Marega, G.A. Rizzi, M.A. Loi, A. Petrozza, E. Menna, Sol R.R.L. (2018) 1800013.
184. Chen, C.; Cheng, M.; Liu, P.; Gao, J.; Kloo, L.; Sun, L. *Nano Energy* **2016,** *23,* 40.

185. Guo, J. J.; Meng, X. F.; Niu, J.; Yin, Y.; Han, M. M.; Ma, X. H.; Song, G. S.; Zhang, F. *Syn. Met.* **2016,** *220,* 462.
186. Lim, I.; Kim, E. K.; Patila, S. A.; Ahn, D. Y.; Lee, W.; Shrestha, N. K.; Lee, J. K.; Seok, W. K.; Cho, C. G.; Han, S. H. *RSC Adv.* **2015,** *5,* 55321.
187. Gratia, P.; Magomedov, A.; Malinauskas, T.; Daskeviciene, M.; Abate, A.; Ahmad, S.; Graetzel, M.; Getautis, V.; Nazeeruddin, M. K. *Angew. Chem. Int. Ed.* **2015,** *54,* 11409.
188. Qin, P.; Paek, S.; Dar, M. I.; Pellet, N.; Ko, J.; Grätzel, M.; Nazeeruddin, M. K. *J. Am. Chem. Soc.* **2014,** *136,* 8516.
189. Zhang, F.; Yi, C.; Wei, P.; Bi, X.; Luo, J.; Jacopin, G.; Wang, S.; Li, X.; Xiao, Y.; Zakeeruddin, S. M.; Grätzel, M. *Adv. Energy Mater* **2016,** *6,* 1600401.
190. Yang, X.; Wang, H.; Cai, B.; Yu, Z.; Sun, L. *J. Ener. Chem.* **2018,** *0 0 0,* 1.
191. Li, H.; Fu, K.; Boix, P. P.; Wong, L. H.; Hagfeldt, A.; Grätzel, M.; Mhaisalkar, S. G.; Grimsdale, A. C. *ChemSusChem* **2014,** *7,* 3420.
192. Paulo, S.; Stoic, G.; Cambarau, W.; Ferrero, E. M.; Palomares, E. *Synth. Met.* **2016,** *222,* 17.
193. Calió, L.; Momblona, C.; Escrig, L. G.; Kazim, S.; Sessolo, M.; Santos, Á. S.; Bolink, H. J.; Ahmad, S. *Sol. Ener. Mat. Sol. Cells* **2017,** *163,* 237.
194. (a) Do Sung, S.; Kang, M. S.; Choi, I. T.; Kim, H. M.; Kim, H.; Hong, M.; Kim, H. K.; Lee, W. I. *Chem. Commun.* **2014,** *50,* 14161.
(b) Huang, C.; Fu, W.; Li, C. Z.; et al. *J. Am. Chem. Soc.* **2016,** *138,* 2528.
195. Li, Y.; Zhang, Y.; Ma, Y.; Ren, T.; Wang, L.; Zhang, J. *Org. Elec.* **2017,** *43,* 96.
196. (a) Zhu, J. Y.; Niu, K.; Li, M.; Lin, M. Q.; Li, J. H.; Wang, Z. K. *Org. Elec.* **2018,** *54,* 9.
(b) Mali, S. S.; Kim, H.; Kim, H. H.; Park, G. R.; Shim, S. E.; Hong, C. K. *Mat. Today Chem.* **2017,** *4,* 53.
(c) Daskeviciene, M.; Paek, S.; Wang, Z.; Malinauskas, T.; Jokubauskaite, G.; Rakstys, K.; Cho, K. T.; Magomedo, A.; Jankauskas, V.; Ahmade, S.; Snaithc, H. J.; Getautis, V.; Nazeeruddin, M. K. *Nano Energy* **2017,** *32,* 551.
197. Wang, Y.; Zheng, X.; Liu, X.; Feng, Y.; Shan, H.; Dong, L.; Fang, G.; Xu, Z. X. *Org. Elect.* **2018,** *56,* 276.
198. Palma, A. L.; Cinà, L.; Pescetelli, S.; Agresti, A.; Raggio, M.; Paolesse, R.; Bonaccorso, F.; Carlo, A. D. *Nano Energy* **2016,** *22,* 349.
199. Selvakumar, D.; Murugadoss, G.; Alsalme, A.; Alkathiri, A. M.; Jayavel, R. *Sol. Energy* **2018,** *163,* 564.
200. Zheng, D.; Yang, G.; Zheng, Y.; Fan, P.; Ji, R.; Huang, J.; Zhang, W.; Yu, J. *Electrochim. Acta* **2017,** *247,* 548.
201. Chen, J.; Park, N. G. *J. Phys. Chem. C* **2018;** https://doi.org/10.1021/acs.jpcc.8b01177.
202. Christians, J. A.; Fung, R. C.; Kamat, P. V. *J. Am. Chem. Soc.* **2014,** *136,* 758.
203. Chavhan, S.; Miguel, O.; Grande, H. J.; Gonzalez-Pedro, V.; Sánchez, R. S.; Barea, E. M.; Mora-Seró, I.; Tena-Zaera, R. *J. Mater. Chem. A* **2014,** *2,* 12754.
204. Qin, P.; Tanaka, S.; Ito, S.; Tetreault, N.; Manabe, K.; Nishino, H.; Nazeeruddin, M. K.; Grätzel, M. *Nat. Commun.* **2014,** *5,* 3834.
205. Jeng, J. Y.; Chen, K. C.; Chiang, T. Y.; Lin, P. Y.; Tsai, T. D.; Chang, Y. C.; Guo, T. F.; Chen, P.; Wen, T. C.; Hsu, Y. *J. Adv. Mater.* **2014,** *26,* 4107.
206. Xie, F.; Chen, C. C.; Wu, Y.; Li, X.; Cai, M.; Liu, X.; Yang, X.; Han, L. *Energy Environ. Sci.* **2017,** *10,* 1942.
207. Wang, F.; Endo, M.; Mouri, S.; Miyauchi, Y.; Ohno, Y.; Wakamiya, A.; Matsuda, Y. K. M. *Nanoscale* **2016,** *8,* 11882.
208. Aitola, K.; Sveinbjörnsson, K.; Correa-Baena, J. P.; Kaskela, A.; Abate, A.; Tian, Y.; Johansson, E. M.; Grätzel, M.; Kauppinen, E. I.; Hagfeldt, A.; et al. *Energy Environ. Sci.* **2016,** *9,* 461.

209. Cheng, M.; Li, Y.; Safdari, M.; Chen, C.; Liu, P.; Kloo, L.; Sun, L. *Adv. Energy Mater.* **2017,** *7,* 1602556.
210. Liu, Z.; He, T.; Liu, K.; Zhi, Q.; Yuan, M. *RSC Adv.* **2017,** *7,* 26202.
211. Tseng, Z. L.; Chen, L. C.; Chiang, C. H.; Chang, S. H.; Chen, C. C.; Wu, C. G. *Sol. Energy* **2016,** *139,* 484.
212. Wu, Q.; Xue, C.; Li, Y.; Zhou, P.; Liu, W.; Zhu, J.; Dai, S.; Zhu, C.; Yang, S. *ACS Appl. Mater. Interfaces* **2015,** *7,* 28466.
213. Dunlap-Shohl, W. A.; Daunis, T. B.; Wang, X.; Wang, J.; Zhang, B.; Barrera, D.; Yan, Y.; Hsu, J. W.; Mitzi, D. B. *J. Mater. Chem. A* **2018,** *6,* 469.
214. Igbari, F.; Li, M.; Hu, Y.; Wang, Z. K.; Liao, L. S. A. *J. Mater. Chem. A* **2016,** *4,* 1326.
215. Hossain, M. I.; Alharbi, F. H.; Tabet, N. *Sol. Energy* **2015,** *120,* 370.
216. Muduli, S. K.; Varrla, E.; Kulkarni, S. A.; Han, G.; Thirumal, K.; Lev, O.; Mhaisalkar, S.; Mathews, N. *J. Power Sour.* **2017,** *371,* 156.
217. Wei, Y.; Yao, K.; Wang, X.; Jiang, Y.; Liu, X.; Zhou, N.; Li, F. *Appl. Sur. Sci.* **2018,** *427,* 782.
218. Hu, Z.; Chen, D.; Yang, P.; Yang, L.; Qin, L.; Huang, Y.; Zhao, X. *Appl. Sur. Sci.* **2018,** *441,* 258.
219. Tang, J.; Jiao, D.; Zhang, L.; Zhang, X.; Xu, X.; Yao, C.; Wu, J.; Lan, Z. *Sol. Energy* **2018,** *161,* 100.
220. Icli, K. C.; Ozenbas, M. *Electrochim. Acta* **2018,** *263,* 338.
221. Yin, X.; Yao, Z.; Luo, Q.; Dai, X.; Zhou, Y.; Zhang, Y.; Zhou, Y.; Luo, S.; Li, J.; Wang, N.; Lin, H. *ACS Appl. Mater. Interfaces* **2017,** *9,* 2439.
222. (a) Cui, E. J.; Meng, F.; Zhang, H.; Cao, K.; Yuan, H.; Cheng, Y.; Huang, F.; Wang, M. *ACS Appl. Mater. Interfaces* **2014,** *6,* 22862.
    (b) Seo, S.; et al. *Nanoscale* **2016,** *8,* 11403.
    (c) You, J.; et al. *Nature Nanotech.* **2016,** *11,* 75.
    (d) Yin, X.; et al. *ACS Nano* **2016,** *10,* 3630.
223. Lee, H.; Huang, Y. T.; Horn, M. W.; Feng, S. P. *Sci. Rep.* **2018,** *8,* 5590.
224. Yanagida, M.; Shimomoto, L.; Shirai, Y.; Miyano, K. *Electrochem* **2017,** *85,* 231.
225. Zhang, H.; Wang, H.; Chen, W.; Jen, A. K. Y. *Adv. Mater.* **2017,** *29,* 1604984.
226. Takahashi, K.; Suzuki, Y. *Jap. J. App. Ph.* **2017,** *56,* 08MC04.
227. Madhavan, V. E.; Zimmermann, I.; Carmona, C. R.; Grancini, G.; Buffiere, M.; Belaidi, A.; Nazeeruddin, M. K. *ACS Energy Lett.* **2016,** *1,* 1112.
228. Ito, S.; Tanaka, S.; Nishino, H. *J. Phys. Chem. Lett.* **2015,** *6,* 881.
229. Rao, H.; Sun, W.; Ye, S.; Yan, W.; Li, Y.; Peng, H.; Liu, Z.; Bian, Z.; Huang, C. *ACS Appl. Mater. Interfaces* **2016,** *8,* 7800.
230. Christians, J. A.; Kamat, P. V. *ACS Nano* **2013,** *7,* 7967.
231. Christians, J. A.; Fung, R. C. M.; Kamat, P. V. *J. Am. Chem. Soc.* **2014,** *136,* 758.
232. Sepalage, G. A.; Meyer, S.; Pascoe, A.; Scully, A. D.; Huang, F.; Bach, U.; Cheng, Y.-B.; Spiccia, L. *Adv. Funct. Mater.* **2015,** *25,* 5650.
233. Huangfu, M.; Shen, Y.; Zhu, G.; Xu, K.; Cao, M.; Gu, F.; Wang, L. *Appl. Surf. Sci.* **2015,** *357,* 2234.
234. Nazari, P.; Ansari, F.; Nejand, B. A.; Ahmadi, V.; Payandeh, M.; Niasari, M. S. *J. Phys. Chem. C* **2017,** *121,* 21935.
235. Xu, L.; Deng, L. L.; Cao, J.; Wang, X.; Chen, W. Y.; Jiang, Z. *Nanoscale Resear. Lett.* **2017,** *12,* 159.
236. Wang, D.; Elumalai, N. K.; Mahmud, M. A.; Wright, M.; Upama, M. B.; Chan, K. H.; Xu, C.; Haque, F.; Conibeer, G.; Uddin, A. *Org. Electr.* **2018,** *53,* 66.

CHAPTER SIX

New Trends in Electrochemical Capacitors

Adam Slesinski, Krzysztof Fic and Elzbieta Frackowiak[1]
Institute of Chemistry and Technical Electrochemistry, Poznań University of Technology, Poznań, Poland
[1]Corresponding author: E-mail: elzbieta.frackowiak@put.poznan.pl

Contents

Abstract

This chapter provides an overview of electrochemical capacitors (ECs) technology. General principles of ion adsorption and interfacial phenomena governing the energy storage in these devices are discussed. True capacitance metrics and appropriate measurement conditions are demonstrated and followed by a discussion of the electrode materials and electrolytes most often applied. Novel trends in ECs, such as pseudocapacitance from electrolytic solutions and hybrid Li-ion capacitors, are also presented.

Advances in Inorganic Chemistry, Volume 72
ISSN 0898-8838
https://doi.org/10.1016/bs.adioch.2018.05.003

1. PERFORMANCE OF ELECTROCHEMICAL CAPACITOR—GENERAL PROPERTIES

Since its discovery in the 19th century and throughout many years, the most efficient and economical way of electrical energy storage has been attributed to the galvanic cell. This kind of cell has evolved over the years, and nowadays dozens of different cell types are available.[1,2] The operation principle of a galvanic cell relies on the chemical reduction and oxidation reactions of two species occurring at different but well-defined potential levels. As a result, the fixed potential difference, called the voltage, is obtained.[3] Depending on the chemistry occurring within the galvanic cell, it can be rechargeable or nonrechargeable. The most popular and rechargeable galvanic cells are mainly: Pb//PbO_2, MH//NiO(OH), Cd//NiO(OH), and Li-ion. One should be careful in using the term battery as it depicts, in general, the number of identical cells interconnected, rather than referring to a single cell of a defined charge storage mechanism. Although the above cells have appeared on the market one after another, bringing about new developments (from which the most important is the progressively increasing energy density), no ultimate replacement occurred, and all of the cell types coexist and share the market. However, an unresolved issue assigned to this kind of cells is related to the power density. These electrochemical cells, owing to typical faradaic charge storage, accept and deliver the charge only at strongly restricted current rates. They are characterized by high energy but limited power. Hence, the capacitors become attractive systems owing to their electrostatic charge storage mechanism; higher power capabilities are available. Typical capacitors, widely used in the electronics, store only a small amount of energy. However, a special type of capacitor, the electrochemical capacitor (EC) based on the formation of an electrical double layer (EDL), allows storing higher energies at comparable power densities. In its simplest form, it comprises two electrodes made of a high-surface-area porous material, most often from activated carbon (AC), immersed in the electrolytic solution and separated by a membrane[4,5] (Fig. 1). The carbon electrodes are placed on the metallic current collectors, and this ensures the electrical connection to the external circuit. Upon application of a potential difference from the external circuit, the separation of ions originating from the electrolyte reservoir takes place. During this process, the electrodes increase their absolute potential values. Upon connection of the load to the EC, the electrodes decrease their absolute potential value to go back to their initial states (Fig. 2). This process,

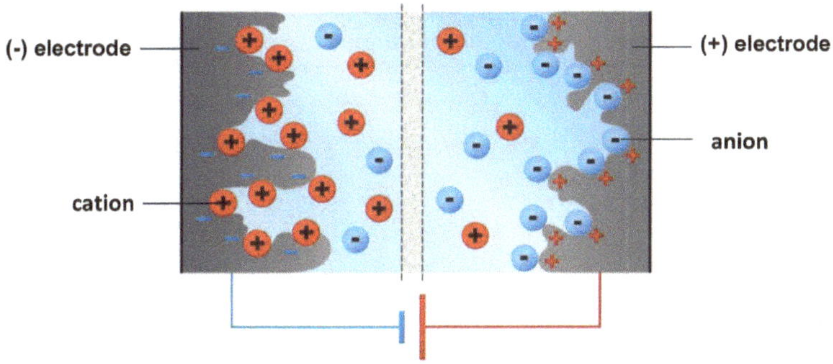

Figure 1 Schematic view of an electrochemical capacitor.

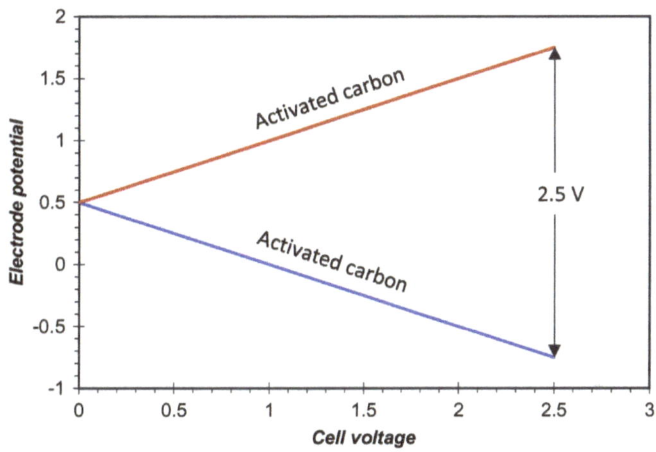

Figure 2 Electrode potential changes upon charging and discharging.

by definition, could be actually repeated an unlimited number of times; however, many issues could be listed that limit the cycle life of such a device. The primary one is aging caused by the operation at elevated voltage values. ECs are often called supercapacitors; however, one should bear in mind that the term *supercapacitor* used to be a registered trademark of Standard Oil Company, which is often neglected. The ECs can be divided into categories, from which the most important one includes the electrolyte solvent type, as it determines the cell performance.[6] Table 1 presents three types of EC electrolytes, which differ by the conductivity range of the electrolyte that mostly imposes the power of the device. The maximum allowable voltage, on the other hand, is governed by the dielectric strength of the electrode/electrolyte interface, which determines the energy of the device to the highest

Table 1 Comparison of Different Solvents Applied in Electrochemical Capacitors

	Aqueous	Organic	Ionic Liquid
Voltage (V)	0.8–1.8	2.5–2.7	3.5
Electrical conductivity (mS/cm)	100	10	8
Energy density (Wh/kg)	10	15	20
Power density (kW/kg)	10	5	1
Price	Low	High	Very high
Safety	High	Low	Medium

degree. Another fact related to the electrolytes is their impact on the environment. One can notice that the currently commercially available capacitors operate using acetonitrile- or carbonate-based electrolytes, which makes them questionable for widespread use. In addition, due to costly chemicals and a complicated assembly process requiring the use of moisture-free carbon, the technology is limited. However, it is currently the only one that has reached successful commercialization. Ionic liquids are special types of electrolytes in which no solvent is used while the molecules remain in the ionic state. As a result, they offer the highest upper voltage limit to be reached. However, the total price of the cells using ionic liquids is even higher than in the case of organic acetonitrile or propylene carbonate.[7,8] Along with the fact of their low power capabilities and high price, even though considered as environment-friendly, their commercial use is restricted. Contrarily, aqueous-based capacitors are very promising in the EC field.[9] Although the voltage of these cells is the lowest because of the thermodynamic stability of water (1.23 V), other properties make them perfect candidates for future ECs. Hence, a lot of research has been done on the development of high voltage cells, which aims at spreading this technology.[10–13]

Owing to the special polarizable character of all ECs, their testing procedures differ from those known from galvanic cell investigations. As the capacitor voltage is not determined thermodynamically, but in fact by the stability of the electrode/electrolyte interface, the selection of the upper voltage limit depends upon the investigator. Scientists around the world use various techniques for selection of the upper voltage limit. Although more time consuming, cycling seems to be a more reliable technique for capacitor life determination at a given upper voltage than floating.[14] The capacitor operating nature is to ramp the voltage, accepting and releasing energy pulses, rather than operate at a constant elevated voltage. On the other hand, floating might be treated as a technique for the accelerated aging test, which gives a fast indication of cycle life. However, obviously in this mode, different aging mechanisms occur and the projection of floating time into cycles is currently obsolete.

The examination of capacitor parameters always includes the determination of capacitance (in Farads). As the different research groups use various electrochemical cells which accept different loadings of electrode material, the use of this unit requires standardization. Various expressions have been introduced, most often F/g or F/cm^3, which are currently the most widely used units that allow direct comparison of the capacitance parameter. However, one should take into account what mass basis or volume do they include, as sometimes the normalization includes the total mass of two electrodes, whereas sometimes the mass of only one electrode is included. Both seem to be correct for symmetric capacitor systems as long as it is clearly stipulated. In case of hybrid systems, this expression loses its sense. This emerges from the fact that both electrodes operate with different charge storage mechanism, where one electrode is a typical double-layer electrode and the second one involves redox reactions. For the EDL electrode, a unit of F/g per electrode mass (or electrodes mass) is appropriate (variable potential), but in the latter case, the capacity should be expressed in mAh/g (fixed potential). Finally, it is difficult to decide on an ultimate unit which describes the outcome of the system properly, should it be F or mAh? Both appear to be improper, especially if the system capacitance varies strongly. The most suitable seems to be the energy in Wh/kg, where neither capacity nor capacitance dominates. Instead, the formula for energy, E, accounts for the voltage limits.

$$E = \frac{C \cdot \left(U_{upper}^2 - U_{lower}^2\right)}{2}$$

Defining the energy in normalized units Wh/kg often leads to confusion, since the mass in the denominator is often not properly defined. This expression is very often used to show the superiority of a certain invention over other technologies. This is extensively found on Ragone plots, where researchers present the energy values of their device coupled with the energy values originating from other authors' devices. This operation is highly doubtful and makes sense only if conducted at identical experimental conditions, which is often difficult to validate. These parameters should not be compared directly on a single graph as they provide an incomparable overview and allow manipulation. Often, the mass basis includes only active material of two electrodes, entire electrodes, or electrodes with a separator.

Care should be taken not only with the selection of upper voltage limit for the capacitor but also with the lower voltage limit. The approach which considers cycling between rated voltage and a certain voltage value only, different than 0 V, is hardly ever mentioned in the literature. From the analytical point of view, cycling between 0 V and the upper voltage limit seems rational, as all available energy and potential range are explored. However, the real application requires a different approach. As the capacitors serve for energy and power delivery, they should operate at the highest possible voltages. In pure EDL capacitors (EDLCs), 75% of the stored energy is localized in the upper half of the voltage. Their deep discharge to 0 V is unnecessary and not recommended, as it can release only one-fourth of the available energy which is actually hardly accessible.[15] Furthermore, most commercial devices require constant power to operate, rather than constant current only, especially when the voltage is variable. Constant power discharge of capacitors assumes that the product of the voltage and current always remains the same. It means that at high voltages, the current is lower than at low voltages, where at extremely low voltages, the current can be exclusively high leading to Joule and Ohmic losses. Fig. 3 presents theoretical curves for pure EDLC during constant power discharge. It implies that high instantaneous capacitance at low voltages is not advantageous. This should be kept in mind while assessing the performance. Most of the articles published on ECs contain information on their long-term performance. Usually, the capacitance values are plotted against cycle number (cycling) or voltage hold time (floating). These capacitance values

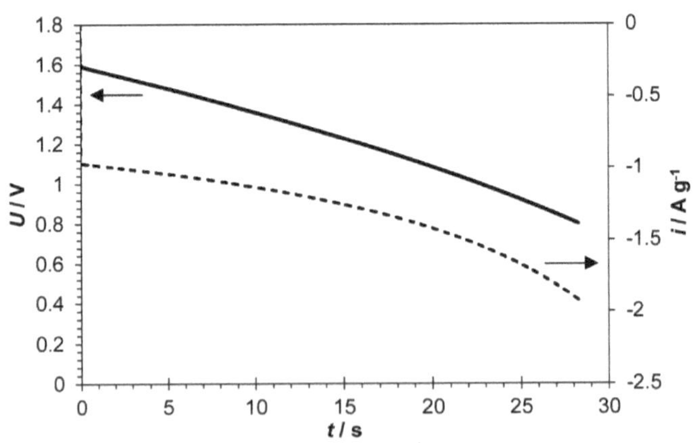

Figure 3 The current waveform and discharge voltage curve profiles during constant power discharge.

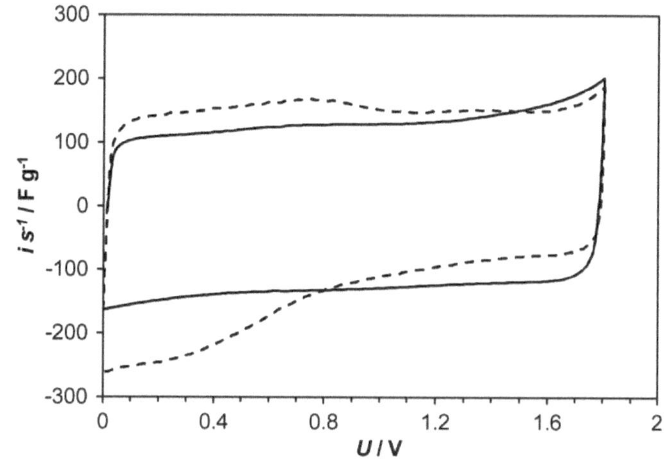

Figure 4 Voltammetry curve (10 mV/s) of two-electrode cell in 1 mol/L Li_2SO_4 recorded before (*solid lines*) and after (*dashed lines*) 2000 cycles in the cycling test.

are usually calculated from the entire voltage range of discharge. As the capacitance of ECs is a function of the voltage, especially during the appearance of redox reactions, expressing single capacitance value is only an average of the instantaneous capacitance values at different instantaneous voltages. In some cases, like long-term capacitor cycling, where capacitance retention is claimed to be excellent, in reality it may not be.[16] The example in Fig. 4 shows the performance of a capacitor before cycling and after cycling at 1.8 V. If the capacitance retention is plotted against the cycle number over the entire voltage range, the system exhibits the same average capacitance before and after the test. Simply, the loss of capacitance at high voltage values is compensated by the capacitance at low voltages. If the capacitance was calculated from the upper half of the voltage range, the capacitor would reach the end of life criteria considerably sooner.

At this point, it is recommended to refer the results to the proper determination of the energy efficiency and discrimination to coulombic efficiency.[17] Coulombic efficiency informs about the total charge which is the ratio of charge delivered and withdrawn, irrespective of the voltage at which this charge is found. Energy efficiency, on the other hand, accounts for at which voltage value this charge is present. The paper states that expressing the coulombic efficiency overestimates the description and rather energy efficiency should be used. But in fact, it reasonably expresses the different properties of the system. The paper stresses the proper calculation

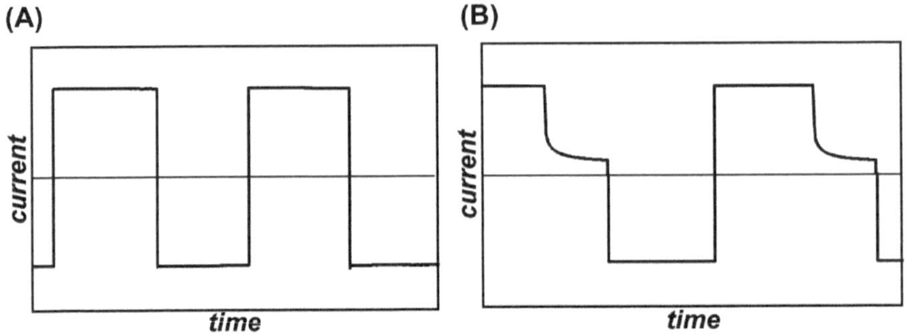

Figure 5 Current waveforms during constant current charging: (A) direct inversion of current upon discharge, (B) discharge followed by the constant voltage period and open circuit voltage time to account for charge redistribution.

of capacitance—the accurate calculation involves the integration of the area under the galvanostatic discharge curve, rather than evaluation of the discharge time and inserting the value into the formula for capacitance. Taking into account the curvature of the discharge curve – the time, i.e., charge gives the overestimated results.

A closely related issue emerges from the fact of charge redistribution within the electrode, which is almost always neglected during galvanostatic cycling.[18] The typical current waveform during performing cycles involves the direct inversion from positive to the negative current of identical magnitude at the point where the maximum cell voltage is reached (Fig. 5A). It means that at a short time during current inversion, its value is doubled. As long as low current regimes are used, this phenomenon is hardly noticeable. It requires attention at high current regimes at which the capacitors are intended to operate. In fact, excessive ohmic losses are observable, which are related to the undistributed charge. It leads to the underestimated energy density values reported at high power regimes than it actually is. A scenario that might be useful in order to account for charge redistribution is to apply a period of constant voltage after charging to minimize the leakage current. Straightaway after constant voltage period, one may directly reverse the current to discharge or introduce a short period of open circuit voltage (OCV) time before the discharge (Fig. 5B).

Ragone plots are often found in the literature to represent the energy of the device at various discharge powers. Accordingly, it seems more rational to present the plot by inserting power values on x-axis and energy at the y-axis. Very often it is shown in the opposite way, which might be

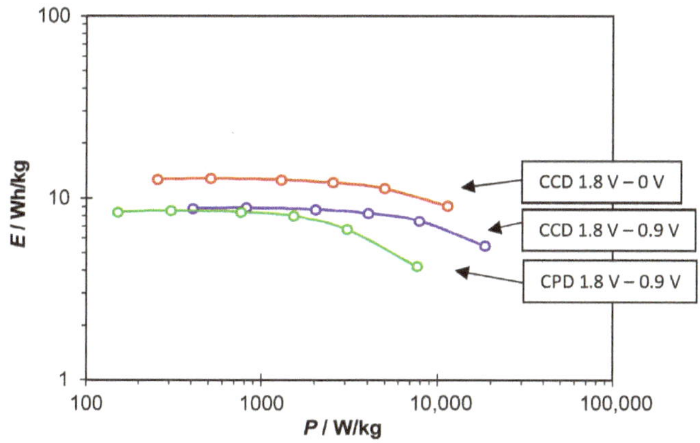

Figure 6 Ragone plot characteristics defined by various techniques. *CCD*, constant current discharge; *CPD*, constant power discharge.

confusing. In addition, numerous papers present Ragone plots, for which points are recalculated from constant current discharge data. This approach is inconsistent with the concept of Ragone plots, where the discharge should be performed at constant power, as it was originally designed for. From a physical point of view, using constant current discharge, the power constantly changes, has its maximum at the upper voltage limit, and tends to zero as the discharge progresses. The calculation of power values from the constant current discharge data, although little more complicated, while mathematically possible, have no physical meaning. With constant current mode, it is possible to discharge to 0 V, showing artificially overestimated values of energy density (Fig. 6). In most real applications, such a deep discharge would be impossible during the constant power discharge mode. Recently, a paper was published on performance testing of supercapacitors, which claims that energy density should be determined at constant power discharge.[19] Furthermore, it was claimed that all known test procedures yield identical capacitance values in case of symmetric systems, whereas in case of hybrid systems, this parameter is strongly dependent on the test technique. Irrespective of the system type, the resistance was the parameter that required special attention, as the resistance determined by various techniques may vary. For example, the formula for power, i.e., $U^2\ (4R)^{-1}$ should be avoided, as the power matched with impedance is in reality not achievable and only gives the first impression of the system. The more realistic approach is to use the simple formula $P = E/t$.

2. MATERIALS

As ECs are devices which store energy by adsorption of ions on the interface, it requires the use of a high-surface-area material. It is in agreement with the well-known formula for capacitance:

$$C = \frac{\varepsilon \cdot S}{d}$$

where ε is the electrical permittivity of the dielectric material (F/m); S, surface area of the interface (m^2); and d, the thickness of the dielectric (m). In case of ECs, the thickness is within the Debye length (10^{-5} m). The second parameter in the formula, ε, for water is close to 80 (in bulk), while it rapidly decreases to 4–8 in the vicinity of the electrode/electrolyte interface, since the water dipoles at the charged surface are perfectly oriented according to the electric field.[20,21] Most of the materials exhibit limited capacitance values. Even the materials, as claimed, of extremely high surface area, cannot exceed 150 F/g in neutral electrolyte or 200 F/g in acidic/basic electrolyte (per mass of active material of one electrode), meaning that the energy density is limited not only by the dielectric strength of the medium but also by the capacitance value. Therefore, a solution seems to be related to the increase of the second factor in the formula—the electric permittivity of the dielectric. Currently, research begins to orientate into the application of high dielectric constant materials which boost the capacitance (barium and strontium titanate, electrically conductive polymers, titanium (IV) oxide, etc.). A secondary effect of using such materials might be related to the increase of dielectric breakdown of the medium as a result of replacement of the electrode/electrolyte interface with an electrode/high-dielectric-constant material/electrolyte one. In fact, the energy density of the final device is expected to increase significantly.

ECs contain electrodes which in most cases are made of carbonaceous materials deposited on metallic current collectors. There are many possible electrode materials that have been proposed in the literature for energy storage purposes, such as carbide-derived carbons, carbon nanotubes (CNTs), graphene, or carbon nanospheres. However, ACs, although known for centuries, still receive significant attention owing to their interesting parameters.[22,23]

2.1 Activated Carbons

ACs are obtained via simple thermal processing of different organic precursors of high elemental carbon content. The most famous ACs are produced from coconut shells, algae, wood, pestles, and other precursors. The requirement for the ideal precursor is its natural origin, ease of processing, and availability in large quantities. Owing to economic and ethical issues, it cannot be food, while preferably only waste. In some cases, it is beneficial if the substrate contains a high content of alkali metals which are extremely useful for further processing. Typically, AC production includes carbonization of a precursor followed by activation. Carbonization is the conversion of a precursor into carbon-containing char by pyrolysis in the flow of an inert atmosphere with simultaneous removal of volatile products. It results in moderate surface area material; however, in order to increase the surface area even more, it is necessary to treat the material by a process called activation. It can occur via a physical or chemical pathway. In the first case, water steam or carbon dioxide are used which act as stimulants for pressure build-up reagents that create microporosity. The latter case involves the preimpregnation of carbonized material with chemicals such as KOH, $ZnCl_2$, or H_3PO_4. These aggressive chemicals at high temperature cause the partial burn-off of the initial material by releasing reactive species which create a secondary microporosity. ACs are very cheap, abundant, and susceptible to modifications, which allow the properties of ACs to be tuned according to special requirements of the application. Modifications of ACs include doping with other elements such as nitrogen, oxygen, and sulfur.

2.2 Graphene, Carbon Nanotubes

Graphene is a material which received much attention. It was first described in the 50s of the 20th century by Rosalind Franklin, but the research on graphene started after its isolation in 2004. It was considered as a significant invention, so that the pioneers were awarded the Nobel Prize. There are claims that this material provides high energy and high power densities.[24,25] From a structural point of view, it is a two-dimensional material which consists of single-graphite sheets. Along with theoretical calculations, it can reach a specific surface area of c. 2700 m^2/g, and even higher values were reported.[26] However, in reality, the value is much lower. Since pure graphene is devoid of any defects in the structure, it is extremely unstable.[27] It tends to adopt more stable conformation by spontaneous adsorption of oxygen, which stabilizes the structure. The initial material, of high electric and thermal conductivity, loses, therefore, its extraordinary properties.[28] Graphene turns into graphene oxide, where the oxygen content usually reaches above 10%. This is the reason why graphene as capacitor electrode material performs worse than expected. Although the material seems to be complicated to handle properly, significant research is devoted to it to extract the special properties of graphene.[29] Several methods have been developed for graphene production, of which high shear rate exfoliation seems to be the most attractive, since it is adaptable and produces graphene with a relatively low number of defects. Graphene is also often obtained via chemical vapor deposition, which is a more sophisticated and more expensive method.[30,31] It is necessary to mention that the price of graphene is very high, irrespective of what production route is applied. All of it makes it questionable whether graphene can really be a material for the future, unless its superior characteristics are uncovered. Nevertheless, the investigation of this material seems to be promising, especially utilizing it as a component of various composites.[32]

Another material of a similar type is CNTs. This material exhibits high electrical conductivity and mechanical strength.[33] CNTs are single- or multi-walled graphene tubes of nanometer diameter. Always their length is much longer than their diameter. They are readily studied as EC electrode materials because they ensure the high power of the device.[34] This is achieved due to the abundance of mesopores in favor of micropores. Therefore, the transport phenomenon is facilitated. As a result of their moderate surface area, with a maximum of 500 m^2/g, they allow only about 40 F/g in a typical EDLC to be obtained. On the other hand, CNTs are successfully used as a support for pseudocapacitive materials (MnO_2, RuO_2, etc.), where their electrical properties and their mesoporosity are preferred. An important drawback of carbon nanotubular materials is their complicated process ability related to debundling during the electrode formation process, even if they constitute only the conductivity enhancer additive at low concentration in the composition. CNTs are obtained via catalytic decomposition of hydrocarbons, during a chemical vapor deposition process.[35] Recently, there is an interest in growing aligned nanotubes directly on current collector substrate.[36] It allows reaching extremely high power of the devices. However, the material loading of such produced coating is too low to start mass and feasible production.

All prepared carbonaceous materials are successfully used as electrode materials in ECs. In order to boost the development, present research tends to improve their performance even more by adequate modifications of the components. One of the modification pathways concerns the electrode material. Various approaches have already been presented, from which doping is most extensively studied.[37–39] It consists of modification using sources of oxygen, sulfur, phosphorus, or nitrogen. Depending on the modification route, the atoms might be attached to the surface of a carbon network in the form of surface chemical groups or inserted into the graphitic layer by substitution of carbon atoms.[40–43] Such heteroatoms incorporated in the carbon matrix modify the physical/chemical properties of the material (conductivity, wettability, redox activity, etc.). The position of such newly formed functional groups in the carbon matrix depends on the energy supplied during the modification process. Usually, in the area of ECs, doping aims at increasing the contribution to the capacitance by provoking pseudocapacitive effects. This, in turn, increases the energy density of the device. Pseudocapacitance is a charge storage mechanism, in which a series of redox reactions occur within a broad potential range.[44] As a result, the capacitance looks like a constant value, which looks like pure double-layer response.

It is well known that the voltage limit of a capacitor operating in aqueous medium is imposed by the electrochemical stability of water. According to the Pourbaix diagram, it is advantageous to operate the capacitor with a significant pH gradient within the cell. In this way, the equilibrium reaction potentials of water decomposition are shifted and the voltage can be extended even to 2.2 V. Various approaches have been undertaken to realize this concept.[45] There are ways to use two different electrolytes for opposite electrodes or semipermeable membranes instead of the porous separator, which results in hindered mixing of the electrolytes.[46,47] However, additional issues arise from considerably increased resistance. Recently, a modification aiming at imposing the surface acidity of AC electrodes in aqueous electrolytes has been proposed.[16] The capacitor electrodes were modified with ammonia gas, which allowed the introduction of the self-controlled pH gradient within the capacitor cell and the formation of a protective layer at the electrode/electrolyte interface. They are responsible for higher overpotentials of solvent decomposition, and finally, the voltage of the cell was increased to 1.8 V.

3. ELECTROLYTES FOR ELECTROCHEMICAL CAPACITORS

The interfacial character of an energy storage process in ECs requires the electrolyte properties to be discussed in line with the electrode materials. In fact, the electrode is the physical place where charge accumulation occurs, but the electrolyte is the source of the ions and plays a decisive role in many parameters and final performance.

3.1 General Issues

As already described, by principle, the ECs store the energy in the form of an EDL, formed at the electrode/electrolyte interface. As previously stated, the capacitance of this interface is proportional to the electrochemically available surface area (but not specific surface area S_{BET} or S_{DFT} determined from gas adsorption[48–50]) and dielectric constant, ε. The thickness of the double layer (in fact, the ion size) has a reciprocal impact. Hence, the electrolyte as a source of ions will determine the amount of charge available for being stored. Taking into account that usually ACs with a well-developed surface area are applied as electrode materials, the number of ions must be of course high. This suggests the first feature of a good electrolyte, i.e., **high concentration.**[51–53] Concentrated solutions usually demonstrate a high conductivity. This parameter controls the **capacitor power** (gravimetric or volumetric).[54] Furthermore, the concentration of the electrolyte will certainly influence the **viscosity, wettability, and thermal capacity.**[55,56] Viscosity of the electrolyte plays an important role because it influences the penetration of the electrode material by the electrolyte.[7] Wettability is related to the viscosity and ensures fast ion transport into the pores and charge accumulation.[57–60] In this case, one should take into account that the wettability of the electrode material with ACs varies remarkably and the electrode material character cannot be neglected. In general, the surface of ACs is hydrophobic, but the presence of surface functionalities (O- and N-based) usually improves this feature.[61] Thermal capacity is responsible for heat management in the operating capacitor. Under real conditions, the resistivity of all components (electrode, binder, electrolyte, etc.) induces heat during the charging/discharging process and temperature increase.[62] The role of the electrolyte, in this case, is to absorb (dissipate) heat generated during operation and maintain the optimal temperature.[62]

From an electrochemical point of view, the kind of electrolyte will determine the **electrochemical stability** (device will operate up to the decomposition voltage) and the final **energy output.**[63]

The interactions of the electrode material with the electrolytic solution will finally affect the cycle life and **long-term capacitor performance.**[64–68]

Therefore, several features of the ECs are governed by the electrolyte properties:

1. Capacitance: by the ion size, concentration, and matching between the ion size and electrode pore width

2. Power density: by the ionic conductivity, ion–electrode interactions (charge propagation), and indirectly by electrochemical stability
3. Energy density: mainly by the electrochemical stability, but also by the concentration
4. Thermal stability: by the boiling and freezing points of the electrolyte and indirectly by the solubility of the salt
5. Internal resistance: by the ionic conductivity, ion mobility, and viscosity
6. Cycle life and long-term performance: by chemical and electrochemical stability.

Thus, the electrolyte is identified as one of the most influential components in the EC performance. All these issues are graphically represented in Fig. 7. The ESR abbreviation stands for Equivalent Series Resistance of an EC. It is the high frequency resistance at the point where the phase angle is zero. It comprises the resistances of current collectors, electrodes, electrolyte, etc.[69]

Although there is no point in the discussion about the importance of the parameters being used for capacitor characterization, from a practical point

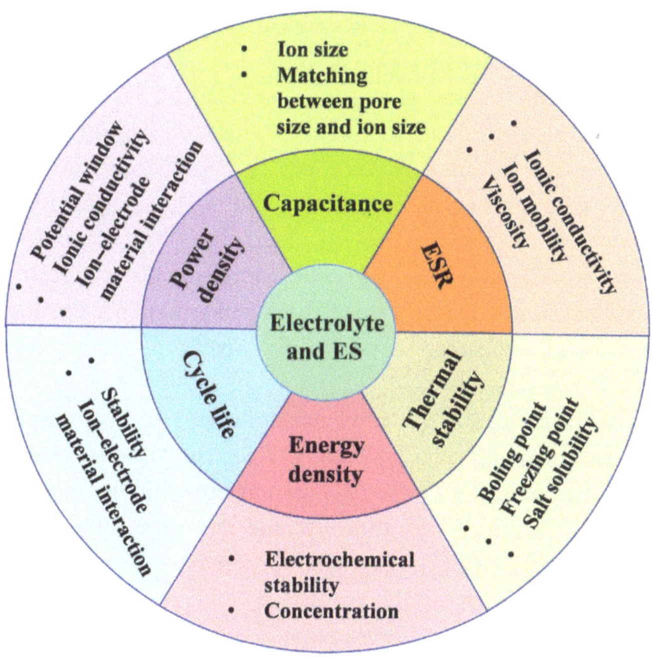

Figure 7 Features of electrolytes and their impact on the electrochemical capacitor performance. *ES*, electrochemical supercapacitor; *ESR*, equivalent series resistance.

of view, energy density and long-term performance seem to be the most important. Worth to say is that both of them depend on or remarkably affect the other features (capacitance, power, leakage currents, self-discharge, etc.).[63,70,71]

The energy density of the EC depends on both, capacitance and maximum operating voltage (to the square).[72] Of course, by selecting the electrolytic solution, one might affect both parameters.

Initially, electrolytes were classified as aqueous (protic) and organic (aprotic).[61] This classification also included ionic liquids, considered as organic electrolytes. The rapid development of ionic liquids during the last decades caused that these compounds are recently considered as an individual class of electrolytes.[7,73,74] Apart from electrolytes in the liquid state, one may also find so-called solid-state-electrolyte (hydro)gels or polyelectrolytes.[75] One should note that this classification is not exhaustive.

3.2 Organic Electrolytes

The most influential feature of the electrolyte in terms of energy is definitely the electrochemical stability (see the formula for energy stored). For this reason, organic electrolytes are usually applied to commercial capacitors. Depending on the solvent (mainly acetonitrile or propylene carbonate), the maximum voltage in this case is 2.5−2.7 V. Of course, such a high voltage requires very special conditions to be preserved during capacitor assembly, i.e., the atmosphere must be oxygen- and moisture-free. Traces of oxygen or water might provoke parasitic reactions and decrease the charging/discharging efficiency. Although the process of moisture removal from microporous carbons, drying of electrolyte, and clean-room assembling is complicated and expensive, all commercially available capacitors are based on organic electrolytes since their performance is the most optimal in terms of energy density and cycle life (negligible corrosion in aprotic media).

From a chemical point of view, acetonitrile and propylene carbonate serve as solvents in these electrolytes. Of course, their exploitation raises several environment-related aspects like toxicity and flammability. Some other compounds were also proposed in the literature (ethylene carbonate, dimethyl carbonate, butylene carbonate, gamma-butyrolactone, gamma-valerolactone, propionitrile, glutaronitrile), but all of them suffer from low conductivity (up to 15 mS/cm, excluding acetonitrile) and low dielectric constant (varying from 26 for propionitrile to 65 for propylene carbonate).[76] Sometimes, adiponitrile or methoxyacetonitrile (low viscosity and good penetration of electrode material) are considered, but their instability at higher voltages inhibits their practical application.

The salts (ion source and charge carrier) used in organic electrolytes should be of high solubility, giving high conductivity, high electrochemical stability, safe (in case of capacitor failure), and preferably inexpensive. Tetraethylammonium tetrafluoroborate ($TEABF_4$) appears to be the best candidate at the moment and is the most often applied salt in commercial capacitors. Here, one should be aware that $TEABF_4$ is highly soluble in acetonitrile while its solubility in propylene carbonate is limited (up to 1 mol/L) and results in high internal resistance of the device. It has been stated that the conductivity of the salts in organic electrolytes depends on the cation and anion in the following order: $TEA^+ > Pr_4N^+ > Bu_4N^+ > Li^+ > Me_4N^+$ for cations and $BF_4^- > PF_6^- > ClO_4^- > CF_3SO_3^-$ for anions. Therefore, $TEABF_4$ appears to be the best candidate as salt for the organic electrolyte.[76–78]

Recently, some other compounds are developed as tentative candidates for capacitor application, essentially because of environmental concerns on acetonitrile. In order to increase the conductivity of propylene carbonate-based electrolytes, several asymmetric tetraalkylammonium and cyclic quaternary ammonium salts have been considered based on 1-ethyl-1-methylpyrrolidinium, triethylmethylammonium, tetramethylenepyrrolidinium, or spiro-(1,1′)-bipyrrolidinium cations. Considerable attention has also been focused on phosphonium-based cations, like tetrakis(diethylamino)-phosphonium, but their performance appeared to be much better in acetonitrile-based electrolytes.[79]

The anion side of the salt usually contains halogen atoms in the structure (BF_4^- or PF_6^-), and for that reason some alternatives were introduced, mainly based on bis(oxalato)borates (BOB). However, the presence of BOB-based anions results usually in lower electrochemical stability of the salt and electrolyte; incorporation of halogen in the anion (e.g., difluoro(oxalato)borate) has been claimed to increase the electrolyte stability.[80–82]

3.3 Ionic Liquids

Ionic liquids (ILs) have also been reported as electrolytes for ECs for several reasons. Chemically, these are electrolytes consisting entirely from ions. Thus, no solvent is necessary. An enormous number of organic compounds, which may be designed prior to requirements, initiated substantial research on ionic liquids as tentative electrolytes. General classification distinguishes three types of ILs: protic (substituted with all organic R^X groups), aprotic (with a labile proton in the structure), and zwitterionic (with merged, usually polymeric structure). Imidazolium-, pyrrolidinium-, ammonium-, sulfonium-, or phosphonium-based cations, together with BF_4^-, PF_6^-, bis(trifluoromethanesulfonyl)imide, bis(fluorosulfonyl)imide, and dicyanamide anions, are the most often reported in electrochemical applications.[7,74,83,84]

Their main advantage over conventional organic electrolytes is their low (negligible) vapor pressure. For that reason, they are claimed as nonvolatile electrolytes. This has an important concern with respect to user safety, since nonvolatile and usually nonflammable electrolytes do not propagate (neither initiate nor catch) fire in case of capacitor failure or damage. Therefore, ionic liquids are quite often used as flame retardants in conventional, organic electrolytes. Unfortunately, their high viscosity and moderate conductivity (usually up to 20 mS/cm) still hinder their further application in ECs.

The electrochemical stability ("electrochemical window") of ILs has been claimed to be very high, reaching even 6 V at glassy-carbon electrodes in a three-electrode experimental setup. However, such high values have never been confirmed on AC electrodes in a real, two-electrode device.[7]

Undoubtedly, ILs bring many opportunities for EC development, since their structure can be easily designed. However, one should be aware that their "exciting" properties (essentially high electrochemical stability) and long-term impact on the environment (as well as decomposition products) still need to be thoroughly studied.

3.4 Aqueous Electrolytes

Unlike organic electrolytes, aqueous solutions display several interesting and promising features as electrolytes in EC application. First of all, they are usually highly conductive (up to 600 mS/cm), nonflammable, quite often nontoxic, have high thermal capacity (usually comparable to pure water), and wide thermal stability.[9,85–87] Furthermore, their formulation and application to capacitors are much simpler than for organic electrolytes. Since they are water based, there is no reason for long-term drying. Therefore, the final cost is significantly reduced.

The capacitance values obtained in aqueous electrolytes are usually remarkably higher than for organic electrolytes and ionic liquids, and might reach 250 F/g. Definitely, there is a significant role of the carbon electrode in providing such capacitance, but one should also be aware that the capacitance value (or rather charge received within the selected potential/voltage range) changes, at least with current load. This change is called charge propagation or capacitance retention. For this reason, the capacitance value should not be considered as an absolute one (although for EDL it is), and in fair reports, the results should always be reported with the current load or scan rate applied.

As usual, the aforementioned advantages are accompanied by some drawbacks: a proton-rich medium might induce corrosion (both on casing, current collector, and electrode material), hence this issue must be considered once there is a selection of the kind of electrolyte and current collectors. For noncorrosive solutions, quite often stainless steel is applied (316L or 304), but sometimes nickel- or titanium-based current collectors are utilized.[67,88,89]

Apart from corrosion-related issues, the major objection against aqueous electrolytes is their moderate electrochemical stability. In theory, the water decomposition voltage limits their operation up to 1.23 V, which is independent of pH and the type of electrode used. In practice, this value can vary significantly and depends on several factors: electrolyte pH and viscosity, type of carbon (micro/meso ratio), and surface functionalities (or heteroatoms in the carbon matrix). Acidic electrolytes (e.g., 1 mol/L H_2SO_4) usually display the lowest electrochemical stability on carbon electrodes (up to 1.0 V), most likely because of the oxidative character of sulfuric acid toward carbon and low oxygen/hydrogen evolution potentials. Moreover, application of acidic electrolytes usually requires current collectors from noble metals (corrosion-resistant), which eliminates them from industrial commercialization.[90]

On the other hand, alkaline electrolytes (e.g., 6 mol/L KOH) display slightly higher electrochemical stability (up to 1.2 V), but long-term performance of the devices operating with alkaline solutions is not satisfactory. Most likely, upon continuous operation, there is a carbon matrix disintegration and progressive electrode damage.[91,92]

Here, it is worth mentioning that alkaline solutions allow the pure capacitance value to be boosted by redox chemistry involving the hydrogen electrosorption process (Fig. 8).[93–97]

This phenomenon, described in the literature as specific for microporous carbons, occurs near to the hydrogen evolution reaction and provides an additional charge to the system according to Reaction (1):

$$\langle C\rangle + H_2O + e^- \leftrightarrow \langle C\rangle H_{ad} + OH^- \quad (1)$$

This reaction is reversible and allows the operational voltage of the capacitor to be slightly extended.

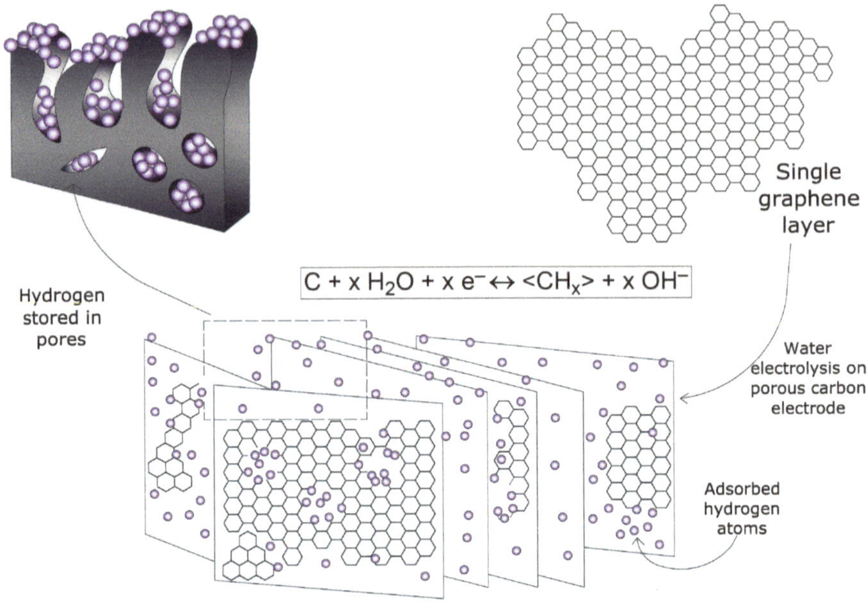

Figure 8 Hydrogen storage mechanism on microporous carbon electrode.

Neutral aqueous electrolytes, such as 0.5–2 mol/L solutions of Li-, Na-, K_2SO_4 and 1–5 mol/L solutions of Li-, Na-, KNO_3, some time ago appeared as the most promising in eco-friendly capacitor applications.[9,85,87,98–100]

First, their pH is neutral—thus, the solution itself is noncorrosive, and stainless steel might be applied as current collector material. Moreover, high overpotentials of oxygen and hydrogen evolution in the neutral medium, as well as strong hydration energies, allow a voltage of 1.8–2.0 V to be reached. This makes these electrolytes an interesting alternative for organic electrolytes. Although the voltage is still c. 1 V lower than for organic electrolytes, the energy should be enhanced by higher capacitance values.

Although much research has recently been done in this field, and aging mechanisms are intensively studied in order to prolong the cycle life of capacitors with neutral, or sulfate- or nitrate-based electrolytes, the capacitance values are not prominent, and the volumetric/gravimetric energy is still moderate.

The capacitance values of ECs might be enhanced by providing additional charge (apart from electrostatic charge/discharge) coming from redox reactions. This additional capacitance, because of its nonelectrostatic origin, is called **pseudocapacitance.**[101,102] Taking into account the high power rates expected from ECs, these reactions must of course be fast and reversible.

Initially, RuO_2 and MnO_2 have been proposed as pseudocapacitive materials.[103] Subsequently, many other materials (mainly transition metal oxides, but also electrically conducting polymers) have been reported as tentative pseudocapacitive materials. Nonetheless, apart from ruthenium- and manganese-based ones, for a large majority of the oxides, the electrochemical characteristics are typically faradaic (with well-defined redox transitions) and for that reason should not be called pseudocapacitive.[91,102]

Pseudocapacitance may originate from an electrode material (as for transition metal oxides), but much more promising results are demonstrated for pseudocapacitance obtained by the application of redox-active electrolytes. In this case, electrodes are not modified by poorly conductive oxides, and their inactive mass does not aggravate the final volumetric/gravimetric capacitance and energy. Moreover, the power rate is not deteriorated by "slow" solid-state reactions; the electrochemical process occurs at the interface between the liquid electrolyte and the solid electrode. The redox-active species might then be rapidly transported from the electrolyte bulk to the interface and undergo the redox process.[71,104]

3.5 Redox-Active Electrolytes

Redox-active electrolytes have been introduced to ECs with the application of 1 mol/L aqueous KI solution to a symmetric carbon/carbon system.[105] The authors demonstrated an increase in the total capacitance versus capacitance of a device operating in 1 mol/L H_2SO_4 solution. The enormous capacitance increase of the positive electrode originates from the following reactions:

$$3\,I^- \leftrightarrow I_3^- + 2e^- \tag{2}$$

$$2\,I^- \leftrightarrow I_2 + 2e^- \tag{3}$$

$$2I_3^- \leftrightarrow 3I_2 + 2e^- \tag{4}$$

$$I_2 + 6\,H_2O \leftrightarrow 2\,IO_3^- + 12\,H^+ + 10\,e^- \tag{5}$$

Interestingly, in a subsequent work, it was demonstrated that the redox activity of the iodide anion depends strongly on the counter ion, and various capacitance values might be recorded for different cations (among alkali

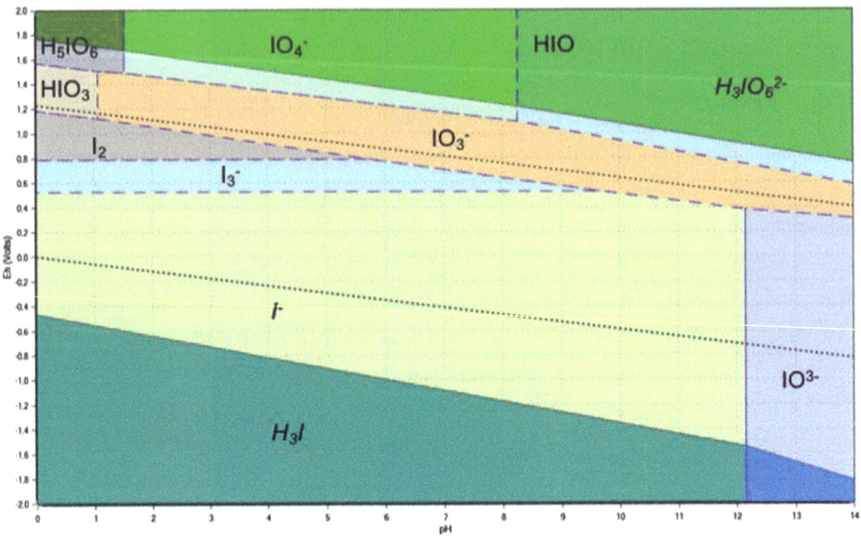

Figure 9 Pourbaix diagram for iodine in water solution at 25°C.

metals, the highest recorded for Rb^+ and the lowest for Li^+) coupled with the iodide anion.[106]

The concentration effect of redox active species has also been verified (from 0.1 to 5 mol/L), and it seems that this parameter influences rather the electrolyte conductivity than the pseudocapacitive effect.[107–109] As noticed in Fig. 9, iodides are active in a wide range of pH (1–9). Hence, they might be mixed with other salts in order to enhance the performance of the system. Unfortunately, the efforts to introduce the shuttle redox-based activity of iodides in an organic medium (and in ionic liquids) are very moderate.

The interactions of iodides, iodine (I_2), and polyiodides (I_3^-) with the carbon electrode surface have been investigated by operando Raman spectroscopy and confirmed that during electrode operation, iodide is firstly converted to iodine and then attacks the carbon surface with weak chemical bond <C_{matrix}>-I formation.[92,109,110] At the same time, there is a competitive reaction between I_2 and I^- leading to polyiodides (I_3^- or I_5^-) formation. At higher electrode potential values, the IO_3^- anion is formed, but the appearance of this species should be avoided. The solubility of KIO_3 is very low, and precipitation product might block the porosity. Of course, blocked porosity has a detrimental effect on the cyclability of the system.[110]

Since the final capacitor capacitance is governed by the electrode of lower capacitance value, the next concept proposed exploiting another redox couple on the negative electrode, based on the redox activity of vanadyl/vanadium species:

$$VOH^{2+} + H^{+} + e^{-} \leftrightarrow V^{2+} + H_2O \quad (6)$$

$$HV_2O_7{}^{3-} + 9H^{+} + 6e^{-} \leftrightarrow 2VO + 5H_2O \quad (7)$$

Iodides on the positive side and vanadyl on the negative side, separated by Nafion 117 membrane, was the first conjugated redox couple reported for capacitor application.[111]

Apart from iodide and vanadyl species, many other redox-active couples were proposed to date (Fig. 10).[115]

Yamazaki et al. introduced the bromine/bromide redox couple for the organic electrolyte.[112] In fact, the pseudocapacitive effect was induced by incorporating bromine in the AC by treatment with brominated water, but the final response was attributed to the bromides present in the electrolyte.[113–115] Complexes with cyanides like $K_3Fe(CN)_6$ and thiocyanates (SCN^{-}) were reported as interesting species also in terms of their conductivity. Apart from well-defined redox response, their high solubility might result in high conductivity values and good power rates. Furthermore, the cycle life of the capacitor operating with KSCN was excellent and reached 100,000 cycles of full charge/discharge ($U_{max} = 1.6$ V).[116,117]

Heteropolyacids like phosphotungstic acid (PWA) and silicotungstic acid (SiWA) are worthy to be considered too. Their redox activity originates from highly mobile protons. These compounds are the perfect media for redox couples, requiring fast proton transfer [like quinone—hydroquinone (HQ)]. Furthermore, high hydrogen overpotential in these electrolytes improves the electrochemical stability to 1.6 V (vs. 1.0 V for conventional acids).[118–121]

Well-known organic redox mediators such as HQ with isomers,[59,122–124] methylene blue,[125] indigo carmine,[124] p-phenylenediamine,[126,127] m-phenylenediamine,[128] lignosulfonates,[129] sulfonated polyaniline,[130] and humic acids[131] have been reported as interesting compounds for the enhancement of carbon electrodes capacitance. Their high activity was unfortunately observed in strongly acidic solutions (excluding phenylenediamines) and aggravates their wide application in commercial devices.

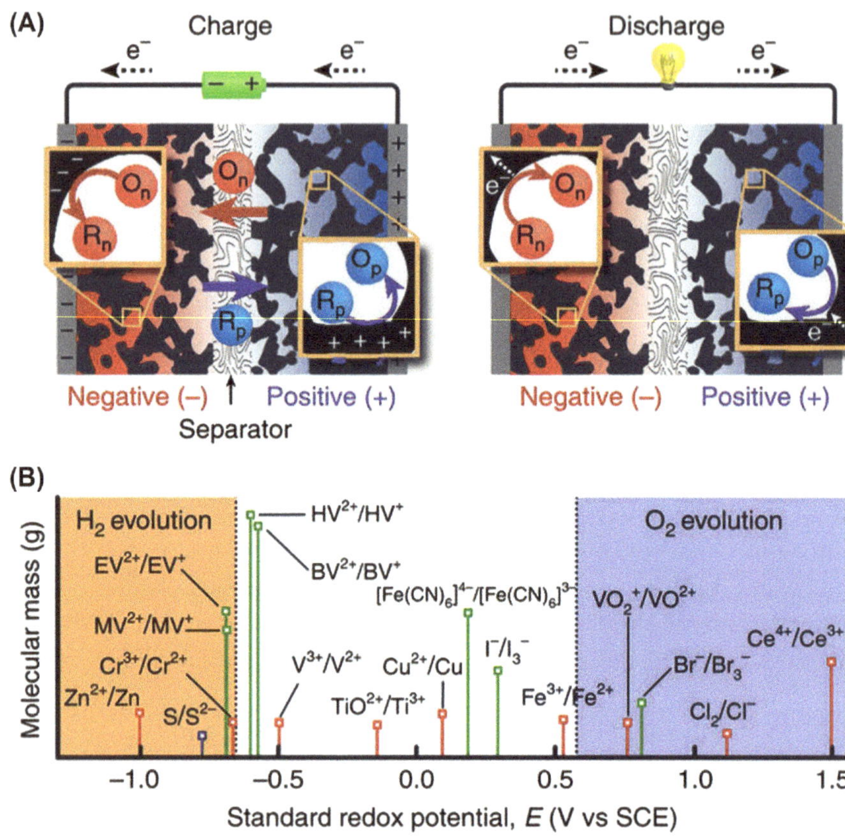

Figure 10 Microscopic processes and candidate couples for redox electric double-layer capacitors. (A) Schematic showing capacitive and faradaic charge-storage processes and (B) redox potentials of the couples vs Saturated Calomel Electrode (SCE).

For organic media, the selection is rather limited. Cluster ions like polyfluorododecaborates $[B_{12}F_xH_{12-x}]^{2-}$, capable of fast charge transfer, have been identified as promising candidates for PC (propylene carbonate)- and ACN (acetonitrile)-based electrolytes.[132] In case of ionic liquids, usually the anion side is modified, and the iodide part is selected as a source of redox activity. Ferrocene-based structures have been recently reported as ionic liquids with bifunctional components.[133]

Finally, redox active electrolytes appeared as an interesting solution for capacitance and energy enhancement. Of course, in introducing redox activity to capacitive systems, one should expect a slight decrease in power density (or response time) since the kinetics of redox processes is usually slower than that of electrostatic attraction.[104,109,134,135]

Redox active electrolytes might also play an interesting role of overcharging protectors (like SCN^-) and play a protective role for carbon electrodes against oxidation. In this way, they impact the cyclability of the system.

Some disadvantages of the redox active electrolytes should also be pointed out. Usually, the systems containing redox species suffer from faster self-discharge and high leakage current. This is related to the shuttling effect of the redox species, hence it has been proposed to separate the electrodes by an ion selective membrane (like Nafion) or to exploit the species becoming insoluble after the redox process (reversibly).

The cyclability of capacitors with redox active electrolytes is usually similar to unmodified electrolytes. However, an inappropriate selection of the redox mediator might aggravate the stability of the system and result in faster capacitance fading.[115]

The rapid development of personal electronics induced the need for water- or liquid-free power sources. Definitely, the electrolyte in a liquid state has many advantages, but in some applications or for some special purposes, they cannot be applied. Hence, solid state electrolytes have been proposed.

3.6 Solid-State Electrolytes and All-Solid-State Devices

The principles of ECs operation, when fully understood, can be successfully used in order to boost their performance. Taking into account these aspects, it was proposed to immobilize the electrolyte. There might be several advantages of electrolyte immobilization. The most important is to promote the enhancement of the water stability window as a result of the introduction of a fixed pH gradient within the capacitor cell. What emerges from this fact is a reduced leakage current. This is possible due to increased viscosity of the electrolyte, which adapts the gel form and decelerates ion mixing. It was shown that it is possible to reach 3 V capacitors operating in propylene carbonate using a gel polymer electrolyte (polyethylene oxide).[136] Other frequently used gel electrolytes are based on the polyvinyl alcohol matrix, usually mixed with H_2SO_4, KOH, and LiCl.[137] Agarose is another gel used in the field, usually in a concentration of c. 1%. The preparation involves heating with the dissolved electrolyte, such as Na_2SO_4, NaCl, or other, using a microwave oven.[138] The gel electrolytes contribute to the production of all-solid-state and flexible capacitors that enlarge their possible applications. Gel electrolytes, apart from their advantages, carry also an important drawback related to increased resistance as a result of high viscosity.[139–141] Such a property must be overcome in order to use these kinds of ECs successfully.

4. LITHIUM-ION CAPACITORS

The ECs (EDL) have attracted much attention during the last decades, which focused mainly on the increase of their specific energy. Nevertheless, it seems that the barrier of ~20 Wh/kg in case of an organic medium and ~10 Wh/kg in case of the aqueous system are difficult to surpass. The incorporation of pseudocapacitive materials increases in addition the energy density; however, the final value is still unsatisfactory for certain applications. Therefore, the idea of combining the advantages of a high energy density electrode coming from already well-mature lithium-ion cell technology (250 Wh/kg) with high power density from an EC (10 kW/kg) seemed to be the solution. In 2001, Amatucci et al.[142] proposed the so-called lithium-ion capacitor (LIC), in which he combined a negative graphite intercalation electrode with a positive one of a capacitive nature. The resulting device allowed for reaching higher energy density than a contemporary EC. The cell consisted out of an AC positive electrode on which reversible adsorption/desorption of ions from the electrolyte occurs and a graphite negative electrode on which intercalation of lithium ions originating from a third lithium electrode, takes place. The energy value of the cell was over 20 Wh/kg at low discharge power. The research progressed gradually and currently two generations of LIC are known. The first type includes the mentioned case, in which a third electrode made of pure lithium acts as a source of lithium ions. However, this type carries a drawback of engaging an auxiliary electrode which does not participate directly in the charge storage mechanism and requires a special arrangement in addition to the graphite and AC electrodes. These are issues that contribute to the increasing mass of the device, which in turn negatively affects the energy density, since it is not possible to fully use the lithium to be inserted into graphite. Moreover, it creates safety issues due to the presence of pure lithium metal, also related to the cell manufacturing that requires an inert atmosphere. The second generation of LIC adopts a source of lithium ions that originates directly from the positive electrode material, which includes a dispersed lithium transition metal oxide within the AC. Since the introduction of this concept by Park et al.,[143] various lithium metal sources, mainly lithium transition metal compounds, have been investigated. This generation is simpler to produce since only two electrodes are involved, and it requires no inert atmosphere during electrode material production. However, it requires separate production of the lithium transition metal compound. Currently, it is synthesized by thermal fusion at high temperature (>600°C) of lithium carbonate salt with a transition metal oxide. The requirements for the suitable lithium metal oxide is that it has relatively low Li^+ extraction potential (below solvent decomposition limit) and high irreversible capacity in order to reduce the dead mass loading after the prelithiation step. Nonetheless, this method seems more viable. There were also studies on hybrid capacitors with an AC electrode as the negative one (rather than as the positive electrode) and graphite or $LiNi_{0.5}Mn_{1.5}O_4$ as the positive electrode. However, this configuration suffered from low voltages. Similar to Li-ion cells, before the LIC is ready for operation, the formation of solid electrolyte interphase from electrolyte decomposition takes place. Lithiation of the graphite electrode can be accomplished by two methods. First, the electrodes are short-circuited, in order for the lithium ions to be inserted into the anode material. Second, Li^+ ions can be extracted from the lithium metal oxide electrochemically and then effectively inserted into the anode. It is possible to calculate the amount of the compound for an appropriate lithium content. The remaining delithiated metal oxide does not act destructively on the AC cathode. However, it may contribute to the reduced conductivity of the electrode since it leaves insulating metal oxide residues. It is very important to stress that the crystallographic phase of the lithium metal compound facilitates lithium extraction. Fig. 11 presents the final assembly of the electrodes. Unlike ECs, but similar to Li-ion cells, lithium-ion capacitors require the use of two current collector materials for specific electrodes. The negative electrode includes a copper current collector and the positive electrode an aluminum current collector, in order to respect the electrochemical stability at the respective electrode potentials. The electrodes are immersed in the solution of the electrolyte, usually 1 M $LiPF_6$ dissolved in a mixture of

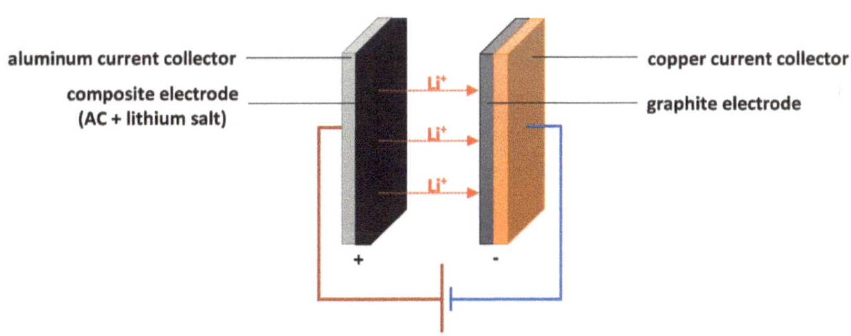

Figure 11 Schematic cell design. *AC*, activated carbon.

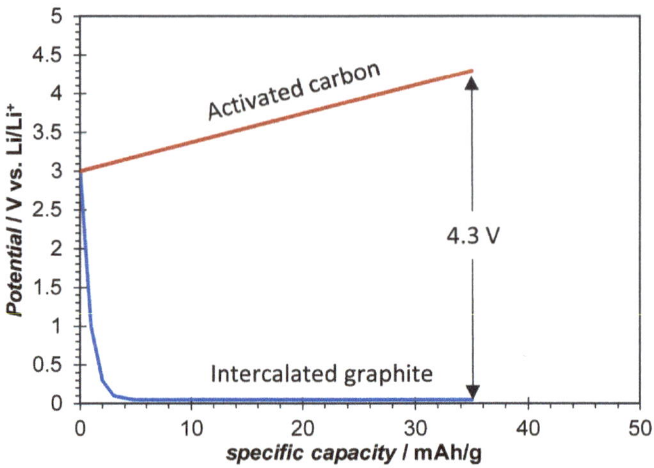

Figure 12 Electrode potential profiles and voltage.

organic solvents EC (ethylene carbonate):DMC (dimethyl carbonate) (usually 50:50 v/v ratio) or 1 M $(C_2H_5)_4N(BF_4)$ in PC.

Typical operating potential profiles are shown in Fig. 12. The graphite electrode reaches the potential close to 0.05 V versus Li/Li^+ at which the lithium intercalates/deintercalates reversibly. At the same time, the capacitive AC electrode increases the potential linearly during adsorption of PF_6^- or BF_4^- ions and decreases during desorption.[144] The maximum voltage of the cell is about 4.3 V, limited by the upper positive potential of the AC electrode. At the point when this voltage is exceeded, decomposition of the electrolyte occurs. On the other hand, the lower voltage limit should be carefully selected. The lower voltage limit cannot be exceeded in order to avoid lithium plating. In this case, short-circuiting of the cell might be detrimental. This will irreversibly lead to a drop in cycle life and capacity. There were trials to increase the energy density by diminishing of the lower voltage limit (4.1–2.0 V) and allowed to reach values of more than 100 Wh/kg. However, the stability of the prelithiated anode was not satisfactory. In general, the deeper the discharge, the worse the observed cycling stability. Fig. 13 presents a comparison of the galvanostatic curves obtained for LIC and EDLC. It is seen that the EDLC operates within lower voltage values than LIC. While LIC has only 1.5 times higher capacitance, owing to its operation at higher voltage values, the energy stored in the device is considerably larger.

The search for another material for irreversible lithium extraction also involves materials such as Li_2RuO_3. The motivation of using such material

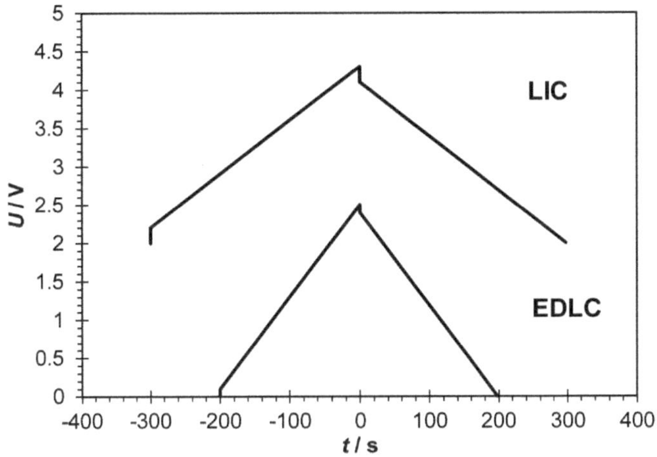

Figure 13 Galvanostatic charge/discharge curve comparison of lithium-ion capacitor (*LIC*) and double-layer capacitor (*EDLC*).

is to boost the capacity of the cell, by the contribution of pseudocapacitance at the positive electrode originating from $Ru^{4+}/Ru^{5+}/Ru^{6+}$. However, the increase in performance is only slight in an organic medium, which does not justify the use of such expensive RuO_2. Li_5FeO_4 is another compound studied as a source for lithium. It contains four lithium ions which can be extracted, giving a theoretical capacity of 700 mAh/g.[145] Li_6CoO_4 is an interesting compound which offers a theoretical capacity of 900 mAh/g. As a result, it is possible to introduce a smaller amount of the lithium-containing compound, while still maintaining respective charge storage in relation to the positive electrode. This might be the way to increase the energy density of the cell owing to the increased irreversibility behavior of Li_6CoO_4.[146] The material Li_5ReO_6 has a low lithium extraction potential of 4.2 V versus Li/Li^+ and almost 400 mAh/g of irreversible capacity.[147] Nickel lithium oxide with a nonstoichiometric ratio has also been used as part of the cathode in LIC, showing almost 100% capacitance retention after 1000 galvanostatic cycles with 0.25 A/g in the voltage range of only 3.8–2.2 V.[148] A different approach has been undertaken in order to combine the properties of a Li-ion cell with a Li-ion capacitor. Namely, the positive electrode consisted of two divided regions, one part made of lithium cobalt oxide while another with AC. A negative electrode was made of graphite covered by lithium metal powder. At low power, the characteristics were almost those of the Li-ion cell, whereas at high power, similar to those of EDLC. The device was called hybrid Li-ion capacitor.[149]

Recently, a new type of lithium source, namely a lithium organic compound 3,4-dihydroxybenzonitrile dilithium salt, was used in LIC for the first time. It presented outstanding performance and high irreversible capacity.[150] It offered very high capacitance retention during 25,000 cycles at 0.5 A/g. Lithium-ion capacitor was also used in aqueous environment; however, with this electrolyte, the interesting properties of the hybrid capacitor cannot be revealed due to relatively low decomposition voltage of water, therefore the voltage of this device is limited to 1.5 V. The resulting energy density is comparable to that of symmetric AC/AC capacitors.[151]

Although Li-ion capacitors are a promising technology for high energy and high power storage, the low abundant lithium metal tends to be replaced by much more available sodium.[152] The sodium-ion capacitors are characterized by slightly higher internal resistance compared to LIC, whereas all the responses in fundamental electrochemistry measurements (such as power or energy densities) are comparable. The most considerable issue emerges from the fact of the significantly worse cyclability of the sodium-ion capacitor, which blocks its success.[153]

Finally, it is worth mentioning that the lithium-ion technology seems especially promising for high-voltage applications but medium power ratings. It requires a lower number of cells in series than ECs (especially those operating in aqueous medium) in order to meet the high voltage requirements. However, their use must be limited to lower power ratings. Nevertheless, it should be noted that the cell operates using an organic solvent which is considered as being flammable and environmental-unfriendly. Moreover, the conductivity of the organic electrolyte is in the range of 11 mS/cm, whereas for aqueous electrolytes, this value might be 10-fold larger. As a rule of thumb, the higher the electrolyte conductivity, the higher the power of the capacitor.[154]

The choice of the capacitor will be determined by its final application. Electrochemical double-layer capacitors are characterized by high power ratings. Therefore, their use is fully justified in peak electric energy harvesting systems, where other technologies are not suitable. However, one must note that it can store only a limited amount of energy. Therefore, its best performance is found when connected in tandem with higher energy density storage technology, such as Li-ion cells. Lithium-ion capacitors were developed to attain the energy density of a Li-ion cell and power density of EDLC. However, it turned out that these parameters cannot be maintained high at the same time. The resulting LIC had energy and power densities being a compromise between these two technologies. Therefore, LIC can perform as a single device in applied energy storage systems. Another aspect is related to cyclability, where the Li-ion cell in general suffers from poor cycle life, especially during high current peaks. This is effectively increased by the use of supporting EDLC, which equalizes the current waveform, thus positively affecting cell cyclability and minimizing their costs.

5. CONCLUDING REMARKS AND PERSPECTIVES

The overview on ECs clearly showed the importance of correct performance measurements, e.g., discharge by constant power for Ragone plot construction, elimination of excessive voltage drop, floating as accelerated ageing test, etc. Appropriate selection of capacitor component properties, i.e., electrode porosity (micro/meso ratio), type of electrolyte (protic, aprotic, redox-active, ILs), and its conductivity, was discussed. Many valuable research were obtained; however, there are still some voids for more work to be done. As highlighted in this chapter, ECs are high-power energy-storage devices with moderate energy densities. The current research orientates in boosting the energy density, nevertheless, it never exceeds the value found in galvanic cells. Therefore, the main application of capacitors is not to store large amounts of energy, but rather to create a buffer which accelerates the energy storage system coupled with another high-energy storage device, e.g., metal-ion batteries. In this way, it is possible to create a system which is able to cover a very broad range of current regimes and offload kinetically limited electrochemical redox reactions incorporated in the operation of the system. Moreover, this operation would improve the stability in terms of long-term operation. Another perspective in EC development seems to be improvement of frequency response which could have significant application for voltage profile smoothing as component in smart energy/conversion of renewable energy. Recently, much attention is devoted to wearable power sources which require development of flexible systems (printed).

Finally, the research should be focused toward biosustainability.[155] The bio-originating electrodes made from waste could be a cheap opportunity for electrodes in electrochemical devices. Nevertheless, it should be herein highlighted that food cannot be taken into account, because there are still many people in this world suffering from a lack of food.

ACKNOWLEDGMENTS

The authors acknowledge the grant of the Ministry of Science and Higher Education, no. 03/31/DSPB/0353.

REFERENCES

1. Ho, J.; Jow, T. R.; Boggs, S. Historical Introduction to Capacitor. *IEEE Electr. Insul. Mag.* **2010,** *26* (1), 20–25.
2. Whittingham, M. S. History, Evolution, and Future Status of Energy Storage. *Proc. IEEE* **2012,** *100* (Special Centennial Issue), 1518–1534.
3. Bard, A. J.; Faulkner, L. R.; York, N.; C; Brisbane, W.; Toronto, S. E. *Electrochemical Methods Fundamentals and Applications,* 1944.
4. Miller, J. R. Introduction to Electrochemical Capacitor Technology. *IEEE Electr. Insul. Mag.* **2010,** *26* (4), 40–47.
5. Kötz, R.; Carlen, M. Principles and Applications of Electrochemical Capacitors. *Electrochim. Acta* **2000,** *45* (15–16), 2483–2498.
6. Zhong, C.; Deng, Y.; Hu, W.; Qiao, J.; Zhang, L.; Zhang, J. A Review of Electrolyte Materials and Compositions for Electrochemical Supercapacitors. *Chem. Soc. Rev.* **2015,** *44* (21), 7484–7539.
7. Salanne, M. Ionic Liquids for Supercapacitor Applications. *Top. Curr. Chem.* **2017.**
8. Eftekhari, A. Supercapacitors Utilising Ionic Liquids. *Energy Storage Materials* **October 2017,** 47–69. Elsevier.
9. Fic, K.; Lota, G.; Meller, M.; Frackowiak, E. Novel Insight into Neutral Medium as Electrolyte for High-Voltage Supercapacitors. *Energy Environ. Sci.* **2012,** *5* (2), 5842–5850.
10. Reber, D.; Kuhnel, R.-S.; Battaglia, C. High-Voltage Aqueous Supercapacitors Based on NaTFSI. *Sustain. Energy Fuels* **2017,** *1* (10), 2155–2161.
11. He, M.; Fic, K.; Frackowiak, E.; Novák, P.; Berg, E. J. Ageing Phenomena in High-Voltage Aqueous Supercapacitors Investigated by in Situ Gas Analysis. *Energy Environ. Sci.* **2016,** *9* (2), 623–633.
12. Blomquist, N.; Wells, T.; Andres, B.; Bäckström, J.; Forsberg, S.; Olin, H. Metal-Free Supercapacitor with Aqueous Electrolyte and Low-Cost Carbon Materials. *Sci. Rep.* **2017,** *7,* 39836.
13. Menzel, J.; Fic, K.; Frackowiak, E. Hybrid Aqueous Capacitors with Improved Energy/power Performance. *Prog. Nat. Sci. Mater. Int.* **2015,** *25* (6), 642–649.
14. Weingarth, D.; Foelske-Schmitz, A.; Kötz, R. Cycle versus Voltage Hold - Which Is the Better Stability Test for Electrochemical Double Layer Capacitors? *J. Power Sources* **2013,** *225,* 84–88.
15. Miller, J. R. Engineering Electrochemical Capacitor Applications. *J. Power Sources* **2016,** *326,* 726–735.
16. Slesinski, A.; Matei-Ghimbeu, C.; Fic, K.; Béguin, F.; Frackowiak, E. Self-Buffered pH at Carbon Surfaces in Aqueous Supercapacitors. *Carbon* **2018,** *129.*
17. Laheäär, A.; Przygocki, P.; Abbas, Q.; Béguin, F. Appropriate Methods for Evaluating the Efficiency and Capacitive Behavior of Different Types of Supercapacitors. *Electrochem. Commun.* **2015,** *60,* 21–25.
18. Graydon, J. W.; Panjehshahi, M.; Kirk, D. W. Charge Redistribution and Ionic Mobility in the Micropores of Supercapacitors. *J. Power Sources* **2014,** *245,* 822–829.
19. Zhao, J.; Gao, Y.; Burke, A. F. Performance Testing of Supercapacitors: Important Issues and Uncertainties. *J. Power Sources* **2017,** *363,* 327–340.

20. Velikonja, A.; Çongadze, E.; Kralj-Iglic, V.; Iglic, A. Charge Dependent Capacitance of Stern Layer and Capacitance of Electrode/electrolyte Interface. *Int. J. Electrochem. Sci.* **2014,** *9* (11), 5885–5894.
21. Gongadze, E.; Velikonja, A.; Perutkova, Š.; Kramar, P.; Maček-Lebar, A.; Kralj-Iglič, V.; Iglič, A. Ions and Water Molecules in an Electrolyte Solution in Contact with Charged and Dipolar Surfaces. *Electrochim. Acta* **2014,** *126,* 42–60.
22. Simon, P.; Gogotsi, Y. Materials for Electrochemical Capacitors. *Nat. Mater.* **2008,** 845–854.
23. Sevilla, M.; Mokaya, R. Energy Storage Applications of Activated Carbons: Supercapacitors and Hydrogen Storage. *Energy Environ. Sci.* **2014,** 7 (4), 1250–1280.
24. Wu, H. P.; He, D. W.; Wang, Y. S.; Fu, M.; Liu, Z. L.; Wang, J. G.; Wang, H. T. Graphene as the Electrode Material in Supercapacitors. In *2010 8th International Vacuum Electron Sources Conference and Nanocarbon;* 2010; pp 465–466.
25. Qi, J. L.; Wang, X.; Lin, J. H.; Zhang, F.; Feng, J. C.; Fei, W.-D. A High-Performance Supercapacitor of Vertically-Oriented Few-Layered Graphene with High-Density Defects. *Nanoscale* **2015,** 7 (8), 3675–3682.
26. Zhang, L.; Zhang, F.; Yang, X.; Long, G.; Wu, Y.; Zhang, T.; Leng, K.; Huang, Y.; Ma, Y.; Yu, A.; et al. Porous 3D Graphene-Based Bulk Materials with Exceptional High Surface Area and Excellent Conductivity for Supercapacitors. *Sci. Rep.* **2013,** *3,* 1408.
27. Luo, G.; Liu, L.; Zhang, J.; Li, G.; Wang, B.; Zhao, J. Hole Defects and Nitrogen Doping in Graphene: Implication for Supercapacitor Applications. *ACS Appl. Mater. Interfaces* **2013,** *5* (21), 11184–11193.
28. Jingyi, Z.; Childress, A. S.; Mehmet, K.; Sushmita, D.; Anurag, S.; Ye, L.; Rao, A. M.; Ramakrishna, P. Defect-Engineered Graphene for High-Energy- and High-Power-Density Supercapacitor Devices. *Adv. Mater.* **2016,** *28* (33), 7185–7192.
29. Yang, W.; Ni, M.; Ren, X.; Tian, Y.; Li, N.; Su, Y.; Zhang, X. Graphene in Supercapacitor Applications. *Curr. Opin. Colloid Interface Sci.* **2015,** *20* (5–6), 416–428.
30. Chen, X.; Zhang, L.; Chen, S. Large Area CVD Growth of Graphene. *Synth. Met.* **2015,** *210,* 95–108.
31. Zhang, Y.; Zhang, L.; Zhou, C. Review of Chemical Vapor Deposition of Graphene and Related Applications. *Acc. Chem. Res.* **2013,** *46* (10), 2329–2339.
32. Gao, Y. Graphene and Polymer Composites for Supercapacitor Applications: A Review. *Nanoscale Res. Lett.* **2017,** *12,* 387.
33. Chen, T.; Dai, L. Carbon Nanomaterials for High-Performance Supercapacitors. *Mater. Today* **2013,** *16* (7–8), 272–280.
34. Rangom, Y.; Tang, X.; Nazar, L. F. Carbon Nanotube-Based Supercapacitors with Excellent Ac Line Filtering and Rate Capability via Improved Interfacial Impedance. *ACS Nano* **2015,** *9* (7), 7248–7255.
35. Notarianni, M.; Liu, J.; Mirri, F.; Pasquali, M.; Motta, N. Graphene-Based Supercapacitor with Carbon Nanotube Film as Highly Efficient Current Collector. *Nanotechnology* **2014,** *25* (43).
36. Wu, G.; Tan, P.; Wang, D.; Li, Z.; Peng, L.; Hu, Y.; Wang, C.; Zhu, W.; Chen, S.; Chen, W. High-Performance Supercapacitors Based on Electrochemical-Induced Vertical-Aligned Carbon Nanotubes and Polyaniline Nanocomposite Electrodes. *Sci. Rep.* **2017,** *7,* 43676.
37. Śliwak, A.; Grzyb, B.; Díez, N.; Gryglewicz, G. Nitrogen-Doped Reduced Graphene Oxide as Electrode Material for High Rate Supercapacitors. *Appl. Surf. Sci.* **2017,** *399,* 265–271.

38. Kesavan, T.; Aswathy, R.; Raj, I. A.; Kumar, T. P.; Ragupathy, P. Nitrogen-Doped Graphene as Electrode Material with Enhanced Energy Density for Next-Generation Supercapacitor Application. *ECS J. Solid State Sci. Technol.* **2015,** *4* (12), M88–M92.
39. Wang, L.; Gao, Z.; Chang, J.; Liu, X.; Wu, D.; Xu, F.; Guo, Y.; Jiang, K. Nitrogen-Doped Porous Carbons As Electrode Materials for High-Performance Supercapacitor and Dye-Sensitized Solar Cell. *ACS Appl. Mater. Interfaces* **2015,** 7 (36), 20234–20244.
40. Li, B.; Dai, F.; Xiao, Q.; Yang, L.; Shen, J.; Zhang, C.; Cai, M. Nitrogen-Doped Activated Carbon for a High Energy Hybrid Supercapacitor. *Energy Environ. Sci.* **2016,** *9* (1), 102–106.
41. Shang, T. X.; Cai, X. X.; Jin, X. J. Phosphorus- and Nitrogen-Co-Doped Particleboard Based Activated Carbon in Supercapacitor Application. *RSC Adv.* **2015,** *5* (21), 16433–16438.
42. Li, Y.; Shang, T.-X.; Gao, J.-M.; Jin, X.-J. Nitrogen-Doped Activated Carbon/graphene Composites as High-Performance Supercapacitor Electrodes. *RSC Adv.* **2017,** 7 (31), 19098–19105.
43. Liu, Y.; Shi, K.; Zhitomirsky, I. Asymmetric Supercapacitor, Based on Composite MnO2-Graphene and N-Doped Activated Carbon Coated Carbon Nanotube Electrodes. *Electrochim. Acta* **2017,** *233,* 142–150.
44. Wang, Y.; Song, Y.; Xia, Y. Electrochemical Capacitors: Mechanism, Materials, Systems, Characterization and Applications. *Chem. Soc. Rev.* **2016,** *45* (21), 5925–5950.
45. Vindt, S. T.; Skou, E. M. The Buffer Effect in Neutral Electrolyte Supercapacitors. *Appl. Phys. Mater. Sci. Process* **2016,** *122* (2), 1–6.
46. Wang, X.; Chandrabose, R. S.; Jian, Z.; Xing, Z.; Ji, X. A 1.8 V Aqueous Supercapacitor with a Bipolar Assembly of Ion-Exchange Membranes as the Separator. *J. Electrochem. Soc.* **2016,** *163* (99), 1853–1858.
47. Lufrano, F.; Staiti, P. Performance Improvement of Nafion Based Solid State Electrochemical Supercapacitor. *Electrochim. Acta* **2004,** *49* (16), 2683–2689.
48. Stoeckli, F.; Centeno, T. A.; Rouquérol, J.; Avnir, D.; Fairbridge, C. W.; Everett, D. H.; Haynes, J. M.; Pernicone, N.; Ramsay, J. D. F.; Sing, K. S. W.; et al. Optimization of the Characterization of Porous Carbons for Supercapacitors. *J. Mater. Chem. A* **2013,** *1* (23), 6865.
49. Lobato, B.; Suárez, L.; Guardia, L.; Centeno, T. A. Capacitance and Surface of Carbons in Supercapacitors. *Carbon* **2017,** *122,* 434–445.
50. Centeno, T. A.; Sereda, O.; Stoeckli, F. Capacitance in Carbon Pores of 0.7 to 15 Nm: A Regular Pattern. *Phys. Chem. Chem. Phys.* **2011,** *13* (27), 12403.
51. Tsay, K. C.; Zhang, L.; Zhang, J. Effects of Electrode Layer Composition/thickness and Electrolyte Concentration on Both Specific Capacitance and Energy Density of Supercapacitor. *Electrochim. Acta* **2012,** *60,* 428–436.
52. Endo, M.; Maeda, T.; Takeda, T.; Kim, Y. J.; Koshiba, K.; Hara, H.; Dresselhaus, M. S. Capacitance and Pore-Size Distribution in Aqueous and Nonaqueous Electrolytes Using Various Activated Carbon Electrodes. *J. Electrochem. Soc.* **2001,** *148* (8), A910.
53. Kim, Y. J.; Horie, Y.; Ozaki, S.; Matsuzawa, Y.; Suezaki, H.; Kim, C.; Miyashita, N.; Endo, M. Correlation between the Pore and Solvated Ion Size on Capacitance Uptake of PVDC-Based Carbons. *Carbon* **2004,** *42* (8–9), 1491–1500.
54. Gogotsi, Y.; Simon, P. True Performance Metrics in Electrochemical Energy Storage. *Science* **2011,** *18,* 917–918.
55. Jiang, J.; Dahn, J. R. Effects of Solvents and Salts on the Thermal Stability of LiC6. *Electrochim. Acta* **2004,** *49* (26), 4599–4604.

56. MacFarlane, D. R.; Tachikawa, N.; Forsyth, M.; Pringle, J. M.; Howlett, P. C.; Elliott, G. D.; Davis, J. H.; Watanabe, M.; Simon, P.; Angell, C. A. Energy Applications of Ionic Liquids. *Energy Environ. Sci.* **2014,** 7 (1), 232–250.
57. Oda, H.; Yamashita, A.; Minoura, S.; Okamoto, M.; Morimoto, T. Modification of the Oxygen-Containing Functional Group on Activated Carbon Fiber in Electrodes of an Electric Double-Layer Capacitor. *J. Power Sources* **2006,** *158* (2 SPEC. ISS.), 1510–1516.
58. Qu, D. Studies of the Activated Carbons Used in Double-Layer Supercapacitors. *J. Power Sources* **2002,** *109* (2), 403–411.
59. Yoshida, A.; Tanahashi, I.; Nishino, A. Effect of Concentration of Surface Acidic Functional Groups on Electric Double-Layer Properties of Activated Carbon Fibers. *Carbon* **1990,** *28* (5), 611–615.
60. Fic, K.; Lota, G.; Frackowiak, E. Effect of Surfactants on Capacitance Properties of Carbon Electrodes. *Electrochim. Acta* **2012,** *60,* 206–212.
61. Beguin, F.; Presser, V.; Balducci, A.; Frackowiak, E. Carbons and Electrolytes for Advanced Supercapacitors. *Adv. Mater.* **2014,** *26* (14), 2219–2251.
62. Bohlen, O.; Kowal, J.; Sauer, D. U. Ageing Behaviour of Electrochemical Double Layer Capacitors. *J. Power Sources* **2007,** *172* (1), 468–475.
63. Burke, A. R&D Considerations for the Performance and Application of Electrochemical Capacitors. *Electrochim. Acta* **2007,** *53* (3 SPEC. ISS.), 1083–1091.
64. Azaïs, P.; Duclaux, L.; Florian, P.; Massiot, D.; Lillo-Rodenas, M. A.; Linares-Solano, A.; Peres, J. P.; Jehoulet, C.; Béguin, F. Causes of Supercapacitors Ageing in Organic Electrolyte. *J. Power Sources* **2007,** *171* (2), 1046–1053.
65. Bittner, A. M.; Zhu, M.; Yang, Y.; Waibel, H. F.; Konuma, M.; Starke, U.; Weber, C. J. Ageing of Electrochemical Double Layer Capacitors. *J. Power Sources* **2012,** *203,* 262–273.
66. He, M.; Fic, K.; Frackowiak, E.; Novák, P.; Berg, E. J. Ageing Phenomena in High-Voltage Aqueous Supercapacitors Investigated by in Situ Gas Analysis. *Energy Environ. Sci.* **2016,** *9* (2), 623–633.
67. Ratajczak, P.; Jurewicz, K.; Béguin, F. Factors Contributing to Ageing of High Voltage Carbon/carbon Supercapacitors in Salt Aqueous Electrolyte. *J. Appl. Electrochem.* **2014,** *44* (4), 475–480.
68. Abbas, Q.; Pajak, D.; Frackowiak, E.; Béguin, F. Effect of Binder on the Performance of Carbon/carbon Symmetric Capacitors in Salt Aqueous Electrolyte. *Electrochim. Acta* **2014,** *140,* 132–138.
69. Zhong, C.; Deng, Y.; Hu, W.; Qiao, J.; Zhang, L.; Zhang, J. A Review of Electrolyte Materials and Compositions for Electrochemical Supercapacitors. *Chem. Soc. Rev.* **2015,** *44* (21), 7484–7539.
70. Miller, J. R.; Burke, A. F. Electrochemical Capacitors: Challenges and Opportunities for Real-World Applications. *Electrochem. Soc. Interface* **2008,** *17* (1), 53–57.
71. Chen, G. Z. Supercapacitor and Supercapattery as Emerging Electrochemical Energy Stores. *Int. Mater. Rev.* **2017,** 173–202.
72. Béguin, F., Frackowiak, E., Eds. *Supercapacitors: Materials, Systems, and Applications;* 2013.
73. Balducci, A.; Bardi, U.; Caporali, S.; Mastragostino, M.; Soavi, F. Ionic Liquids for Hybrid Supercapacitors. *Electrochem. Commun.* **2004,** *6* (6), 566–570.
74. Balducci, A. Electrolytes for High Voltage Electrochemical Double Layer Capacitors: A Perspective Article. *J. Power Sources* **2016,** *326,* 534–540.
75. Le Bideau, J.; Viau, L.; Vioux, A. Ionogels, Ionic Liquid Based Hybrid Materials. *Chem. Soc. Rev.* **2011,** *40* (2), 907–925.

76. Kim, I. T.; Egashira, M.; Yoshimoto, N.; Morita, M. Effects of Electrolytic Composition on the Electric Double-Layer Capacitance at Smooth-Surface Carbon Electrodes in Organic Media. *Electrochim. Acta* **2010,** *55* (22), 6632–6638.
77. Ue, M.; Ida, K.; Mori, S. Electrochemical Properties of Organic Liquid Electrolytes Based on Quaternary Onium Salts for Electrical Double-Layer Capacitors. *J. Electrochem. Soc.* **1994,** *141* (11), 2989.
78. Yamada, H.; Moriguchi, I.; Kudo, T. Electric Double Layer Capacitance on Hierarchical Porous Carbons in an Organic Electrolyte. *J. Power Sources* **2008,** *175* (1), 651–656.
79. Kurig, H.; Jänes, A.; Lust, E. Substituted Phosphonium Cation Based Electrolytes for Nonaqueous Electrical Double-Layer Capacitors. *J. Mater. Res.* **2010,** *25* (8), 1447–1450.
80. Jiang, M.; Zhu, J.; Chen, C.; Lu, Y.; Pampal, E. S.; Luo, L.; Zhu, P.; Zhang, X. Superior High-Voltage Aqueous Carbon/carbon Supercapacitors Operating with in Situ Electrodeposited Polyvinyl Alcohol Borate Gel Polymer Electrolytes. *J. Mater. Chem. A* **2016,** *4* (42), 16588–16596.
81. Wu, P.; Huang, J.; Meunier, V.; Sumpter, B. G.; Qiao, R. Complex Capacitance Scaling in Ionic Liquids-Filled Nanopores. *ACS Nano* **2011,** *5* (11), 9044–9051.
82. Nambu, N.; Takahashi, R.; Takehara, M.; Ue, M.; Sasaki, Y. Electrolytic Characteristics of Fluoroethylene Carbonate for Electric Double-Layer Capacitors at High Concentrations of Electrolyte. *Electrochemistry* **2013,** *81* (10), 817–819.
83. Merlet, C.; Rotenberg, B.; Madden, P. A.; Salanne, M. Computer Simulations of Ionic Liquids at Electrochemical Interfaces. *Phys. Chem. Chem. Phys.* **2013,** *15* (38), 15781.
84. Balducci, A.; Dugas, R.; Taberna, P. L.; Simon, P.; Plée, D.; Mastragostino, M.; Passerini, S. High Temperature Carbon-Carbon Supercapacitor Using Ionic Liquid as Electrolyte. *J. Power Sources* **2007,** *165* (2), 922–927.
85. Bichat, M. P.; Raymundo-Piñero, E.; Béguin, F. High Voltage Supercapacitor Built with Seaweed Carbons in Neutral Aqueous Electrolyte. *Carbon* **2010,** *48* (15), 4351–4361.
86. Khomenko, V.; Raymundo-Piñero, E.; Frackowiak, E.; Béguin, F. High-Voltage Asymmetric Supercapacitors Operating in Aqueous Electrolyte. *Appl. Phys. Mater. Sci. Process* **2006,** *82* (4 SPEC. ISS.), 567–573.
87. Gao, Q.; Demarconnay, L.; Raymundo-Piñero, E.; Béguin, F. Exploring the Large Voltage Range of Carbon/carbon Supercapacitors in Aqueous Lithium Sulfate Electrolyte. *Energy Environ. Sci.* **2012,** *5* (11), 9611.
88. Frąckowiak, E.; Ratajczak, P.; Béguin, F. Electrochemical Capacitors Based on Carbon Electrodes in Aqueous Electrolytes. In *Large Lithium Ion Battery Technology and Application Symposium, LLIBTA 2015 and Large EC Capacitor Technology and Application Symposium, ECCAP 2015-Held at AABC Europe 2015;* 2016; pp 285–312.
89. Ratajczak, P.; Jurewicz, K.; Skowron, P.; Abbas, Q.; Béguin, F. Effect of Accelerated Ageing on the Performance of High Voltage Carbon/carbon Electrochemical Capacitors in Salt Aqueous Electrolyte. *Electrochim. Acta* **2014,** *130,* 344–350.
90. Fic, K.; Meller, M.; Menzel, J.; Frackowiak, E. Around the Thermodynamic Limitations of Supercapacitors Operating in Aqueous Electrolytes. *Electrochim. Acta* **2016,** *206,* 496–503.
91. Chen, G. Z. Understanding Supercapacitors Based on Nano-Hybrid Materials with Interfacial Conjugation. *Prog. Nat. Sci.: Materials International* **2013,** *23* (3), 245–255.
92. Fic, K.; Meller, M.; Frackowiak, E. Interfacial Redox Phenomena for Enhanced Aqueous Supercapacitors. *J. Electrochem. Soc.* **2015,** *162* (5), A5140–A5147.

93. Jurewicz, K.; Frackowiak, E.; Béguin, F. Towards the Mechanism of Electrochemical Hydrogen Storage in Nanostructured Carbon Materials. *Appl. Phys. A* **2004,** *78* (7), 981–987.
94. Béguin, F.; Kierzek, K.; Friebe, M.; Jankowska, A.; Machnikowski, J.; Jurewicz, K.; Frackowiak, E. Effect of Various Porous Nanotextures on the Reversible Electrochemical Sorption of Hydrogen in Activated Carbons. *Electrochim. Acta* **2006,** *51* (11), 2161–2167.
95. Béguin, F.; Friebe, M.; Jurewicz, K.; Vix-Guterl, C.; Dentzer, J.; Frackowiak, E. State of Hydrogen Electrochemically Stored Using Nanoporous Carbons as Negative Electrode Materials in an Aqueous Medium. *Carbon* **2006,** *44* (12), 2392–2398.
96. Babel, K.; Janasiak, D.; Jurewicz, K. Electrochemical Hydrogen Storage in Activated Carbons with Different Pore Structures Derived from Certain Lignocellulose Materials. *Carbon* **2012,** *50* (14), 5017–5026.
97. Lota, G.; Fic, K.; Jurewicz, K.; Frackowiak, E. Correlation of Hydrogen Capacity in Carbon Material with the Parameters of Electrosorption. *Cent. Eur. J. Chem.* **2011,** *9* (1), 20–24.
98. Chae, J. H.; Chen, G. Z. 1.9 v Aqueous Carbon-Carbon Supercapacitors with Unequal Electrode Capacitances. *Electrochimica Acta* **2012,** *86,* 248–254.
99. Redondo, E.; Goikolea, E.; Mysyk, R. The Decisive Role of Electrolyte Concentration in the Performance of Aqueous Chloride-Based Carbon/carbon Supercapacitors with Extended Voltage Window. *Electrochim. Acta* **2016,** *221,* 177–183.
100. Suo, L.; Borodin, O.; Gao, T.; Olguin, M.; Ho, J.; Fan, X.; Luo, C.; Wang, C.; Xu, K. "Water-in-Salt" electrolyte Enables High-Voltage Aqueous Lithium-Ion Chemistries. *Science* **2015,** *350* (6263), 938–943.
101. Conway, B. E.; Birss, V.; Wojtowicz, J. The Role and Utilization of Pseudocapacitance for Energy Storage by Supercapacitors. *J. Power Sources* **1997,** *66* (1–2), 1–14.
102. Brousse, T.; Belanger, D.; Long, J. W. To Be or Not To Be Pseudocapacitive? *J. Electrochem. Soc.* **2015,** *162* (5), A5185–A5189.
103. Conway, B. E.; Pell, W. G. Double-Layer and Pseudocapacitance Types of Electrochemical Capacitors and Their Applications to the Development of Hybrid Devices. *J. Solid State Electrochem.* **2003,** *7* (9), 637–644.
104. Akinwolemiwa, B.; Peng, C.; Chen, G. Z. Redox Electrolytes in Supercapacitors. *J. Electrochem. Soc.* **2015,** *162* (5), A5054–A5059.
105. Lota, G.; Frackowiak, E. Striking Capacitance of Carbon/iodide Interface. *Electrochem. Commun.* **2009,** *11* (1), 87–90.
106. Lota, G.; Fic, K.; Frackowiak, E. Alkali Metal Iodide/carbon Interface as a Source of Pseudocapacitance. *Electrochem. Commun.* **2011,** *13* (1), 38–41.
107. Menzel, J.; Fic, K.; Meller, M.; Frackowiak, E. The Effect of Halide Ion Concentration on Capacitor Performance. *J. Appl. Electrochem.* **2014,** *44* (4), 439–445.
108. Menzel, J.; Fic, K.; Frackowiak, E. Hybrid Aqueous Capacitors with Improved Energy/power Performance. *Prog. Nat. Sci. Mater. Int.* **2015,** *25* (6), 642–649.
109. Frackowiak, E.; Meller, M.; Menzel, J.; Gastol, D.; Fic, K. Redox-Active Electrolyte for Supercapacitor Application. *Faraday Discuss* **2014,** *172,* 179–198.
110. Meller, M.; Menzel, J.; Fic, K.; Gastol, D.; Frackowiak, E. Electrochemical Capacitors as Attractive Power Sources. *Solid State Ionics* **2014,** *265,* 61–67.
111. Frackowiak, E.; Fic, K.; Meller, M.; Lota, G. Electrochemistry Serving People and Nature: High-Energy Ecocapacitors Based on Redox-Active Electrolytes. *ChemSusChem* **2012,** *5* (7), 1181–1185.
112. Yamazaki, S.; Ito, T.; Yamagata, M.; Ishikawa, M. Non-Aqueous Electrochemical Capacitor Utilizing Electrolytic Redox Reactions of Bromide Species in Ionic Liquid. *Electrochim. Acta* **2012,** *86,* 294–297.

113. Gastol, D.; Walkowiak, J.; Fic, K.; Frackowiak, E. Enhancement of the Carbon Electrode Capacitance by Brominated Hydroquinones. *J. Power Sources* **2016,** *326,* 587–594.
114. Barpanda, P.; Fanchini, G.; Amatucci, G. G. Structure, Surface Morphology and Electrochemical Properties of Brominated Activated Carbons. *Carbon* **2011,** *49* (7), 2538–2548.
115. Chun, S. E.; Evanko, B.; Wang, X.; Vonlanthen, D.; Ji, X.; Stucky, G. D.; Boettcher, S. W. Design of Aqueous Redox-Enhanced Electrochemical Capacitors with High Specific Energies and Slow Self-Discharge. *Nat. Commun.* **2015,** *6.*
116. Gorska, B.; Bujewska, P.; Fic, K. Thiocyanates as Attractive Redox-Active Electrolytes for High-Energy and Environmentally-Friendly Electrochemical Capacitors. *Phys. Chem. Chem. Phys.* **2017,** *19* (11), 7923–7935.
117. Ma, G.; Li, J.; Sun, K.; Peng, H.; Mu, J.; Lei, Z. High Performance Solid-State Supercapacitor with PVA-KOH-K 3[Fe(CN)6] Gel Polymer as Electrolyte and Separator. *J. Power Sources* **2014,** *256,* 281–287.
118. Lian, K. K.; Li, C. Asymmetrical Electrochemical Capacitors Using Heteropoly Acid Electrolytes. *ECS Transactions* **2008,** *16,* 127–132.
119. Lian, K.; Li, C. Heteropoly Acid Electrolytes for Double-Layer Capacitors and Pseudocapacitors. *Electrochem. Solid state Lett.* **2008,** *11* (9), A158.
120. Lian, K.; Li, C. M. Solid Polymer Electrochemical Capacitors Using Heteropoly Acid Electrolytes. *Electrochem. Commun.* **2009,** *11* (1), 22–24.
121. Gao, H.; Tian, Q.; Lian, K. Polyvinyl Alcohol-Heteropoly Acid Polymer Electrolytes and Their Applications in Electrochemical Capacitors. *Solid State Ionics* **2010,** *181* (19–20), 874–876.
122. Shul, G.; Bélanger, D. Self-Discharge of Electrochemical Capacitors Based on Soluble or Grafted Quinone. *Phys. Chem. Chem. Phys.* **2016,** *18* (28), 19137–19145.
123. Nazar, L. F.; Meller, M.; Fic, K.; Frackowiak, E. Quinone-Decorated Carbon Materials for Capacitive Energy Storage Applications. In *Materials Research Society Symposium Proceedings;* 2014.
124. Roldán, S.; González, Z.; Blanco, C.; Granda, M.; Menéndez, R.; Santamaría, R. Redox-Active Electrolyte for Carbon Nanotube-Based Electric Double Layer Capacitors. *Electrochim. Acta* **2011,** *56* (9), 3401–3405.
125. Roldán, S.; Granda, M.; Menéndez, R.; Santamaría, R.; Blanco, C. Supercapacitor Modified with Methylene Blue as Redox Active Electrolyte. *Electrochim. Acta* **2012,** *83,* 241–246.
126. Zhang, Z. J.; Zhu, Y. Q.; Chen, X. Y.; Cao, Y. Pronounced Improvement of Supercapacitor Capacitance by Using Redox Active Electrolyte of P-Phenylenediamine. *Electrochim. Acta* **2015,** *176,* 941–948.
127. Hatzell, K. B.; Beidaghi, M.; Campos, J. W.; Dennison, C. R.; Kumbur, E. C.; Gogotsi, Y. A High Performance Pseudocapacitive Suspension Electrode for the Electrochemical Flow Capacitor. *Electrochim. Acta* **2013,** *111,* 888–897.
128. Yu, H.; Fan, L.; Wu, J.; Lin, Y.; Huang, M.; Lin, J.; Lan, Z. Redox-Active Alkaline Electrolyte for Carbon-Based Supercapacitor with Pseudocapacitive Performance and Excellent Cyclability. *RSC Adv.* **2012,** *2* (17), 6736.
129. Lota, G.; Milczarek, G. The Effect of Lignosulfonates as Electrolyte Additives on the Electrochemical Performance of Supercapacitors. *Electrochem. Commun.* **2011,** *13* (5), 470–473.
130. Chen, L.; Chen, Y.; Wu, J.; Wang, J.; Bai, H.; Li, L. Electrochemical Supercapacitor with Polymeric Active Electrolyte. *J. Mater. Chem. A* **2014,** *2* (27), 10526–10531.
131. Wasiński, K.; Walkowiak, M.; Lota, G. Humic Acids as Pseudocapacitive Electrolyte Additive for Electrochemical Double Layer Capacitors. *J. Power Sources* **2014,** *255,* 230–234.

132. Ionica-Bousquet, C. M.; Casteel, W. J.; Pearlstein, R. M.; GirishKumar, G.; Pez, G. P.; Gómez-Romero, P.; Palacín, M. R.; Muñoz-Rojas, D. Polyfluorinated Boron Cluster - [B12F11H]2- - Based Electrolytes for Supercapacitors: Overcharge Protection. *Electrochem. Commun.* **2010,** *12* (5), 636−639.
133. Xie, H. J.; Gélinas, B.; Rochefort, D. Redox-Active Electrolyte Supercapacitors Using Electroactive Ionic Liquids. *Electrochem. Commun.* **2016,** *66,* 42−45.
134. Guan, L.; Yu, L.; Chen, G. Z. Capacitive and Non-Capacitive Faradaic Charge Storage. *Electrochim. Acta* **2016,** *206,* 464−478.
135. Zhong, C.; Hu, W. *Electrolytes for Electrochemical Supercapacitors,* 2016.
136. Ramasamy, C.; Palma, J.; Anderson, M. A 3-V Electrochemical Capacitor Study Based on a Magnesium Polymer Gel Electrolyte by Three Different Carbon Materials. *J. Solid State Electrochem.* **2014,** *18* (10), 2903−2911.
137. Peng, X.; Liu, H.; Yin, Q.; Wu, J.; Chen, P.; Zhang, G.; Liu, G.; Wu, C.; Xie, Y. A Zwitterionic Gel Electrolyte for Efficient Solid-State Supercapacitors. *Nat. Commun.* **2016,** *7,* 11782.
138. Moon, W. G.; Kim, G. P.; Lee, M.; Song, H. D.; Yi, J. A Biodegradable Gel Electrolyte for Use in High-Performance Flexible Supercapacitors. *ACS Appl. Mater. Interfaces* **2015,** 7 (6), 3503−3511.
139. Dvořák, P. Overview of Non-Aqueous Electrolytes for Supercapacitors. *Energy* **2010,** *4* (3), 2−6.
140. Yamagata, M.; Soeda, K.; Yamazaki, S.; Ishikawa, M. Alginate Gel Containing an Ionic Liquid and Its Application to Non-Aqueous Electric Double Layer Capacitors. *Electrochem. Solid State Lett.* **2011,** *14* (11), A165.
141. Yang, P.; Mai, W. Flexible Solid-State Electrochemical Supercapacitors. *Nano Energy* **2014,** *8,* 274−290.
142. Amatucci, G. G.; Badway, F.; Du Pasquier, A.; Zheng, T. An Asymmetric Hybrid Nonaqueous Energy Storage Cell. *J. Electrochem. Soc.* **2001,** *148* (8), A930.
143. Park, M.-S.; Lim, Y.-G.; Kim, J.-H.; Kim, Y.-J.; Cho, J.; Kim, J.-S. A Novel Lithium-Doping Approach for an Advanced Lithium Ion Capacitor. *Adv. Energy Mater.* **2011,** *1* (6), 1002−1006.
144. Sivakkumar, S. R.; Pandolfo, A. G. Evaluation of Lithium-Ion Capacitors Assembled with Pre-Lithiated Graphite Anode and Activated Carbon Cathode. *Electrochim. Acta* **2012,** *65,* 280−287.
145. Park, M.-S.; Lim, Y.-G.; Hwang, S. M.; Kim, J. H.; Kim, J.-S.; Dou, S. X.; Cho, J.; Kim, Y.-J. Scalable Integration of Li_5FeO_4 towards Robust, High-Performance Lithium-Ion Hybrid Capacitors. *ChemSusChem* **2014,** 7 (11), 3138−3144.
146. Lim, Y.-G.; Kim, D.; Lim, J.-M.; Kim, J.-S.; Yu, J.-S.; Kim, Y.-J.; Byun, D.; Cho, M.; Cho, K.; Park, M.-S. Anti-Fluorite Li6CoO4 as an Alternative Lithium Source for Lithium Ion Capacitors: An Experimental and First Principles Study. *J. Mater. Chem. A* **2015,** *3* (23), 12377−12385.
147. Jezowski, P.; Fic, K.; Crosnier, O.; Brousse, T.; Beguin, F. Lithium Rhenium(vii) Oxide as a Novel Material for Graphite Pre-Lithiation in High Performance Lithium-Ion Capacitors. *J. Mater. Chem. A* **2016,** *4* (32), 12609−12615.
148. Jezowski, P.; Fic, K.; Crosnier, O.; Brousse, T.; Béguin, F. Use of Sacrificial Lithium Nickel Oxide for Loading Graphitic Anode in Li-Ion Capacitors. *Electrochim. Acta* **2016,** *206,* 440−445.
149. Zheng, J.-S.; Zhang, L.; Shellikeri, A.; Cao, W.; Wu, Q.; Zheng, J. P. A Hybrid Electrochemical Device Based on a Synergetic Inner Combination of Li Ion Battery and Li Ion Capacitor for Energy Storage. *Sci. Rep.* **2017,** *7,* 41910.
150. Jeżowski, P.; Crosnier, O.; Deunf, E.; Poizot, P.; Béguin, F.; Brousse, T. Safe and Recyclable Lithium-Ion Capacitors Using Sacrificial Organic Lithium Salt. *Nat. Mater.* **2017,** *17* (2), 167−173.

151. Hanna, O.; Luski, S.; Brousse, T.; Aurbach, D. Aqueous Energy-Storage Cells Based on Activated Carbon and $LiMn_2O_4$ Electrodes. *J. Power Sources* **2017,** *354,* 148–156.
152. Ajuria, J.; Redondo, E.; Arnaiz, M.; Mysyk, R.; Rojo, T.; Goikolea, E. Lithium and Sodium Ion Capacitors with High Energy and Power Densities Based on Carbons from Recycled Olive Pits. *J. Power Sources* **2017,** *359,* 17–26.
153. Babu, B.; Shaijumon, M. M. High Performance Sodium-Ion Hybrid Capacitor Based on $Na_2Ti_2O_4(OH)_2$ Nanostructures. *J. Power Sources* **2017,** *353,* 85–94.
154. Shukla, A. K.; Banerjee, A.; Ravikumar, M. K.; Jalajakshi, A. Electrochemical Capacitors: Technical Challenges and Prognosis for Future Markets. *Electrochim. Acta* **2012,** *84,* 165–173.
155. Fic, K.; Platek, A.; Piwek, J.; Frackowiak, E. Sustainable Materials for Electrochemical Capacitors. *Mater. Today* **2018,** *2* (4), 437–454.

CHAPTER SEVEN

Integrated and Sustainable Solutions for Li-ion Energy Storage Systems

Michał Świętosławski*, Monika Bakierska*, Joanna Pacek*,§, Krystian Chudzik*, Marcelina Lis*, Weronika Marszałowicz*, Rafał Knura* and Marcin Molenda*,§,1

*Faculty of Chemistry, Jagiellonian University, Kraków, Poland
§MarCelLi Adv Tech Sp. z o.o., Kraków, Poland
[1]Corresponding author: E-mail: molendam@chemia.uj.edu.pl

Contents

Abstract

This paper describes the most prominent directions in electrode materials production and development. The batteries' market demand for such solutions is bigger than ever before, and the current trends clearly suggest that this direction will not change in the near future. In this review on electrode materials, we present the most interesting technologies which in our opinion could fill the gap between limitations of current Li-ion systems and sustainable development of automotive and energy industries.

1. INTRODUCTION

The historical forecast for worldwide demand for lithium-ion batteries (LIBs) prepared by Avicienne Energy[1] in 2013 presented a target value of close to 160 GWh/year in 2025. The global installed LIB manufacturing capacity was a bit more than 100 GWh/year in 2017.[2] At the same time,

Advances in Inorganic Chemistry, Volume 72
ISSN 0898-8838
https://doi.org/10.1016/bs.adioch.2018.05.007

the forecast for the worldwide demand for LIBs for the year 2025 increased to 600 GWh/year.[3] Actually, this value is close to four times higher than expected in 2013 and seems to be underestimated. According to Bloomberg [2] the forecast for worldwide demand for LIBs for the year 2030 will be close to 1500 GWh/year. Such rapid growth of the LIB market over the next years will be mainly driven by the electric vehicle market (xEV) and stationary energy storage systems (ESS) for renewable energy (photovoltaic, wind energy) and Smart Grid solutions. It is obvious that to fulfill the Li-ion manufacturing capacity demand, the huge amount of raw materials has to be secured at the lowest possible price. Actually, most of the LIB production is based on toxic and expensive cobalt compounds [$LiCoO_2$ (LCO), $LiMn_{1/3}Ni_{1/3}Co_{1/3}O_2$ (NMC), and $Li(Ni_{0,85}Co_{0,1}Al_{0,05})O_2$ (NCA)], and cobalt mainly determines the high material cost (>80 k$/t in March 2018). On the other hand, to reduce the cost of the final LIB value, the supply chain of raw materials has to be secured with the lowest possible carbon footprint. Besides the economics and logistics, another obligatory factor has to be taken into consideration, which concerns the operational safety of the LIBs. Unfortunately, cobalt-based cathode materials do not reveal sufficient safety, such that their large-scale application is limited.

The quest for new, safer, and cheaper cobalt-free cathode materials leads to three-dimensional (3D) materials—$LiMn_2O_4$ (LMO) spinel, $LiMXO_4$ (M = metal *3d*, X = S, P, As, Mo, W) systems—in particular, $LiFePO_4$ (LFP) and Li_2MSiO_4 (M = Fe, Mn, Ni) silicate systems. On the other hand, the demand for lowering the carbon footprint and securing the supply necessitates the finding of an alternative for natural graphite or artificial graphite from fossil raw materials (oil or coal based). It seems that a reasonable alternative for fossil raw materials could be renewable natural polysaccharides from plant starch.[4] However, to fulfill all the demands mentioned above, the development and integration of materials and processes must be performed at every stage discussed in this chapter.

Application of nanomaterials as electrode materials in LIBs leads to power increase and achievement of operating parameters close to the theoretical values. The large surface area of nanomaterials significantly increases the charge transfer in the electrode, which in turn increases the current density in the cell. Decreasing the active material particle size significantly shortens the diffusion path of lithium ions, which decreases the polarization of the electrode. On the other hand, nanometric electrode materials reveal an increase in the reactivity toward electrolytes, i.e., the surface reactivity in particular. This obviously affects the stability and safety of the cell. The

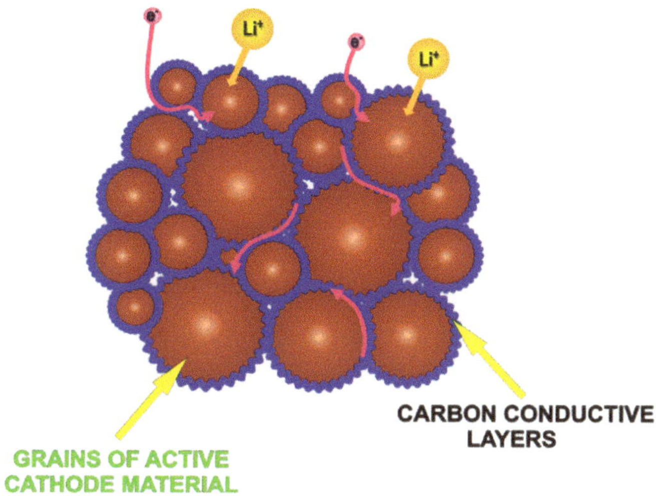

Figure 1 Functional model of nanocomposite electrode.

solution is to use protective coatings on nanomaterials with greater chemical stability (e.g., carbon), and the formation of nanocomposites wherein the coating does not limit the electrical conductivity of the system and must ensure the permeability of lithium ions. The solution which integrates the above demands is a nanocomposite with conductive carbon layers (CCLs) (Fig. 1) permeable for lithium ions and covering each of the nanometric active material grains. This provides a stable electrical connection of nanometric active material grains with the current collector via a 3D network of nanometric carbon layers, which improves the homogeneity of the nanocomposite, increases the chemical stability of the material toward the electrolyte and can decrease the amount of carbon additives in the final electrode. As CCL precursors, water soluble hydrophilic polymers (e.g., poly(N-vinyl formamide) (PNVF) or polyacrylamide (PAAm))[5,6] or gelatinized starch[7] can be used. The physicochemical properties of CCL nanocomposites ensure integration of the following functions:

- high macroscopic electrical conductivity of the nanocomposite via a 3D CCL conductive network,
- easy diffusion paths (permeability) for lithium ions through the unique structure of the CCL,
- high chemical stability of the nanocomposite toward the electrolyte,
- durability against volume changes of the active material in the nanocomposite.

2. MODIFIED LITHIUM MANGANESE SPINELS

The low cost and high availability of manganese raw materials, contributed to the increased interest in its compounds for use as Li-ion electrode materials. The main representative for LIBs of manganese-based materials is lithium manganese spinel ($LiMn_2O_4$, LMO) and its derivatives. Present research focuses on using it as the cathode of a high-power LIB for hybrid electric vehicles. As such, except for the desired performance, their preparation must be economical in terms of mass production. Therefore, all of the spinels mentioned below were synthesized by a sol-gel method based on hydrolysis and condensation processes and followed by mild calcination. Synthesis like this enables us to obtain homogeneous, pure, and nanosized products even at relatively low temperatures. Another advantage of this process is the ability to control the amount of doping atoms that are introduced into the spinel structure. It is suitable for widescale use, as it exploits only widely available substrates such as acetates and ammonia.[8,9]

Basic lithium manganese spinel, in contrast with layered oxides like LCO and lithium nickel oxide ($LiNiO_2$), is characterized by superior safety, competitive capacity of approximately 148 mAh/g, and high working potential due to its 3D structure which offers fast Li^+ ions transport.[10–14] Electrochemical performance of this material depends on the crystallinity, purity, and particle size.[13] A major drawback is that the stoichiometric LMO suffers from capacity fading during charging/discharging processes, which limits its lifespan and rate performance.[15–17] The drop in capacity of LMO cell properties has been generally combined with Jahn-Teller distortion of high spin Mn^{3+} ions that leads to an LMO phase transition from cubic (*Fd-3m*) to orthorhombic (*Fddd*) at room temperature and increased surface reactivity in the delithiated state, leading to manganese dissolution in the liquid electrolyte.[15–17] A standard liquid electrolyte used in cells with spinel-type electrode materials is a lithium hexafluorophosphate solution in ethylene carbonate and dimethyl carbonate (1.0 M $LiPF_6$ in EC/DMC 50/50 v/v).[18] In order to deal with these issues, a few approaches have been proposed. First of them is coating of the active material, which can reduce the electrode and electrolyte interfacial area and restrain manganese release.[19,20] Another method is the substitution of Mn^{3+} ions by Al^{3+}, Cr^{3+}, Co^{3+}, Mg^{2+}, Ni^{2+},[21] or oxygen sublattice modification with sulfur.[9,22]

The partial exchange of manganese by the other cations mentioned above improves the structural stability of LMO but negatively affects the capacity of the Li-ion cell. As it has been shown, performance decline has no place when oxygen in the anion sublattice is alternated with larger, isoelectronic sulfur atoms since disturbance in the structure framework suppresses the phase transition related to Jahn-Teller distortion near room temperature. Furthermore, it may increase the capacity and coulombic efficiency of the material. On the other hand, sulfur-doped lithium manganese spinel ($LiMn_2O_{4-y}S_y$, LMOS) showed slightly lower electrical conductivity than the pure LMO.[9,22] Reports indicate that a surface carbon coating of cathode materials may improve the electrochemical characteristics, as well as restrain reactions between the electrode and electrolyte.[23,24]

Doping with sulfur at a level of 0.01 mol per 1 mol of LMO leads to phase-pure material (Fig. 2A). Analysis of the X-ray diffraction (XRD) pattern shows that the lattice constant is slightly higher, which proves that sulfur substitution for oxygen in LMOS is homogeneous. In Fig. 3A, we can observe that substitution significantly increases the specific capacity of the material. Furthermore, the addition of such a small amount of sulfur does not decrease the material's conductivity, which is reflected in a good rate capability.[25]

Most of the research carried out on modified spinels comes down to doping with just one type of cation or anion, although co-doping can further improve structural and electrochemical properties of the electrode. One of the propositions is to simultaneously execute nickel and sulfur substitution ($LiMn_{1,9}Ni_{0,1}O_{3,99}S_{0,01}$, LMNOS).[26,27] Sulfur in this case affects the spinel in the same way as before. Also nickel inclusion instead of Mn^{3+} ions can contribute to structural and chemical stabilization and even lead to elimination of manganese solubility in the standard liquid electrolyte.[16] Research carried out in our group also confirmed that the addition of excessive nickel affects the electrochemical performance by decreasing the electrical conductivity at constant, reduced activation energy of the process. In addition, it was demonstrated that the LMNOS sample exhibits a higher capacity (10 mAh/g more) (Fig. 3B), as well as an extraordinary rate capability, capacity retention, and reversibility, which is much better than stoichiometric LMO since it loses only 19% of the initial capacity after 650 cycles under 5C (1C current corresponds to current needed for full charging or discharging of the cell; 5C = full charge/discharge in 1/5 h) current (Fig. 3(c)).[28] An obstacle that occurs while using nickel-doped material is the thick solid electrolyte interface that results from electrolyte decomposition on the cathode surface at a high voltage due to the Ni^{2+}/Ni^{4+} redox couple potential (Fig. 3).[29]

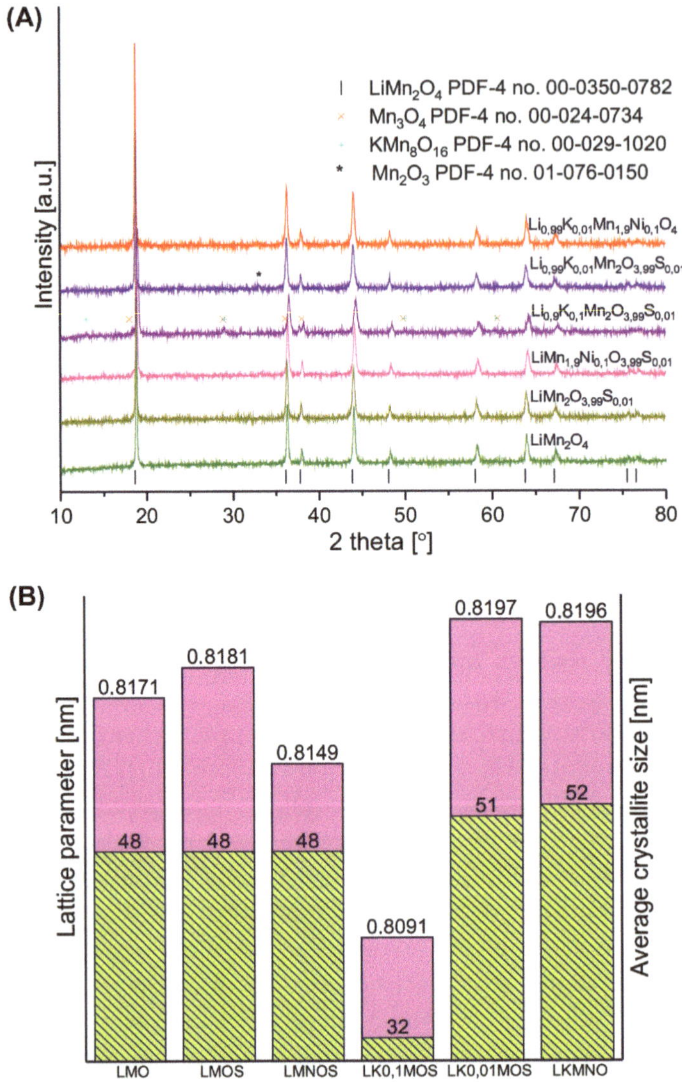

Figure 2 X-ray diffraction patterns of doped spinel materials (A); lattice constants and average crystallite sizes (B) appropriate *Powder Diffraction Files*™ numbers are added in the legend.

Another mixed metal spinel, which can be useful as cathode material for high-power Li-ion energy storage, is a K- and S-doped spinel. Material with only potassium dopant provides high working voltage and prevents the cell from overcharging. Moreover, when the initial specific capacity was set at about 130 mAh/g, a capacity retention larger than 95% was observed after

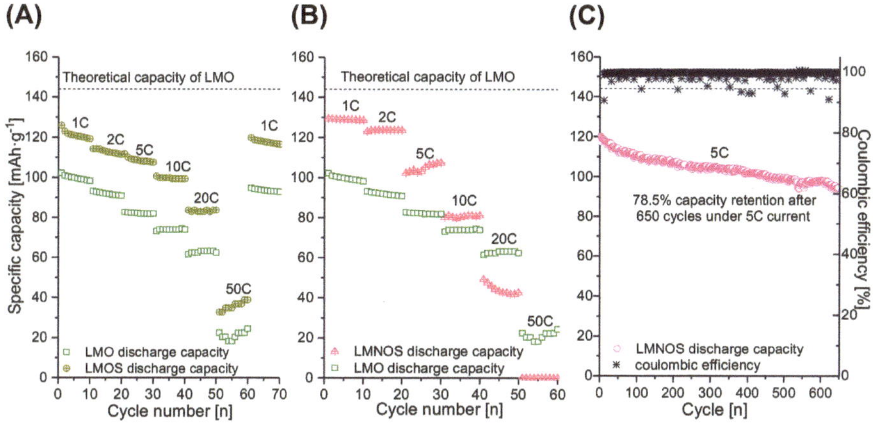

Figure 3 Discharge capacities of LMO and LMOS (A); discharge capacities of LMO and LMNOS (B); discharge capacities of LMNOS in long-term tests (C).

90 cycles.[30] Introduction of potassium and sulfur into the structure of the spinel resulted in forming a material with a chemical formula of $L_{0,99}K_{0,01}Mn_2O_{3,99}S_{0,01}$ (LK0,01MOS). This amount of potassium built into the lithium sublattice causes expansion of the unit cell (increasing lattice constant) (Fig. 2B) and increases the Li^+ ions mobility, which is reflected in increased Li^+ diffusion coefficients (Fig. 4B). From studies carried out by our group, it was seen that this combination of dopants allows the material to transfer higher currents and limits cell polarization. The capacity of this material is 5% lower than for the LMOS material as a consequence of reduction in lithium content and potassium blocking of some neighboring lithium ions. Long-term cell tests performed on this material confirmed that it can withstand even extreme conditions[31] (Fig. 4A,B).

Increasing the amount of potassium in K- and S-doped spinel to levels that lead to material with a chemical formula of $L_{0,9}K_{0,1}Mn_2O_{3,99}S_{0,01}$ (LK0,1MOS) did not enhance the previously mentioned traits of LK0,01MOS. In terms of the lattice constant of spinel, this combination of dopants is significantly lower, as well as the average crystallite size. Electrochemical performance of LK0,1MOS is deteriorated, the material exhibits lower discharge capacities and diffusion coefficient (Fig. 4A,B). Furthermore, XRD studies exhibit formation of impurities in LK0,1MOS material at much higher levels than in materials with different combinations of dopants (Fig. 2A). From the data presented, it can be concluded that an amount of potassium as high as 0.1 mol K per 1 mol of LMOS is too high to obtain electrochemically efficient and phase-pure material.

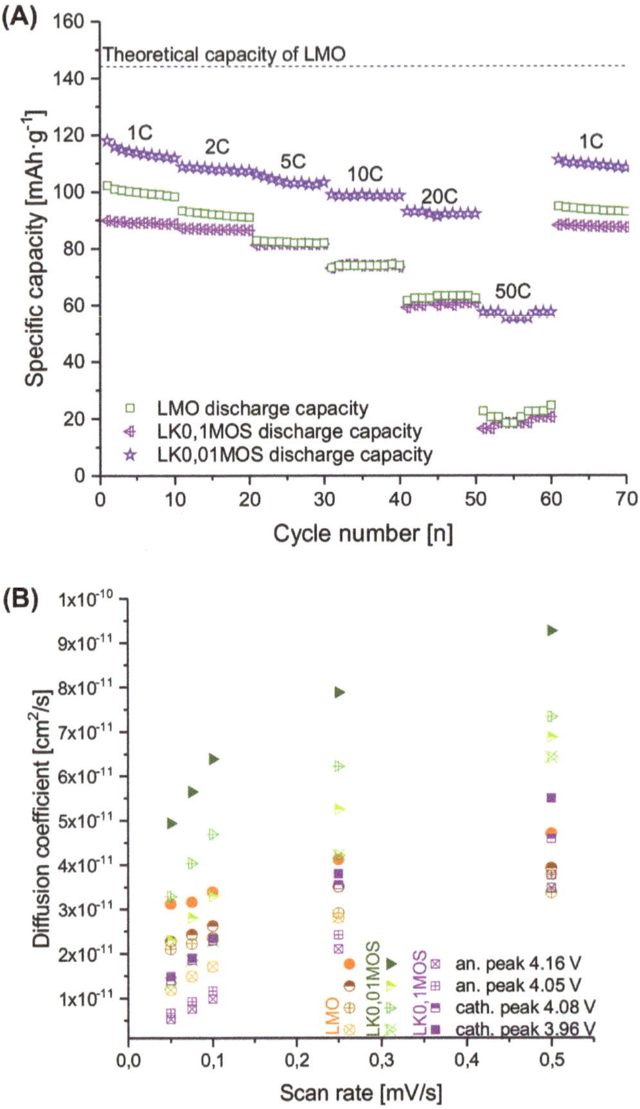

Figure 4 Discharge capacities of LMO; LK0,1MOS; and LK0,01MOS (A) and dependence of the lithium ions diffusion coefficient on cyclic voltammetry scan rate for LMO; LK0,1MOS; and LK0,01MOS (B). *an.*, anode; *cath.*, cathode.

A different approach was to change both lithium and manganese sublattices with potassium and nickel, respectively, at the same time. As our studies showed, spinel materials with the general formula $Li_{1-x}K_xMn_{2-y}Ni_yO_4$ (LKMNO) lead to unexpected results. Introduction of these cations into

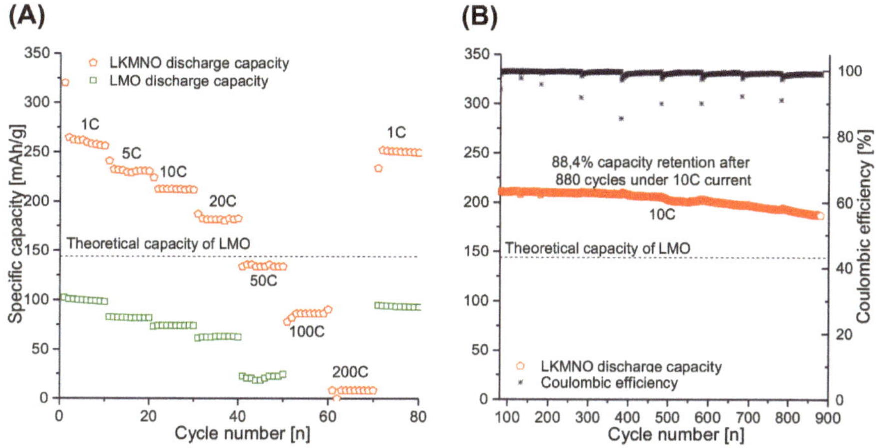

Figure 5 Discharge capacities of LMO and LKMNO ($Li_{0.99}K_{0.01}Mn_{1.9}Ni_{0.1}O_2$) (A); discharge capacities of LKMNO in long-term tests (B).

the structure leads to almost a 70% increase in capacity in contrast to stoichiometric LMO, showing at least 250 mAh/g under a current load of 1C without any optimization of cathode composition (Fig. 5A). In addition, this material works in a high potential range (4.0–4.7 V) and is characterized by coulombic reversibility of more than 99%. Another benefit is cell operation efficiency under high current conditions (100–200 C), which translates into current transfer without any damage.[32] These extraordinary properties of the LKMNO material are currently intensively investigated in our group.

To sum up the above considerations, it can be inferred that introduction of a small amount of dopants has a big impact on the material properties, as well as on the cell performance. Another conclusion is the fact that doping with different atoms does not have an additive, but rather a synergistic effect.

3. LITHIUM IRON PHOSPHATE ($LIFEPO_4$) AND C/$LIFEPO_4$ COMPOSITES

Lithium iron phosphate ($LiFePO_4$, LFP) was first presented as a suitable cathode material by Padhi et al. in 1997.[33] While working in an electrochemical cell, $LiFePO_4$ is cyclically transformed into its deintercalated form, $FePO_4$, during charging and back to $LiFePO_4$ during discharging. The flat potential during charge/discharge cycles, independent of the composition of the material, indicates that a two-phase reaction occurs.[34–36]

This process is possible due to the ability of LFP to conduct electric current and to intercalate and deintercalate lithium ions to and from its structure via one-dimensional lithium ion diffusion channels.[37] However, the material features some main drawbacks, viz. low electronic conductivity ($<10^{-9}$ Scm^{-1} at RT)[35,36,38] with energy activation of ca. 0.65 eV,[39] as well as a low ionic diffusion coefficient ($<10^{-14}$ cm^2/s),[35,36,38] and existence of impurities preventing lithium diffusion that limits its energy-related characteristics that have to be overcome in order to utilize lithium iron phosphate as competitive cathode material.[39]

Although lithium iron phosphate may not have the best characteristics associated with energy and power densities, its superior features related to economic and environmental aspects make it very desirable for energy and automotive industries.[35]

Power and energy densities of electrical cells are two of the most crucial parameters when considering their future applications. Both depend on cell voltage, as well as the capacity of the electrode and lithium ions diffusion rate for energy and power density, respectively. The material's voltage against that of lithium, greatly depend on the type of transition metal cation. In the case of lithium iron phosphate, its voltage against lithium is relatively low, ca. 3.45 V, in comparison to other types of Li-ion cathodes or even very similar polyanionic compounds like $LiMnPO_4$, ca. 4.1 V, and $LiCoPO_4$, 4.8 V. Nevertheless, the shortcomings related to these systems make them harder to apply in LIB systems.[33,34,40]

Another very significant factor when considering energy density is the gravimetric capacity of the material. Due to the fact that LFP consists of one lithium atom per formula unit and that the phosphate anion has a significantly larger molar mass than, e.g., a pair of oxygen atoms, lithium iron phosphate has an inferior theoretical gravimetric capacity of 170 mAh/g[33,38] compared to other widely exploited cathode materials, e.g., layered oxide cathodes. On the other hand, there are cathode materials currently in use with far lower theoretical capacity. It should also be noticed that significant research indicate that due to its remarkable stability, it is possible to deintercalate all lithium ions from the material to reach its theoretical capacity, whereas such an action would irreversibly damage the structure of other types of cathode materials (e.g., $LiCoO_2$).[34,38]

The key factor when considering the cell's power density apart from the voltage is the lithium ion diffusion rate due to its direct relation with the cell's available current. As was previously mentioned, LFP features a low ionic diffusion coefficient, as well as enables lithium ion diffusion in only one direction due to the strong difference between the diffusion activation energy along the b and c directions.[37] However, there are approaches that allow mitigating those drawbacks and some of them will be discussed further on.

From an economical point of view, lithium iron phosphate is one of the most prominent materials proposed as Li-ion cathode. It is the result of two main aspects: availability of materials and simplicity of synthesis along with the elimination of its high-energy-requiring stages.

Apart from lithium that is essential in all Li-ion cathode materials and available only at selected locations on Earth, $LiFePO_4$ consists of elements that are abundant and inexpensive.[35] It is worth noting that the cost of iron is much lower than the cost of other transition metals used in Li-ion cathodes, such as cobalt or nickel, whereas its abundance is greater and nonrestricted to specific locations.[41]

Furthermore, it is possible to obtain lithium iron phosphate using a wide variety of wet chemical methods (e.g., hydrothermal, solvothermal, or precipitation processes). Such an approach allows for minimization of high-energy-requiring stages (e.g., long-term ball milling, high temperature treatment) and better control over the synthesis as well as the structure of the final product, which is also crucial when considering power density and cyclability.[40,42]

Finally, LFP is proven to work efficiently for thousands of cycles with very low maintenance, especially when combined with a lithium titanate oxide anode ($Li_4Ti_5O_{12}$). This is a result of its stable polyanionic structure, as well as low volume changes, ca. 6.81%, during cycling and may contribute to less frequent replacement of used cells.[33,43,44]

The environmental impact of lithium iron phosphate may be considered in relation to its synthesis in accordance to principles of green chemistry, as well as with respect to safety issues and effects that it may cause when misused. The 12 principles of green chemistry proposed by Anstas and Warner[45] give practical insight into whether a chemical process is environmentally friendly and allows for its quantitative assessment. These principles refer to reducing the amounts of waste, derivatives, and auxiliary compounds, assuring environmental benignness of substances used during the process; designing routes that maximize quantity and quality of the end-product while minimizing additional resources (e.g., energy); as well as real-time control and prevention from pollution. Just to demonstrate the principles of green chemistry in practice, one can consider the following example. Elimination of the synthesis stages requiring high energy is not only an economical advantage but also a great environmental benefit. When the majority of available energy comes from nonrenewable energy sources, reduction of energy demand during large-scale production might alleviate the overall impact of the material on the environment. The same applies to the use of widely available and environmentally benign solvents (e.g., water), as well as other aspects of the chemical process.[45]

When considering safety issues, it should be noted that a polyanionic structure with its strong 3D bonding greatly influences the material's thermal and chemical stability, therefore making it less prone to cause damage in its surroundings when short-circuited, overcharged, or otherwise mistreated. It has been established that $LiFePO_4$ has one of the highest breakdown temperatures among cathode materials, c. 320°C, whereas in the case of other positive electrodes, the value does not exceed 260°C. As a result, LFP is less susceptible to release oxygen after mishandling.[44] Also, a lack of toxic elements such as cobalt or nickel in the material's structure guarantees its environmental benignness and reduces the risk of dangerous pollution in case of damaging the battery casing.[46]

Our present studies on lithium iron phosphate focuses on two main areas: minimizing particle size and coating the surface of grains with a CCL.[38] Decreasing the particle size leads to reducing the length of diffusion, consequently, improving the reversibility of the Li-ion intercalation/deintercalation process. $LiFePO_4$ as a material with one-dimensional atomic migration channels in which the diffusion constant strongly depends on the particle size therefore requires optimization of preparation procedures in order to prepare nanosized material grains. Modification of active materials with carbon coating improved the lattice electronic conductivity and did not change the structure parameters, thus pathways for electron transport are provided.[47] Coating LFP with conductive materials allows to enhance electrical conductivity in the electrode. Furthermore, it provides for grain growth and prevents potential electrode/electrolyte reactions. The procedure of coating with carbon has been confirmed as an effective method to improve the electronic conductivity of LFP.[48] Based on the developed technology,[49] composite cathodes CCL/$LiFePO_4$ were successfully prepared in a simple and cheap process of wet impregnation of the polymer carbon precursor followed by controlled pyrolysis. The formed carbon nanocoatings significantly improved the electrical conductivity of the $LiFePO_4$ cathode material from $\sim 10^{-9}$ S /cm to the level of $\sim 10^{-2}$ S / cm at room temperature. Attempts have been made to find optimal parameters for the preparation of CCLs onto electrode materials, with different morphologies for LIBs.[6]

Table 1 Comparison of Average Grain Size, Carbon Content, Surface Area, and Electrical Conductivity Values for C/$LiFePO_4$ and Conductive Carbon Layer (*CCL*)/$LiFePO_4$ Composites

Sample	Average Grain Size [nm]	Carbon Content [%]	Specific Surface Area [m^2/g]	σ at 25° C [10^{-3} S/cm]
C/$LiFePO_4$	33	1.4	14.3	8.5
2.5% CCL/$LiFePO_4$	34	2.5	11.3	16.6
5% CCL/$LiFePO_4$	34	5.0	-	8.8

The usability of the carbon-coating procedure was proven on commercial LFP. $LiFePO_4$ was purchased from Pulead Technology Industry Co. Ltd. It contained residual carbon (carbon content ~1.4 wt%) after synthesis (Table 1). The wet impregnation method, mentioned above, was used to prepare carbon-coated CCL/$LiFePO_4$ nanocomposites. The homogeneous carbon dispersion in CCL/LFP nanocomposites effectively enhances its electrochemical performance, in particular the electrical conductivity of the material and its capacity retention.[50] The specific discharge capacity versus cycle number is shown in Fig. 6A—C). It was performed as a function of cycle number at various C rates. The discharge capacity for the current C/5 grows with the cycle number, and for the 50th discharge cycle, a reversible capacity of about 150 mAh/g and 125 mAh/g was obtained for samples 2.5% CCL/$LiFePO_4$ and 5% CCL/$LiFePO_4$, respectively. By way of comparison, the C/$LiFePO_4$ composite (as produced with residual carbon), without the CCL, achieved nothing but 100 mAh/g. The voltage plateau at 3.4 V in the discharge curve was observed for all investigated composites (Fig. 6D—F). It is typical for a two-phase reaction of lithium extraction/insertion. The polarization of the electrode was low and did not increase during further cycles. Moreover, the tests suggested that the CCLs contribute to increase the chemical stability and improve the overall electrochemical performance of the nanocomposite cathode material based on $LiFePO_4$.[50]

The next attempt was based on the reduction of particle size and improvement of its electrochemical parameters with CCLs. Nanosized $LiFePO_4$ was synthesized via a modified Delacourt et al.[51] method in compliance with a patent-pending procedure.[42] The described technology of carbon coating was successfully applied to improve the electrochemical performance of nanostructured LFP material (nano-LFP). The XRD patterns of nano-LFP and CCL/nano-LFP are shown in Fig. 7. All diffraction lines indicated in the XRD pattern are well matched to the pure phase of

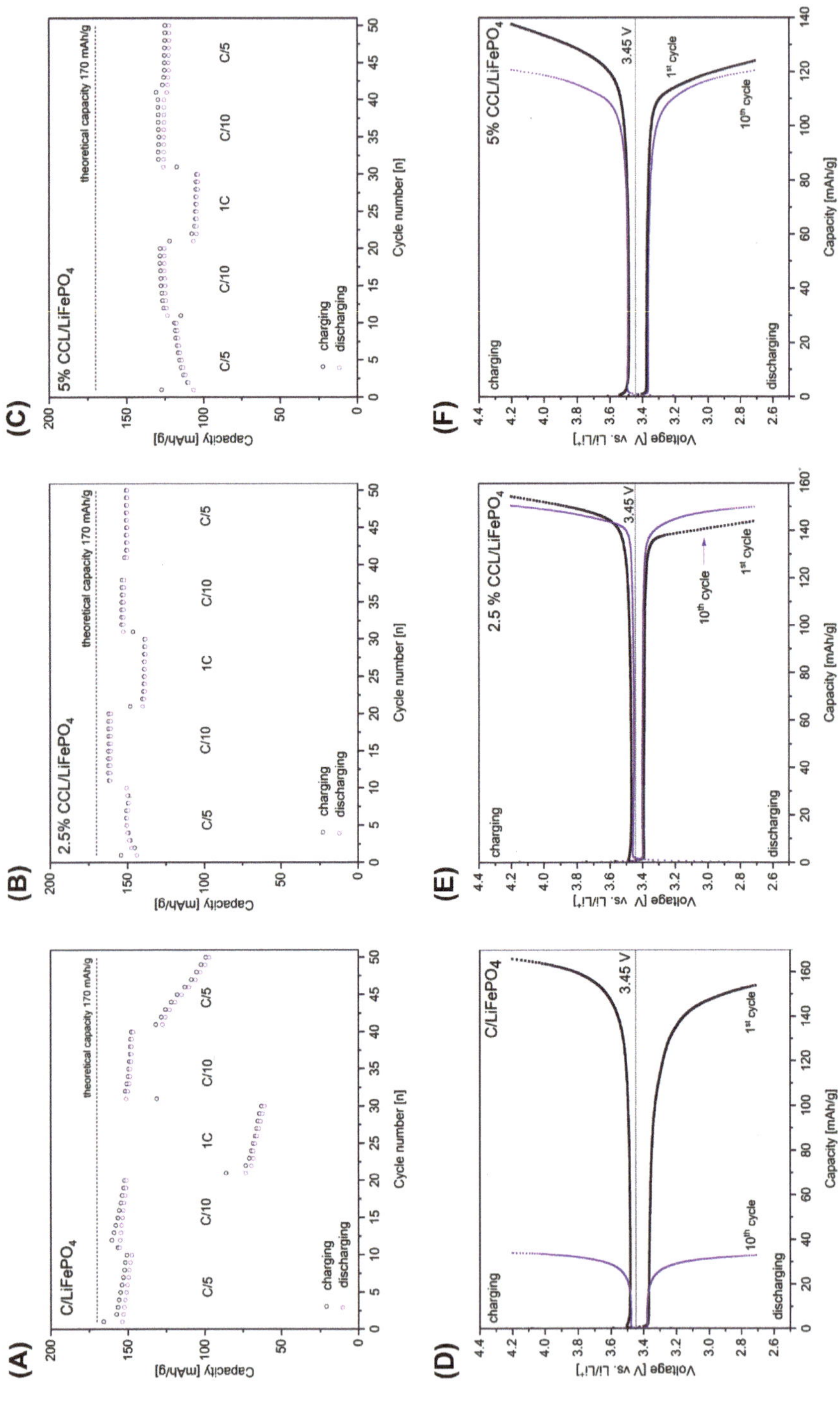

Figure 6 (A–F) Electrochemical properties of C/LFP and (*CCL*)/LFP composites.

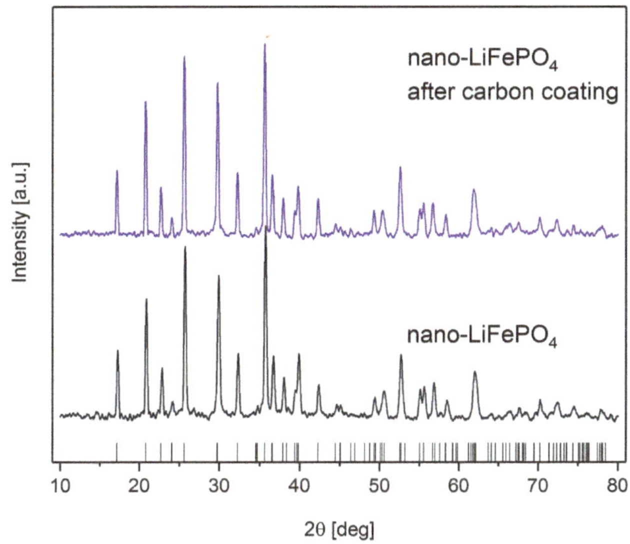

Figure 7 X-ray diffraction patterns of nano-LFP and conductive carbon layer (*CCL*)/ nano-LFP composites.

$LiFePO_4$ with orthorhombic structure (*Pnma*). Any possible impurities were not detected in the XRD pattern. The carbon-coating process did not result in structural changes of the pristine nano-LFP. The nano-$LiFePO_4$ coated with the CCL exhibited very good cyclability and rate performance at different current rates (Fig. 8). Under the C/2 current rate, the reversible discharge capacity close to the theoretical value was found.

It has to be pointed out that the achievement of about 100% of the theoretical capacity is due to the small and homogeneous size distribution combined with conductive carbon nanocoating. The presented technology is very valuable and the process of producing carbon-coated $LiFePO_4$ cathode materials according to the described method requires optimization in relation to cathode materials of different morphology. Moreover, the process of carbon coating does not influence the growth of grain size (Table 1), and the prepared CCL may be characterized as a thin, stick, and elastic layer with good carbon dispersion. Besides, it was found that the CCL increases the chemical stability of the active material surface during electrochemical tests. The process of CCL preparation must be considered from an economic and environmental point of view, if we consider to scale up from a laboratory process to mass production. The technology may be defined as cheap (according to the reduced manufacturing cost, resulting from lower costs of raw materials and consumed energy) and scalable without difficulty

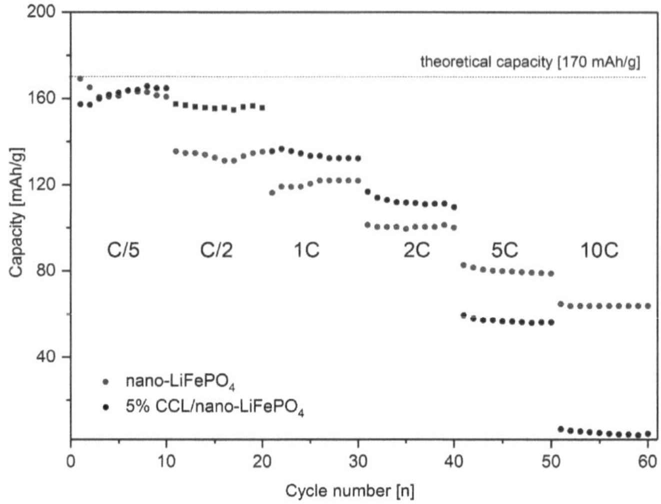

Figure 8 Electrochemical properties of nano-LFP and conductive carbon layer (*CCL*)/nano-LFP composites.

(due to common equipment and the simple and easy controlled procedure used). In addition, the technology is consistent with Green Chemistry rules: the estimated numerical value on an eco-scale for laboratory processes is more than 80 points and corresponds to a process classified as excellent. The CCL composite production technology is characterized by its lower environmental impact because of the low emission, which improves the economics of the process. Most importantly, it provides high-performance cathode materials with improved electrical properties, increased chemical stability, durability, and longer product life. The proposed technology can be used in a large-scale process, and the final product is dedicated for application in safe premium-class LIBs for the automotive sector of hybrid and electric vehicles (*x*EV) and ESS market segments.

4. POLYSILICATE CATHODE MATERIALS

Another interesting group of cathode materials for LIBs is the family of polysilicates with the general formula of Li_2MSiO_4, where M is one of the 3d metals. The structure of these materials is similar to polyphosphates, known mainly by the aforementioned LFP material. The successful commercialization of LFP resulted in the first attempts to develop a battery system based on a lithium iron polysilicate cathode. The first experimental studies on these materials were published by the group of Thomas from the Uppsala University in 2004.[52] The obtained Li_2FeSiO_4 ($Pmn2_1$) revealed an initial capacity of 165 mAh/h and a reversible capacity after few cycles of 140 mAh/g working, with a potential close to 3 V versus Li. This work initiated broader interest in polysilicate materials capable of reversible insertion of Li-ions.

In general, the crystal structure of Li_2MSiO_4 (Fig. 9) is an analog of the lithium phosphate(V) structure.[53] Li_2MSiO_4 consists of LiO_4, MO_4, and SiO_4 tetrahedra in a way that none of the tetrahedra walls are shared. The joined tetrahedra form a distorted layer structure in which they can be combined by the tops or edges and can be oriented along the same or in opposite direction to the adjacent ones. Li_2MSiO_4 has a few different polymorphs: a low-temperature α phase with $Pmn2_1$ space group and two high-temperature phases: β ($Pmnb$) and γ ($P2_1/n$).[54–57] Because the formation energies of all three polymorphs are quite similar, most preparation routes lead to the formation of a mixture of at least two of the mentioned polymorphs. There

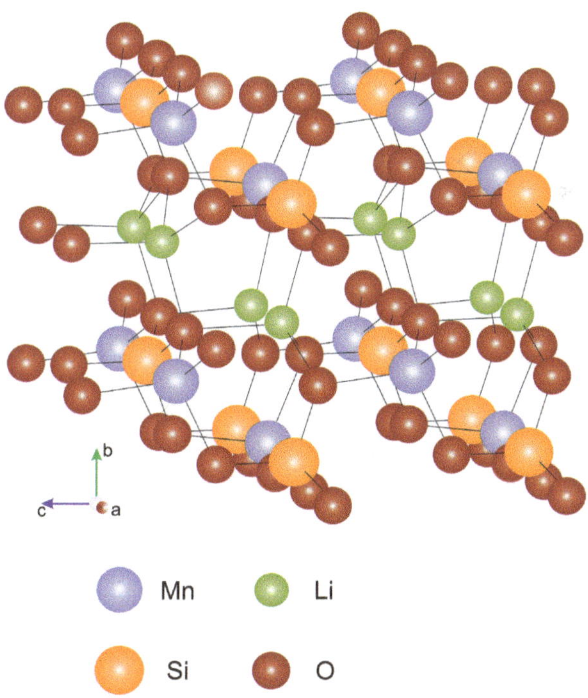

Figure 9 The structure of lithium polysilicate materials.

is also a *Pn* polymorph of the Li_2MSiO_4 structure which was obtained by ion-exchange from the sodium manganese silicate. Nevertheless, all of the polymorphic forms should be electrochemically active in an LIB cell. The most popular synthetic techniques for the preparation of Li_2MSiO_4 include: solid-state,[52,55,58–61] hydrothermal,[54,62–66] solvothermal,[67,68] spray-pyrolysis,[69] and sol-gel processes.[70–75,77,78] Due to the fact that Li_2MSiO_4 is characterized by a low ionic conductivity (only one-dimensional Li^+ diffusion paths) and extremely low electronic conductivity (the presence of polyanion results in greater separation of electron carriers which hinders polaron hopping), its synthesis requires a similar approach as for the production of LFP systems.[79] To reduce the influence of a low effective chemical diffusion coefficient, the material needs to be obtained in the form of nanoparticles. Increasing the specific surface and lowering the diffusion length allow us to perform the electrochemical reaction at an acceptable rate (up to 0.5 C). The prepared nanopowder cannot be used for electrode preparation due to the extremely low electronic conductivity of such materials. The material needs to be modified, usually by coating with conductive material. Simple mixing or milling the active material with conductive additives, as it is realized for other cathode materials like layered oxides or spinels, is not sufficient due to the disproportion in grain size of the active material and conductive additives. The most popular, at least for LFP systems, is modification by conductive carbon coating.

The most interesting feature connected to this particular structure is the possibility of conducting a reversible two-step electrochemical reaction:

$$Li_2M^{2+}SiO_4 \leftrightarrow LiM^{3+}SiO_4 + Li^+ + e^- \leftrightarrow \square_2M^{4+}SiO_4 + 2Li^+ + 2e^-$$

This ability leads to high capacities of up to 333 mAh/g for the manganese-based system. Depending on the chosen 3d metal, the electrochemical properties of materials change significantly. The calculated reaction potentials and theoretical capacities for different polysilicates are summarized in Table 2. The possibility to use both of the lithium ions in an electrochemical reaction makes lithium silicates with manganese, cobalt, and nickel especially interesting. LFS operating at lower potential in a one-electron reaction seems to be a closer analog to lithium iron phosphate than to other described materials. That was the reason to focus work in our group on the Li_2MnSio_4 (LMS), Li_2CoSiO_4 (LCS), and Li_2NiSiO_4 (LNS) systems.

We decided to implement the sol-gel method for polysilicate nanomaterial synthesis due to good homogeneity and grain size control provided

Table 2 Reaction Potentials and Theoretical Capacities of Different Polysilicate Cathode Materials

	$M^{2+} \leftrightarrow M^{3+}$	$M^{3+} \leftrightarrow M^{4+}$	Theoretical Capcity [mAh/G]
Li_2FeSiO_4 (LFS)	3.2 V	–	167
Li_2MnSiO_4 (LMS)	4.2–4.4 V	4.6–4.7 V	333
Li_2CoSiO_4 (LCS)	4.3 V	5.1 V	325
Li_2NiSiO_4 (LNS)	4.6 V	5.1 V	325.5

by this technique. For modification of the materials into an electrode composite, the carbon-coating process, previously developed in our laboratories, was chosen.[5,79,80] The first attempts proved that polysilicates synthesis is very demanding and both stoichiometry and homogeneity of gel precursors play a crucial role. Our efforts, as well as those from other groups, to synthesize Li_2NiSiO_4 were unsuccessful. On the other hand, the manganese- and cobalt-based systems, despite the initial difficulties, were successfully synthesized as phase-pure materials.

Detailed description of the synthesis of Li_2CoSiO_4 was published.[84] We found that the obtained gel precursor needed to be heat treated under oxidizing conditions to prevent the reduction of cobalt ions to metallic cobalt. Such conditions limit the possibility of nanomaterial formation. Due to the fact that the oxidation of the precursor's organic matrix occurs at temperatures lower than the formation of Li_2CoSiO_4, the grain size control during calcination is hindered. On the other hand, heat treatment in neutral or reductive atmosphere (like in the case of LMS) leads to organic matrix carbonization, which prevents polysilicate nanograins from growing and sintering. The sol-gel synthesis of Li_2CoSiO_4, followed by additional purification steps, can lead to the formation of the phase-pure β_I-Li_2CoSiO_4 or γ-Li_2CoSiO_4. Studies on the electrical conductivity of these polymorphs clearly confirmed the poor electrical conductivity of both LCS types ($<10^{-12}$ S/cm) and proved the need to prepare such materials in the form of well-defined conductive nanocomposites (Fig. 10). Unfortunately, the low stability of LCS in the reductive environment made formation of conductive carbon layers on the LCS surface impossible [LMS-LCS SSI]. The pure LCS, or its mixture with conductive additives, does not provide good electrochemical properties.[57,65,81] Difficulty in synthesis and poor electrode formation ability, together with high cobalt prices and its negative environmental impact, make this material less promising than it was previously suspected.

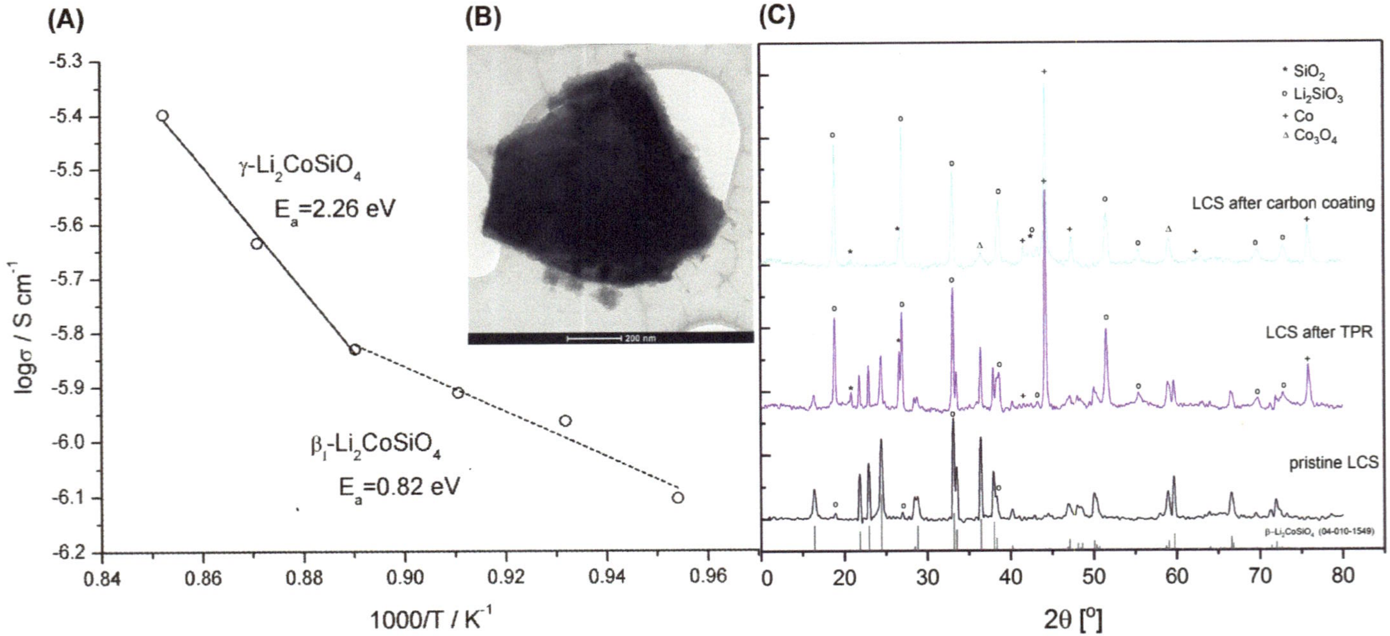

Figure 10 Li_2CoSiO_4: electrical conductivity measurements (A); transmission electron microscopy micrograph (B); X-ray diffraction patterns for pristine LCS and LCS after carbon coating and after heat treatment in reductive environment (C).

Fortunately, manganese, which is the second most abundant heavy metal (after iron) and much more environmentally friendly than cobalt or nickel, can reversibly change its oxidation state from +2 to +4 in the polysilicate structure. In the course of our studies on polysilicate cathode materials, we optimized the preparation route of nano-Li_2MnSiO_4 (*Pmn2₁*), as well as well-defined nanocomposites with *Conductive Carbon Layers* deposited on LMS nanograins (*CCL*/Li_2MnSiO_4).[71,72,82] The careful sol-gel synthesis (Pechini-type), followed by heat treatment under inert atmosphere, ensures production of a carbon/lithium manganese polysilicate composite with homogeneous 5–10 nm grains of Li_2MnSiO_4 covered with amorphous carbon. Formation of the carbon matrix is a result of heat treatment conditions, which cause carbonization of the organic gel precursor (built with citric acid and ethylene glycol). At first, the goal of our studies was to use this phenomenon for a one-step electrode composite formation. Unfortunately, Pechini's chemistry restrains this possibility. Even the highest ratio of inorganic/organic compounds in the gel precursor, which leads to synthesis of the phase-pure Li_2MnSiO_4, causes production of the C/LMS composite with at least 30 wt% of carbon. Such a high carbon loading in a composite, despite its high electronic conductivity (0.36 S/cm), limits Li^+ diffusivity to/from the active material and results in extremely poor electrochemical properties.[83] To usefully apply the LMS as cathode in a Li-ion system, the technology of carbon coating from hydrophilic polymers was applied. To prepare LMS for coating, the primary carbon (from gel matrix) was burned at temperatures close to 400°C.[72] This procedure results in partial oxidation of LMS grains to mainly lithium manganese spinel and lithium silicate. This reaction is reversible, and heat treatment of the partially oxidize LMS in a reductive environment can restore the Li_2MnSiO_4 structure. Both H_2-TPR (H_2-temperature-programmed reduction) and carboreduction with the organic CCL precursor rebuild LMS, so the partially oxidize LMS can, without other preparation, be used as the substrate for the composite precursor. Such a material is impregnated with a polymer (mixture of poly-N-vinylformamide and pyromellitic acid) in water solution. Subsequent pyrolysis under appropriate conditions assures formation of the CCL/LMS composite with ~25-nm LMS grains covered with well-defined, up to 5-nm-thick, amorphous carbon coating. The produced CCLs are characterized by high electrical conductivity (~10^{-2} S/cm), good adhesion to the active material grains, and good Li^+ diffusivity.

Our further studies (Fig. 11) on LMS and CCL/LMS composites confirmed that stabilization of the surface of nanograined LMS is not

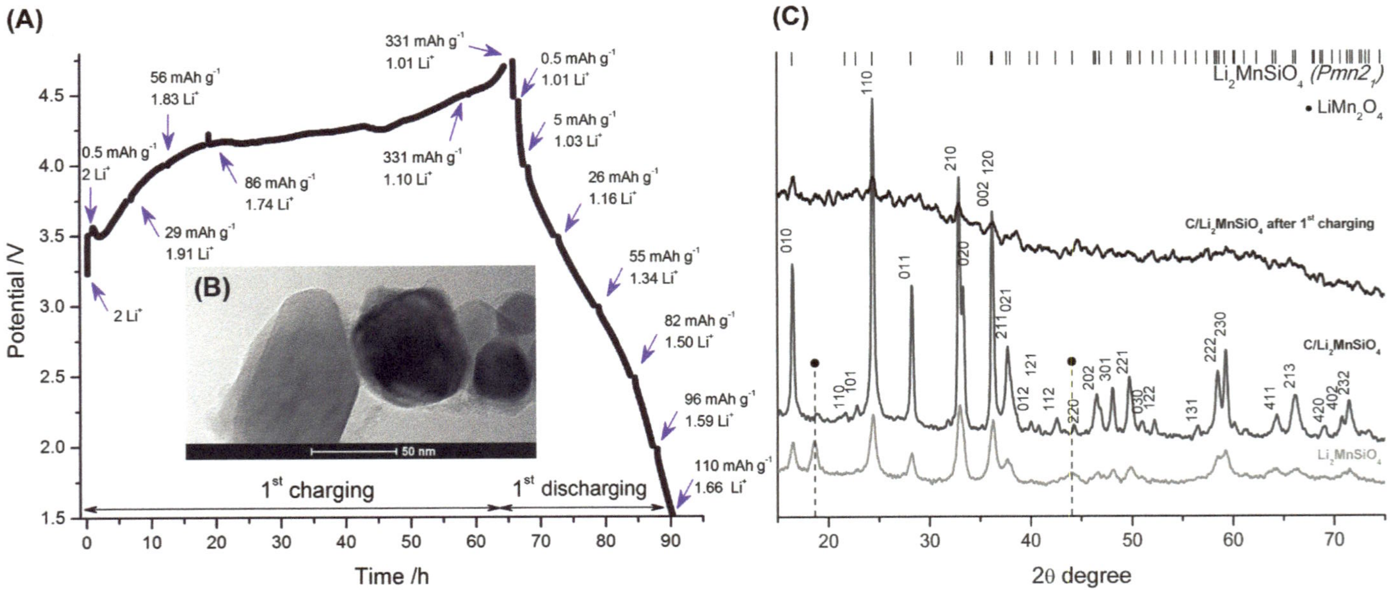

Figure 11 Li_2MnSiO_4: potential curve (A); transmission electron microscopy micrograph (B); X-ray diffraction patterns for pristine LMS and LMS after carbon coating and after initial charge/discharge cycle (C).

sufficient to stabilize the LMS structure.[84,85] Amorphization of LMS occurs during the initial charging/discharging cycles and results in poor cycling behavior of the cell. The amorphous LMS can still deliver a reversible capacity of 185 mAh/g, but the capacity retention after 10 cycles is only 67%.[72] The destruction of LMS is caused by Jahn-Teller distortion, related to the presence of Mn^{+3} during the electrochemical reaction.[86–88] This explanation was later developed by Lee in theoretical studies on the stability of LMS and LFS polymorphs.[89] The proposed hypothesis suggests that the amorphization of the LMS structure is related not only to the distortion of MnO_4 tethraedra during manganese ions oxidation but also to exfoliation process which can occur for the delithated $MnSiO_4$ structure. Current studies being carried out in our group focus on the experimental confirmation of Lee's hypothesis. The in-situ XRD experiment together with ex-situ transmission electron microscopy observations[90] and X-ray photoelectron spectroscopy (XPS) studies of manganese oxidation states that LMS's different states of charge will be applied to analyze the mechanism of the Li_2MnSiO_4 structure amorphization. These studies are conducted in collaboration with the Academic Centre for Materials and Nanotechnology of AGH University of Science and Technology.

5. CARBON AEROGELS FROM NATURAL POLYMERS AS ANODE MATERIAL

Thus far, different anode materials including a wide variety of carbonaceous materials, titanium-based oxides, and alloying- and conversion-type compounds have been extensively investigated in rechargeable LIBs.[91–94] Among these, advanced material carbons are regarded as the leading and still unrivalled anode materials.[95,96] Since 1991 when Sony Co. introduced the first LIB into the market, graphite materials were the most commonly used anodes.[97,98] In fact, in 2016, around 89% of the commercial LIBs was based on graphite.[99] This indispensability of graphitic carbons in commercial LIBs arose from their specified characteristics, which is a compromise between the basic requirements for active materials used in Li-ion technology, namely operation reliability (good reversibility for intercalation/deintercalation of lithium ions, low operation potential, relatively high specific capacity, and long cycle life) as well as financial affordability and environmental benignity.[95,100,101] Unfortunately, the increasing market demands for various electronic devices utilizing ESS led to the great consumption of natural resources of graphite which are limited and nonrenewable. Moreover, the act of extracting graphite from the earth can impoverish the natural surroundings and also be hazardous.[95,102] At the same time, the manufacturing process of synthetic (artificial) graphite requires complicated procedures involving harsh conditions, which makes the production expensive and time-consuming.[103] On the other hand, popular graphitic anode materials cannot fully satisfy the increasing challenges for large-scale and high-energy applications, particularly in targeting the automotive industry and Smart Grid technology.[104,105] All things considered, to address an urgent necessity of more efficient energy storage and conversion technologies and to achieve global energy sustainability, there is a crucial need for new carbon-based anode materials with desired properties, produced from natural and renewable resources via facile, low-cost, and environmentally friendly methods. Of great interest is the use of naturally abundant and highly accessible biomass precursors that possess favorable, hierarchical nanostructures, controllable morphology, and rich porosity.[100,102,106] Starch, next to the cellulose, chitin, and lignin, is one of the most available worldwide, renewable, and inexpensive biopolymers which can be successfully converted to sophisticated green carbons that exhibit tailored properties and enhanced functionality, designed for use as anode materials in Li-ion technology.[4,102,107] Starch consists of two glucose polymers: amylose (linear polymer) and amylopectin (branched polymer), and thus offers distinct advantage over other natural carbon sources that occur with natural inhomogeneity (different amounts of amylose and amylopectin depending on the plant material from which it was obtained). Furthermore, the length of the polysaccharides chain is also varied by the botanical origin of starch.[108] These prominent variables in the composition of starch modulate its structure and properties, and hence affect and control the specification of starch-derived carbon materials.

Taking into account all of these issues, we started to work on hierarchically arranged, nanostructured carbon materials [carbon aerogels (CAGs)] that are obtained from the abundant, renewable, and nontoxic precursors based on naturally occurring polymers.[4,7,109] Generally, carbon aerogels, a relatively novel class of electrode materials, are synthesized as the result of polycondensation of resorcinol and formaldehyde.[110,111] Possible precursors of organic aerogels, often described in the literature, may also include: cresols, melamine, isocyanate, phenol, furfural, and many other compounds.[112,113] However, in these synthetics, often harmful materials are replaced in our research with benign alternatives to promote green chemistry. These ultralight carbon materials exhibit many extraordinary features for energy storage applications, such as determined porosity, large surface area, chemical and thermal stability, high electrical conductivity, and high capacitance. The structure of these materials is well defined and continuous with covalently bonded carbon atoms.[110,114,115] Moreover, it is notable that during the preparation process of carbon aerogels, by changing various parameters of synthesis and thermal treatment, as well as by changing the source of carbon precursors, we can functionalize nanostructured carbon materials depending on the purpose of their use.

Herein, we describe nanostructured carbon aerogels derived from natural starches of different botanical origin such as potato (PS), maize (MS), and rice (RS), which can be applied in LIBs or supercapacitors.[4,7,109] These CAG materials are synthesized via sol-gel polymerization of the aforementioned plant polysaccharides, followed by the drying process and the carbonization of the obtained samples. There are four main stages of their preparation, schematically presented in Fig. 12, that have to be differentiated: the gelatinization process, the solvent exchange (used to reduce the capillary forces responsible for the pore texture destruction),[116] the alcogels drying, and the controlled pyrolysis of the organic aerogels. Each one of them needs to be optimized in order to produce carbon materials with

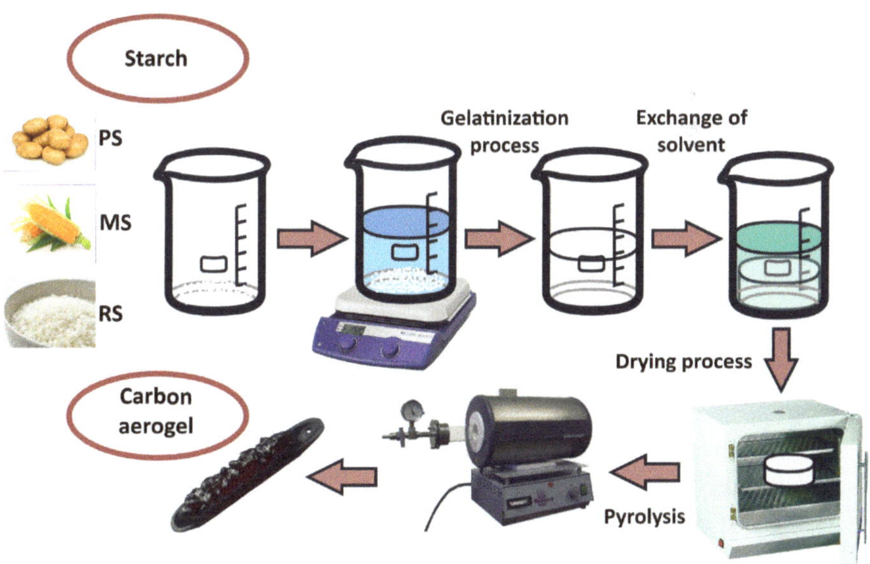

Figure 12 The preparation process of carbon aerogel materials. *MS*, maize; *PS*, potato; *RS*, rice.

desirable properties. Nevertheless, the major challenge in the preparation of such aerogels is to eliminate the liquid solvent from the gel, without collapsing the already existing structure and thereby avoiding the subsequent shrinkage of the gel. The most common techniques in the literature involve supercritical and freeze drying processes.[117,118] However, the classical method that we use (drying the gel at an elevated temperature and under atmospheric pressure after exchange of the solvent), seems to be just as effective but more simple and economically affordable. As mentioned before, the influence of the botanical origin of the substrate on the final product's morphology and electrochemical behavior is also observed. Therefore, our recent studies on this topic are focused on optimization of carbon aerogels (CAG_PS, CAG_MS, CAG_RS) production and their characterization in terms of structural, textural, and electrical features. After determining the gelatinization temperature of various types of starches (which are 65°C, 71°C, and 79°C for PS, MS, and RS, respectively) by differential scanning calorimetry, the proper conditions for the starch aerogels carbonization process were selected by thermal analysis methods (TG/DTG/SDTA).[4,7,109] The studied aerogels were pyrolysed under a flow of argon at 600°C, 700°C, 800°C, and 900°C for 6 h. On the macroscopic scale, it was noted that after pyrolysis, carbon aerogels maintain the morphology of organic aerogels (Fig. 12). Furthermore, the observed structures of these materials are very similar. As it can be seen in Fig. 13A–C for all XRD patterns, there are two broad reflexes located at around 23 degrees and 43 degrees 2θ, corresponding to the graphene-like domains. The intensities of these reflections increase with the increase in carbonization temperature, which indicates the increase in degree of graphitization. In order to point out the exact nature of the carbonaceous samples, the Raman spectroscopy measurements were conducted. Based on all selected Raman spectra (Fig. 13D), as well as the values of the intensity ratio of the D to G bands (I_D/I_G are equal to 1.53 for CAG_PS_700°C and CAG_MS_700°C, and 1.61 for CAG_RS_700°C samples), the carbon aerogels can be characterized as a disordered, nanocrystalline graphitic phase. Afterward, the textural properties of the carbon aerogels were determined. The nitrogen adsorption-desorption isotherms (N_2-BET) show (Fig. 13F) that the carbon materials' specific surface area decreases with decreasing carbonization temperature from 800°C to 600°C for CAG_PS and CAG_MS, as well as from 900°C to 600°C for CAG_RS. Furthermore, the most developed surface area was achieved for CAG_RS_900°C. The shape of the isotherms (Fig. 13E) suggests the microporous character of the studied powders. The

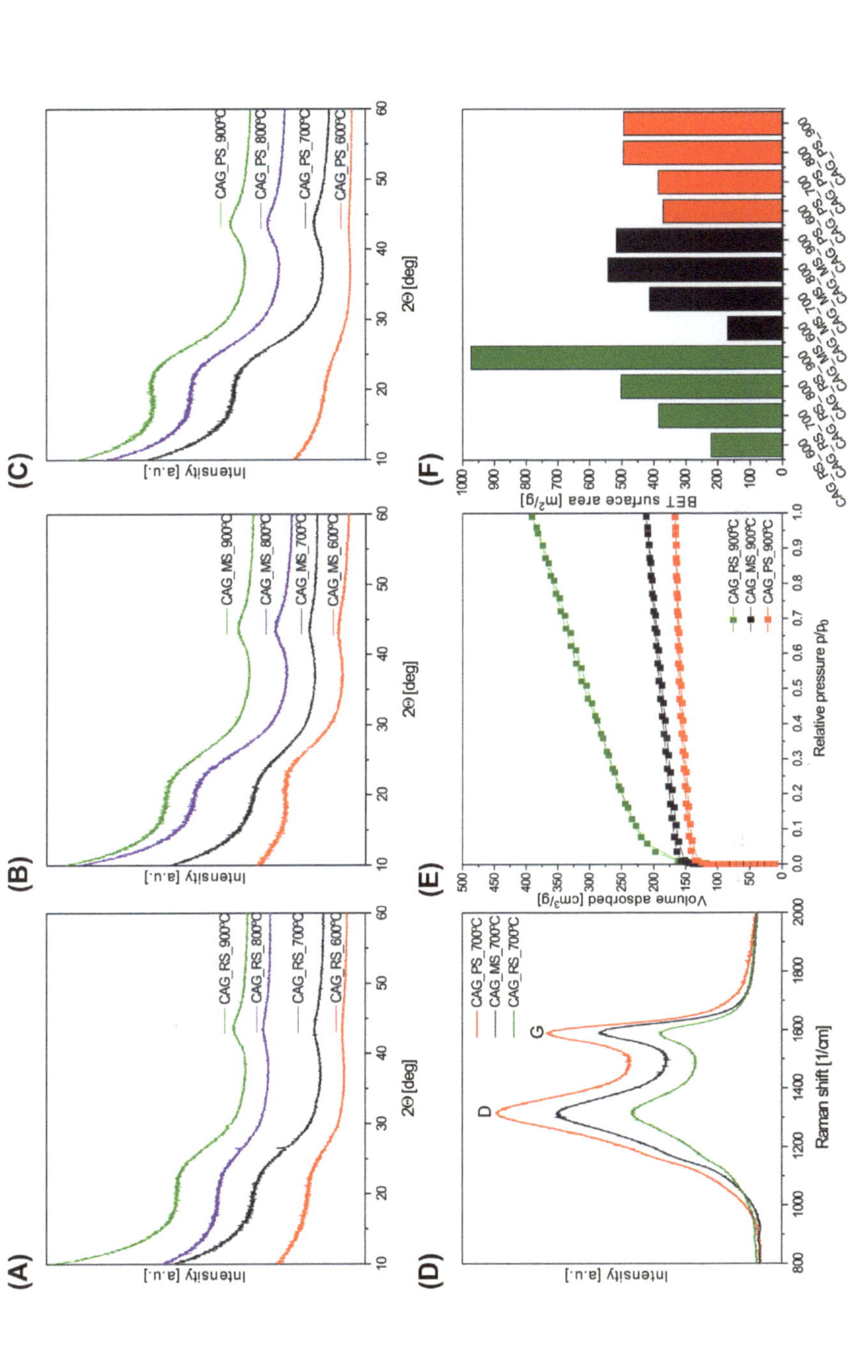

Figure 13 The X-ray diffraction patterns (A)–(C), Raman spectra (D), N2 adsorption–desorption isotherms (E), and surface area (F) of selected carbon aerogels (*CAGs*) samples. *MS*, maize; *PS*, potato; *RS*, rice.

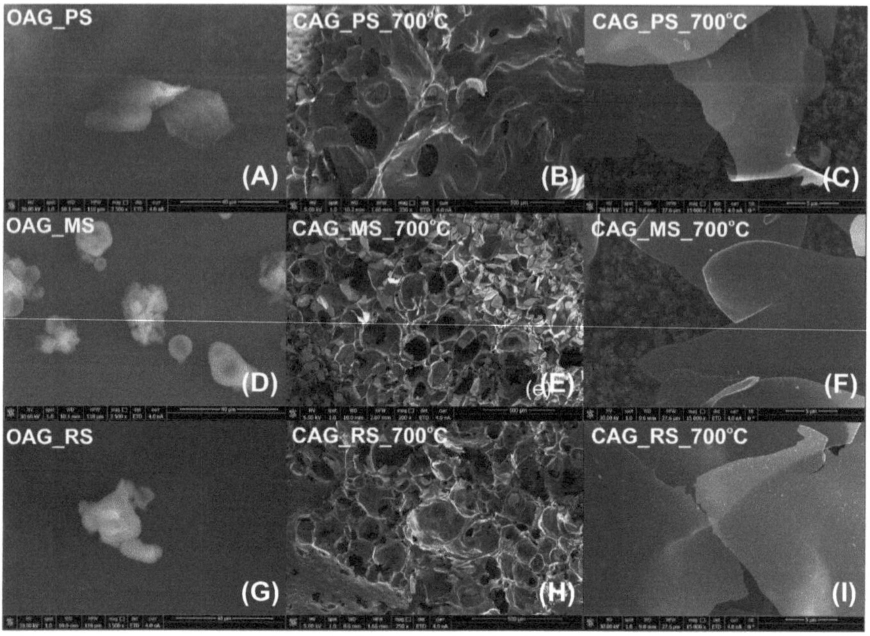

Figure 14 Scanning electron microscopy images of organic aerogels (*OAGs*) (A), (D), (G); selected carbon aerogels (*CAGs*) after pyrolysis (B), (E), (H); and the same CAGs samples after further grinding (C), (F), (I). *MS*, maize; *PS*, potato; *RS*, rice.

analyses of the scanning electron microscopy images lead to the conclusion that carbon materials before pyrolysis (Fig. 14A,D,G) are extremely porous. Aerogels after thermal treatment (Fig. 14B,E,H) look like ultrathin, stretched layers; however, after the grinding process (Fig. 14C,F,I), holes and irregular edges appear in their structure. Finally, our materials were investigated in terms of electrochemical properties. All examined carbon aerogels are considered to be good electrical conductors (Fig. 15A,B). The values of the electrical conductivity increase with the pyrolysis temperature, which confirms the connection between the thermal treatment conditions and the degree of graphitization. Moreover, carbonaceous samples exhibit very low values of the activation energy, about 0.1 and 0.01 eV for CAG_600°C and CAG_700°C, respectively. The results of galvanostatic charging/discharging tests, after 10 cycles of electrochemical activation (C/2), under a different current load (Fig. 15C), show very good electrochemical performance, stability, and reversibility of the selected electrodes. Depending on the origin of starch, the specific capacity of the Li-ion cell is miscellaneous, and its highest values are exhibited by Li/Li^+/CAG_MS as

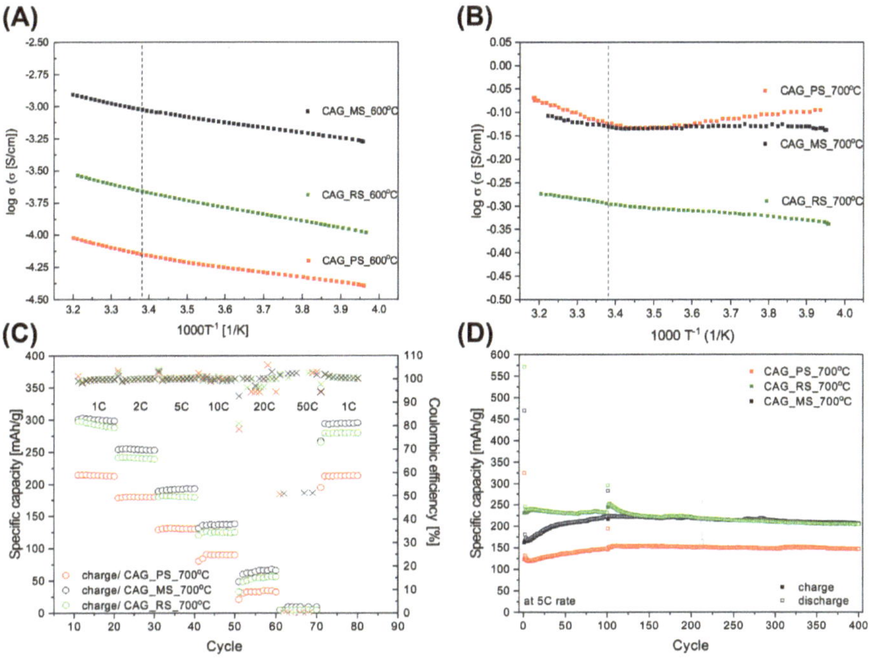

Figure 15 Electrical conductivity (A), (B); the galvanostatic charge/discharge tests under different current loads (C); and the long-term cycling performance at 5C rate (D) of selected carbon aerogels (*CAGs*) samples. *MS*, maize; *PS*, potato; *RS*, rice.

well as Li/Li^+/CAG_RS. Long-term charging/discharging cycles under 5C current rate (Fig. 15D) confirm the much better operation parameters of Li/Li^+/CAG_MS and Li/Li^+/CAG_RS cells than for the cell with CAG_PS.

To sum up, although carbon aerogels are known for about 28 years, the commonly used carbon aerogel precursors are still poisonous and expensive.[110,112,113] Therefore, actual studies have been largely focused on carbon precursors that are easily accessible, renewable, low-cost, as well as environmentally friendly.[4,7,109,119,120] One of the representatives is the most abundant, nontoxic, naturally occurring polysaccharide starch, which, for these reasons, is used in our technology. Thanks to its application, our materials are not only competitive in terms of their properties but also in the context of economics and ecology. Besides changes in the botanical origin of starch, thermal treatment or synthesis conditions allow us to control the carbon aerogel features such as surface development, degree of graphitization, or electrochemical behavior. As it can be seen, all these parameters make our hierarchically arranged, bio-derived, and nanostructured carbons very unusual and useful for energy storage applications.

Further progress of our research includes the development of new technology that allows us to obtain carbonaceous materials (carbogels, CBGs) through the direct controlled pyrolysis of organic alcogels prepared from starch in an inert atmosphere.[109] This means the complete elimination of the time-consuming drying process during sample preparation, making it better suited for industrial application. Moreover, the role of the carbon precursors will be played not only by starch but also by cellulose or even various mixtures of different types of the mentioned polysaccharides. In the first stage of the planned studies, the physicochemical properties of the carbogel precursors will be examined with simultaneous determination of the content of individual polysaccharide fractions. Then, a series of nanostructured CBGs model systems (based on various amounts of amylose and amylopectin) will be obtained and characterized in order to find the interconnection between chemical composition of the precursors and eligible features of the carbon materials. At this stage, the method of synthesis of carbogels will be fully optimized. After that, all the important variables and their correlations with structural, textural, and electrochemical features of carbonaceous materials will be specified, and finally, the mechanism of CBGs formation will be established. To conclude, our work in this matter is aimed to design novel systems with specific, strictly defined properties for LIBs, as well as supercapacitors. Due to their competitive attributes, low production cost, high availability of their precursors, environmental friendliness, and simplicity of synthesis, carbogels are prominent alternatives to the currently used carbon materials.

6. CONCLUSIONS AND PERSPECTIVES

The presented work summarizes the most prominent directions in electrode materials development and production for rapid growth of the LIB industry. The two main types of cathode materials, spinel and polyanionic structures, are considered to be the most important in Li-ion system development in near future. The proposed technologies of modified spinel materials, CCL composites, polyanionic cathodes (LFP, LMS, etc.), as well as carbon aerogel anodes can fill the gap between current technological limitations and sustainable development of mankind. Of course, currently the most important cathode materials which are layered cobalt oxides

LCO and its derivatives like NMC and NCA will still play an important role in the battery industry. Unfortunately, limited safety of such cathodes together with high prices of cobalt (>80 k$/t in March 2018, +54% relative to 2017) and the environmental hazards associated with its use will restrain application of layered oxides in Li-ion cells in the near future. According to Avicenne Energy analysis[1] during the upcoming decade, the market for the new cathode materials in the LIB industry will be dominated by high-voltage spinels and polyphosphates. Widespread application of those materials in commercial Li-ion systems will assure sustainable development of battery industry and reduce its dependence on limited cobalt resources. Current scientific trends in the field of ESS suggest that further development of lithium-based batteries will focus on lithium—sulfur (Li-S) and lithium—air (Li-Air) systems. It is anticipated that commercial Li-S batteries will be available at the end of 2020s and Li-Air batteries in following decade. The dynamics of battery market development in recent years is so fast that all of the predictions concerning LIBs are notoriously understated and the attempts to predict which of those technologies will be dominating in 10 years from now are pointless.

ACKNOWLEDGMENTS

Authors acknowledge the financial support from the following granting agencies:
National Centre for Research and Development in Poland, research grant no. LIDER/463/L-6/14/NCBR/2015 concerning work on modified spinel materials.
The research and development work concerning conductive carbon layers (CCL) and CCL/nano-LFP composites was co-financed by the European Regional Development Fund as part of the Smart Growth Operational Programme for 2014 — 2020 under grant no. POIR.01.02.00-00-0006/16-01.
National Science Centre of Poland, research grant no. 2014/13/B/ST5/04531 concerning work on polysilicate cathodes and research grant no. 2015/19/B/ST8/01077 concerning work on carbon aerogels.

REFERENCES

1. Avicenne ENERGY. *The Worldwide Battery Market 2012—2025,* 2013. Report.
2. Curry, C. *Lithium-ion Battery Costs and Market;* Bloomberg New Energy Finance, 5 July 2017.
3. European Commission Analysis, Available: Online: (https://ec.europa.eu/commission/) 12 February 2018.
4. Bakierska, M.; Molenda, M.; Majda, D.; Dziembaj, R. *Procedia Eng.* **2014,** *98,* 14—19.
5. Molenda, M.; Dziembaj, R.; Drozdek, M.; Podstawka, E.; Proniewicz, L. M. *Solid State Ion.* **2008,** *179,* 197—201.
6. Molenda, M.; Świętosławski, M.; Drozdek, M.; Dudek, B.; Dziembaj, R. *J. Nanomat.* **2014,** *2014,* 103418.

7. Bakierska, M.; Chojnacka, A.; Świętosławski, M.; Natkański, P.; Gajewska, M.; Rutkowska, M.; Molenda, M. *Materials* **2017,** *10,* 1336.
8. Hench, L. L.; West, J. K. *Chem. Rev.* **1990,** *90,* 33–72.
9. Molenda, M.; Dziembaj, R.; Podstawka, E.; Łasocha, W.; Proniewicz, L. M. *J. Phys. Chem.* **2006,** *67,* 1347–1350.
10. Scrosati, B. *Electrochim. Acta* **2000,** *45,* 2461–2466.
11. Fergus, J. W. *J. Power Sources* **2010,** *195,* 939–954.
12. Myung, S. T. *J. Power Sources* **2015,** *283,* 219–236.
13. Croguennec, L.; Palacin, M. R. *J. Am. Chem. Soc.* **2015,** *137,* 3140–3156.
14. Lee, H.-W.; Muralidharan, P.; Ruffo, R.; Mari, C. M.; Cui, Y.; Kim, D. K. *Nano Lett.* **2010,** *10,* 3852–3856.
15. Ishizawa, N.; Tateishi, K.; Oishi, K.; Kishimoto, S. *Am. Min.* **2014,** *99,* 1528–1536.
16. Yang, L.; Takahashi, M.; Wang, B. *Electrochim. Acta* **2006,** *51,* 3228–3234.
17. Wang, L. F.; Ou, C. C.; Striebel, K. A.; Chen, J. S. *J. Electrochem. Soc.* **2003,** *150,* A905–A911.
18. Bakierska, M.; Świętosławski, M.; Chojnacka, A.; Majda, D.; Dziembaj, R.; Molenda, M. *Mater. Technol.* **2016,** *31,* 614–622.
19. Molenda, M.; Dziembaj, R.; Podstawka, E.; Proniewicz, L. M.; Piwowarska, Z. *J. Power Sources* **2007,** *174,* 613–618.
20. Jiang, R.; Cui, C.; Ma, H.; Ma, H. *Chen T. J. Electroanal. Chem.* **2015,** *744,* 69–76.
21. Molenda, J.; Marzec, J.; Świerczek, K.; Ojczyk, W.; Ziemnicki, M.; Molenda, M.; Drozdek, M.; Dziembaj, R. *Solid State Ion.* **2004,** *171,* 215–227.
22. Molenda, M.; Dziembaj, R.; Majda, D.; Dudek, M. *Solid State Ion.* **2005,** *176,* 1705–1709.
23. Dominko, R.; Gaberscek, M.; Bele, M.; Mihailovic, D.; Jamnik, J. *J. Eur. Ceram. Soc.* **2007,** *27,* 909–913.
24. Doeff, M. M.; Wilcox, J. D.; Kostecki, R.; Lau, G. *J. Power Sources* **2006,** *163,* 180–184.
25. Molenda, M.; Bakierska, M.; Majda, D.; Świętosławski, M.; Dziembaj, R. *Solid State Ion.* **2015,** *272,* 127–132.
26. Sun, Y. K.; Oh, S. W.; Yoon, C. S.; Bang, H. J.; Prakash, J. *J. Power Sources* **2006,** *161,* 19–26.
27. Raja, M. W.; Mahanty, S.; Basu, R. N. *J. Power Sources* **2009,** *192,* 618–626.
28. Bakierska, M.; Świętosławski, M.; Gajewska, M.; Kowalczyk, A.; Piwowarska, Z.; Chmielarz, L.; Dziembaj, R.; Molenda, M. *Materials* **2016,** *9,* 366.
29. Bakierska, M.; Świętosławski, M.; Gajewska, M.; Majda, D.; Drozdek, M.; Molenda, M. *J. Nanomat.* **2016,** *2016,* 1497983.
30. Atwater, T. B.; Salkind, A. J. *J. Pow. Sources* **2006,** *162,* 773–779.
31. Bakierska, M.; Świętosławski, M.; Chudzik, K.; Lis, M.; Molenda, M.. 2018, 317, 190–193.
32. Bakierska, M.; Molenda, M.; Świętosławski, M.; Bielecka, P. *LKMNO Cathode Materials and Method of Production Thereof.* WO2017203422A1, 2017.
33. Padhi, A. K.; Nanjundaswamy, K. S.; Goodenough, J. B. *J. Electrochem. Soc.* **1997,** *144,* 1188–1194.
34. Doeff, M. M. In *Batteries for Sustainability;* Brodd, R. J., Ed.; Springer: New York, 2013; pp 5–49.
35. Goren, A.; Costa, C. M.; Silva, M. M.; Lanceros-Mendez, S. *Composites Part B* **2015,** *83,* 333–345.
36. Yang, Z.; Dai, Yu; Wang, S.; Yu, J. *J. Mater. Chem. A* **2016,** *4,* 18210–18222.
37. Morgan, D.; Van der Ven, A.; Ceder, G. *Electrochem. Solid State Lett.* **2004,** *7,* A30–A32.
38. Świder, J.; Molenda, M.; Kulka, A.; Molenda, J. *Funct. Mater. Lett.* **2016,** *9,* 1641007.
39. Zaghib, K.; Mauger, A.; Julien, C. M. *J. Solid State Electrochem.* **2012,** *16,* 835–845.
40. Gong, Z.; Yang, Y. *Energy Environ. Sci.* **2011,** *4,* 3223–3242.

41. Gaines, L.; Cuenca, R. *Costs of Lithium-ion Batteries for Vehicles;* Argonne National Laboratory: Argonne, IL, 2000.
42. Molenda, J.; Zając, W.; Kulka, A. *Process of Fabrication of Crystalline Nanometric Lithium Transition Metal Phosphate.* WO 2015/137836, 2015.
43. Zaghib, K.; Dontigny, M.; Guerfi, A.; Trottier, J.; Hamel-Paquet, J.; Gariepy, V.; Galoutov, K.; Hovington, P.; Mauger, A.; Groult, H.; Julien, C. M. *J. Power Sources* **2012,** *216,* 192–200.
44. Julien, C. M.; Mauger, A.; Groult, H.; Zaghib, K. *Thin Solid Films* **2014,** *572,* 200–207.
45. Anastas, P. T.; Warner, J. C. *Green Chemistry: Theory and Practice;* Oxford University Press: New York, 1998.
46. Toprakci, O.; Toprakci, H. A. K.; Ji, L.; Zhang, X. *KONA Powder. Part. J.* **2010,** *28,* 50–73.
47. Xie, H.; Zhou, Z. *Electrochim. Acta* **2006,** *51,* 2063–2067.
48. Wang, J.; Sun, X. *Energy Environ. Sci.* **2012,** *5,* 5163–5185.
49. Molenda, M.; Dziembaj, R.; Kochanowski, A.; Bortel, E.; Drozdek, M.; Piwowarska, Z.. Process for the preparation of conductive carbon layers on powdered supports, Pat. PL 216549 B1 (2014); Pat. US 8,846,135 B2(2014); Pat. JP 5476383 B2 (2014); Pat. KR 10-2011-7004142 (2015), Pat. EP 2326745 A2 (2017).
50. Molenda, M.; Świder, J.; Świętosławski, M.; Kochanowski, A. *Polimery* **2017,** *62* (7–8), 532–538.
51. Delacourt, C.; Poizot, P.; Levasseur, S.; Masquelier, C. *Electrochem. Solid State Lett.* **2006,** *9,* A352–A355.
52. Nyten, A.; Abouimrane, A.; Armand, M.; Gustafsson, T.; Thomas, J. *Electrochem. Commun* **2005,** *7,* 156–160.
53. Arroyo-deDompablo, M. E.; Amador, U.; Gallardo-Amores, J. M.; Morán, E.; Ehrenberg, H.; Dupont, L.; Dominko, R. *J. Power Sources* **2009,** *189,* 638–642.
54. Dominko, R.; Bele, M.; Gaberscek, M.; Meden, a; Remskar, M.; Jamnik, J. *Electrochem. Commun.* **2006,** *8,* 217–222.
55. Zaghib, K.; Aitsalah, a; Ravet, N.; Mauger, a; Gendron, F.; Julien, C. *J. Power Sources* **2006,** *160,* 1381–1386.
56. Nishimura, S.; Hayase, S.; Kanno, R.; Yashima, M.; Nakayama, N.; Yamada, A. *J. Am. Chem. Soc.* **2008,** *130,* 13212–13213.
57. Lyness, C.; Delobel, B.; Armstrong, a. R.; Bruce, P. G. *Chem. Commun.* **2007,** *46,* 4890.
58. Liu, W.; Xu, Y.; Yang, R. Synthesis. *J. Alloys Compd.* **2009,** *480,* L1–L4.
59. Mancini, A.; Barbieri, V. R.; Neuefeind, J. C.; Page, K.; Malavasi, L. *J. Mater. Chem. A* **2014,** *2,* 17867–17874.
60. Politaev, V. V.; Petrenko, A. A.; Nalbandyan, V. B.; Medvedev, B. S.; Shvetsova, E. S. *J. Solid State Chem.* **2007,** *180,* 1045–1050.
61. Boulineau, A.; Sirisopanaporn, C.; Dominko, R.; Armstrong, R.; Bruce, P. G.; Masquelier, C. *Dalton Trans.* **2010,** *39,* 6310–6316.
62. Jiang, X.; Xu, H.; Liu, J.; Yang, J.; Mao, H.; Qian, Y. *Nano Energy* **2014,** *8,* 1–8.
63. Gummow, R. J.; He, Y. *J. Power Sources* **2014,** *253,* 315–331.
64. Santamaría-Perez, D.; Amador, U.; Tortajada, J.; Dominko, R.; Arroyo-deDompablo, M. E. *Inorg. Chem.* **2012,** *51,* 5779–5786.
65. Gong, Z. L.; Li, Y. X.; Yang, Y. *J. Power Sources* **2007,** *174,* 524–527.
66. Ferrari, S.; Capsoni, D.; Casino, S.; Destro, M.; Gerbaldi, C.; Bini, M. *Phys. Chem. Chem. Phys.* **2014,** *16,* 10353–10366.
67. Kuezma, M.; Devaraj, S.; Balaya, P. *J. Mater. Chem.* **2012,** *22,* 21279.
68. Bhuvaneswari, D.; Kalaiselvi, N. *Dalton Trans.* **2014,** *43,* 18097–18103.
69. Sasaki, H.; Nemoto, A.; Moriya, M.; Miyahara, M.; Hokazono, M.; Katayama, S.; Akimoto, Y.; Nakajima, A.; Hirano, S. *Ceram. Int.* **2015,** *41,* 680–685.

70. Świetosławski, M.; Molenda, M.; Natkański, P.; Kuśtrowski, P.; Dziembaj, R.; Gajewska, M. Sol-Gel Synthesis. *Funct. Mater. Lett.* **2014,** 7, 1440001.
71. Molenda, M.; Świętosławski, M.; Rafalska-Łasocha, A.; Dziembaj, R. *Funct. Mater. Lett.* **2011,** *4,* 135.
72. Świętosławski, M.; Molenda, M.; Furczoń, K.; Dziembaj, R. *J. Power Sources* **2013,** 2–6.
73. Deng, C.; Zhang, S.; Fu, B. L.; Yang, S. Y.; Ma, L. *Mater. Chem. Phys.* **2010,** *120,* 14–17.
74. Deng, C.; Zhang, S.; Yang, S. Y. *J. Alloys Compd.* **2009,** *487,* 18–23.
75. Aravindan, V.; Karthikeyan, K.; Ravi, S.; Amaresh, S.; Kim, W. S.; Lee, Y. S. *J. Mater. Chem.* **2010,** *20,* 7340.
76. Deleted in review.
77. Bhaskar, A.; Deepa, M.; Rao, T. N.; Varadaraju, U. V. *J. Electrochem. Soc.* **2012,** *159,* 1954–1960.
78. Kokalj, A.; Dominko, R.; Mali, G.; Meden, A.; Gaberscek, M.; Jamnik, J. *Chem. Mater.* **2007,** *19,* 3633–3640.
79. Molenda, M.; Dziembaj, R.; Piwowarska, Z.; Drozdek, M. *J. Therm. Anal. Calorim.* **2007,** *88,* 503–506.
80. Molenda, M. *Funct. Mater. Lett.* **2011,** *4,* 129–134.
81. Devaraju, M. K.; Truong, Q. D.; Honma, I. *RSC Adv.* **2013,** *3,* 20633.
82. Molenda, M.; Świętosławski, M.; Dziembaj, R. *Composites and Their Properties.* Chapter 4; InTech, 2012.
83. Świętosławski, M.; Molenda, M.; Zaitz, M. M.; Dziembaj, R. *ECS Trans.* **2012,** *41,* 129–137.
84. Świętosławski, M.; Molenda, M.; Grabowska, M.; Wach, A.; Kuśtrowski, P.; Dziembaj, R. *Solid State Ion.* **2014,** *263,* 99–102.
85. Molenda, M.; Świętosławski, M.; Wach, A.; Majda, D.; Kuśtrowski, P.; Dziembaj, R. *Solid State Ion.* **2014,** 4–7.
86. Li, Y.-X.; Gong, Z.-L.; Yang, Y. *J. Power Sources* **2007,** *174,* 528–532.
87. Dominko, R.; Bele, M.; Kokalj, a.; Gaberscek, M.; Jamnik, J. *J. Power Sources* **2007,** *174,* 457–461.
88. Wu, S.; Zhang, J.; Zhu, Z.; Yang, Y. *Curr. Appl. Phys.* **2007,** 7, 611–616.
89. Lee, H. *Chem. Mater.* **2014,** *26,* 3896–3899.
90. Świętosławski, M.; Molenda, M.; Gajewska, M. *Funct. Mater. Lett.* **2016,** *9,* 1641003.
91. Roy, P.; Srivastava, S. K. *J. Mater. Chem. A.* **2015,** *3,* 2454–2484.
92. Rahman, M. A.; Wong, Y. C.; Song, G.; Wen, C. *J. Porous Mater.* **2015,** *22,* 1313–1343.
93. Goriparti, S.; Miele, E.; De Angelis, F.; Di Fabrizio, E.; Zaccaria, R. P.; Capiglia, C. *J. Power Sources* **2014,** *257,* 421–443.
94. Ji, L.; Lin, Z.; Alcoutlabi, M.; Zhang, X. *Energy Environ. Sci.* **2011,** *4,* 2682–2699.
95. Xing, B.; Zhang, C.; Cao, Y.; Huang, G.; Liu, Q.; Zhang, C.; Chen, Z.; Yi, G.; Chen, L.; Yu, J. *Fuel Process. Technol.* **2018,** *172,* 162–171.
96. Mondal, A. K.; Kretschmer, K.; Zhao, Y.; Liu, H.; Fan, H.; Wang, G. *Micropor. Mesopor. Mater.* **2017,** *246,* 72–80.
97. Xu, J.; Han, Z.; Wu, J.; Song, K.; Wu, J.; Gao, H.; Mi, Y. *Mater. Chem. Phys.* **2018,** *205,* 359–365.
98. Li, W.; Chen, M.; Wang, C. *Mater. Lett.* **2011,** *65,* 3368–3370.
99. Pillot, C.. The Rechargeable Battery Market and Main Trends 2016–2025, Presentation at 33rd Annual International Battery Seminar & Exhibit, 20 March 2017, Fort Lauderdale, FL, USA.
100. Fromm, O.; Heckmann, A.; Rodehorst, U. C.; Frerichs, J.; Becker, D.; Winter, M.; Placke, T. *Carbon* **2018,** *128,* 147–163.
101. Park, D. Y.; Park, D. Y.; Lan, Y.; Lim, Y. S.; Kim, M. S. *J. Ind. Eng. Chem.* **2009,** *15,* 588–594.
102. Gao, Z.; Zhang, Y.; Song, N.; Li, X. *Mat. Res. Lett.* **2017,** *5,* 69–88.

103. Campbell, B.; Ionescu, R.; Favors, Z.; Ozkan, C. S.; Ozkan, M. *Sci. Rep.* **2015,** *5,* 14575.
104. Cui, C.; Sun, X.; Li, X.; Li, C.; Niu, Y. *RSC Adv.* **2015,** *5,* 55348–55352.
105. Fang, M. D.; Ho, T. H.; Yen, J. P.; Lin, Y. R.; Hong, J. L.; Wu, S. H.; Jow, J. J. *Materials* **2015,** *8,* 3550–3561.
106. Imtiaz, M.; Zhu, C.; Li, Y.; Pak, M.; Zada, I.; Bokhari, S. W.; Chen, Z.; Zhang, D.; Zhu, S. J. *Alloys Compd.* **2017,** *724,* 296–305.
107. Nowak, A. P.; Wicikowska, B.; Lisowska-Oleksiak, A. *Solid State Ion.* **2014,** *263,* 131–139.
108. BeMiller, J.; Whistler, R. *Starch: Chemistry and Technology,* 3rd ed.; Elsevier: New York, 2009.
109. Bakierska, M.; Molenda, M.; Chojnacka, A.; Świętosławski, M. *Carbogel Anode Materials and Method for Their Preparation,* 2017. WO2017153855A1.
110. Pekala, R. W. *J. Mater. Sci.* **1989,** *24,* 3221–3227.
111. Halama, A.; Szubzda, B.; Pasciak, G. *Electrochim. Acta* **2010,** *55,* 7501–7505.
112. Pekala, R. W.; Farmer, J. C.; Alviso, C. T.; Tran, T. D.; Mayer, S. T.; Miller, J. M.; Dunn, B. *J. Non-cryst. Solids* **1998,** *225,* 74–80.
113. Zhu, Y.; Hu, H.; Li, W. C.; Zhang, X. *J. Power Sources* **2006,** *162,* 738–742.
114. Kumar, T. P.; Kumari, T. S. D.; Stephan, A. M. *J. Indian Inst. Sci.* **2009,** *89,* 393–424.
115. Shariff, A. M.; Beshir, D. M.; Bustam, M. A.; Maitra, S. T. *Indian Ceram. Soc.* **2010,** *69,* 1–6.
116. Job, N.; Théry, A.; Pirard, R.; Marien, J.; Kocon, L.; Rouzaud, J. N.; Beguin, F.; Pirard, J. P. *Carbon* **2005,** *43,* 2481–2494.
117. García-González, C. A.; Camino-Rey, M. C.; Alnaief, M.; Zetzl, C.; Smirnova, I. *J. Supercrit. Fluids* **2012,** *66,* 297–306.
118. Tamon, H.; Ishizaka, H.; Yamamoto, T.; Suzuki, T. *Carbon* **2000,** *38,* 1099–1105.
119. García-González, C. A.; Alnaief, M.; Smirnova, I. *Carbohydr. Polym.* **2011,** *86,* 1425–1438.
120. Zeng, X.; Wu, D.; Fu, R.; Lai, H.; Fu, J. *Electrochim. Acta* **2008,** *53,* 5711–5715.

CHAPTER EIGHT

Advances in Functionalization of Inorganic Porous Materials for Environmental Catalysis

Lucjan Chmielarz[1], Małgorzata Rutkowska and Andrzej Kowalczyk
Faculty of Chemistry, Jagiellonian University, Kraków, Poland
[1]Corresponding author: E-mail: Lucjan.Chmielarz@uj.edu.pl

Contents

Abstract

Various methods of silica and silica–alumina inorganic porous materials functionalization (zeolites, mesoporous silica sieves, and cationic layered clay minerals), for selected processes of environmental catalysis (NO reduction by ammonia, N_2O reduction and decomposition, NH_3 oxidation to N_2, and synthesis of ethers from alcohols), are presented and critically discussed. In the case of zeolites, various strategies for the preparation of materials with hierarchical mesoporous–microporous structures enabling faster internal diffusion of reactants are presented and discussed as catalysts for nitrogen oxide and ammonia conversions. Various methods of catalytically active metals deposition in the form of monomeric cations and aggregated oxide species on mesoporous silica supports, resulting in their uniform distribution, are presented as effective methods for the synthesis of catalysts for the selected environmental processes. Two different strategies for intercalation of cationic layered clay minerals with inorganic

Advances in Inorganic Chemistry, Volume 72
ISSN 0898-8838
https://doi.org/10.1016/bs.adioch.2018.05.005

pillars are presented and critically discussed as effective methods for the preparation of catalysts for NO_x conversion and production of ethers from alcohols. The presented analyses and discussion are mainly based on studies and experience of the authors of this chapter.

1. INTRODUCTION

Heterogeneous catalysis is of paramount importance in many areas of the chemical[1] and energy[2] industries, including environmental protection.[3] Catalysts are responsible for the production of over 60% of all chemicals and are used in some 90% of all chemical processes worldwide.[4,5] Manufacturing of catalysts alone accounts for over $10 billion in sales worldwide in four major sectors: refining, chemicals, polymerization, and exhaust emission catalysts.[5] The total value of the products obtained by using catalysts is estimated to be about $10 trillion.[5]

Environmental catalysis has continuously grown in importance over the last few decades and is focused mainly on the removal of nitrogen oxides (NO_x) from stationary and mobile sources,[6] conversion of volatile organic compounds (VOCs),[7] selective oxidation of ammonia to dinitrogen,[8] oxidation of CO to CO_2,[9] combustion of carbon soots,[10] production for environmentally friendly fuels[11] including hydrogen,[12] elimination of sulfur[13] and nitrogen[14] from crude oil, and elimination of organic pollutants from waste water.[15]

Looking to the future, new catalytic materials with an atomically tailored structure of active sites guaranteeing conversion of reactants directly to products without generating any by-products is one of the more important challenges of modern catalysis, including environmental catalysis. Another very important issue is the use of suitable supports, which by interaction with the active phase may modify its catalytic properties.[16] On the other hand, catalytic supports characterized by high surface area with designed porous structure are important for the generation of a large number of well-defined active sites but may also influence high selectivity to desired reaction products without limitation of internal diffusion of reactants.[17]

Inorganic porous materials have been intensively studied with regard to various applications including catalysts and catalyst supports. According to the IUPAC definition, porous solids are divided into three classes: microporous (pore size < 2 nm), mesoporous (2–50 nm), and macroporous (>50 nm) materials.[18] Moreover, the term *nanoporous materials* is also increasingly being used. However, it is not clearly defined and loosely refers to pores in the nanometer size range.

Among porous materials used, zeolite catalysts play a very important role. Zeolites are three-dimensional, microporous, crystalline alumina silicates with well-defined structures. In zeolites the silicon and aluminum atoms are tetrahedrally coordinated with each other through shared oxygen atoms forming very-well-defined and uniform porous structures. Zeolites are especially interesting for heterogeneous catalysis because (1) the presence of exchangeable cations allows the introduction of cations with various catalytic properties; (2) the generation of acid sites by introduction of protons into cation exchange sites is possible; (3) micropore diameters of zeolites are in the range similar to the size of chemical molecules and can therefore be used as shape-selective catalysts; (4) they have high thermal and hydrothermal stability. Therefore, it is not surprising that zeolite-based catalysts are used in various large-scale industrial processes, including the oil refining and petrochemical industries (e.g., FCC),[19]and highly selective processes for conversions of organic molecules, viz., hydroxylation (e.g., phenol), alkylation (e.g., ethylbenzene, cumene), oximation (e.g., cyclohexanone oxime), and epoxidation (e.g., propylene oxide).[20] Moreover, zeolites are the preferred catalysts for selective catalytic reduction of nitrogen oxides with ammonia (NH_3-SCR) produced in diesel engines because they can withstand high operating temperatures, which is essential in automotive NH_3-SCR applications.[20] Zeolite-based catalysts are highly effective for conversion of NO_x under steady-state operation.

The main directions of functionalization of zeolites for catalysis are synthesis of zeolites with hierarchical porous structures containing both micropores and mesopores for conversion of bulky molecules and limitation of internal diffusion restrictions, as well as synthesis of zeolites with homogeneous, strictly defined active sites.

One of the important drawbacks of zeolites is their relatively small pore size, which limits conversion of bulky molecules on active sites located inside the pores. Therefore, development of the method for the synthesis of mesoporous silica materials by researches from Mobile Oil Company[21] aroused hope for the extension of classical petrochemical and refinery reactions performed in the presence of zeolites for the conversion of much larger molecules over mesoporous silica-based catalysts. Unfortunately, due to limited thermal stability and surface acidity of mesoporous silica sieves, the application of these materials in conversion of larger organic molecules in petrochemical and refinery processes was impossible. Mesoporous silica materials, characterized by the uniform porous structure, high surface area, and porosity, are typically synthesized by surfactant-directed methods.[21] The size and arrangement of pores can be adjusted by using various types of surfactants and also by controlling the other synthetic parameters.[22] Since 1992, when the first successful synthesis of this type of materials was reported,[23] many different types of mesoporous silicas have been developed, and methods of their synthesis were reported in the scientific literature.[22–24] In contradiction to zeolites, mesoporous silica materials have walls built of amorphous silica, which results in their limited thermal and hydrothermal stability. Moreover, pure silica mesoporous materials do not exhibit the presence of acidic sites, as well as ion-exchange properties, which limits the spectrum of methods for deposition of catalytically active components, ensuring their uniform distribution in the pore system. Nevertheless, these drawbacks of mesoporous silica sieves, they are still very interesting materials for potential applications in various areas, including catalysis. The main directions of studies on silica mesoporous sieves are focused on the synthesis of materials with increased thermal and hydrothermal stability,[25] synthesis of mesoporous materials with crystal walls or containing crystal domains incorporated into silica walls,[26] and functionalization of mesoporous silica for catalysis by controlled deposition of the atomically tailored structure of active sites.[27]

Another, important group of porous materials can be obtained by modifications of cationic clay minerals. Such materials are characterized by unique structures, which contain parallel-ordered negatively charged silica–alumina or silica–magnesia layers. The negative charge of clay layers is compensated by interlayer cations, typically Na^+, K^+, Ca^{2+}, or Mg^{2+}, which in the case of clay minerals can be relatively easily exchanged for other cations, including also large oligocations and bulky organic cations.[28] Ion-exchange properties are key factors determining the possibility of clay minerals intercalation with different pillars. Among them, intercalation of cationic layered minerals with metal oxide pillars (e.g., Al_2O_3, TiO_2, ZrO_2) is considered as one of the most important methods.[29] Clay minerals modified by this way, called pillared interlayered clays (PILCs), are characterized by a relatively high surface area (150–350 m^2/g), developed microporous structure, ion-exchange properties, and surface acidity.

In 1995, a new method of clay minerals intercalation with silica pillars, based on the surfactant-directed method, resulting in materials called porous clay heterostructures (PCHs), was developed.[30] In subsequent years, fast progress in methods of PCH synthesis resulted in the development of procedures for intercalation of a wide range of cationic layered clay minerals, as well as intercalation of multicomponent pillars (e.g., silica–titania[31] or silica–alumina[32]). The PCH materials are characterized by high surface area (up to 1000 m^2/g), combined microporous and mesoporous structure, relatively high thermal and hydrothermal stability, cation-exchange properties, and surface acidity, which depends on both the type of parent clay mineral used for PCH synthesis, as well as chemical composition of pillars introduced into the interlayer space of clay.[33] Recent trends in the studies of intercalated clay minerals, especially PCHs, are focused on controlling the distribution of pillars in the interlayer space and therefore also designing pore size, scale-up of the pillaring process, and controlled deposition of catalytically active components on modified clay minerals.

In the next part, selected recent trends in the functionalization of three types of inorganic silica-containing materials, viz., zeolites, mesoporous silica, and intercalated cationic clay minerals, for catalysis, especially environmental catalysis, are reviewed and discussed.

2. ZEOLITES

Systematic research focused on the synthesis of zeolites was initiated by Richard M. Barrer in the 1930s.[34] Continuation of his pioneering work by researchers of Union Carbide resulted in the development of technology for the first synthetic zeolites (e.g., zeolites A, X, and Y) preparation.[35] In subsequent years, zeolites quickly found many applications in various areas, including the chemical industry. The current worldwide consumption of synthetic zeolites is estimated to be about 1.7–2 million metric tons per year,[36–38] whereas the natural zeolites consumption is approximately 2.5 million metric tons per year.[36] The major industrial areas of synthetic zeolite applications are the production of detergents, adsorbents, and catalysts (mainly for oil refining and petrochemical industrial processes). According to the estimation,[38] the global zeolite market, including both natural and synthetic zeolites, was valued at around \$3.5 billion in 2014 and is expected to reach \$4.5 billion in 2020.

Synthesis of zeolites with the bimodal microporous and mesoporous structures and controlled deposition of catalytically active components belong to the most important recent trends in studies of the zeolite functionalization for catalysis. This great demand for zeolites in the market, together with the high potential of their application in the new fields of industrial development, is the driving force for studies on their modification and functionalization techniques.

One of the recent trends in zeolites modification is the combination of their main advantages, such as crystallinity, acidity, and high hydrothermal stability, with the presence of mesopores. The main idea of the synthesis of mesoporous zeolites was connected with the problem of intracrystalline diffusion limitations and hindered accessibility of the internal pore system by bulky molecules. Such a modification can be visualized as increasing the external surface area, making it available for large molecules.[24,39] A general systematization of hierarchical zeolites synthetic methods divides them into two kinds of approaches: (1) constructive, so-called "*bottom-up*" methods, when mesopores are created simultaneously with micropores during zeolite crystallization (e.g., zeolitization of mesoporous material, double templating with soft and hard templates, mesotemplate-free method, assembly of zeolite nanocrystals, and deposition of zeolite seeds onto the surface of mesoporous materials) and (2) destructive, so-called "*top-down*" methods, basing on the postsynthesis modification of porosity (e.g., desilication, dealumination, recrystallization of zeolites, delamination, and pillaring of 2D zeolites).[40–46] It is very important to emphasize that the modification of the porous structure in many cases influences the Si/Al ratio of the zeolite, which means that the concentration and strength of acidic sites (important factors in catalysis) can be tuned. Moreover, both porosity and acidity of zeolite influence the form and amount of introduced transition metals by ion exchange or impregnation. Thus, the properties of modern redox catalysts can be modified and adjusted for particular reactions by the generation of mesoporosity and functionalization with transition metals.

The first literature reports on zeolite modification by steaming (dealumination) comes from 1967, when McDaniel and Maher published a method to increase the stability of the Y zeolite.[47] This process, leading to so-called ultra-stable Y (USY), was very important for the petroleum industry (cracking of heavy oil fractions). At that time, McDaniel and Maher did not know that the steam treatment they applied could be used for the modification of zeolites porosity. Nowadays, zeolite Y is one of the most intensively studied hierarchical zeolites for the application in fluid catalytic cracking and hydrocracking.[48–50]

Among the various methods used for the mesoporous zeolites synthesis, desilication seems to be one of the most simple, cheap, and widely applied methods. Desilication (controlled leaching of framework silica) in alkaline medium results in the formation of mesopores or macropores, depending on the Si/Al ratio (a high concentration of aluminum prevents the extraction of silicon from the framework). Optimum duration and conditions (concentration of leaching agent, temperature, stirring speed) are the key factors to obtain the relevant parameters of mesoporosity.[51–53] The first patented desilication of the zeolite (mordenite) was connected with increased benzene adsorption capacity after this postsynthesis treatment (D.A. Young in the late 1960s[54]). Since that time, many studies concerning the mechanism of zeolites dissolution in alkaline media have been performed. One of the most interesting studies from the last few years is related to the possibility of tailoring the mesopore size by the addition of quaternary ammonium cations into the leaching solution. Pérez-Ramírez et al.[55] developed this novel desilication method in 2009, proposing that the tetrapropyl ammonium cations (TPA^+) or tetrabutyl ammonium cations (TBA^+) assisted desilication of ZSM-5. These so-called "*pore-directing agents*" act as pore-growth moderators, enhancing the mesopore surface area development, without a severe decrease in the micropore volume. Good control of pore formation during desilication with organic compounds, in comparison to sodium hydroxide, enables the slower desilication kinetics of this process.[56]

During the last few decades, desilicated zeolites were tested in various reactions, including processes of environmental catalysis, such as NH_3-SCR (selective catalytic reduction of ammonia, $DeNO_x$ process),[57–59] NH_3-SCO (selective catalytic oxidation of ammonia),[57,58] N_2O decomposition,[59] synthesis of dimethyl ether (DME) from methanol,[60] ethyl levulinate biodiesel production,[61] catalytic fast pyrolysis of lignocellulosic biomass,[62] or catalytic degradation of low-density polyethylene.[63]

One of the most intensively examined zeolites in this field has been ZSM-5. Its modification by desilication (treatment with NaOH/TPAOH mixture for 2 hours, $TPA^+/OH^- = 0.2$) resulted in the introduction (by ion exchange) of copper in various amounts and with different properties.[57] Divalent copper cations introduced into microporous–mesoporous ZSM-5 were much easily reducible to Cu^+. The reduction of $Cu^{2+} \rightarrow Cu^+$ is connected with their catalytic activity in the NH_3-SCR and NH_3-SCO processes. It was proved that the desilicated sample presented much higher activity in NO reduction and NH_3 oxidation in comparison to the conventional ZSM-5 (Fig. 1). The rapid decrease in NO conversion (Fig. 1) corresponds to the maximum conversion of ammonia. At lower temperatures, preferably NO reduction with ammonia to N_2 occurs, whereas at higher temperatures ammonia is oxidized to NO (conversion drop), which in the

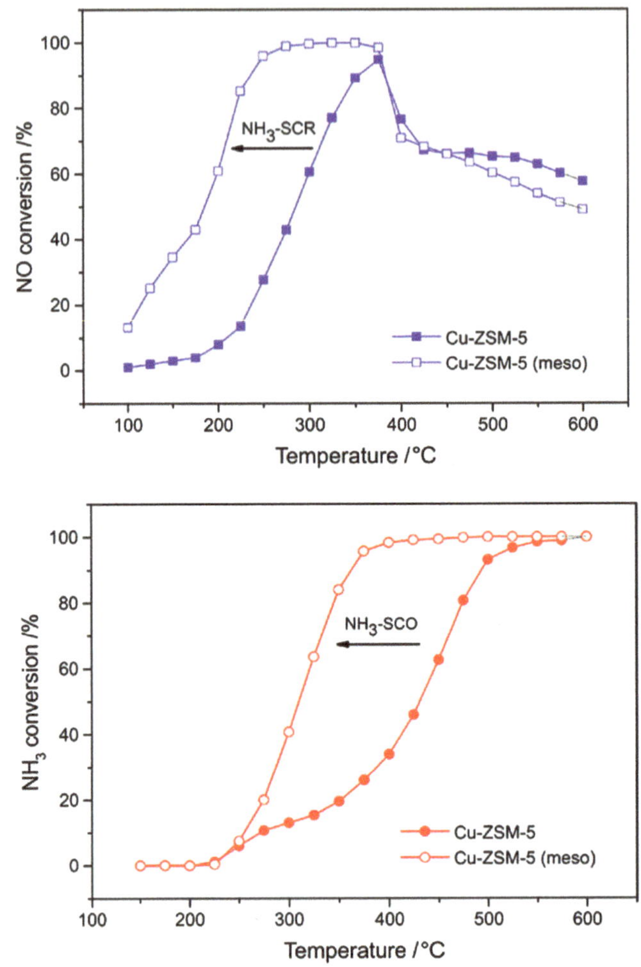

Figure 1 Results of the catalytic tests in the NH_3-SCR and NH_3-SCO processes performed in the presence of conventional Cu-ZSM-5 and mesoporous Cu-ZSM-5 obtained by desilication.[57] *SCO*, selective catalytic oxidation; *SCR*, selective catalytic reduction.

next step is reduced to N_2 (so-called internal SCR mechanism). A comparison of NO and NH_3 conversions over the mesoporous Cu-ZSM-5 samples desilicated for 1 or 2 h with NaOH and TPAOH bases ($TPA^+/OH^- = 0.2$), or exclusively with NaOH, is presented in Fig. 2.

Similar results were obtained in the case of iron-modified desilicated ZSM-5 and mesoporous ZSM-5 produced by a direct synthetic method using amphiphilic organosilanes as a mesopore-directing agent[58] or copper-modified CHA (chabazite) treated with Na_2H_2EDTA.[59] In the former

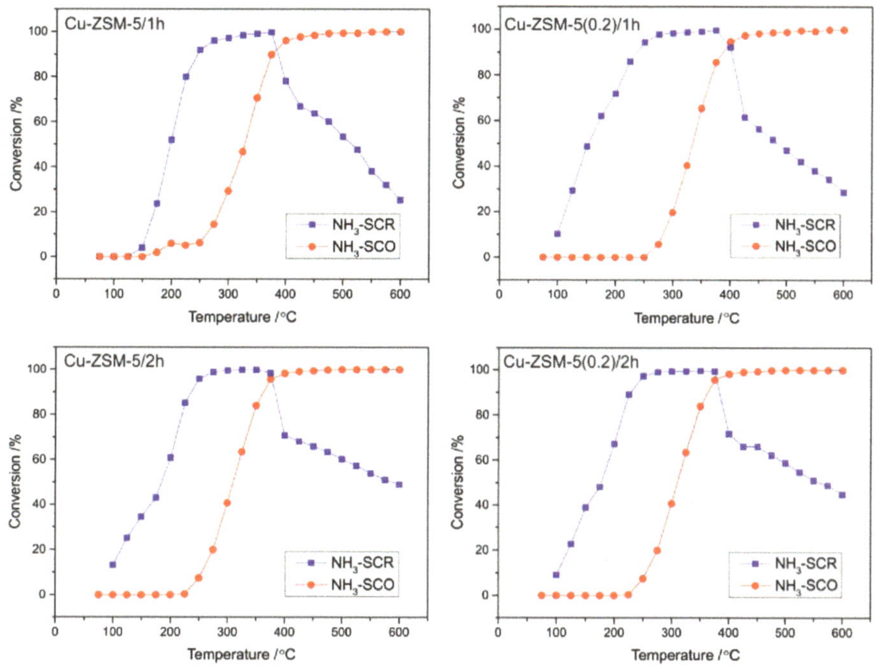

Figure 2 Comparison of NO and NH_3 conversion in the NH_3-SCR and NH_3-SCO processes over the Cu-ZSM-5 samples desilicated at different conditions.[57] *SCO*, selective catalytic oxidation; *SCR*, selective catalytic reduction.

case of mesoporous ZSM-5 modified by ion exchange, iron was present in the form of relatively well-dispersed species (monomeric Fe^{3+} cations and small oligomeric species), whereas in the case of conventional microporous ZSM-5 Fe_2O_3 clusters were found. For copper-modified CHA zeolite, the bulky CuO aggregates were present only in microporous CHA, whereas monomeric cations and oligomeric Cu_xO_y species were dominant in mesoporous CHA, which increased its activity in N_2O decomposition.[59]

By modification of the zeolite porous structure, the accessibility of the acidic sites (ion-exchange positions) can be increased. It influences the dispersion of introduced transition metals and prevents surface metal aggregation (unwanted in many catalytic reactions). Moreover, the stabilization of the cations in the ion-exchange positions is disturbed, which may influence the redox properties of the catalytically active sites. Despite the changes in functionalization of conventional and mesoporous zeolites for the application in redox catalysis, desilication directly influences the concentration and strength of the acidic sites. Thus, this modification method can be

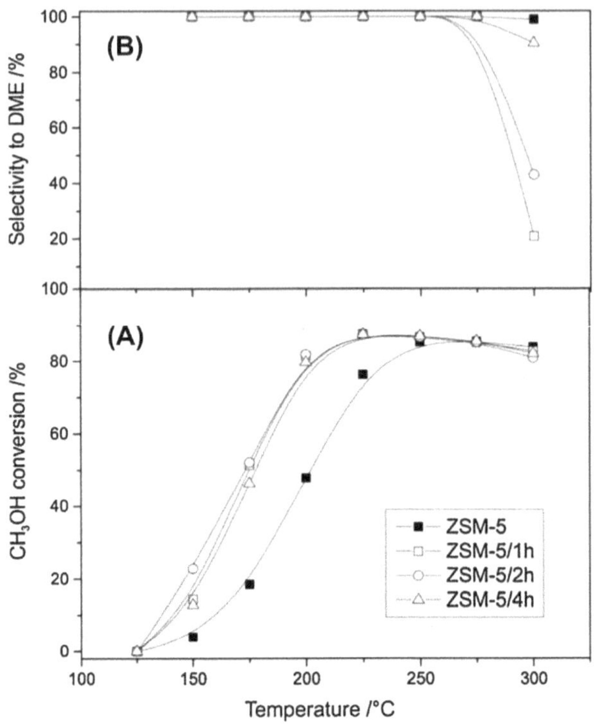

Figure 3 Temperature dependence of (A) CH_3OH conversion and (B) selectivity to DME over ZSM-5 and desilicated zeolites.[60] *DME*, dimethyl ether.

also used for tailoring of the catalytic properties for reactions proceeding on acidic sites. Desilication of ZSM-5 for 1, 2, and 4 h (0.1 M NaOH) increased the total concentration of acidic sites (acidity increased with increasing alkaline treatment duration), which resulted in their catalytic activation in DME synthesis from methanol (Fig. 3).[60] DME is one of the most promising eco-friendly alternative fuels characterized by a high cetane number (about 55—60), high oxygen content (34.8% by mass), and lack of C—C bonds.[64] DME synthesis from methanol is reported to be an acid-catalyzed exothermic reaction.[64,65] Zeolites due to strong acidic sites are one of the most promising catalysts of this reaction. However, fast coking and blocking of micropores by tar deposit are the main drawbacks to be overcome for their practical application. It is important to notice that in the case of the desilicated ZSM-5 samples, the selectivity for DME decreased in comparison to the parent ZSM-5, which is connected with their increased acidity (Fig. 3B). This effect is less important when the treatment duration increases,

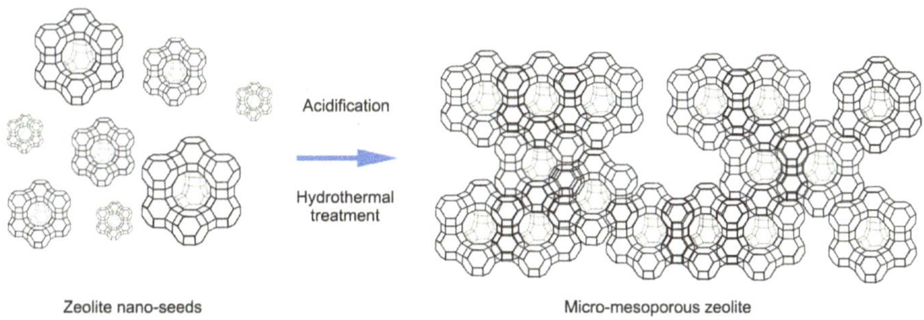

Scheme 1 The concept of mesotemplate-free method.

which suggests balancing of the acidity factor by the generated mesoporosity (enhanced with the increasing treatment duration) and prevented fast blocking of the active sites.

Another interesting method for the synthesis of mesoporous zeolites, besides desilication, is a mesotemplate-free method. This method is based on the formation of the porous structure without the use of the mesopore-directing agent. Scheme 1 shows the concept of the mesotemplate-free synthesis method, whereas Fig. 4 presents STEM micrographs of microporous–mesoporous ZSM-5 prepared using this method. In this synthetic route, zeolite nanoparticles or zeolite seeds are used as a silica source (these particles are obtained by the standard zeolite synthesis route, but for shorter time). The major difference in comparison with other templating methods lies in the fact that in this case, no structure-directing agent, such as organic surfactants (supramolecular templating method) or polymer and carbon black beads (solid templating method), is added. This leads to important economic and ecological advantages in comparison to the syntheses with the use of templates to direct formation of mesopores. In the

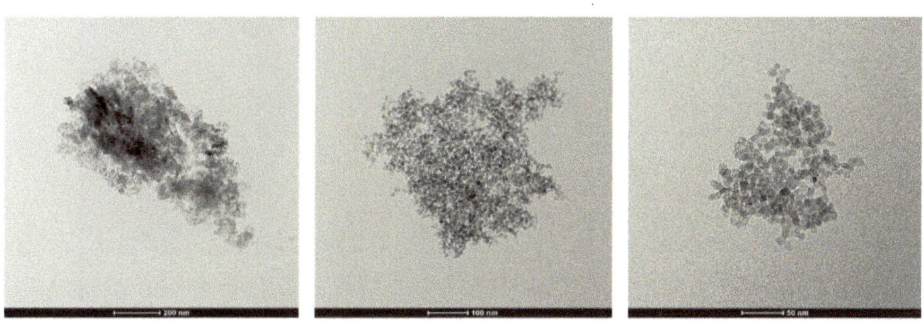

Figure 4 *BFTEM* (bright field transmission electron microscope) images of the microporous–mesoporous ZSM-5 sample with different magnifications.[67]

mesotemplate-free method, the synthesis is performed under controlled conditions, which favor nucleation over crystal growth (e.g., acidification of nanoseeds solution, varying the crystallization duration and temperature).[66–69]

Small changes in the synthesis temperature of the zeolite nanoparticles have a direct impact on the final acidic properties of the bimodal material, which was confirmed by *in situ* IR spectroscopic studies.[70,71] Rutkowska et al.[69] analyzed the influence of hydrothermal treatment duration before and after acidification of ZSM-5 nanoparticles, on the catalytic activity in the DME synthesis. Optimal synthesis conditions, balancing structural properties of mesoporous ZSM-5, such as porosity, acidity, and crystallinity, lead to materials with high catalytic stability and increased resistance for carbon deposit formation. The same mesotemplate-free method was applied for the synthesis of mesoporous beta. Such zeolitic support, modified with Fe, Cu, and Co, was found to be an active catalyst for N_2O decomposition.[72,73] Another method used for the synthesis of microporous–mesoporous catalytic materials is based on deposition of zeolite nanoparticles on the surface of mesoporous materials such as Santa Barbara Amorphous (SBA)-15[72] or Mesostructured Cellular Foam (MCF)[74] by an impregnation method (Scheme 2).

Comparison of the catalytic properties of conventional Fe-beta zeolite modified with iron, mesoporous beta obtained by mesotemplate-free method (Fe-Beta/meso), SBA-15 impregnated with beta nanoparticles at different conditions (Fe-Beta/SBA-15, Fe-Beta(A)/SBA-15), and SBA-15 modified with iron in the process of N_2O decomposition is presented in Fig. 5. Calculation of the reaction rate (the samples differ in the surface

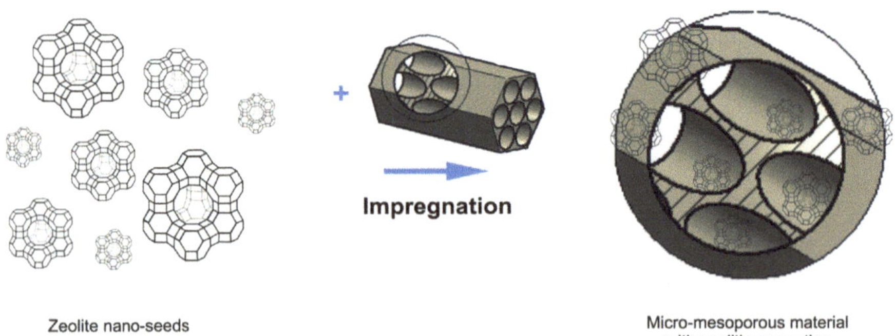

Scheme 2 The concept of the microporous–mesoporous material preparation based on deposition of zeolite nanoparticles on the surface of mesoporous silica materials.

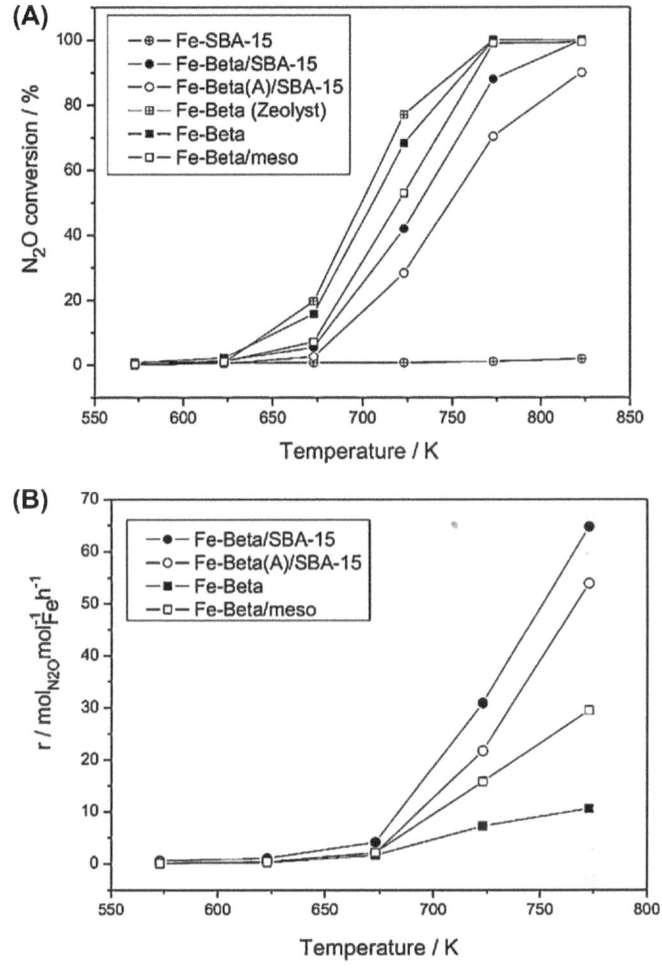

Figure 5 Temperature dependence of (A) N_2O conversion and (B) reaction rate for iron-exchanged Beta zeolite, SBA-15 mesoporous silica, and microporous–mesoporous samples with zeolitic properties.[70]

area and in the iron content) enabled the selection of the most catalytically active species, which were generated in the microporous–mesoporous hybrid materials obtained by zeolite nanoparticles deposition on mesoporous SBA-15.[72]

As it was presented previously, zeolites are still very important materials for catalysis, including environmental catalysis. One of the important recent trends in their functionalization for catalysis is focused on the formation of materials with the hierarchical porous structure allowing for the increased internal diffusion of reactants. It was shown that in many cases, the formation of such bimodal porous structures improves the catalytic properties of zeolitic materials in various processes, including environmentally important ones.

3. MESOPOROUS MICELLE-TEMPLATED SILICA MATERIALS

The first successful preparation of mesoporous micelle-templated silica (MTS) materials using alkyl quaternary ammonium surfactants was described in the patent literature in 1971 by Sylvania Electric Products, Inc.[75] However, due to a lack of analysis, its unique porous structure was not fully recognized. Twenty years later, in 1992, a similar material was obtained by scientists at the Mobil Oil Corporation, who discovered the remarkable mesoporous nature of this novel type of silica.[21] MCM-41 (Mobil Composition of Matter No. 41), which was synthesized in a first series of the mesoporous silica M41S family, is characterized by highly ordered hexagonal array of unidimensional pores with a very narrow mesopore size distribution.[21,76] Other related members of the M41S family are MCM-48 and MCM-50, which are characterized by cubic and lamellar mesostructures, respectively. The discovery of ordered mesoporous silica of the M41S family has been a breakthrough in Materials Science and Engineering. Ordered mesoporous silica materials are generally formed using a micelle-templating process, following either an electrostatically driven cooperative assembly pathway or nonionic route in the presence of uncharged surfactants as structure-directing agents.[77] Therefore, since 1992 there has been impressive progress in the development of many new mesoporous silica synthesized by the surfactant-directed method.[78] Significant research efforts have been focused on achieving control over the characteristics of mesoporous silica, with special emphasis on pore size and morphology. It was shown that the pore geometry and size can be designed by using suitable templates and controlling the conditions of the ordered mesoporous silica synthesis.[22,79] An important disadvantage of ordered mesoporous silica is their limited thermal, hydrothermal, and mechanical stability. It is related to amorphous and relatively thin walls separating channels. Therefore, the next challenge has been the development of methods to increase the stability of ordered mesoporous silica. Among various reported methods, the hydrophobization of the silica surface by grafting of hydrophobic organic species was one of the used procedures for increasing the hydrothermal stability of silica materials.[80] On the other hand, development of methods for the synthesis of ordered mesoporous materials with thick silica walls (e.g., SBA-15), at last, partially solved the problem of their mechanical stability.[81]

A very important issue is the functionalization of ordered mesoporous silica materials for catalysis focused mainly on deposition of catalytically active metals in the form of well-defined and uniform species. Pure calcined mesoporous silica does not exhibit ion-exchange properties, and therefore, deposition of metal cations or cationic aggregated species by ion-exchange methods is impossible. On the other hand, simple impregnation methods are not suitable for deposition of well-distributed and uniform metal species. Moreover, in the case of deposition of larger amounts of metal species, such methods may result in a partial clogging of the pore system. Therefore, intensive studies that focused on development of effective methods for deposition of catalytically active species uniformly distributed on the mesoporous silica supports have been conducted for more than 2 decades. The most promising methods, according to the authors' opinion, seem to be (1) selective grafting of metalorganic complexes on silica surface, followed by their thermal decomposition, resulting in well-dispersed monomeric metal cations—*molecular designed dispersion* (*MDD*); (2) grafting of surface anionic organic functional species able to bind monomeric or oligomeric metal species, followed by sample calcination, resulting in well-distributed metal species—*deposition of oligomeric metal species* (*DOMS*); and (3) deposition of catalytically active cationic species by exchange of surfactants in the prepared (noncalcined) mesoporous silica materials—*template ion-exchange method (TIE)*.

3.1 Molecular Designed Dispersion

The MDD method was developed for uniform deposition of monomeric metal species on support surface and therefore obtaining the catalysts with a designed distribution of catalytically active components.[82–87] The MDD technique consists of two main steps.[88,89] In a first step, a metalorganic complex of metal reacts with the hydroxyl group of the silica support, resulting in the formation of a bond between the silica surface and the metalorganic complex. Depending on the used metalorganic complex, its anchoring to the silica surface may occur by covalent or hydrogen bonding.[90] In a second step, the anchored surface complex is decomposed in an oxygen-containing atmosphere at elevated temperatures (calcination process), resulting in the supported catalysts with a monomeric and uniformly distributed active component.

One of the most promising precursors of catalytically active metals deposited on the silica surface by the MDD method is their acetylacetonate complexes. Molecules of metal acetylacetonates are characterized by a relatively small size, which is important in the case of the deposition of metal species onto microporous or mesoporous supports. Moreover, the important feature of acetylacetonate complexes of metals is their high reactivity toward silanol groups of support. The interaction of metal acetylacetonate molecules with the silica surface may occur according to two different mechanisms. First, π-electrons of the acetylacetonate ligand can be involved in hydrogen bonding with the surface hydroxyl group.[88,89] Second, the surface hydroxyl group may react with the $M^{n+}(acac)_n$ molecule, resulting in the formation of a covalent bond between the metal and surface oxygen, and loss of a ligand as pentanedione (Hacac). Scheme 3 presents the possible mechanisms of acetylacetonate complexes of copper and iron reactions with the surface silanol groups. As can be seen in the case of copper acetylacetonate, the interaction with the silica surface is possible by the formation of hydrogen bonds between the surface silanol group and π-electrons of the acetylacetonate ligand. On the other hand, for deposition of iron using the

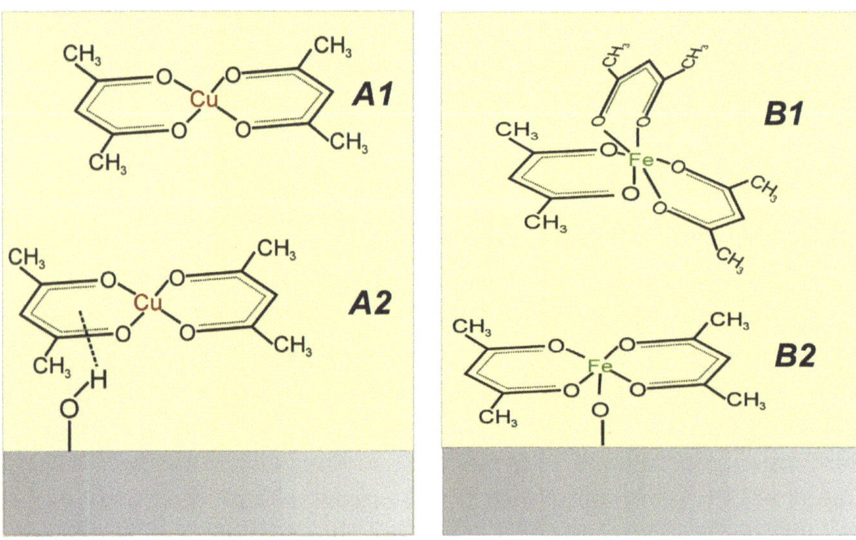

Scheme 3 Mechanisms of interaction of acetylacetonate complexes of copper and iron with surface silanol groups. (A1) $Cu(acac)_2$ molecule; (A2) $Cu(acac)_2$ molecule bonded to the surface silanol group by hydrogen bound; (B1) $Fe(acac)_3$ molecule; and (B2) $Fe(acac)_2$ molecule bonded to surface by coordination of surface oxygen (dehydroxylated silanol group)—ligand exchange mechanism.[88]

acetylacetonate complex as the precursor, the formation of a covalent bond between surface and iron cation was reported.[90] In this case, one acetylacetonate ligand is liberated from the complex and reacts with a proton, resulting in pentanedione (Hacac). Simultaneously, surface oxygen forms a bond with iron. It should be noted that in the case of iron deposition, the formation of a covalent bond with the surface stabilizes the location of metal species on the silica surface during the calcination process. On the other hand, the higher surface mobility of the copper complex bound to the surface by a much weaker hydrogen bond is expected and may result in sintering of copper during the high-temperature calcination process. The mechanism of $M^{n+}(acac)_n$ interaction with the silica surface depends on the stability and geometry of the complex,[88,89] as well as on the properties of the silica support used.[91,92]

Chmielarz et al.[93] compared the efficiency of copper and iron deposition on MCM-41, SBA-15, and MCF mesoporous silica using MDD and classical impregnation methods. It was shown that in the case of all studied silica supports, iron was deposited mainly in the form of monomeric cations and small oligomeric species when the MDD method was used, whereas the impregnation method resulted in a significant contribution of iron in the form of bulky iron oxide aggregates. Such effect was not found for copper-modified silica materials, where independent of the method used for metal deposition, both monomeric and aggregated copper oxide species were identified. The obtained results are in full agreement with the suggested nature of metal complex interaction with the support surface (c.f. Scheme 3). In the case of the metal complex strongly bound to the surface by a covalent bond, their surface mobility is restricted during calcination, which significantly limits the aggregation process. On the other hand, the metal complex attached to the surface by hydrogen bond can relatively easily move on the surface with the formation of more aggregated species. Mesoporous silicas modified with copper and iron were studied as catalysts for selective reduction of NO with ammonia (NH_3-SCR or $DeNO_x$).[93] Examples of the results obtained for iron-modified silica are presented in Fig. 6. It can be seen that independent of the mesoporous silica support used for the deposition of iron by the MDD method, the resulting catalysts are more active and more selective to N_2 (desired reaction product) in comparison to silica doped with iron by incipient wetness impregnation. It should also be mentioned that the samples modified with iron by the MDD method contained slightly lower loading of this metal compared to the catalysts obtained by the impregnation method, e.g., the iron content

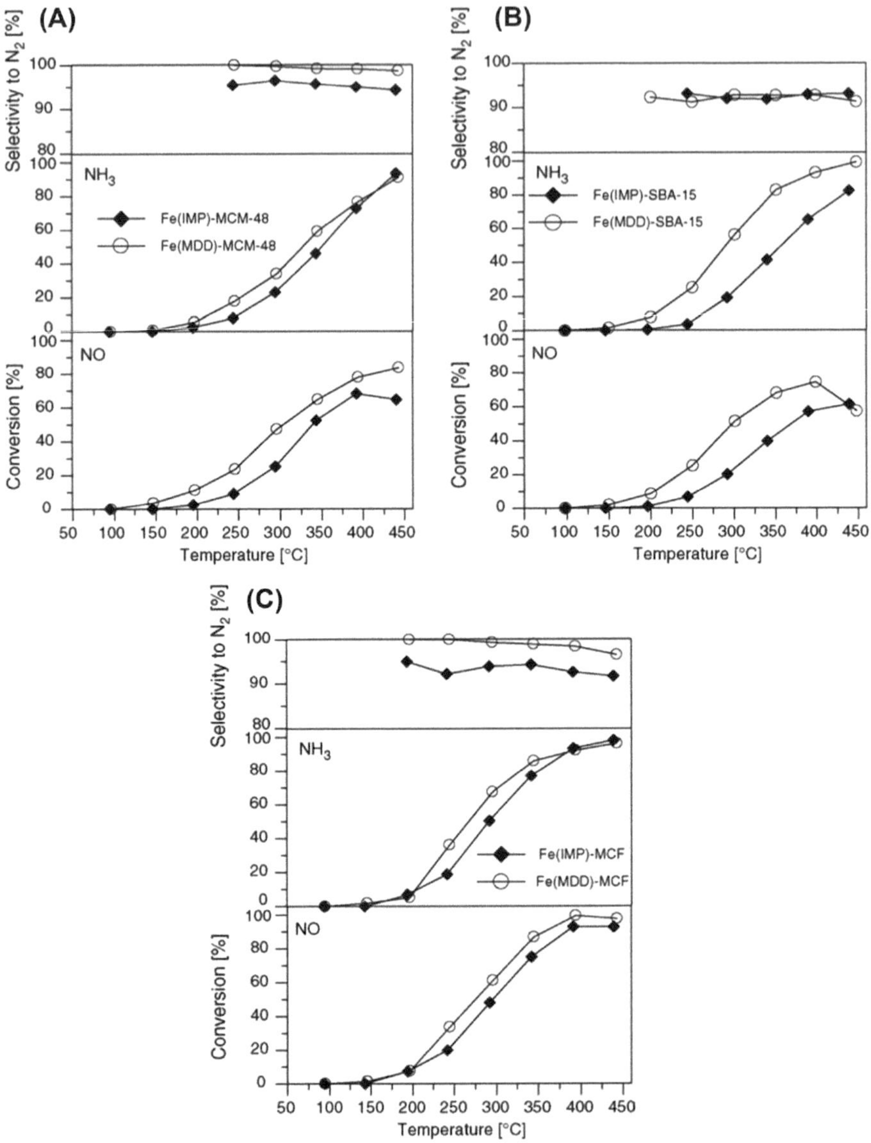

Figure 6 Results of the catalytic tests in the NH_3-SCR process performed in the presence of various mesoporous silicas modified with iron by MDD and impregnation methods.[91] *MDD*, molecular designed dispersion; *SCR*, selective catalytic reduction.

in Fe(IMP)-SBA-15 and Fe(MDD)-SBA-15 was 3.8 and 2.7 wt%, respectively.[93] Moreover, it was shown that modification of the SBA-15 mesoporous silica surface with titanium or aluminum before deposition of iron or copper (all metals introduced by the MDD method) significantly increased

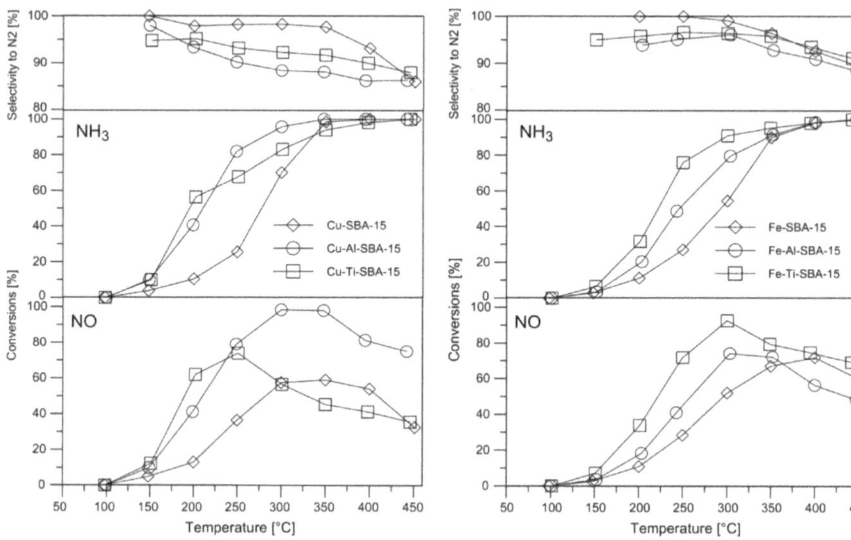

Figure 7 Results of the catalytic tests in the NH_3-SCR process performed in the presence of SBA-15 mesoporous silica and its modifications with titanium-or aluminum-doped iron and copper (all metals deposited by MDD method).[88] *MDD*, molecular designed dispersion; *SCR*, selective catalytic reduction.

their catalytic activity in the NH_3-SCR process.[90] It can be seen (Fig. 7) that modification of SBA-15 with aluminum before the deposition of copper resulted in more effective catalytic activation of the sample than in the case of titanium-modified silica. However, it should also be noted that modification of SBA-15 with titanium or aluminum resulted mainly in activation of the catalysts in the low-temperature range, whereas at higher temperatures, the side process of competitive direct ammonia oxidation by oxygen present in the reaction mixture resulted in a decrease in efficiency of NO conversion. In the case of iron-modified catalysts, codeposition of titanium resulted in a more efficient activation of the catalysts in NH_3-SCR. Similar to the Cu-containing catalysts, also for iron-modified SBA-15, modification of the silica support with titanium and aluminum resulted in its catalytic activation in the low-temperature range, whereas at higher temperatures resulted in intensification of the side process of direct ammonia oxidation. The activation effect of introducing titanium or aluminum is related to the formation of surface acid sites for ammonia sorption. Therefore, in the low-temperature range there is an excess of ammonia molecules on the catalyst surface (ammonia reservoir), which by the sequence of desorption—adsorption cycles can be supplied to active sites located possibly on the surface copper or iron species. On the other hand, ammonia desorption is more preferable than its adsorption, and under such conditions, ammonia chemisorbed on strong acidic sites is possibly oxidized by oxygen present in the reaction mixture to dinitrogen and nitrogen oxides.

Segura et al.[94,95] studied the catalytic activity of SBA-15 modified with titanium or vanadium species, as well as the silica support modified together with both these highly dispersed metal oxide species deposited by the MDD method. It should be mentioned that commercial catalysts for the NH_3-SCR process are based on the $V_2O_5-TiO_2$ metal oxide system.[96] The advantage of titanium and vanadium deposition on SBA-15 is to obtain a high–surface area catalyst with easily accessible active components. Results from the catalytic studies of the NH_3-SCR process for SBA-15 modified with titanium or vanadium species by the MDD method are presented in Fig. 8. In general, it can be seen that SBA-15 modified with only one active component, titanium or vanadium, effectively operates only in the high-temperature region and strongly depends on the loading of metal species. In the case of the catalyst with higher vanadium loading, the low-temperature activity in NH_3-SCR increased, but at higher temperatures the efficiency of NO reduction decreased due to the side process of direct ammonia oxidation. Simultaneous presence of titanium and vanadium deposited on SBA-15 resulted in a significant activation of the silica-based catalysts in the NH_3-SCR process (Fig. 9). It should be noted that the loading of active components is a crucial point that influences the catalytic performance of the studied samples. The catalysts with lower content of vanadium and titanium presented only limited activity, which could be explained by the necessity of contact between titanium and vanadium species to effectively convert NO molecules. Possibly, in the case of the samples with lower loading of these active components, they are separated on the silica surface and cannot cooperate in the conversion of NO by ammonia. On the other hand, in the case of the samples with higher loading of active components, they are in much better contact and may effectively cooperate in the NH_3-SCR process. Another possible explanation is the formation of more aggregated vanadium and titanium oxide species, with higher catalytic activity in the studied reaction. However, it seems that also in this case the contact between titanium and vanadium oxide phases is necessary to obtain active catalysts for the NH_3-SCR process. It should be noted that for the most active $V_2O_5-TiO_2$ SBA-15 catalyst, a significant decrease in the efficiency of NO conversion, related to the side process of direct ammonia oxidation, is observed at higher temperatures (Fig. 9).

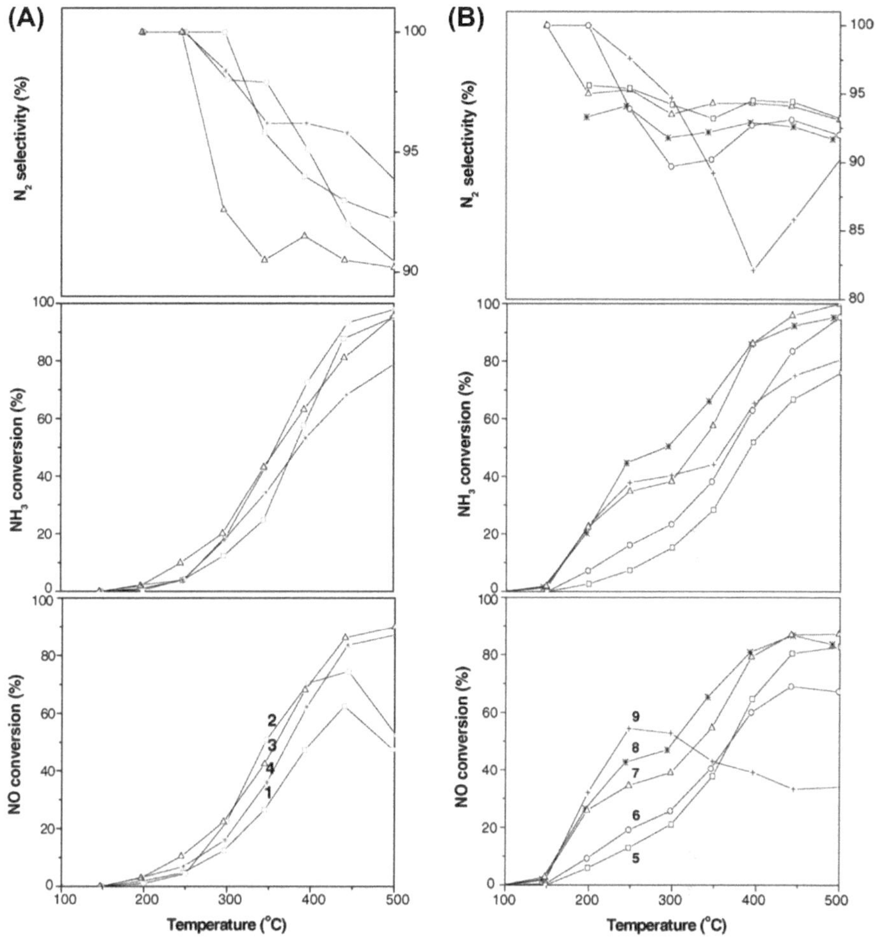

Figure 8 (A) Catalytic activity data of TiO_x/SBA-15 catalysts in NH_3-SCR having (1) 0.17, (2) 0.37, (3) 0.55, and (4) 1 mmol Ti/g. (B) Catalytic activity data of VO_x/SBA-15 catalysts having (5) 0.11, (6) 0.25, (7) 0.40, (8) 0.45, and (9) 0.77 mmol V/g.[92] *SCR*, selective catalytic reduction.

This effect is also a characteristic of the commercial NH_3-SCR catalysts based on the $V_2O_5-TiO_2$ metal oxide system.[96] Thus, it seems that this is a characteristic feature of the catalytic systems containing V_2O_5 and TiO_2 together.

Mesoporous silica materials modified with rhodium using the MDD method were also studied as catalysts for decomposition of nitrous oxide. This is a very important environmental process resulting in the conversion of strong greenhouse gas N_2O to N_2 and O_2. Chmielarz et al.[97] studied

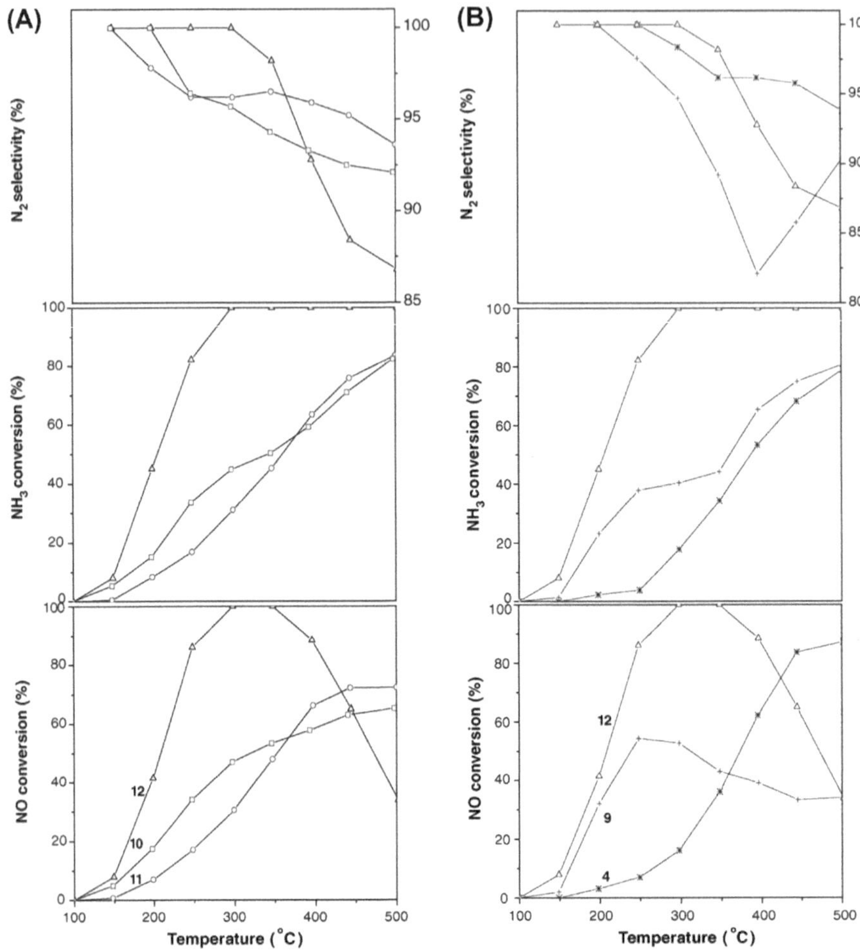

Figure 9 (A) Catalytic activity data of the VO_x—TiO_x/SBA-15 catalysts in NH_3-SCR with different metal contents (10) 0.37 Ti, 0.37 V, (11) 0.37 Ti, 0.27 V, and (12) 0.77 Ti, 0.77 V mmol/g. (B) Catalytic activity data of (4) TiO_x/SBA-15, (9) VO_x/SBA-15, and (12) VO_x—TiO_x/SBA-15 catalysts having about the same metal content.[92] *SCR*, selective catalytic reduction.

the catalytic performance of rhodium deposited on the SBA-15 mesoporous silica support by the MDD method. First of all, it was shown that the use of the MDD method resulted in much better dispersion of rhodium on the silica surface in comparison to the impregnation method (Fig. 10) Moreover, as can be seen, rhodium dispersion on SBA-15 decreased with an increase in its loading. Thus, again it was proved that the MDD method is very useful for the preparation of catalysts containing a highly dispersed active phase.

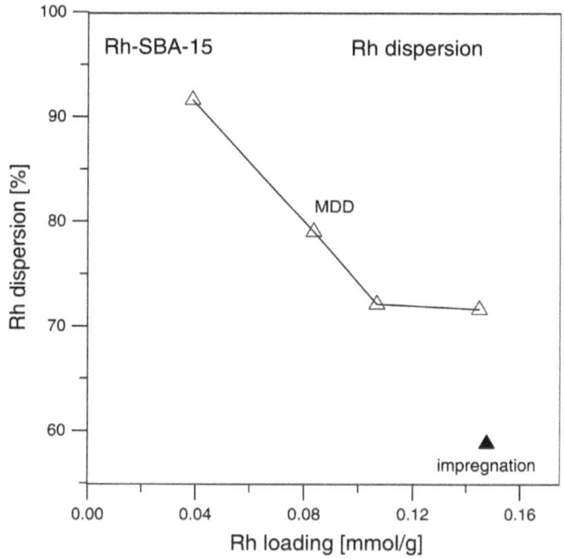

Figure 10 Correlation between loading and dispersion of rhodium in the SBA-15 samples.[95] Dispersion in defined a parentage of Rh atoms exposed on the catalyst surface (demined by H_2 chemisorption) to the Rh loading in catalysts (determined by chemical analysis). *MDD*, molecular designed dispersion.

SBA-15 modified with rhodium was found to be an active catalyst for catalytic decomposition of N_2O; however, its activity increased with an increase in noble metal loading (Fig. 11). Additional introduction of iron or copper (also by the MDD method) into Rh-SBA-15 resulted in a decrease in catalytic activity, whereas activation of the catalyst was observed after deposition of aluminum by the MDD method. This interesting effect is possibly related to the generation of acidic sites by surface Al species, which play a role of N_2O adsorption sites. It should be also noted that the efficiency of N_2O decomposition decreased in the reaction mixture containing water vapor; however, also in this case the activity of the Rh-SBA-15 catalyst is better than many other catalytic systems reported in the literature (Fig. 11).

In conclusion, the MDD method is an effective way for deposition of metals on the silica surface in the form of highly dispersed species. As it was shown previously, such species were found to be active in the conversion of NO and N_2O. The disadvantage of the MDD method is its relatively high cost mainly due to expensive precursors of active components (metal-organic complexes) and multistage procedures of metal species deposition.

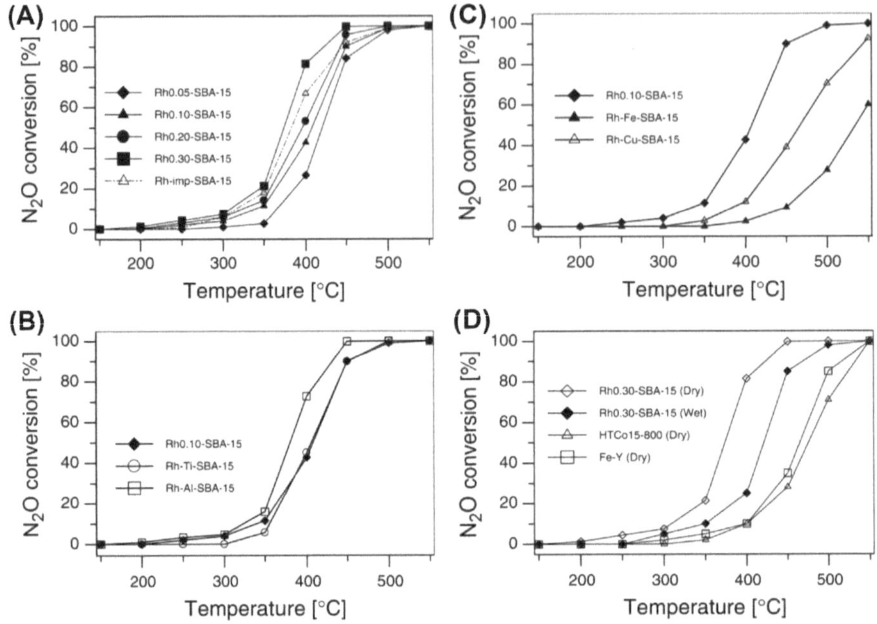

Figure 11 Results of catalytic tests for a series of rhodium-modified SBA-15 (A), Rh-Ti-SBA-15 and Rh-Al-SBA-15 (B), Rh-Fe-SBA-15 and Rh-Cu-SBA-15 (C), as well as the comparison of catalytic activity of Rh0.30-SBA-15 in a flow of wet and dry reaction mixtures with activity of the reference catalysts Fe-Y and HTCo15-800 (D).[95]

3.2 Deposition of Oligomeric Metal Species

Many catalytic processes (e.g., selective oxidation of ammonia to N_2 or VOCs incineration) need transition metals, which play the role of catalytically active components, in the form of aggregated species.[98,99] To perform processes with high selectivity, such aggregated species should be characterized by a similar structure and size as well as should be uniformly distributed on the support surface. Also in this case, mesoporous silica materials seem to be very interesting supports for production of catalysts. However, the design and synthesis of such catalysts should take into account some issues: (1) the size of deposited oligomeric species should be small enough for their effective deposition inside the pores of mesoporous silica materials and (2) pure silica mesoporous materials do not exhibit ion-exchange properties, and any reactive surface sites are able to directly bind oligomeric metals species. The strategy of DOMS on silica surface is based on deposition of functionalized alkoxy silicates by their selective reaction with the surface silanol (≡Si–OH) groups, optionally followed by the conversion of anchored

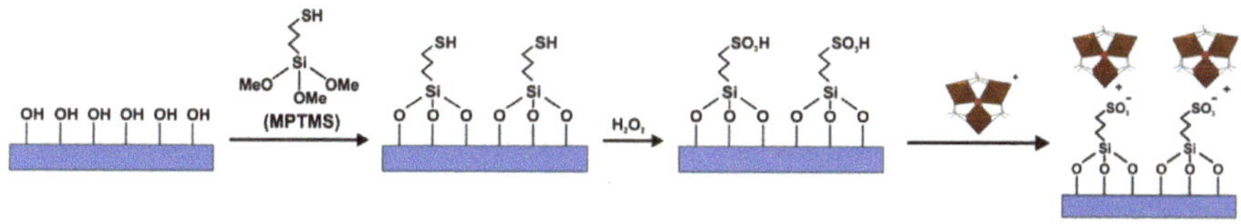

Scheme 4 Procedure used for deposition of oligomeric iron species via anchoring and oxidation of 3-(mercaptopropyl)trimethoxysilane (MPTMS) by the DOMS method.[98] *DOMS*, deposition of oligomeric metal species.

functional groups into their anionic or cationic forms. This procedure results in generation of ion-exchange properties in mesoporous silica, and therefore, in the next step, oligomeric species (e.g., oligocations) can be deposited by the ion-exchange method. Finally, silica materials modified in this way should be calcined to remove organic species from the samples and thermally decompose oligomeric species to stable metal oxide aggregates. An example of such a procedure is presented in Scheme 4.[100] 3-(Mercaptopropyl)trimethoxysilane (MPTMS) was anchored on the surface of mesoporous silica by reaction with the surface silanol groups. In the next step, —SH groups were oxidized to —SO_3H using a solution of H_2O_2 as an oxidizing agent. Then, trinuclear iron acetate oligocations, $[Fe_3O(OAc)_6\cdot AcOH\cdot 2H_2O]^+$, were introduced by the ion-exchange method. Finally, after drying, the samples were calcined.[101] Detailed characterization of the samples obtained in this way proved that iron species were uniformly distributed on the catalyst surface. The majority of iron, as it was expected, was present in the form of small FeO_x aggregates; however, also, monomeric iron cations were identified. The catalytic performance of iron-modified SBA-15, obtained by the DOMS and MDD procedures in the process of selective reduction of NO with ammonia ($DeNO_x$), is compared in Fig. 12. It can be seen that the catalysts obtained by the DOMS method are significantly more active in comparison to MCM-41 modified with iron by the MDD method. A decrease in the efficiency of the $DeNO_x$ process at higher temperatures is related to the side reaction of direct oxidation of ammonia by oxygen present in the reaction mixture. Fig. 12 (right side) presents the turnover-number (TON) of the NO conversion. It was assumed that iron ions play the role of active centers located on the catalyst surface. It can be seen that iron in the form of aggregated species is more catalytically active in the $DeNO_x$ process in comparison to monomeric iron cations dispersed on the silica surface.

Mesoporous silicas, SBA-15 and Hexagonal Mesoporous Sieves (HMS), modified with Fe, Cr, and mixed Fe—Cr trinuclear metal acetate oligocations,

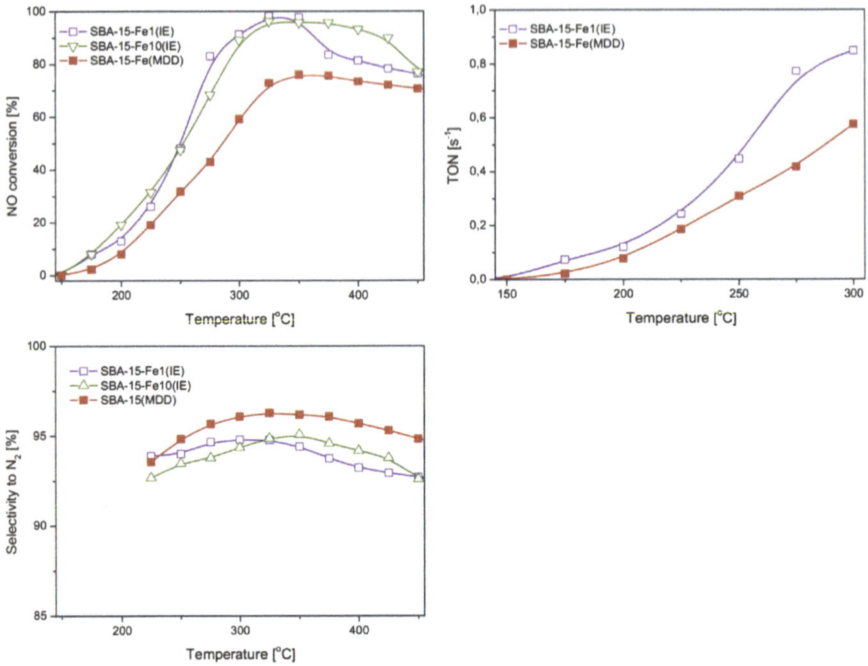

Figure 12 Results of the catalytic tests in the $DeNO_x$ process for iron-modified SBA-15 by DOMS (IE-series) and MDD methods (left side) and turn over numbers (TON) determined for SBA-15-Fe1(IE) and SBA-15-Fe(MDD) (right side).[98] DOMS, deposition of oligomeric metal species; MDD, molecular designed dispersion.

$[Fe_3O(OAc)_6 \cdot AcOH \cdot 2H_2O]+$, $[Cr_3O(OAc)_6 \cdot AcOH \cdot 2H_2O]+$, $[Fe_2CrO(OAc)_6 \cdot AcOH \cdot 2H_2O]+$, and $[FeCr_2O(OAc)_6 \cdot AcOH \cdot 2H_2O]+$, deposited by the DOMS method, after calcination were tested as catalysts for the selective oxidation of ammonia to dinitrogen.[102] It was shown that MPTMS was successfully anchored and oxidized on the surface of mesoporous silica, and in the following step, trinuclear metal acetate oligocations were effectively deposited by the ion-exchange method. Calcination of SBA-15 modified with oligomeric species resulted in a decrease in specific surface area by about 25% and shift of the maximum of the pore size distribution in the direction of lower pore diameter. It shows that a significant part of the oligomeric species was deposited inside the pores of SBA-15. A different effect was observed for HMS, which is characterized by smaller pores than SBA-15 (pore diameter: 5.3 nm for SBA-15 and 2.9 nm for HMS). In the case of HMS, deposition of oligomeric species resulted in a decrease in specific surface area by about 5%, while no change in pore

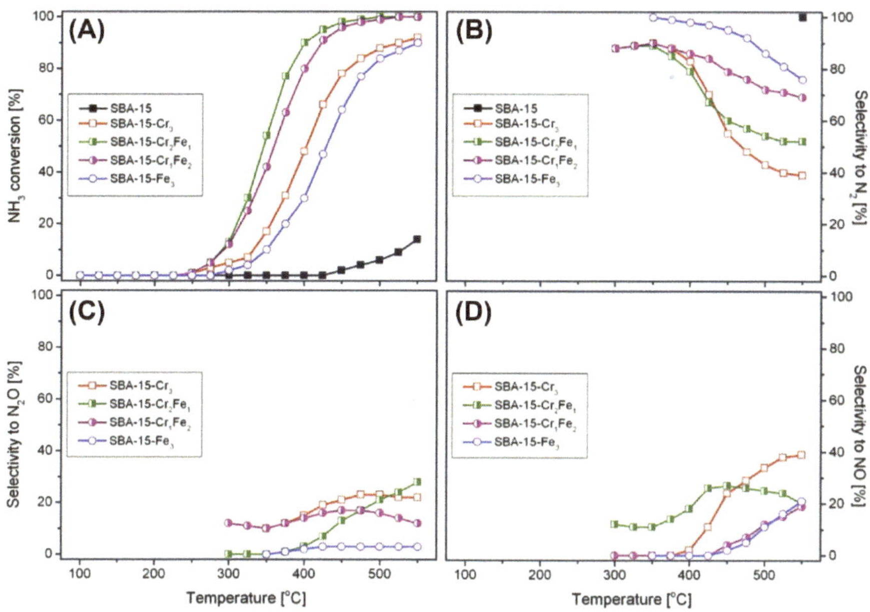

Figure 13 Results of (A) NH_3 conversion and (B) selectivity to N_2, (C) N_2O, and (D) NO obtained for the SBA-15 samples modified with Cr—Fe species.[100]

diameter was observed. This effect could be related to the limited diffusion rate of oligocations inside the HMS pores, and therefore, the majority of oligomeric species was deposited on the outer surface of mesoporous silica grains. The results of catalytic tests on the process of selective oxidation of ammonia to dinitrogen obtained for SBA-15 modified with oligomeric Fe, Cr, and Fe—Cr species are presented in Fig. 13.[102] It can be seen that SBA-15 nonmodified with metals is inactive in the process of ammonia oxidation. Deposition of iron or chromium species resulted in a very significant catalytic activation of SBA-15; however, the Cr-modified sample presented slightly higher catalytic activity in the low-temperature range in comparison with the Fe-containing catalyst. A very interesting effect was observed for the samples modified with mixed Fe—Cr oligocations, which were more catalytically active than the catalyst modified with monometallic Fe- or Cr-oligomeric species. Thus, these results clearly show the synergetic effect of the simultaneous presence of both iron and chromium. It was shown that reduction of Fe—Cr oxide species occurred at a temperature lower than for monometallic Fe or Cr oxide species (Fig. 14). It seems possible that

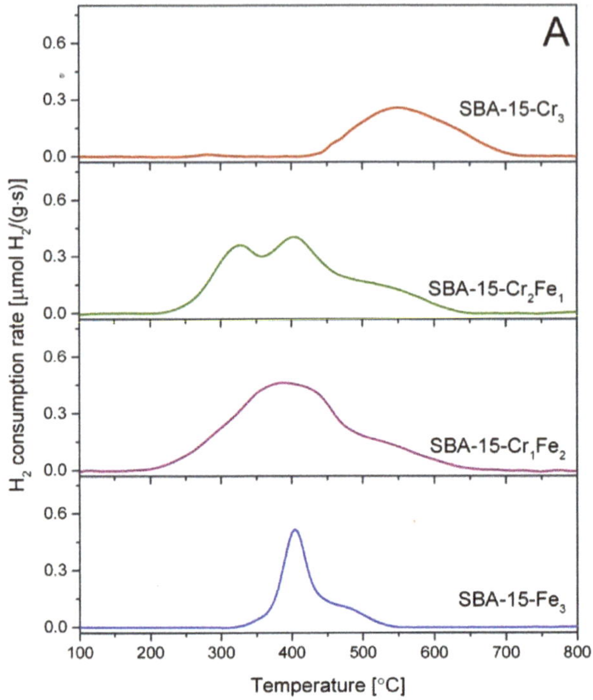

Figure 14 *H_2-TPR* (temperature programmed reduction with hydrogen) results obtained for SBA-15 modified with Fe, Cr, and Fe–Cr species.[100]

the high catalytic activity of bimetallic Fe–Cr species in the selective oxidation of ammonia to dinitrogen is related to their improved redox properties.

Macina et al.[103] reported a high catalytic activity of mesoporous silica materials modified with aggregated aluminum species deposited by the DOMS method as effective catalysts for conversion of methanol to DME. It is known that this process is catalyzed by acidic sites of medium strength.[64,104–106] Thus, the concept to design a highly active catalyst for the methanol to DME conversion process was based on deposition of aluminum oligomeric species, generating acidic properties on the high–surface area mesoporous supports. In this case, mesoporous silicas of the SBA-15 and MCF type, with pore diameters of 5.9 and 11.8 nm, respectively, were modified with aluminum Keggin cations, $[Al_{12}(AlO_4)(OH)_{24}(H_2O)_{12}]^{7+}$ (diameters about 1 nm), by the DOMS method. The ion-exchange step (deposition of Keggin cations) was proceeded by anchoring of MPTMS on the silica surface followed by oxidation of –SH

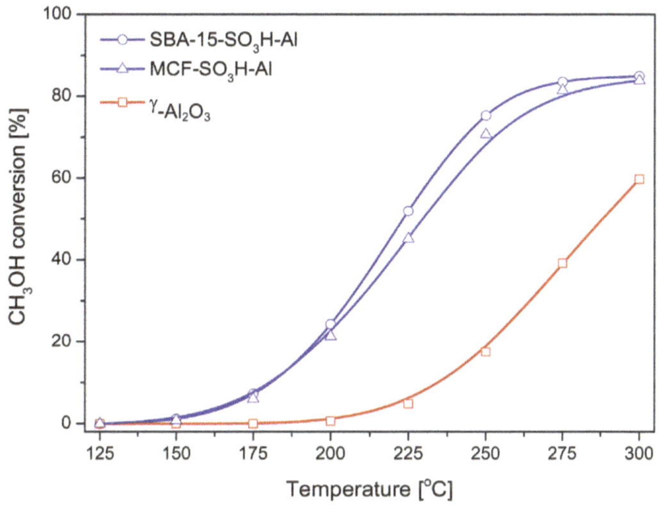

Figure 15 Temperature dependence of CH_3OH conversion to DME over silica supports modified with alumina species and γ-Al_2O_3.[101] *DME*, dimethyl ether.

to $-SO_3H$ by H_2O_2 (c.f. Scheme 4). After calcination, SBA-15 and MCF modified with aluminum Keggin cations were tested as catalysts for the synthesis of DME from methanol. Results of the catalytic tests are presented in Fig. 15. It can be seen that mesoporous silica modified with aluminum species are much more active in the conversion of methanol to DME than γ-Al_2O_3, which is one of the most important commercial catalysts for this process. Such promising catalytic results were obtained due to the uniform distribution of aluminum species on the high−surface area silica supports by the DOMS method.[103] It was shown by the temperature programmed desorption of ammonia (NH_3-TPD) method (Fig. 16) that the number of acidic sites formed by deposition of aluminum species on mesoporous silica is much higher than that on γ-Al_2O_3. Therefore, assuming that catalytic activity is related to the number of acidic sites, the results of the catalytic tests are not surprising.

The above presented procedure of DOMS includes anchoring and oxidation of MPTMS, as it was presented in Scheme 4. However, also, other species can be deposited on the silica surface to generate ion-exchange properties. An example can be anchoring of 2-cyanoethyltriethoxysilane followed by the oxidation of the −CN group to −COOH by a solution of HCl (Scheme 5).[107] Mesoporous silica modified with carboxylic groups was also reported to be able to accommodate various

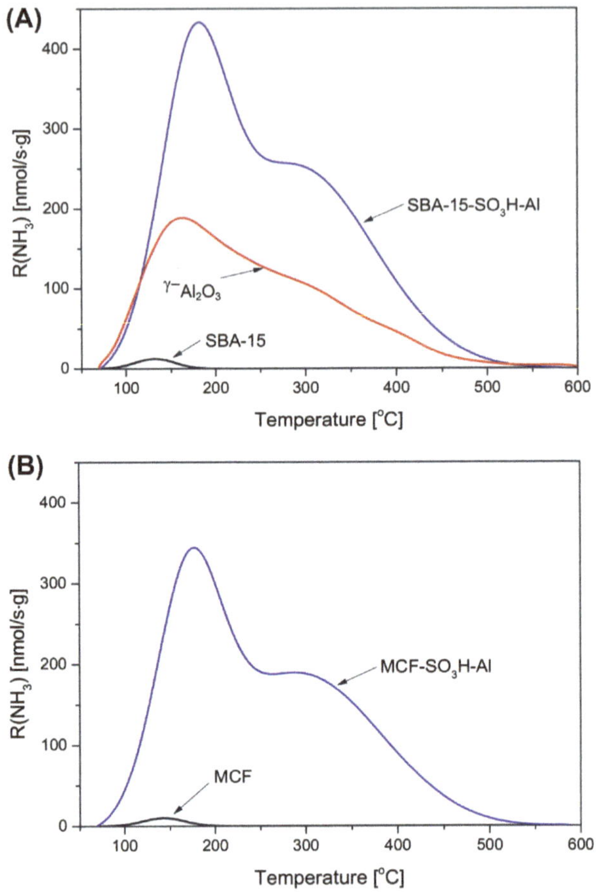

Figure 16 *NH_3-TPD* (temperature programmed desoprtion of ammonia) profiles of the pure silica supports and their modifications with alumina species and γ-Al_2O_3.[101]

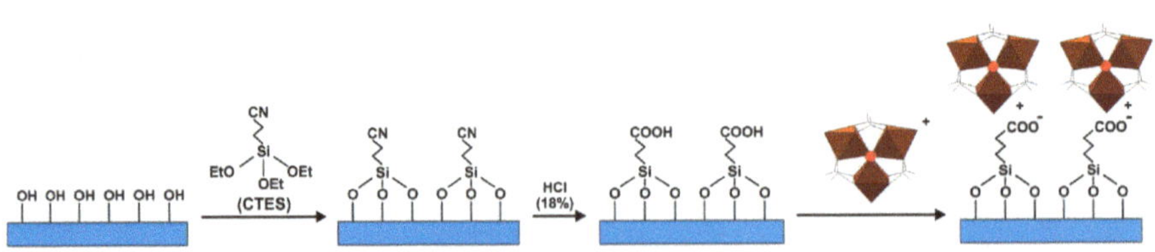

Scheme 5 Procedure used for deposition of oligomeric iron species via deposition and oxidation of 2-cyanoethyltriethoxysilane (CTES) by the DOMS method.[106] *DOMS*, deposition of oligomeric metal species.

oligomeric species including Keggin aluminum, $[Al_{12}(AlO_4)(OH)_{24}(H_2O)_{12}]^{7+}$, and trinuclear iron acetate $[Fe_3O(OAc)_6 \cdot AcOH \cdot 2H_2O]^+$ oligocations.[100]

3.3 Template Ion-Exchange Method

The methods presented previously, viz., MDD and DOMS, result in the deposition of the catalytically active phase in highly dispersed forms. To some extend it is also possible to control the aggregation of the deposited metal oxide species. However, both these methods are rather complex and consist of many steps and, therefore, are relatively expensive. An interesting alternative method is TIE, which is much cheaper and faster in comparison with MDD and DOMS. The TIE method proposed in 1990s[108] is based on extraction of organic templates from synthesized mesoporous silica in polar solvent containing cations. Originally, this method was used for the recovery of organic templates from freshly synthesized molecular sieves.[108] Then, the application of the TIE method was extended to the deposition of metal species into mesoporous silica materials.[109,110] The applicability of TIE for deposition of metal species into the prepared mesoporous silica depends on the nature of the interactions between the inorganic silica network and organic template, which can be coulombic (S^+I^-) and hydrogen bonding ($S^\circ I^\circ$ or $S^+X^-I^+$), where S° and S^+ represent the neutral and cationic surfactants, respectively, I for the inorganic silica network, and X^- for the anionic species.[111,112] Templates bound to the silica framework by hydrogen bonding can be relatively easily extracted using alcohol,[112] whereas for the removal of templates bound to the silica network by coulombic forces, an exchange reagents are required.[108] The number of scientific reports presenting the application of the TIE method for simultaneous removal of organic surfactants and disposition of metal species is limited. Therefore, it seems that the potential of the TIE method is not fully exploited. Scientific reporters have been focused on deposition of iron,[111,113,114] manganese,[109,110,115,116] chromium,[117] cobalt,[118–120] copper,[121,122] nickel,[123,124] calcium,[125] and titanium[126] species into MCM-41. Ethanol was used as a polar solvent for deposition of iron and chromium species, whereas in the case of other metals, water was used as a solvent. The metal loadings into MCM-41 obtained by TIE reported in the literature[109–126] are rather low (typically, no more than few weight percent). The concept of the TIE method is presented in Scheme 6.

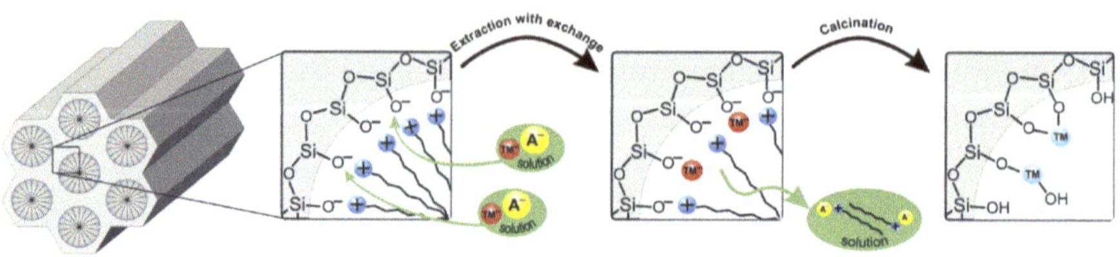

Scheme 6 Graphic presentation of the TIE method concept.[126] *TIE*, template ion-exchange method.

Kowalczyk et al.[101] reported studies on the deposition of monomeric and aggregated iron species into the prepared MCM-41 and Ti-MCM-41 using the TIE method. Methanol solutions of $FeCl_2$ and $[Fe_3(OAc)_6O(H_2O)_3]NO_3$ were used as precursors for the deposited monomeric and aggregated iron species. It was shown that in the case of using a solution of $FeCl_2$, iron was deposited nearly exclusively in the form of highly dispersed species (mainly monomeric iron cations with a small contribution of oligomeric FeO_x species). An increase in iron loading in MCM-41 above 8.3 wt% resulted in an increase in the contribution of aggregated iron oxide species. In the case of Ti-MCM-41, the contribution of aggregated FeO_x species was higher in comparison with that of MCM-41. Deposition of $[Fe_3(OAc)_6O(H_2O)_3]^+$ besides small oligomeric and monomeric iron species also resulted in the formation of Fe_2O_3 nanorods on the outside of the MCM-41 silica grains. This effect is probably related to the limited diffusion of relatively large $[Fe_3(OAc)_6O(H_2O)_3]^+$ oligocations (size of about 1 nm) in the relatively small pores of MCM-41 (diameter of 2.8 nm).

MCM-41 and Ti-MCM-41 modified with iron species by the TIE method were tested as catalysts for the SCR of NO with ammonia (NH_3-SCR, $DeNO_x$) and SCO of ammonia to dinitrogen (NH_3−SCO). As it can be seen from Figs. 17 and 18, the catalysts obtained by the TIE method were found to be active and selective in the high-temperature NH_3-SCR and NH_3-SCO processes. In general, all the studied catalysts were active in the NH_3-SCR process; however, the samples containing well-dispersed iron species presented better catalytic efficiency at higher temperature compared with the catalysts containing iron in the form of Fe_2O_3 crystallites. It was explained by the high catalytic activity of such iron oxide crystallites in direct oxidation of ammonia by oxygen present in the reaction mixture. It is fully supported by the results of the ammonia oxidation studies, in which the samples containing Fe_2O_3 crystallites were significantly more active in comparison to the other studied catalysts. The

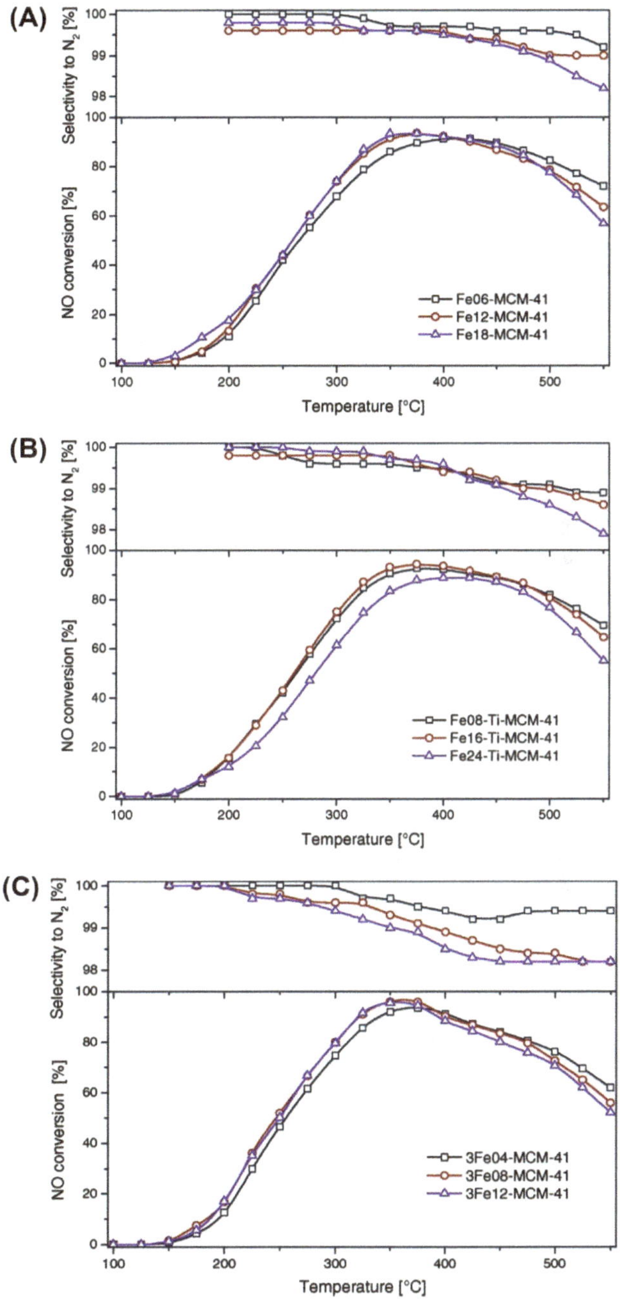

Figure 17 Results of the catalytic tests in the NH_3-SCR process for (A) MCM-41 and (B) Ti-MCM-41 modified with monomeric iron cations and (C) MCM-41 modified with oligomeric iron species.[99] *SCR*, selective catalytic reduction.

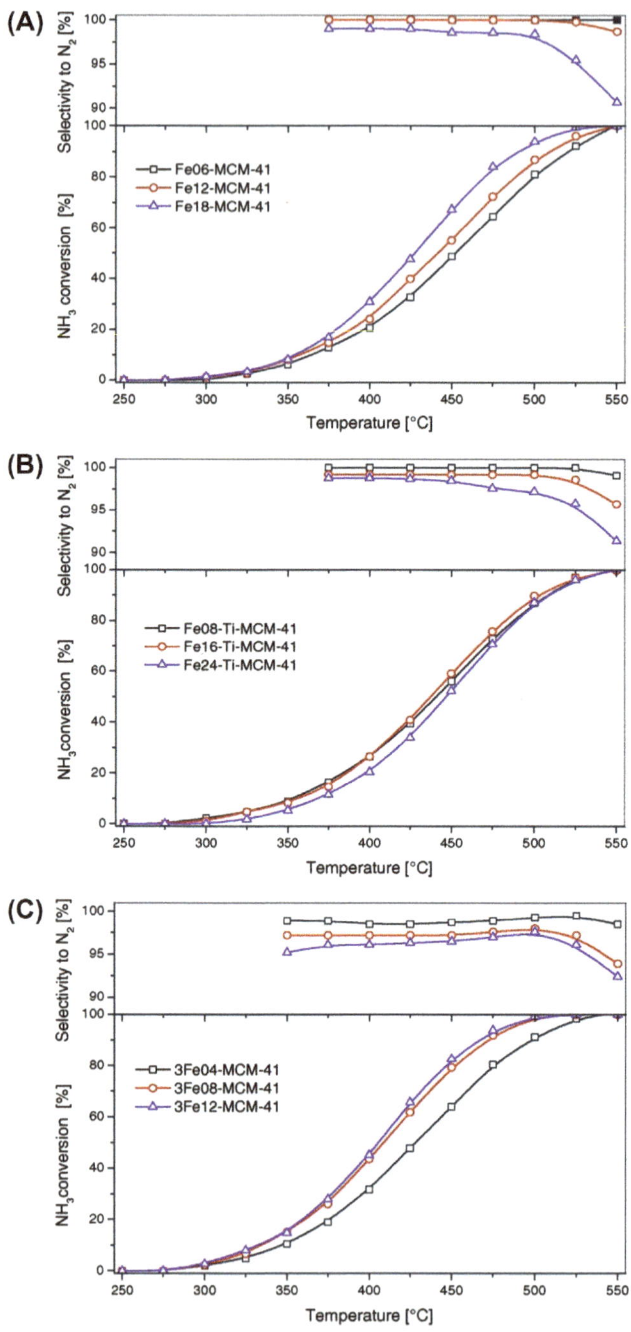

Figure 18 Results of the catalytic tests in the NH_3-SCO process for (A) MCM-41 and (B) Ti-MCM-41 modified with monomeric iron cations and (C) MCM-41 modified with oligomeric iron species.[99] *SCO*, selective catalytic oxidation.

silica-based catalysts of the NH_3-SCO process showed a very high selectivity to dinitrogen at temperatures as high as 500–550°C and, therefore, seemed to be very promising for conversion of NH_3 in hot waste gases.

The TIE method was used not only for the deposition of a single metallic component on MCM-41 but also for the synthesis of bimetallic catalysts based on mesoporous silica of the MCM-41 type.[127] In this case, copper, iron, and manganese were deposited on MCM-41 using methanol solutions of $CuCl_2$, $FeCl_2$, and $MnCl_2$, respectively. It was shown that introduction of single transition metals into MCM-41 by the TIE method resulted in deposition of iron and manganese in the form of highly dispersed species (mainly monomeric cations), whereas in the case of copper, CuO crystallites were formed. Such aggregated copper oxide species were not detected in the bimetallic samples containing copper and iron, as well as copper and manganese. MCM-41 modified with transition metals by the TIE method was found to be active catalysts for the NH_3-SCO process; however, catalytic performance of the studied samples strongly depended on the type of deposited metal (Fig. 19). The samples modified with iron and manganese effectively catalyzed ammonia oxidation to dinitrogen only in the high-temperature range. On the other side, MCM-41 modified with copper was catalytically active at significantly lower temperatures; however, selectivity to dinitrogen decreased at elevated temperatures. It was shown that simultaneous incorporation of copper and iron, as well as copper and manganese, resulted in catalysts for the low-temperature NH_3-SCO process with significantly increased selectivity to dinitrogen in comparison with the monometallic copper-containing catalyst. This interesting effect was explained by synergistic effects between transition metals deposited on the surface of MCM-41.

In conclusion, the TIE method is effective and relatively cheap for deposition of catalytically active components in the form of highly dispersed metal species. In the case of introducing aggregated oligomeric metal oxide species, this method is less effective due to limited mobility of large metal oligocations in relatively small pores of MCM-41. Another drawback of the TIE method is their applicability only for mesoporous silica synthesized with the use of cationic surfactants.

4. CATIONIC LAYERED CLAY MINERALS

Minerals are very important raw materials for the preparation of catalysts. This is related to a great variety of minerals with different chemical

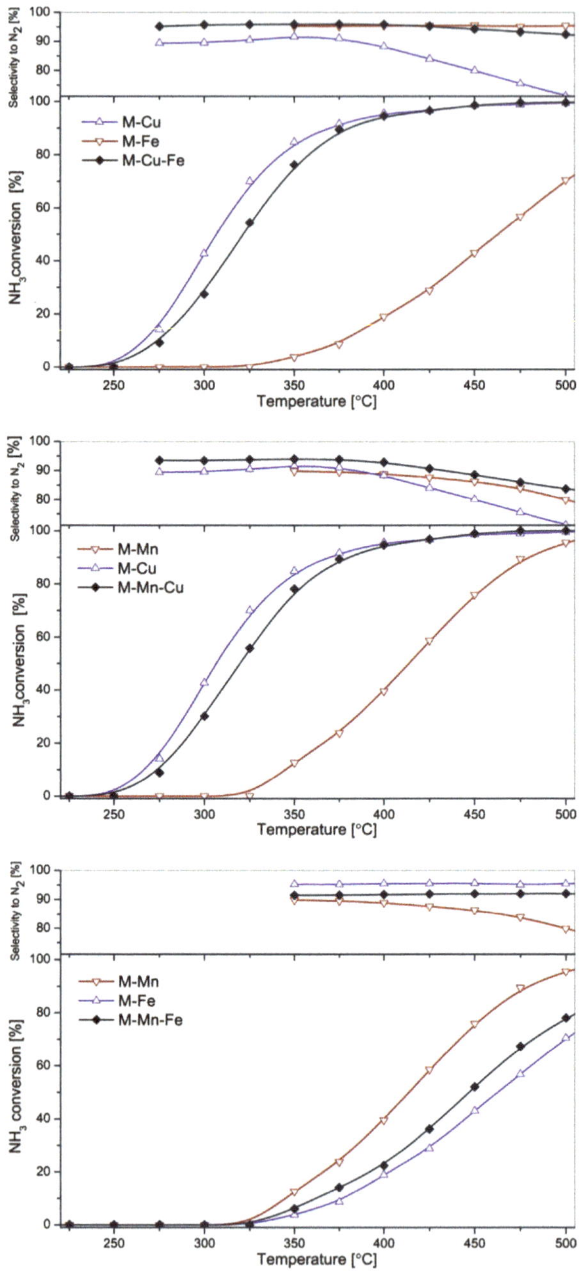

Figure 19 Results of catalytic tests of MCM-41 modified with transition metals (monometallic and bimetallic samples) in the process of selective catalytic oxidation of ammonia to dinitrogen.[126]

composition, structure, porous nature, and morphology, and therefore, such minerals can be tailored for specific catalytic applications. Another very important issue is high availability of minerals and also their rather low cost. On the other hand, the major drawback of minerals is the presence of different types of impurities (e.g., coexistence of other minerals) that can influence their catalytic properties.

Cationic layered clay minerals belong to the most intensively studied minerals that can be used as raw materials for catalyst preparation.[28–33] It is related to their unique structure, which can be easily modified to adjust their properties to the specific catalytic needs. There is a great variety of clay minerals with different layered structures, containing various intralayer and interlayer chemical elements.

Fig. 20 presents the structure of montmorillonite, which is probably the most intensively studied cationic clay mineral. Each montmorillonite layer consists of three sublayers. The central octahedral sublayer is sandwiched between two tetrahedral sublayers. Such structure is called 2:1 (two tetrahedral sheets and one octahedral sheet) or T-O-T (tetrahedral-octahedral-tetrahedral sheets). Si^{4+} cations are located in the tetrahedral sublayer, whereas Al^{3+} cations are in octahedral sublayers. In montmorillonite, all tetrahedral positions are occupied by Si^{4+} cations, whereas only two-thirds of the octahedral positions are occupied by Al^{3+} cations with vacancies in one-third of the octahedral positions. Clay minerals with such layered structure are classified as dioctahedral (two octahedral positions occupied and one vacancy). Moreover, in montmorillonite, a part of the trivalent Al^{3+} cations in the octahedral sheets is replaced by divalent Mg^{2+} ions. Such substitution results in negative charging of the montmorillonite layers. This negative charge is compensated by common cations (e.g., Na^+, K^+, Mg^{2+}), which together with water molecules are located in the interlayer space of the clay mineral. What is very important, these interlayer cations can be relatively easily exchanged for other cations; thus montmorillonite exhibits ion-exchange properties, which gives a great possibility for various modifications.

Probably, the most important method for functionalization of layered clay minerals for catalysis is their intercalation with inorganic pillars. In general, this method is based on deposition of bulky inorganic species into the interlayer space of clay, resulting in an increase of interlayer distance and, therefore, opening up the interlayer space for reacting molecules. There are two main procedures for the pillaring of clay minerals. The first one, resulting in materials called PILCs, is based on intercalation of bulky metal

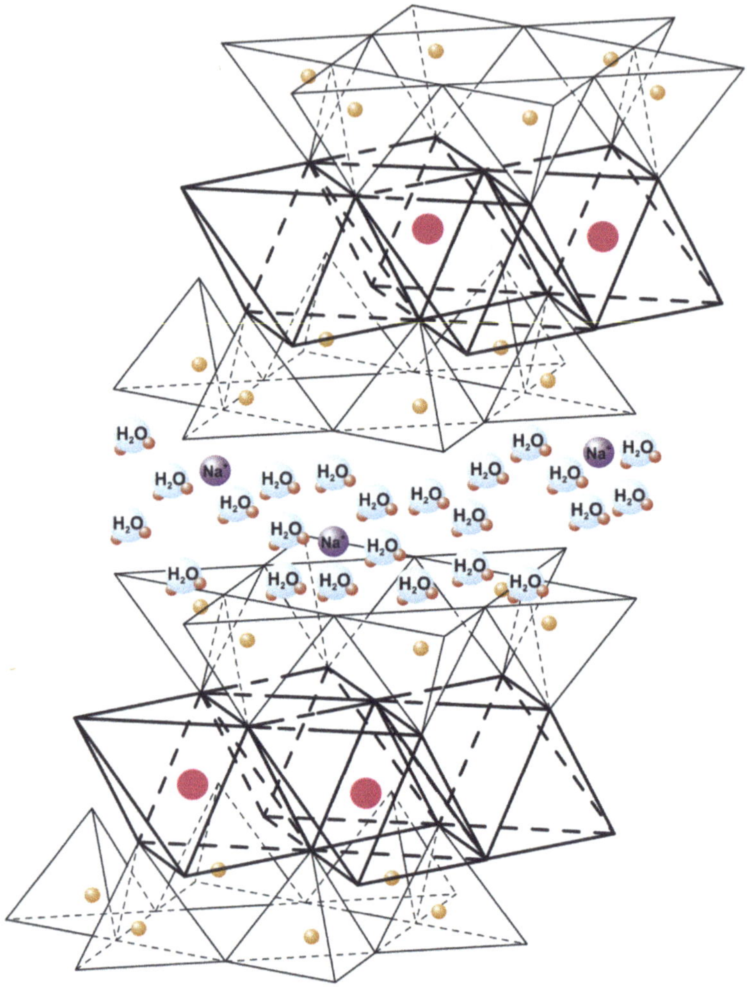

Figure 20 Model structure of montmorillonite.

oligocations into the interlayer space of the clay mineral by ion-exchange procedure, followed by thermal decomposition of oligocations, resulting in stable metal oxide pillars that permanently open the layered structure of the clay mineral (Scheme 7). The second procedure, resulting in materials called *porous clay heterostructures* (PCHs), is based on deposition of cationic alkyl ammonium surfactants and neutral alkylamine cosurfactants, which in the interlayer space of clay form micelles. In the next step, silica pillars are created by in situ polymerization of the silica source around such micelles

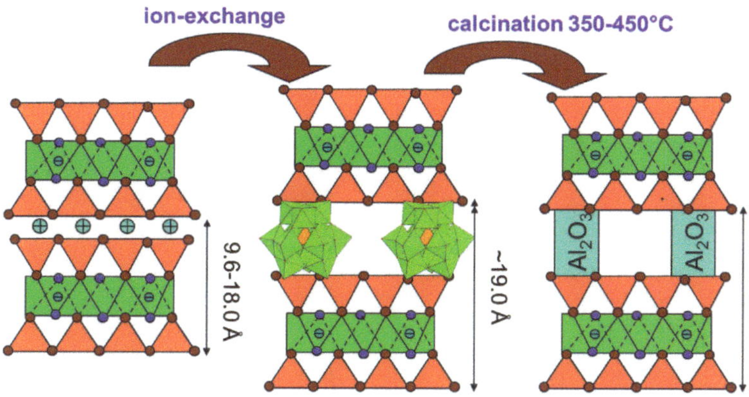

Scheme 7 Intercalation of cationic layered clay mineral with alumina pillars. Aluminum Kegging oligocations, $[Al_{12}(AlO_4)(OH)_{24}(H_2O)_{12}]^{7+}$, are used as precursors of Al_2O_3 pillars.

in the interlayer space of clay. Finally, organic surfactants and cosurfactants should be removed from the prepared PCHs by calcination or extraction with suitable solvents (Scheme 8).

4.1 Pillared Interlayered Clays

The procedure used for the intercalation of cationic layered clay mineral, presented in Scheme 7, results in deposition of alumina pillars into its interlayer space. In the first step, interlayer cations present in the parent clay mineral (or its sodium form) are exchanged for metal polyoxocations by ion-exchange method. Thus the driving force necessary for effective deposition of oligocations into the interlayer space of clay is its cation-exchange capacity (CEC). The problem of CEC in various clay minerals is discussed in many specialized reports.[28,128] Apart from aluminum oligocations,[27,29] other polyoxocations (e.g., zirconium[25,26] or titanium[27,29]) can also be used as precursors for interlayer pillars. Intercalation of cationic layered minerals with metal oxide pillars (e.g., Al_2O_3, TiO_2, ZrO_2) results in materials characterized by a relatively high surface area (150–350 m^2/g), developed microporous structure (slit-like interlayer pores), as well as ion-exchange properties and surface acidity. Ion-exchange properties are very important because they can be used for deposition of metal cations into PILCs to generate catalytic redox sites. On the other hand, the surface acidic sites are important in many catalytic processes, including cracking of hydrocarbons[129] or synthesis of ethers from alcohols.[130] The drawback of PILCs obtained from montmorillonite is their limited stability of the

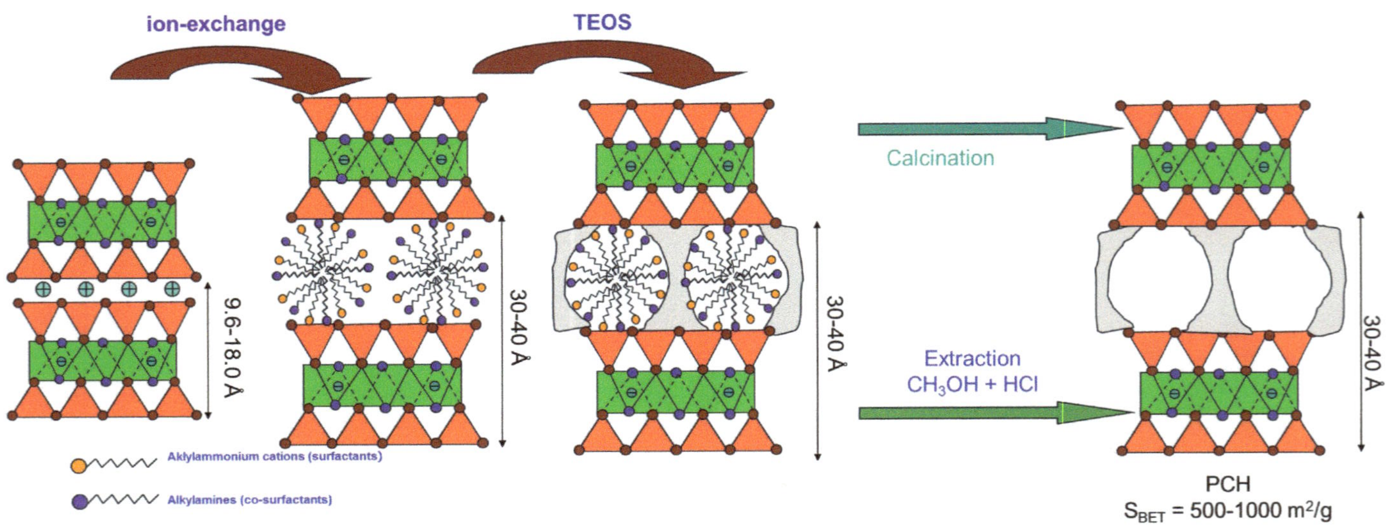

Scheme 8 Synthesis of PCH materials. Tetraethyl orthosilicate (TEOS) was used as a precursor of interlayer silica pillars. *PCH*, porous clay heterostructure.

pillared structure, which can collapse at temperatures above 450°C. Therefore, such pillared clays can be used only for low- and medium-temperature processes. More stable PILCs can be obtained by intercalation of vermiculite[131] and phlogopite[132] with alumina pillars, but in these cases, the pillaring procedures are more complex and intercalation is less effective. Because PILCs has been intensively studied since 1970s, there are many scientific reports presenting various aspects of these interesting materials, among them are some review papers.[28,133,134]

PILCs and their modifications were tested as catalysts of various chemical processes, including environmental processes. One of them is NH_3-SCR. Chmielarz et al.[25,26,135] reported a high catalytic activity of montmorillonite-based PILCs intercalated with ZrO_2 pillars and doped with manganese and carbon deposits (Fig. 21). It can be seen that the initial acid treatment of clay (A2) decreased its catalytic activity in the NH_3-SCR process. It is possibly related to the partial leaching of Al^{3+} cations and, therefore, decreasing the number of surface acidic sites responsible for chemisorption and activation of ammonia molecules. Intercalation of montmorillonite with ZrO_2 resulted in a slight activation in the NH_3-SCR process; however, still the sample obtained from the acid-treated clay (M2) was less active in comparison with the catalyst obtained from the sodium form of montmorillonite (M1). Modifications of PILCs (M1 and M2) with carbon increased their catalytic activity and improved selectivity to N_2; however, in this case the catalyst obtained from the acid-treated clay (M2C) was significantly more active in comparison with the sample obtained from the sodium form of montmorillonite. It is possibly related to the lower loading of carbon deposits inside the porous system of M2C, which in contradiction to M1C, only slightly decreased the diffusion of the reactants inside the pores. A very significant activation of the catalysts was observed after introduction of manganese. The best catalytic results were obtained for the PILC sample obtained from acid-treated clay modified with carbon deposits and manganese. It was suggested that the presence of a small amount of carbon aggregates in the pore system of PILCs resulted in deposition of highly dispersed (small clusters and/or individual cations) manganese species, active in the process of selective reduction of NO to dinitrogen. Detailed studies on the catalytic stability of the M2C and M2CMn14 samples in the reaction mixtures containing optional water vapor or SO_2, which are typical components of flue gases emitted by electric power stations powered by coal, showed their relatively high resistance to these poisons.[135]

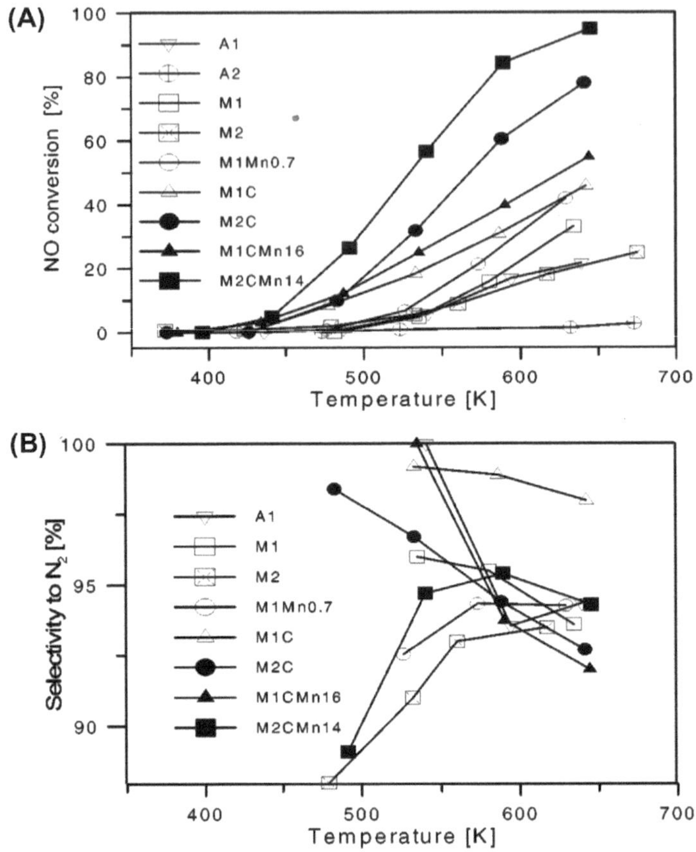

Figure 21 Results of NH_3-SCR tests over PILC-based catalysts: A1-sodium form of montmorillonite, A2 (montmorillonite treated with a solution of HCl [20%]), M1 and M2 (PILCs intercalated with ZrO_2 obtained from A1 and A2, respectively), M1C and M2C (PILCs modified with carbon deposit), M1Mn*X* and M2Mn*X* (PILCs modified with manganese; *X* denotes the amount of introduced manganese, wt%).[25] *PILC*, pillared interlayered clay; *SCR*, selective catalytic reduction.

Very promising catalytic properties in the NH_3-SCR process were reported for montmorillonite intercalated with titania pillars and additionally modified with highly dispersed copper species (introduced by the ion-exchange method).[27,136] In this case, the procedure applied for intercalation of the TiO_2 pillars into montmorillonite resulted in the formation of a delaminated pillared structure (nonparallel ordering of clay layers, "house of card" structure). For comparison, a very similar procedure used for deposition of Al_2O_3 pillars resulted in an intercalated clay composed of parallel layers. Delaminated pillared structure, which is more chaotic, contains

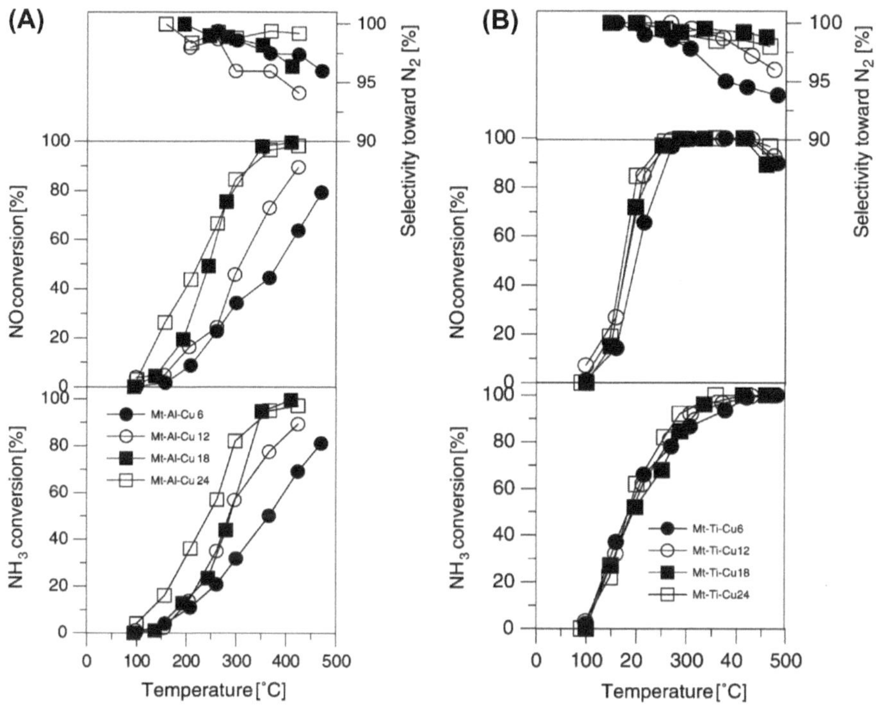

Figure 22 Results of catalytic tests in NH3-SCR process for montmorillonites pillared with alumina (A) or titania (B) modified with copper introduced by the ion-exchange method.[27] *SCR*, selective catalytic reduction.

both large pores (macropores and mesopores) suitable for fast internal diffusion of reactants and micropores with specific sites for their conversion. Thus Ti-PILCs belong to the category of catalysts with the hierarchical porous structure. Results of catalytic tests presented in Fig. 22 show that the conversion of NO over Ti-PILCs modified with copper occurred in a broad temperature range with a very high selectivity to N_2. The content of copper introduced into Ti-PILCs only slightly influenced the performance of the catalysts. A decrease in NO conversion observed at temperatures above 400°C is a result of the side process of direct oxidation of ammonia by oxygen present in the reaction mixture. Such promising catalytic results were an inspiration for extending the studies on Ti-PILCs, modified with dispersed iron species deposited by the ion-exchange method.[29] It was reported that iron species introduced to delaminated Ti-PILCs are slightly less catalytically active in NH_3-SCR than copper but still more active than the majority of catalysts of this process reported in the literature.

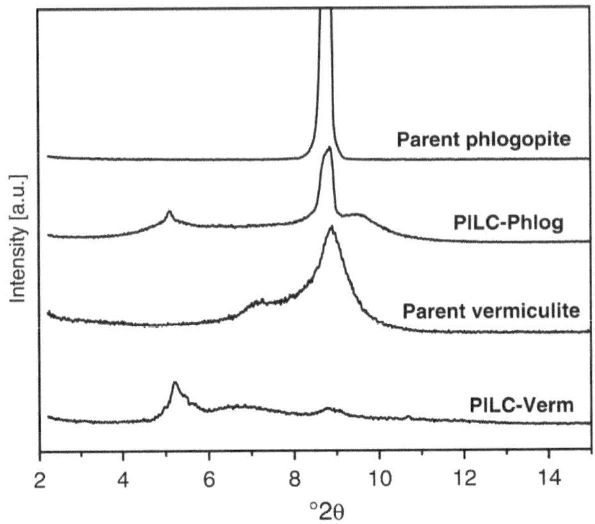

Figure 23 XRD diffractograms recorded for the parent clays and their pillared forms.[136] *PILC*, pillared interlayered clay.

As mentioned previously, the pillared structure of Al-PILCs obtained from phlogopite and vermiculite is more stable compared to PILCs based on montmorillonite.[132,133] Such Al-PILCs with extended thermal stability were tested as catalytic supports for the process of selective oxidation of ammonia to dinitrogen (NH_3-SCO).[137] First of all, it was shown that intercalation of phlogopite and vermiculite with Al_2O_3 pillars is less effective compared with intercalation of montmorillonite. As it can be seen in Fig. 23, the reflection characteristic of nonpillared minerals at 2θ about 8.8° are still present in the diffractograms recorded for intercalated clays, especially for PILC-Phlog. The reflection characteristic of intercalated clays, at 2θ about 5.1−5.2 degrees, shows that only part of the clay layers was successfully pillared (increase in basal spacing from 10 to about 17.5−17.6 Å, taking into account the thickness of clay layer of about 9.6 Å, the interlayer distance increased from about 0.4 to 7.9−8.0 Å). Al-PILCs obtained from phlogopite and vermiculite and doped with highly dispersed copper or iron species (introduced by ion-exchange method) were found to be active and selective catalysts for high-temperature oxidation of ammonia to N_2 (Fig. 24). Intercalated vermiculite modified with copper was operated at temperatures lower than other catalysts. Moreover, the conversion of ammonia over this sample proceeded nearly exclusively to N_2, what in the case of high-temperature ammonia oxidation is rather a unique

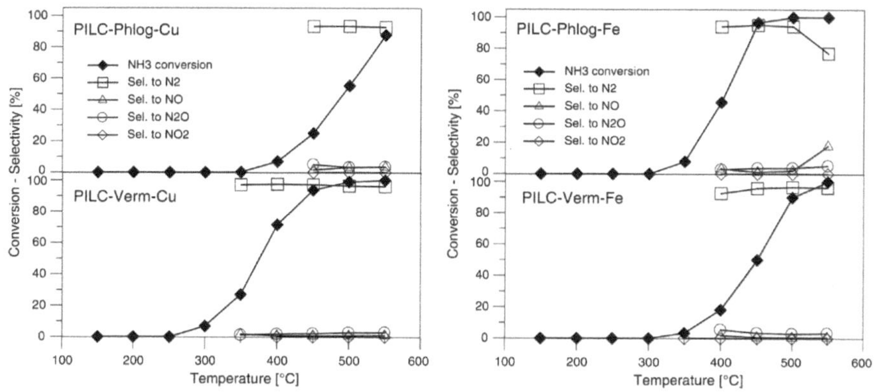

Figure 24 Results of oxidation of ammonia in the presence of copper and iron-modified clay–based catalysts.[136]

feature. Such promising catalytic properties of the PILC-Verm-Cu sample are probably related to its relatively high specific surface area and porosity, high density of surface acidic sites necessary for ammonia chemisorption and activation, as well as specific properties of highly dispersed copper species in the selective oxidation of ammonia to dinitrogen.

4.2 Porous Clay Heterostructures

The synthesis of PCHs is based on the formation of the stable MCM-like porous silica structure between layers of the clay mineral (Scheme 8). PCH materials are characterized by the combined microporous and mesoporous structure, high surface area (up to 1000 m^2/g), and relatively high thermal and hydrothermal stability. In addition, PCH materials exhibit surface acidity and ion-exchange properties, which depend mainly on the type of the parent clay used for their synthesis and additional components incorporated into silica pillars. Various features of cationic layered clay minerals used for the synthesis of PCHs influence the nature and quality of the final product; however, the most important seems to be the ability for exchange of interlayer cations and layer charge density of host clays, as well as their plate dimension. Exchange of interlayer inorganic cations, present in host clay, for alkyl ammonium cations, is a key step necessary for the formation of PCHs. Thus, only clay minerals exhibiting ion-exchange properties can be used for the synthesis of PCH materials. As it was reported by Galarneau et al.,[30] clay minerals with various layer charge density, including rectorite, fluorohectorite, magadiite, and vermiculite, can be successfully converted into PCHs. Moreover, it was shown that the interlayer pore sizes were invariant of the type of host clay. The size of layers (plate dimension) in the parent clay is an important feature that influences the porous structure of the obtained PCHs. Benjelloun et al.[138] reported that for large-sized host montmorillonite, with a plate dimension of 1000 nm, the formation of micellar templates occurs mainly in the interlayer space, and the face-to-face orientation (parallel ordering of clay layers) is favored. On the other hand, for small-size synthetic saponite with a plate dimension of 50 nm, the stacking of clay layers surrounding the micelles does not occur in parallel due to the favored edge-to-edge and edge-to-face orientations (delaminated structure).

As mentioned previously, the ion-exchange properties of the catalytic supports are very important for deposition of catalytically active metallic species in highly dispersed form. In the case of PCH materials, ion-exchange properties are determined by CEC of the parent clay used for their synthesis.[138,139] Similarly, the surface acidic properties, which are very important in many catalytic processes, in PCHs are also mainly related to the surface acidity of the host clay used. However, surface acidity of clay minerals can to some extent be modified by their acid treatment. Chmielarz et al.[140,141] reported that treatment of vermiculite and phlogopite with nitric acid solution increased not only their surface area but also surface acidity; however, an increase in the duration of acid treatment from 2 to 24 h resulted in a decrease in surface acidity of vermiculite and phlogopite. This effect was explained by gradual leaching of Al^{3+} cations from the clay layers responsible for the presence of acidic sites. The successful synthesis of PCHs from such acid-treated vermiculite was reported.[142] Another strategy for the generation of acidic sites in PCHs is based on incorporation of "acidic" elements into interlayer silica pillars. Successful incorporation of Ti^{4+},[31,130,143] Al^{3+},[31,130] and Zr^{4+}[130] in the form of monomeric cations or small aggregated species into the structure of interlayer silica pillars was reported.

PCHs were reported to be effective catalysts and catalytic supports for various chemical reactions, including environmentally important processes. Chmielarz et al.[144] reported the synthesis of silica-intercalated PCHs from montmorillonite, followed by treatment of the freshly calcined solid product with gaseous ammonia. In calcined PCHs, the negative charge of the clay layers is compensated by protons formed during thermal decomposition of organic templates. Protons can migrate into vacant cationic sites in the clay layers (octahedral vacancies in dioctahedral clays or/and defects in trioctahedral clays) and in consequence reduce the CEC of PCHs.[138] It was shown that this effect can be limited by conversion of protons for NH_4^+ cations by treatment of the freshly calcined PCHs with ammonia ($H^+NH_3 = NH_4^+$). NH_4^+ cations are too large to migrate into the clay layers, and therefore, the CEC potential of the sample is protected. The ion-exchange method using aqueous solutions of copper and iron nitrates resulted in a higher loading of copper introduced into the ammonia-treated PCH (1.43 wt%) compared with its protonated form (0.59 wt%). This effect was not observed for the Fe-containing samples (1.41 and 1.43 wt% for NH_4^+ -PCH and H^+-PCH, respectively). In the copper-containing catalysts, monomeric Cu^{2+} cations and oligonuclear $(CuO)_n$ species dominated, whereas bulky CuO clusters were present in a significantly lower quantity. For the Fe-containing samples, the transition metal was found in the form of monomeric Fe^{3+} cations and oligonuclear $(FeO)_n$ species. The catalysts based on the PCH supports have been found to be active, selective, and stable for the NH_3-SCR process (Fig. 25). The Cu-containing catalysts are more effective at lower temperatures than PCHs modified with iron. Moreover, the Cu-containing catalyst obtained from the NH^{4+} form of PCH (PCH-NH3-Cu) was significantly more active compared with the catalyst obtained from the protonated form of PCH (PCH-Cu). At higher temperatures, the NO conversion decreased due to the direct oxidation of ammonia by oxygen present in the reaction mixture. High selectivity of

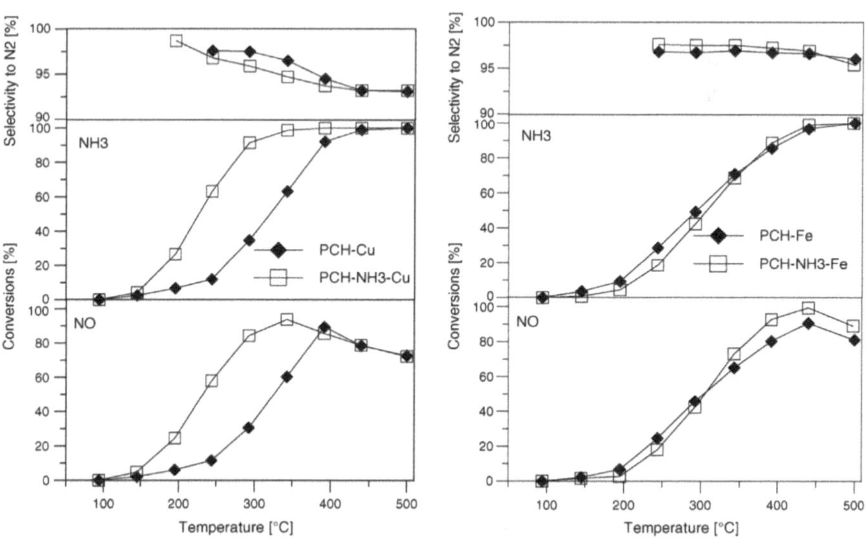

Figure 25 Results of activity tests in the NH_3-SCR process for PCH-based catalysts.[143] *PCH*, porous clay heterostructure; *SCR*, selective catalytic reduction.

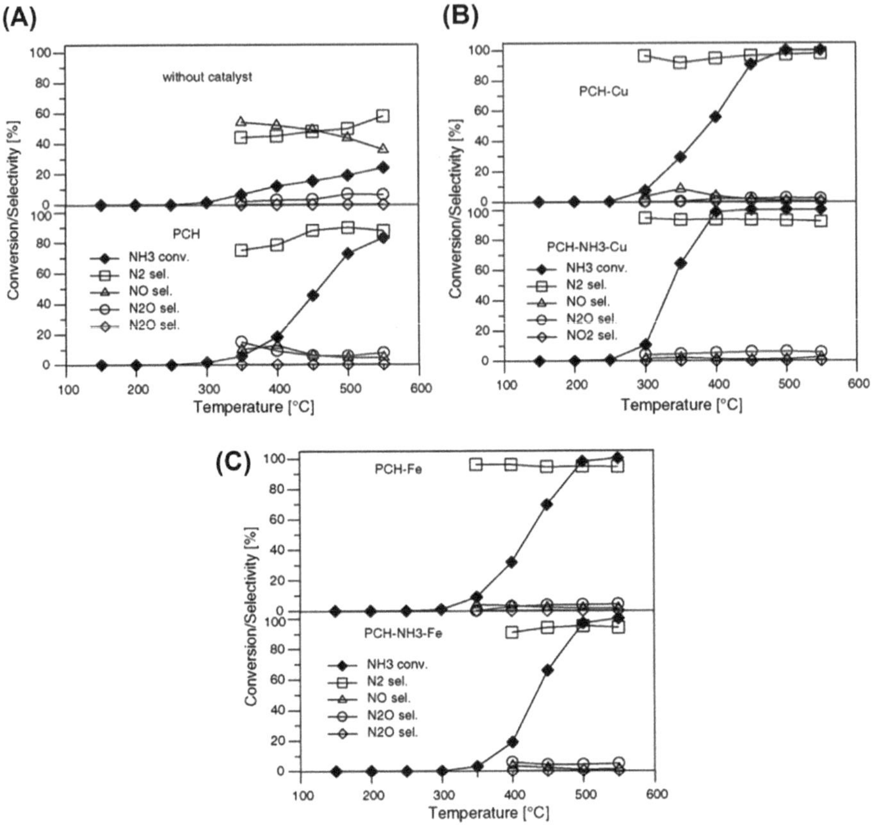

Figure 26 Results of activity tests in the process of selective oxidation of ammonia to dinitrogen for PCH-based catalysts.[144] *PCH*, porous clay heterostructure.

the NH_3-SCR process for dinitrogen in the high-temperature range (Fig. 25) suggests that ammonia is oxidized mainly to N_2. It was proven by additional experiments, in which PCHs modified with copper and iron were tested as catalysts for the process of the selective oxidation of ammonia.[145] The results of these studies presented in Fig. 26 show that among the studied samples, the best catalytic results were obtained for ammonia-treated PCHs modified with copper (PCH-NH3-Cu). Taking into account the very high selectivity to dinitrogen, obtained in the presence of this sample at elevated temperatures, it seems to be a promising candidate as a commercial catalyst for selective oxidation of ammonia in hot flue gases.

As mentioned previously, the type of clay mineral used for the synthesis of PCHs determines their properties, such as surface acidity, ion-exchange

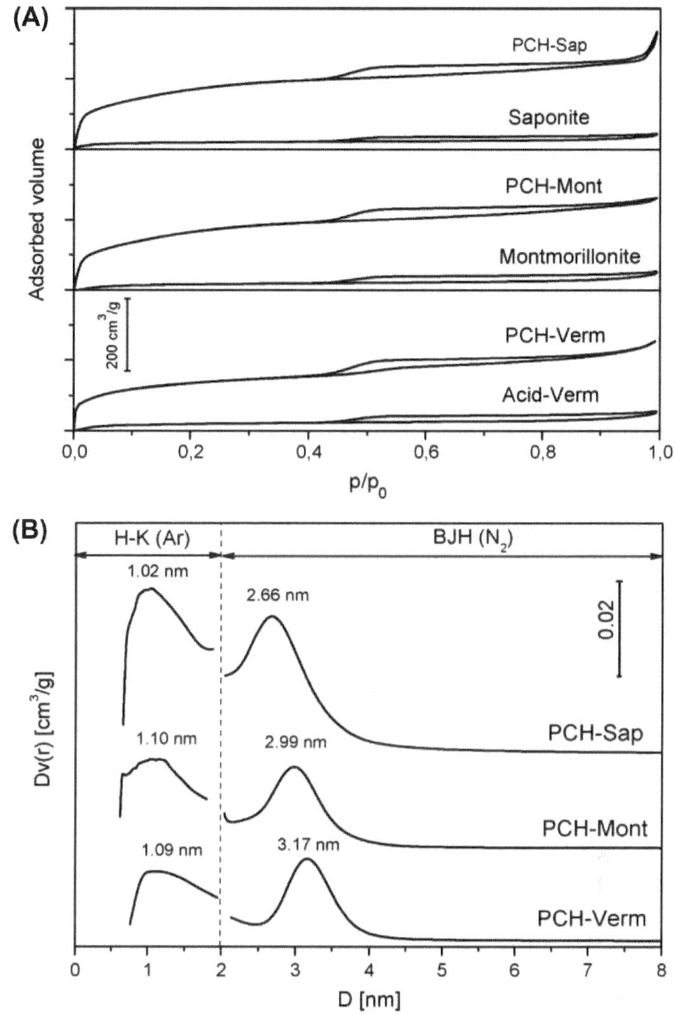

Figure 27 Nitrogen adsorption–desorption isotherms (A) as well as H–K(Ar) and BJH (N_2) pore size distribution (B) of the PCH materials.[145] *PCH*, porous clay heterostructure.

capacity, or the nature of the porous structure. Chmielarz et al.[146] compared PCHs obtained from montmorillonite, acid-treated vermiculite, and synthetic saponite as supports of the catalyst for the NH_3-SCR process. As it can be seen from Fig. 27, intercalation of layered clay minerals resulted in the formation of materials with uniform microporous–mesoporous structures. The PCH obtained from saponite is characterized by the presence of mesopores (2.66 nm) smaller than for materials produced from montmorillonite (2.99 nm) and vermiculite (3.17 nm). On the other hand, the Brunauer–Emmett–Teller (BET)-specific surface areas determined for PCH-Sap, PCH-Mont, and PCH-Verm were 906, 816, and 612 m^2/g, respectively. Differences in the textural parameters of PCHs depend on the properties of the parent clay minerals, including CEC and size of layers (plate dimension). PCHs and their modifications with copper and iron were tested as catalysts for the NH_3-SCR process (Fig. 28). Among PCHs not modified with transition metals, the best catalytic results with respect to high NO conversion and selectivity to N_2 were obtained for the sample obtained from acid-treated vermiculite (PCH-Verm). This could be explained by the relatively high content of iron (13.7 wt% in reference to Fe_2O_3), which is a natural component of vermiculite and is known to be catalytically active in the NH_3-SCR process. For comparison, the iron content (as Fe_2O_3) in parent montmorillonite and synthetic saponite was 3.0 and 0.0 wt%, respectively. Introduction of copper or iron significantly increased the catalytic activity of PCHs. In both series of the samples modified with iron and copper, the best catalytic results were obtained for the catalysts based on Sap-PCHs. It is possibly related to higher specific surface area and porosity of this support in comparison with Verm-PCH and Mont-PCH, and the nature of deposited iron and copper species. It was found that in the case of the most active catalysts, there were significant contributions of iron in the form of small aggregated metal oxide species (PCH-Sap-Fe) and monomeric Cu^{2+} cations (PCH-Sap-Cu), which are known to be active in the low-temperature NH_3-SCR process. In Fig. 28, for comparison, results of the catalytic tests for Fe- and Cu-modified γ-Al_2O_3 and SBA-15 are shown. It can be seen that these reference catalysts are not only efficient in the NH_3-SCR process but also very active in the side process of direct ammonia oxidation by oxygen present in the reaction mixture. Therefore, the overall catalytic performance of PCH-based catalysts is much better than the reference catalysts. It clearly shows a very important role of the catalytic supports in designing efficient catalysts for the NH_3-SCR process. It should be added that the studied PCH-based catalysts were stable in the tests performed in the presence of SO_2 and water vapor, which are typical components of flue gases produced during fossil fuel combustion.

Very promising results for the NH_3-SCR process were obtained in the presence of PCHs intercalated with silica–titania pillars and doped with copper and iron.[31,33] It was shown that incorporation of titanium into silica pillars significantly increased the surface acidity of PCHs by generation of

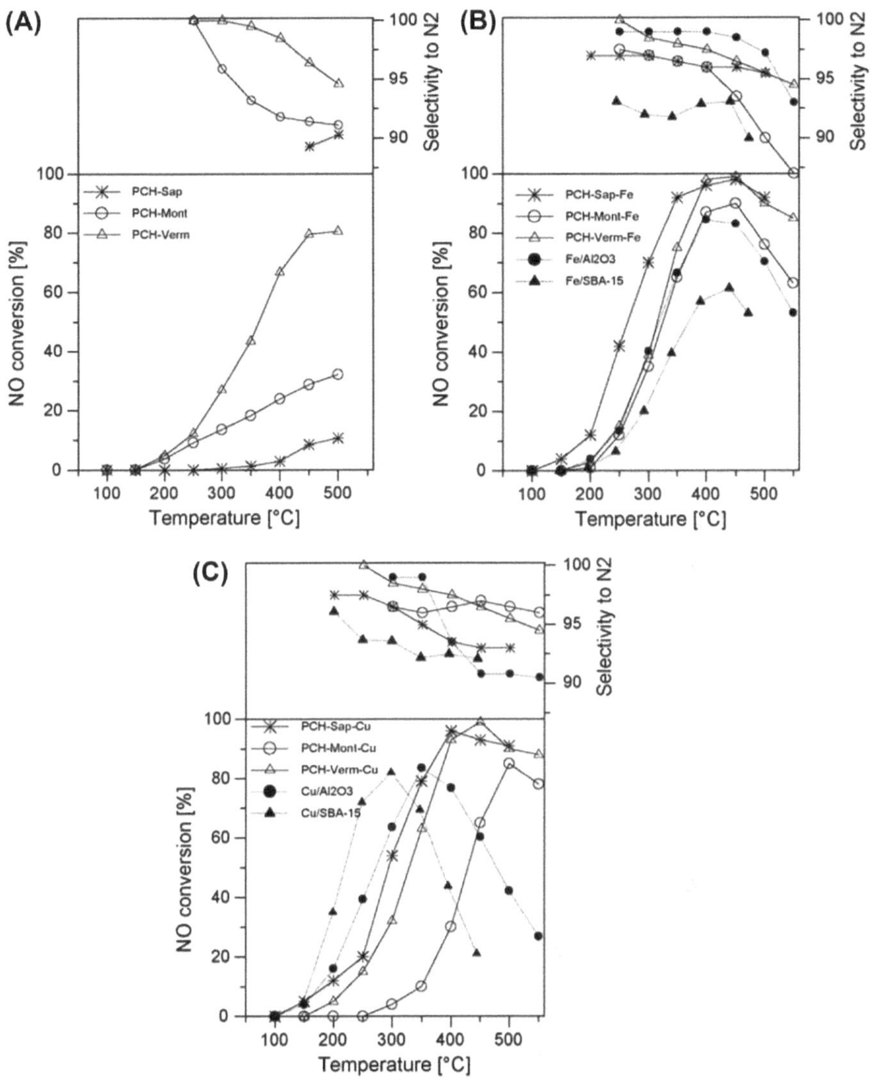

Figure 28 Results of catalytic tests in the NH_3-SCR process for the PCH supports (A) and samples modified with iron (B) and copper (C).[145] *SCR*, selective catalytic reduction.

both Lewis and Brønsted acidic sites. Titanium introduced into the PCH samples also influenced the type of transition metal surface species formed during deposition of copper and iron. Transition metals deposited on the surface of the PCH supports were present in the form of monomeric cations and small oligonuclear metal oxide clusters. Silica titania–pillared PCHs modified with transition metals were significantly more active than the

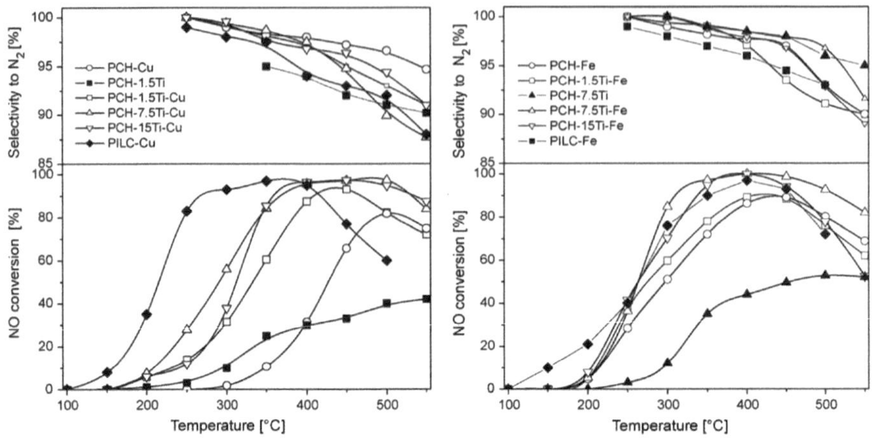

Figure 29 Results of catalytic tests for PCH intercalated with silica and silica–titania pillars modified with copper and iron.[33] *PCH*, porous clay heterostructure; *PILC*, pillared interlayered clay.

catalysts based on PCHs intercalated with pure silica pillars (Fig. 29). The side process of ammonia oxidation, especially in the presence of the Cu-doped catalysts with high loading of titanium (PCH-7.5Ti-Cu and PCH-15Ti-Cu), was observed at temperatures as high as 500°C. Therefore, these materials seem to be very promising candidates as catalysts of the high-temperature NH_3-SCR process, in contradiction to PILC-Cu (alumina-intercalated montmorillonite doped with copper) active in the low-temperature NH_3-SCR process (c.f. Fig. 29). In addition, the catalysts based on PCHs intercalated with silica and silica–titania were reported to be only slightly deactivated in the presence of water vapor, as well as SO_2.

PCHs intercalated with the multicomponent silica–titania, silica–alumina, or silica–zirconia pillars were also studied as catalysts for the synthesis of DME and diethyl ether (DEE) from methanol and ethanol, respectively.[130] Moreover, the PCH samples intercalated with the pure silica pillars were doped with aluminum using the TIE method. Dehydration of ethanol results in ethylene, which is one of the main feed stocks for the petrochemical industry and DEE. DEE has excellent properties as transportation fuel, with respect to a cetane number above 125.[147] On the other hand, blending of DEE with ethanol was reported to improve the cold start problem in ethanol fueled cars.[148]

In the case of all the studied PCH-based samples, the deposited metal (Al, Ti, and Zr) species were present in highly dispersed forms and resulted

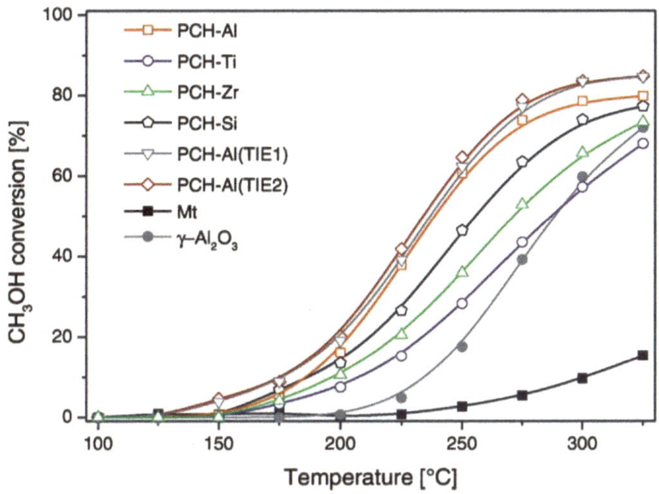

Figure 30 Results of catalytic tests in the process of methanol to dimethyl ether dehydration over the PCH-based catalysts.[130] *PCH*, porous clay heterostructure.

in generation of acidic sites catalytically active in dehydration of methanol and ethanol.[130] As it can be seen from Figs. 30 and 31, the best catalytic results in both the studied processes were obtained for PCHs doped with aluminum, especially introduced by using the TIE method, which were more catalytically active in the process of methanol to DME conversion than γ-Al_2O_3 (one of the most important commercial catalysts of this process). It was suggested that the high catalytic activity of the aluminum-doped PCH samples is related to the nature and strength of the acidic sites active in the process of alcohol dehydration.

In conclusion, layered clay minerals are still very promising materials for catalysis. Intercalation of such minerals with inorganic pillars results in thermally stable, high—surface area porous materials, which still exhibit surface acidity and ion-exchange properties and, therefore, can be tailored for specific needs of acid and redox catalyzed reactions.

5. CONCLUSIONS

Heterogeneous catalysis is a very important tool for the elimination of various pollutants from flue gases emitted by stationary sources, such as electric power stations or industrial plants, as well as vehicles. On the other hand, heterogeneous catalysts are used in the conversion of crude oil to liquid fuels, as well as the production of fuel additives. Another important issue

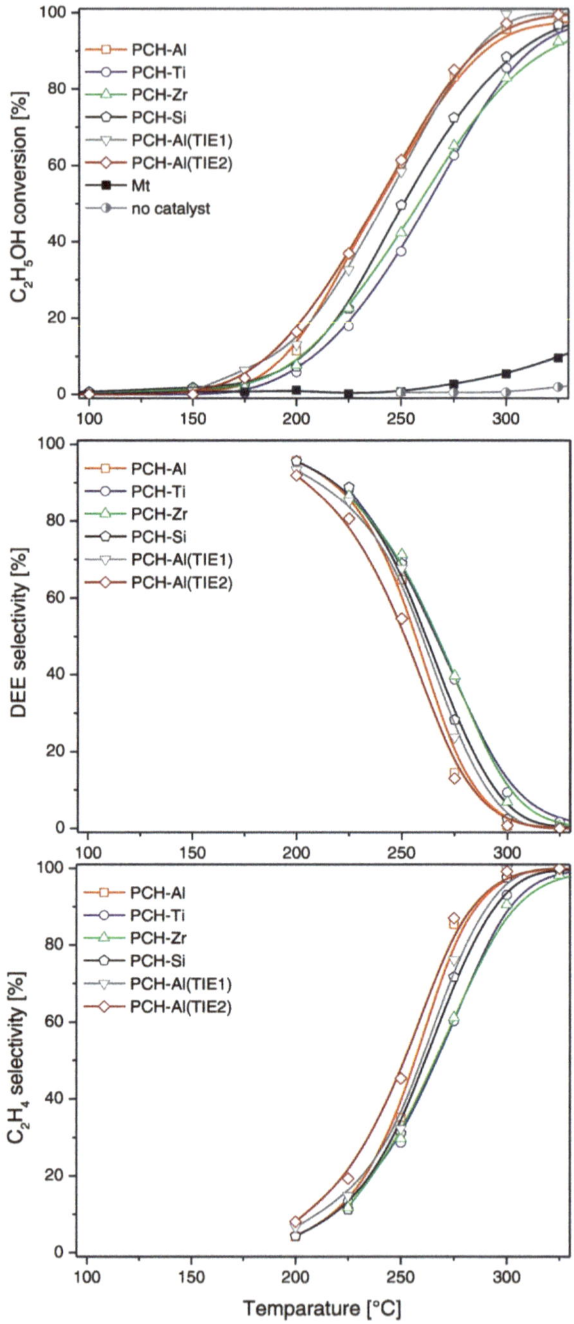

Figure 31 Results of catalytic tests in the process of ethanol to diethyl ether and ethylene dehydration over the PCH-based catalysts.[130] *DEE*, diethyl ether; *PCH*, porous clay heterostructure.

is catalytic conversion of biomass to biofuels and various industrially important chemicals. Thus, heterogeneous catalysis is strongly coupled with technologies used for fuel and energy production and conversion. The progress in the development of the synthesis of catalysts with improved activity, selectivity, and stability is one of the most important trends in science and technology since many decades.

Various methods used for the synthesis of zeolites with the hierarchical mesoporous–microporous structure result in catalysts with improved rate of internal diffusion of reactants. It was shown that zeolites with such hierarchical porous structure presented improved catalytic performance in various processes, including the synthesis of DME from methanol. Moreover, it was shown that the type of deposited metal species, especially their aggregation, depends on the porous structure of zeolitic materials.

Mesoporous silicas, due to their extremely high surface area and uniform porous structure, are very promising as catalytic supports. However, the leak of ion-exchange properties in pure silica mesoporous materials and, therefore, also problems with the uniform deposition of catalytically active metal species limit their application in industrial catalysis. It was shown that catalytically active metals can be deposited as uniform species (monomeric cations or oligomeric metal oxide species) on mesoporous silica by grafting of metal species precursors, followed by their thermal decomposition into stable forms. Moreover, metal species can be uniformly deposited on prepared mesoporous silica (non-calcined) by exchange of alkyl ammonium surfactants for monomeric or oligomeric metal cations. This method, called TIE, is much cheaper in comparison to grafting but results in less homogeneous deposition of metal species.

Cationic layered clay minerals, due to their relatively low cost and high availability in nature, are very promising raw materials for catalyst preparation. Intercalation of clay minerals with inorganic pillars results in porous materials with a relatively high surface area, which are characterized by the presence of surface acidic sites and ion-exchange properties. It was shown that various clay minerals can be intercalated with monometallic and multicomponent pillars. Such pillared clays were found to be active catalysts for various catalytic processes, including reduction of NO by ammonia and synthesis of DME and DEE from methanol and ethanol, respectively. Very promising catalytic results in NH_3-SCR were obtained for montmorillonite intercalated with silica–titania pillars by the surfactant-directed method, additionally doped with copper or iron by the ion-exchange method. In this case, NO was selectively reduced to N_2 by ammonia in a very-broad-temperature range. On the other hand, montmorillonite intercalated with silica—alumina pillars was very active in the synthesis of DME from methanol. The efficiency of this catalyst was higher than the commercial catalyst based on γ-Al_2O_3.

One of the most important expected innovations in environmental catalysis is related to the development of the process of low-temperature NO_x reduction. Typically, NH_3-SCR catalyst, in the form of monolith, is located upstream of an electrostatic precipitator (ESP) and therefore operates with dusty gas stream (high-dust SCR). Particles of dust may clog channels in the monolith and deactivate the catalytic system. To avoid such problem the monolithic NH_3-SCR catalyst could be located downstream of the ESP unit (low-dust NH_3-SCR). However, temperature of flue gases downstream of ESP is about 250°C or even lower.[149] Therefore, for such low-dust NH_3-SCR process effective catalysts operating below 250°C are necessary. It seems that novel, porous catalysts with high loadings of uniformly distributed active component are very promising candidates for the low-temperature NH_3-SCR process. Another important problem of environmental catalysis is related to selective oxidation of ammonia to dinitrogen in flue gases in a broad temperature range, especially at temperatures above 400°C, where oxidation of ammonia may result not only in dinitrogen but also in toxic nitrogen oxides. Thus, for controlling the selectivity of ammonia oxidation in the high-temperature range, development of new type of the catalytic systems is necessary. Again, it seems that deposition of uniform catalytically active species on high—surface area supports could be an important step in development of such catalysts. In the field of DME and diethyl ether synthesis, an important problem seems to be the development of high-performance catalysts with increased resistance for the poisoning by coke deposition. In this case the porosity, strength, location, and distribution of the catalytically active acid sites seem to be externally important. It was shown that these features can be controlled in the synthesis of the aforementioned catalysts.

ACKNOWLEDGMENTS

The studies related to the catalytic functionalization of mesoporous silica materials for catalysis were carried out within the frame of project 2012/05/B/ST5/00269 from the National Science Centre (Poland). The studies on cationic layered clay minerals were carried out within the frame of project 2016/21/B/ST5/00242 from the National Science Centre (Poland).
Parts of the presented research was done with equipment purchased within the frame of the European Regional Development Fund (Polish Innovation Economy Operational Program - contract no. POIG.02.01.00-12-023/08).

REFERENCES

1. Farrauto, R. J.; Dorazio, L.; Bartholomew, C. H. *Introduction to Catalysis and Industrial Catalytic Processes;* Wiley—AIChE: New Jersey, 2016.
2. Barbaro, P.; Bianchini, C. *Catalysis for Sustainable Energy Production;* Wiley-VCH Verlag GmbH: Weinheim, 2009.
3. Ertl, G.; Knözinger, H.; Weitkamp, J. *Environmental Catalysis;* Wiley-VCH: Verlag GmbH: Weinheim, 2008.
4. Bartholomew, C. H.; Farrauto, R. J. *Fundamentals of Industrial Catalytic Processes,* 2nd ed.; Wiley—AIChE: New York, 2005.
5. Raveendran Shiju, N.; Guliants, V. V. *Appl. Catal. A Gen.* **2009,** *356,* 1—17.
6. Liu, Z.; Woo, S. I. *Catal. Rev.* **2006,** *48,* 43—89.
7. Basąg, S.; Kocoł, K.; Piwowarska, Z.; Rutkowska, M.; Baran, R.; Chmielarz, L. *React. Kinet. Mech. Catal.* **2017,** *121,* 225—240.
8. Devadas, M.; Krocher, O.; Elsener, M.; Wokaun, A.; Soger, N.; Pfeifer, M.; Demel, Y.; Mussmann, L. *Appl. Catal. B Environ.* **2006,** *67,* 187—196.
9. Santra, A. K.; Goodman, D. W. *Electrochim. Acta* **2002,** *47,* 3595—3609.
10. Chmielarz, L.; Jabłońska, M. *RSC Adv.* **2015,** *54,* 43408—43431.
11. Sharma, Y. C.; Singh, B.; Korstad, J. *Fuel* **2011,** *90,* 1309—1324.
12. Holladay, J. D.; Hu, J.; King, D. L.; Wang, Y. *Catal. Today* **2009,** *139,* 244—260.
13. Wang, A.; Wang, Y.; Kabe, T.; Chen, Y.; Ishihara, A.; Qian, W. *J. Catal.* **2001,** *199,* 19—29.
14. Oyama, S. T.; Gott, T.; Zhao, H.; Lee, Y.-K. *Catal. Today* **2009,** *143,* 94—107.
15. Garrido-Ramírez, E. G.; Theng, B. K. G.; Mora, M. L. *Appl. Clay Sci.* **2010,** *47,* 182—192.
16. Breysse, M.; Geantet, C.; Afanasiev, P.; Blanchard, J.; Vrinat, M. *Catal. Today* **2008,** *130,* 3—13.
17. Rutkowska, M.; Chmielarz, L. In *Advances in Materials Science Research,* Vol. 22, Wythers, M. C., Ed.; Nova Science Publishers Inc.: New York, 2016; pp 137—156.
18. Sing, K. S. W.; Everett, D. H.; Haul, R. A. W.; Moscou, L.; Pierotti, R. A.; Rouquerol, J.; Siemieniewska, T. *Pure Appl. Chem.* **1985,** *57,* 603—619.
19. Naber, J. E.; de Jong, K. P.; Stork, W. H. J.; Kuipers, H. P. C. E.; Post, M. F. M. *Stud. Surf. Sci. Catal.* **1994,** *84,* 2197—2219.
20. Yilmaz, B.; Müller, U. *Top. Catal.* **2009,** *52,* 888—895.
21. Kresge, C. T.; Leonowicz, M. E.; Roth, W. J.; Vartuli, J. C.; Beck, J. S. *Nature* **1992,** *359,* 710—712.
22. Meynen, V.; Cool, P.; Vansant, E. F. *Micropor. Mesopor. Mater.* **2009,** *125,* 170—223.
23. Linssen, T.; Cassiers, K.; Cool, P.; Vansant, E. F. *Adv. Colloid Interface Sci.* **2003,** *103,* 121—147.
24. Corma, A. *Chem. Rev.* **1997,** *97,* 2373—2419.
25. Chmielarz, L.; Dziembaj, R.; Grzybek, T.; Klinik, J.; Łojewski, T.; Olszewska, D.; Papp, H. *Catal. Lett.* **2000,** *68,* 95—100.
26. Chmielarz, L.; Dziembaj, R.; Grzybek, T.; Klinik, J.; Łojewski, T.; Olszewska, D.; Węgrzyn, A. *Catal. Lett.* **2000,** *70,* 51—56.
27. Chmielarz, L.; Kuśtrowski, P.; Zbroja, M.; Rafalska-Łasocha, A.; Dudek, B.; Dziembaj, R. *Appl. Catal. B Environ.* **2003,** *45,* 103—116.
28. Chmielarz, L.; Kowalczyk, A. In *Comprehensive Guide for Mesoporous Materials. Synthesis and Characterization;* Aliofkhazraei, M., Ed.; Nova Science Publishers Inc.: New York, 2015; pp 27—54.
29. Chmielarz, L.; Kuśtrowski, P.; Zbroja, M.; Łasocha, W.; Dziembaj, R. *Catal. Today* **2004,** *90,* 43—49.
30. Galarneau, A.; Barodawalla, A.; Pinnavaia, T. J. *Nature* **1995,** *374,* 529—531.

31. Chmielarz, L.; Gil, B.; Kuśtrowski, P.; Piwowarska, Z.; Dudek, B.; Michalik, M. *J. Solid State Chem.* **2009,** *182,* 1094–1104.
32. Chmielarz, L.; Kowalczyk, A.; Wojciechowska, M.; Boroń, P.; Dudek, B.; Michalik, M. *Chem. Pap.* **2014,** *68,* 1219–1227.
33. Chmielarz, L.; Piwowarska, Z.; Kuśtrowski, P.; Gil, B.; Adamski, A.; Dudek, B.; Michalik, M. *Appl. Catal. B Environ.* **2009,** *91,* 449–459.
34. Barrer, R. M. *J. Chem. Soc.* **1948,** *33,* 127–132.
35. Lauriente, D. H.; Inoguchi, Y. *The Chemical Economics Handbook;* SRI consulting: Stanford, 2005.
36. Bellussi, G. *Stud. Surf. Sci. Catal.* **2004,** *154 A,* 53–65.
37. Maesen, T. In *Introduction to Zeolite Science and Practice;* Čejka, J., van Bekkum, H., Corma, A., Schueth, F., Eds., 3rd rev. ed;; Elsevier: USA, 2007; pp 1–12.
38. Market Research Store, Published 22-Dec-2015, http://www.marketresearchstore.com/news/global-zeolite-market-140
39. Čejka, J.; Mintova, S. *Catal. Rev.* **2007,** *49,* 457–509.
40. van Donk, S.; Janssen, A. H.; Bitter, J. H.; de Jong, K. P. *Catal. Rev.* **2003,** *45,* 297–319.
41. Tao, Y.; Kanoh, H.; Abrams, L.; Kaneko, K. *Chem. Rev.* **2006,** *106,* 896–910.
42. Pérez-Ramírez, J.; Christensen, C. H.; Egeblad, K.; Christensend, C. H.; Groen, J. C. *Chem. Soc. Rev.* **2008,** *37,* 2530–2542.
43. Chal, R.; Gérardin, C.; Bulut, M.; van Donk, S. *ChemCatChem* **2011,** *3,* 67–81.
44. Hua, Z. L.; Zhou, J.; Shi, J. L. *Chem. Commun.* **2011,** *47,* 10536–10547.
45. Serrano, D. P.; Escola, J. M.; Pizarro, P. *Chem. Soc. Rev.* **2013,** *42,* 4004–4035.
46. Rutkowska, M.; Díaz, U.; Palomares, A. E.; Chmielarz, L. *Appl. Catal. B Environ.* **2015,** *168,* 531–539.
47. Barrer, R. M. *Molecular Sieves;* Society of Chemical Industry: London, 1968.
48. García-Martínez, J.; Lia, K.; Krishnaiaha, G. *Chem. Commun.* **2012,** *48,* 11841–11843.
49. García-Martínez, J.; Johnson, M.; Valla, J.; Lib, K.; Ying, J. Y. *Catal. Sci. Technol.* **2012,** *2,* 987–994.
50. Li, K.; Valla, J.; García-Martínez, J. *Chem. Cat. Chem.* **2014,** *6,* 46–66.
51. Groen, J. C.; Jansen, J. C.; Moulijn, J. A.; Pérez-Ramírez, J. J. *Phys. Chem. B* **2004,** *108,* 13062–13065.
52. Groen, J. C.; Peffer, L. A. A.; Moulijn, J. A.; Pérez-Ramírez, J. *Chem. Eur. J.* **2005,** *11,* 4983–4994.
53. Verboekend, D.; Pérez-Ramírez, J. *Catal. Sci. Technol.* **2011,** *1,* 879–890.
54. Young, D.A. US Patent No. 3 326 797, 1967.
55. Pérez-Ramírez, J.; Verboekend, D.; Bonilla, A.; Abelló, S. *Adv. Funct. Mater.* **2009,** *19,* 3972–3979.
56. Abelló, S.; Bonilla, A.; Pérez-Ramírez, J. *Appl. Catal. A Gen.* **2009,** *364,* 191–198.
57. Rutkowska, M.; Pacia, I.; Basąg, S.; Kowalczyk, A.; Piwowarska, Z.; Duda, M.; Tarach, K. A.; Góra-Marek, K.; Michalik, M.; Díaz, U.; Chmielarz, L. *Micropor. Mesopor. Mater.* **2017,** *246,* 193–206.
58. Góra-Marek, K.; Brylewska, K.; Tarach, K. A.; Rutkowska, M.; Jabłońska, M.; Choi, M.; Chmielarz, L. *Appl. Catal. B Environ.* **2015,** *179,* 589–598.
59. Rutkowska, M.; Duda, M.; Kowalczyk, A.; Chmielarz, L. *CR. Chim.* **2017,** *20,* 850–859.
60. Rutkowska, M.; Macina, D.; Mirocha-Kubień, N.; Piwowarska, Z.; Chmielarz, L. *Appl. Catal. B Environ.* **2015,** *174,* 336–343.
61. Nandiwale, K. Y.; Niphadkar, P. S.; Deshpande, S. S.; Bokade, V. V. *J. Chem. Technol. Biotechnol.* **2014,** *89,* 1507–1515.
62. Li, J.; Li, X.; Zhou, G.; Wang, W.; Wang, C.; Komarneni, S.; Wang, Y. *Appl. Catal. A Gen.* **2014,** *470,* 115–122.
63. Zhou, Q.; Wang, Y.-Z.; Tang, C.; Zhang, Y.-H. *Polym. Degrad. Stab.* **2003,** *80,* 23–30.

64. Rownaghi, A. A.; Rezaei, F.; Stante, M.; Hedlund, J. *Appl. Catal. B Environ.* **2012,** *119–120,* 56–61.
65. Sung, D. M.; Kim, Y. H.; Park, E. D.; Yie, J. E. *Catal. Commun.* **2012,** *20,* 63–67.
66. Van Oers, C. J.; Stevens, W. J. J.; Bruijn, E.; Mertens, M.; Lebedev, O. I.; Van Tendeloo, G.; Meynen, V.; Cool, P. *Micropor. Mesopor. Mater.* **2009,** *120,* 29–34.
67. Stevens, W. J. J.; Meynen, V.; Bruijn, E.; Lebedev, O. I.; Van Tendeloo, G.; Cool, P.; Vansant, E. F. *Micropor. Mesopor. Mater.* **2008,** *110,* 77–85.
68. Van Oers, C. J.; Kurttepeli, M.; Mertens, M.; Bals, S.; Meynen, V.; Cool, P. *Micropor. Mesopor. Mater.* **2014,** *185,* 204–212.
69. Rutkowska, M.; Macina, D.; Piwowarska, Z.; Gajewska, M.; Díaz, U.; Chmielarz, L. *Catal. Sci. Technol.* **2016,** *6,* 4849–4862.
70. Van Oers, C. J.; Góra-Marek, K.; Sadowska, K.; Mertens, M.; Meynen, V.; Datka, J.; Cool, P. *Chem. Eng. J.* **2014,** *237,* 372–379.
71. Van Oers, C. J.; Góra-Marek, K.; Prelot, B.; Datka, J.; Meynen, V.; Cool, P. *Langmuir* **2014,** *30,* 1880–1887.
72. Rutkowska, M.; Chmielarz, L.; Macina, D.; Piwowarska, Z.; Dudek, B.; Adamski, A.; Witkowski, S.; Sojka, Z.; Obalová, L.; Van Oers, C.; Cool, P. *Appl. Catal. B Environ.* **2014,** *146,* 112–122.
73. Rutkowska, M.; Piwowarska, Z.; Micek, E.; Chmielarz, L. *Micropor. Mesopor. Mater.* **2015,** *209,* 54–65.
74. Rutkowska, M.; Chmielarz, L.; Macina, D.; Dudek, B.; Van Oers, C. J.; Cool, P. *JCAMS* **2013,** *1,* 48–55.
75. Chiola, V.; Ritsko, J.E.; Vanderpool, C.D.. US Patent No. 3 556 725, 1971.
76. Beck, J. S.; Vartuli, J. C.; Roth, W. J.; Leonowicz, M. E.; Kresge, C. T.; Schmitt, K. D.; Chu, C. T. W.; Olson, D. H.; Sheppard, E. W.; McCullen, S. B.; Higgins, J. B.; Schlenker, J. L. *J. Am. Chem. Soc.* **1992,** *114,* 10834–10843.
77. Kumar, S.; Malik, M. M.; Purohit, R. *Mater. Today: SAVE Proc.* **2017,** *4,* 350–357.
78. Weitkamp, J.; Hunger, M.; Rymsa, U. *Micropor. Mesopor. Mater.* **2001,** *48,* 25–270.
79. Taguchi, A.; Schuth, F. *Micropor. Mesopor. Mater.* **2005,** *77,* 1–45.
80. Bu, J.; Rhee, H.-K. *Catal. Lett.* **2000,** *65,* 141–145.
81. Zhang, F.; Yan, Y.; Yang, H.; Meng, Y.; Yu, C.; Tu, B.; Zhao, D. *J. Phys. Chem. B* **2005,** *109,* 8723–8732.
82. Baltes, M.; Collart, O.; Van Der Voort, P.; Vansant, E. F. *Langmuir* **1999,** *15,* 5841–5845.
83. White, G. M. *Catal. Today* **1993,** *18,* 73–109.
84. Van Der Voort, P.; Possemiers, K.; Vansant, E. F. *J. Chem. Soc. Faraday Trans.* **1996,** *92,* 843–848.
85. Van Der Voort, P.; Babich, I. V.; Grobet, P. J.; Verberckmoes, A. A.; Vansant, E. F. *J. Chem. Soc. Faraday Trans.* **1996,** *92,* 3635–3642.
86. Van Der Voort, P.; Mitchell, M. B.; Vansant, E. F.; White, J. M. G. *Interface Sci.* **1997,** *5,* 169–197.
87. Van Der Voort, P.; Baltes, M.; Vansant, E. F.; White, J. M. G. *Interface Sci.* **1997,** *5,* 199–206.
88. Babich, I. V.; Plyuto, Y. V.; Van Der Voort, P.; Vansant, E. F. *J. Coll. Interface Sci.* **1997,** *189,* 144–150.
89. Kenvin, J. C.; White, M. G.; Mitchell, M. B. *Langmuir* **1991,** *7,* 1198–1205.
90. Chmielarz, L.; Kuśtrowski, P.; Dziembaj, R.; Cool, P.; Vansant, E. F. *Micropor. Mesopor. Mater.* **2010,** *127,* 133–141.
91. Van Veen, J. A. R.; Jonkers, G.; Hesselink, W. H. *J. Chem. Soc. Faraday Trans.* **1998,** *85,* 389–413.
92. Van Veen, J. A. R.; De Jong-Versloot, P. C.; Van Kessel, G. M. M.; Fels, F. J. *Thermochim. Acta* **1989,** *152,* 359–370.

93. Chmielarz, L.; Kuśtrowski, P.; Dziembaj, R.; Cool, P.; Vansant, E. F. *Appl. Catal. B Environ.* **2006,** *62,* 369–380.
94. Segura, Y.; Chmielarz, L.; Kuśtrowski, P.; Cool, P.; Dziembaj, R.; Vansant, E. F. *Appl. Catal. B Environ.* **2005,** *61,* 69–78.
95. Segura, Y.; Cool, P.; Kuśtrowski, P.; Chmielarz, L.; Dziembaj, R.; Vansant, E. F. *J. Phys. Chem. B* **2005,** *109,* 12071–12079.
96. Busca, G.; Lietti, L.; Ramis, G.; Berti, F. *Appl. Catal. B Environ.* **1998,** *18,* 1–36.
97. Chmielarz, L.; Kuśtrowski, P.; Drozdek, M.; Rutkowska, M.; Dziembaj, R.; Michalik, M.; Cool, P.; Vansant, E. F. *J. Porous Mater.* **2011,** *18,* 483–491.
98. Basąg, S.; Piwowarska, Z.; Kowalczyk, A.; Węgrzyn, A.; Baran, R.; Gil, B.; Michalik, M.; Chmielarz, L. *Appl. Clay Sci.* **2016,** *129,* 122–130.
99. Jabłońska, M.; Chmielarz, L.; Węgrzyn, A.; Góra-Marek, K.; Piwowarska, Z.; Witkowski, S.; Bidzińska, E.; Kuśtrowski, P.; Wach, A.; Majda, D. *Appl. Clay Sci.* **2015,** *114,* 273–282.
100. Macina, D.; Piwowarska, Z.; Góra-Marek, K.; Tarach, K.; Rutkowska, M.; Girman, V.; Błachowski, A.; Chmielarz, L. *Mater. Res. Bull.* **2016,** *78,* 72–82.
101. Kowalczyk, A.; Piwowarska, Z.; Macina, D.; Kuśtrowski, P.; Rokicińska, A.; Michalik, M.; Chmielarz, L. *Chem. Eng. J.* **2016,** *295,* 167–180.
102. Macina, D.; Opioła, A.; Rutkowska, M.; Basąg, S.; Piwowarska, Z.; Michalik, M.; Chmielarz, L. *Mater. Chem. Phys.* **2017,** *187,* 60–71.
103. Macina, D.; Piwowarska, Z.; Tarach, K.; Góra-Marek, K.; Ryczkowski, J.; Chmielarz, L. *Mater. Res. Bull.* **2016,** *74,* 425–435.
104. Xu, M.; Lunsford, J. H.; Goodman, D. W.; Bhattacharyya, A. *Appl. Catal. A Gen.* **1997,** *149,* 289–301.
105. Takeguchi, T.; Yanagisawa, K.; Inui, T.; Inoue, M. *Appl. Catal. A Gen.* **2000,** *192,* 201–209.
106. Yaripour, F.; Baghaei, F.; Schmidt, I.; Perregaard, J. *Catal. Commun.* **2005,** *6,* 147–152.
107. Macina, D.. Doctoral Thesis, Jagiellonian University, Krakow, 2017.
108. Whitehurst, D.D.. US Patent No. 5 143 879, 1992.
109. Yonemitsu, M.; Tanaka, Y.; Iwamoto, M. *Chem. Mater.* **1997,** *9,* 2679–2681.
110. Yonemitsu, M.; Tanaka, Y.; Iwamoto, M. *J. Catal.* **1998,** *178,* 207–213.
111. Tanev, P. T.; Pinnavaia, T. J. *Chem. Mater.* **1996,** *8,* 2068–2074.
112. Bourlinos, A. B.; Karakassides, M. A.; Petridis, D. *J. Phys. Chem. B* **2000,** *114,* 4375–4380.
113. Wang, Y.; Zhang, Q.; Shishido, T.; Takehira, K. *J. Catal.* **2002,** *209,* 186–196.
114. Zhang, Q.; Guoa, Q.; Wang, X.; Shishido, T.; Wang, Y. *J. Catal.* **2006,** *239,* 105–116.
115. Zhang, Q.; Wang, Y.; Itsuki, S.; Shishido, T.; Takehira, K. *J. Mol. Catal. A: Chemical.* **2002,** *188,* 189–200.
116. Derylo-Marczewska, A.; Gac, W.; Popivnyak, N.; Zukocinski, G.; Pasieczna, S. *Catal. Today* **2006,** *114,* 293–306.
117. Wang, Y.; Ohishi, Y.; Shishido, T.; Zhang, Q.; Yang, W.; Guo, Q.; Wan, H.; Takehira, K. *J. Catal.* **2003,** *220,* 347–357.
118. Ohtsuka, Y.; Takahashi, Y.; Noguchi, M.; Arai, T.; Takasaki, S.; Tsubouchi, N.; Wang, Y. *Catal. Today* **2004,** *89,* 419–429.
119. Lang, N.; Aouine, M.; Tuel, A. *Stud. Surf. Sci. Catal.* **2005,** *158,* 439–446.
120. Blanco, A. A. G.; Amaya, M. G.; Jalil, M. E. R.; Nazzarro, M.; Oliva, M. I.; Sapag, K. *Top. Catal.* **2011,** *54,* 190–200.
121. Liu, C. C.; Teng, H. *Appl. Catal. B Environ.* **2005,** *58,* 69–77.
122. Canepa, A. L.; Chanquia, C. M.; Eimer, G. A.; Casuscelli, S. G. *Appl. Catal. A Gen.* **2013,** *462–463,* 8–14.
123. Tanaka, M.; Itadani, A.; Kuroda, Y.; Iwamoto, M. *J. Phys. Chem. C* **2012,** *116,* 5664–5672.

124. Lehmann, T.; Wolff, T.; Hamel, C.; Veit, P.; Garke, B.; Seidel-Morgenstern, A. *Micropor. Mesopor. Mater.* **2012,** *151,* 113–125.
125. Pirouzmand, M.; Nikzad-Kojanag, B.; Seyed-Rasulzade, S. K. *Catal. Commn.* **2015,** *69,* 196–201.
126. Iwamoto, M.; Tanaka, Y. *Catal. Surv. Jpn.* **2001,** *5,* 25–36.
127. Kowalczyk, A.; Borcuch, A.; Michalik, M.; Rutkowska, M.; Gil, B.; Sojka, Z.; Chmielarz, L. *Micropor. Mesopor. Mater.* **2017,** *240,* 9–21.
128. Brigatti, M. F.; Galan, E.; Theng, B. K. G. In *Handbook of Clay Science;* Bergaya, F., Theng, B. K. G., Lagely, G., Eds.; Elsevier Science: Amsterdam, 2006; pp 19–86.
129. Burch, R. *Catal. Today* **1988,** *2,* 185–186.
130. Chmielarz, L.; Kowalczyk, A.; Skoczek, M.; Rutkowska, M.; Gil, B.; Natkański, P.; Radko, M.; Motak, M.; Dębek, R.; Ryczkowski, J.. Appl. Clay Sci.. DOI: 10.1016/j.clay.2017.12.015
131. del Rey-Perez-Caballero, F. J.; Poncelet, G. *Micropor. Mesopor. Mater.* **2000,** *41,* 169–181.
132. del Rey-Perez-Caballero, F. J.; Poncelet, G. *Micropor. Mesopor. Mater.* **2000,** *37,* 313–327.
133. Vaccari, A. *Appl. Clay Sci.* **1999,** *14,* 161–198.
134. Gil, A.; Korili, S. A.; Trujillano, R.; Vicente, M. A. *Appl. Clay Sci.* **2011,** *53,* 97–105.
135. Chmielarz, L.; Dziembaj, R.; Łojewski, T.; Węgrzyn, A.; Grzybek, T.; Klinik, J.; Olszewska, D. *Solid State Ionics* **2001,** *141–142,* 715–719.
136. Chmielarz, L.; Kuśtrowski, P.; Zbroja, M.; Gil-Knap, B.; Datka, J.; Dziembaj, R. *Appl. Catal. B Environ.* **2004,** *53,* 47–61.
137. Chmielarz, L.; Kuśtrowski, P.; Piwowarska, Z.; Michalik, M.; Dudek, B.; Dziembaj, R. *Top. Catal.* **2009,** *52,* 1017–1022.
138. Benjelloun, M.; Cool, P.; Linssen, T.; Vansant, E. F. *Micropor. Mesopor. Mater.* **2001,** *49,* 83–94.
139. Meier, P. L.; Kahl, G. *Clays Clay Miner.* **1999,** *47,* 386–388.
140. Chmielarz, L.; Kowalczyk, A.; Michalik, M.; Dudek, B.; Piwowarska, Z.; Matusiewicz, A. *Appl. Clay Sci.* **2010,** *48,* 156–162.
141. Chmielarz, L.; Wojciechowska, M.; Rutkowska, M.; Adamski, A.; Węgrzyn, A.; Kowalczyk, A.; Dudek, B.; Boroń, P.; Michalik, M.; Matusiewicz, A. *Catal. Today* **2012,** *191,* 25–31.
142. Pinto, M. L.; Marques, J.; Pires, J. *Sep. Purif. Technol.* **2012,** *98,* 337–343.
143. Chmielarz, L.; Piwowarska, Z.; Kuśtrowski, P.; Węgrzyn, A.; Gil, B.; Kowalczyk, A.; Dudek, B.; Dziembaj, R.; Michalik, M. *Appl. Clay Sci.* **2011,** *53,* 164–173.
144. Chmielarz, L.; Kuśtrowski, P.; Dziembaj, R.; Cool, P.; Vansant, E. F. *Catal. Today* **2007,** *119,* 181–186.
145. Chmielarz, L.; Kuśtrowski, P.; Drozdek, M.; Dziembaj, R.; Cool, P.; Vansant, E. F. *Catal. Today* **2006,** *114,* 319–325.
146. Chmielarz, L.; Kuśtrowski, P.; Piwowarska, Z.; Dudek, B.; Gil, B.; Michalik, M. *Appl. Catal. B Environ.* **2009,** *88,* 331–340.
147. Takahara, A.; Saito, M.; Inaba, M.; Murata, K. *Catal. Lett.* **2015,** *105,* 249–252.
148. Ciftci, A.; Varisli, D.; Tokay, K. C.; Sezgi, N. A.; Dogu, T. *Chem. Eng. J.* **2012,** *207–208,* 85–93.
149. Mladenović, M.; Paprika, M.; Marinković, A. *Renew. Sustain. Energy Rev.* **2018,** *82,* 3350–3364.

CHAPTER NINE

Abatement of Volatile Organic Compounds Emission as a Target for Various Human Activities Including Energy Production

Piotr Kuśtrowski[1], Anna Rokicińska and Tomasz Kondratowicz
Faculty of Chemistry, Jagiellonian University, Kraków, Poland
[1]Corresponding author: E-mail: piotr.kustrowski@uj.edu.pl

Contents

Abstract

In this review, the issue of volatile organic compounds (VOCs) emission is discussed in detail. Various sources of VOCs release, especially related to stationary and mobile fuel combustion, are presented. Successive changes in the distribution of fuels toward a utilization of growing amounts of biocomponents are underscored. The impact of

Advances in Inorganic Chemistry, Volume 72
ISSN 0898-8838
https://doi.org/10.1016/bs.adioch.2018.05.004

different kinds of VOCs emitted to the atmosphere on the environment and human health is discussed within the context of tightening regulations. Main remediation methods developed for the elimination of VOCs are shown with a special emphasis on the catalytic total oxidation. We present advantages of this approach in comparison to the thermal oxidation performed at significantly higher temperatures in the absence of a catalyst. The most important features of catalysts studied in the VOCs combustion, as well as the speculated reaction mechanisms, are elaborated. The catalytic performance and stability of noble metal–containing materials are compared critically to transition metal oxides. Finally, combining adsorption with subsequent catalytic conversion is revealed as a very promising way to abate amounts of emitted VOCs.

1. INTRODUCTION

Nowadays, the feedstock for the most important chemical technologies and energy production is mainly based on fossil fuels. In 2010, oil provided for about 34% of the total energy consumption, coal 30%, and natural gas 24%.[1] Nevertheless, taking into account the increasing demand for energy and limited availability of these resources, alternative methods of energy production still need to be developed. Undoubtedly, the significance of renewable sources (hydroelectricity, solar, and wind energy) will have to grow. Furthermore, biomass produced by photosynthesis will become a more relevant raw material for the production of fuels and chemicals in the near future. The application of biomass for energy production on a large scale exhibits some benefits for the environment due to the closed carbon cycle, but it should be kept in mind that processing and burning of fossil fuels and biomass can cause the emission of various harmful pollutants. Among them, nitrogen oxides (NO_x), SO_2, CO, volatile organic compounds (VOCs), semi-VOCs (including polycyclic aromatic hydrocarbons [PAHs], as well as polychlorinated biphenyls, dibenzodioxins, and dibenzofurans), and particulate matter (especially those with the aerodynamic diameter less than 2.5 μm [$PM_{2.5}$]) can be distinguished. In this review, special emphasis will be put on the emission of VOCs and possible ways of their elimination.

Generally, VOCs represent a large family of organic compounds, which can easily evaporate and be released to the atmosphere. Obviously, many different definitions for VOCs have been formulated. However, in practice, all carbon compounds exhibiting a boiling point less than or equal to 250°C at a pressure of 101.3 kPa belong to VOCs. Among them, alkanes, olefins, aromatic hydrocarbons, and heteroatom-containing organic compounds are included. The definition of VOCs proposed by the US Environmental Protection Agency indicates one more characteristic feature of these compounds: "volatile organic compound means any compound of carbon, excluding carbon monoxide, carbon dioxide, carbonic acid, metallic carbides or carbonates, and ammonium carbonate, which participate in atmospheric photochemical reactions."[2] From this point of view, methane, which is the most abundant organic compound present in the atmosphere (with an average concentration of about 1.8 ppm),[3] is undoubtedly one of the VOCs, despite many definitions that exclude CH_4 from this group of pollutants.

2. PROFILE OF VOLATILE ORGANIC COMPOUNDS EMISSION

2.1 Main Sources of Emission

A major part of VOCs released to the atmosphere is formed by natural sources (so-called biogenic emission). In this case, isoprene, terpenes, esters, alcohols, and carbonyl compounds that evaporate from oil cells of plants dominate among the emitted VOCs. Nevertheless, human activities have significantly changed the complex physical and biogeochemical equilibria established in the biosphere. Many anthropogenic sources of VOCs emission have appeared, in particular, since the explosion of the industrial revolution in the 19th century in Europe and North America. Obviously, the scale and profile of VOCs emission strongly depend on the population number, industrialization, as well as prosperity and regional changes between, for example, parts of the world or rural and urban areas.

In Fig. 1, the emission of VOCs from various anthropogenic sources in the United States, over the period of 1970–2016, is demonstrated. This analysis is possible due to the availability of detailed data published systematically by the US Environmental Protection Agency.[4] It is observed that transportation and industrial production make the main contribution in the release of VOCs. Very important is that the level of emission has decreased since 1970 due to improved technologies and changes in legislation. The implementation of very efficient catalysts for the limitation of pollutants emitted by vehicles is here the best example. Against this background, stationary fuel combustion makes a rather slight contribution but is still more than 0.5 Tg of VOCs produced annually by this source. Apart from that, the intensity of VOCs emission

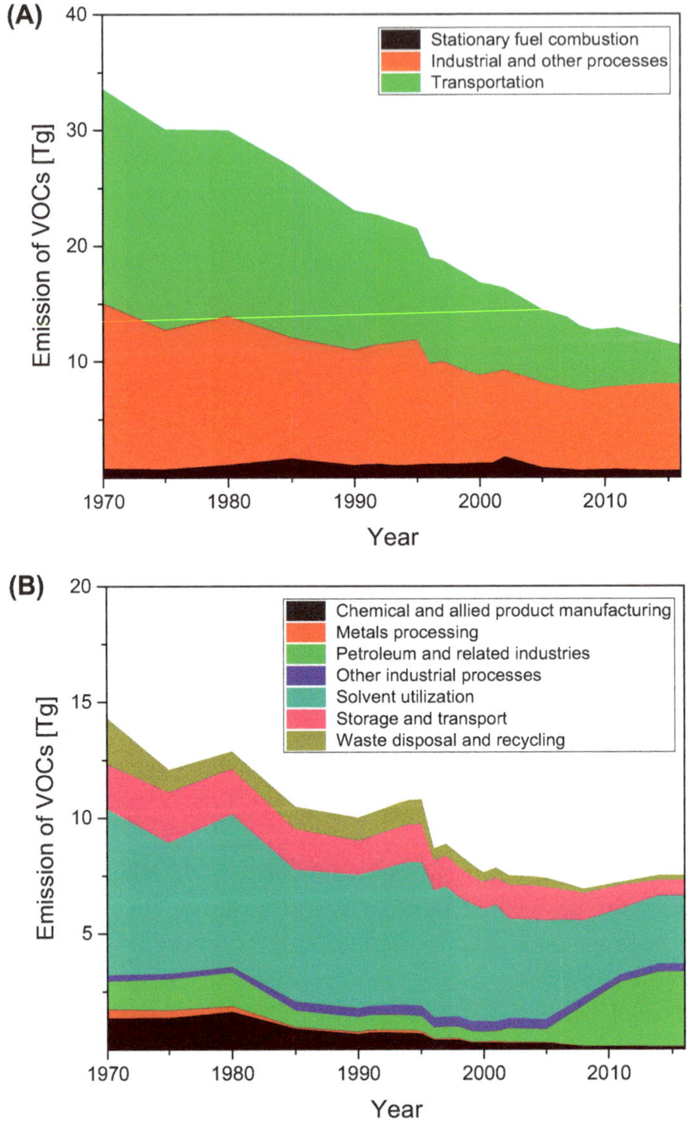

Figure 1 Emission of VOCs from main anthropogenic sources in the United States over the period of 1970–2016 (A) with specified contributing industrial processes (B) according to data published by the US Environmental Protection Agency.[4]

from the stationary fuel combustion in other parts of the world fluctuates and is sometimes significantly higher.

Among the industrial sources of VOCs release in the United States, crude oil production and solvent utilization have presently the highest significance (Fig. 1B). It should not, however, be forgotten that chemical and metallurgical production, waste storage and treatment, as well as food manufacturing are also considered as important sources of VOCs. Furthermore, surface coating materials used in building construction and vehicle production should be enumerated here. It was shown[5] that automotive paint manufacturing releases about 0.02–0.09 kg of VOCs per kilogram of a produced paint. The automotive surface coating materials exhibit a higher VOCs emission (especially toluene, ethylbenzene and xylenes) than target marine or architectural surface coating materials. It may be surprising to know that even innocent New Year's Eve fireworks are responsible for the formation of $PM_{2.5}$ and VOCs.[6]

Various indoor human activities release an additional amount of VOCs. For example, cooking contributes meaningfully to the appearance of different carbon-containing species in the atmosphere. It should, however, be kept in mind that the kind and amount of emitted compounds depend strongly on various factors (including fuel and ingredients used, cooking intensity, process temperature, or ventilation). The Chinese cuisine is known to produce huge amounts of harmful pollutants, making a contribution of 12%–20% in the overall emitted ambient aerosol in this region of the world.[7] Furthermore, the use of household chemicals (e.g., cleaning agents, air fresheners, or personal care products), electronic equipment (especially printers and photocopiers), ventilation systems, and even smoking tobacco or reading newspapers can be included in this kind of VOCs producing activities.

2.2 Emission From Stationary Fuel Combustion

The burning of various types of carbon-containing raw materials is often considered to be the most widely distributed VOCs emission source in the world. Especially, uncontrolled combustion of biomass results in an emission of huge amounts of aromatic VOCs (mainly benzene, toluene, and xylenes) and semi-VOCs. Open field burning of crop residues, which is often practiced as a disposal and fertilization method to limit a number of fungal diseases and pest eggs, clearly illustrates this effect.[8] Unfortunately, the extent of this agricultural practice is so high that only in the continental United States, approximately 1.2 mln ha of croplands is burned annually. It is not surprising that this way of VOCs emission varies seasonally. For instance, in China, where open field burning of crop residues accounts for about 25% of the total emission related to biomass burning, two enhancement periods of VOCs release are observed in May/June and October due to intensified cultivation.[9]

The composition of products emitted during the burning of crop residues (mainly $PM_{2.5}$, VOCs, semi-VOCs, CO, and CO_2) is strongly influenced by the kind of biomass and soil, as well as the moisture content.[10] Inomata et al.[9] observed that the emission of VOCs during combustion of wet wheat straw is suppressed compared with that of dry wheat due to more smoldering conditions. Such process of biomass combustion can take place in geographical regions characterized by relatively high temperature and humidity (for example subtropical and tropical areas), which leads to a higher content of moisture in inner parts of pilled-up crop residues and/or parts directly in contact with the ground. On the contrary, Aurell et al.[11] demonstrated that biomass piles (timber-harvested Douglas fir from western Oregon) emit significantly higher amounts of pollutants (including $PM_{2.5}$, benzene, propylene, acetone, acrolein, vinyl acetate, toluene, 1,3-butadiene, 2-butanone, styrene, or PAHs) after exposure to moisture. Nevertheless, it should be underlined that especially during a dry season, plants release considerable amounts of VOCs to the atmosphere. This phenomenon is commonly known and can lead to serious implications, including accelerating forest fires, which, for example, occur often in Mediterranean countries. VOCs accumulated in a fire area create a flammable atmosphere that can lead consequently to an eruptive fire. In the case of Mediterranean plant species, a predominant role in the generation of the described effect is played by α-pinene derivatives and benzene.[12]

Furthermore, it should be kept in mind that during biomass combustion, apart from typical VOCs, very toxic nitrogen-containing compounds (such as hydrogen cyanide, isocyanic acid, and acetonitrile) are also evolved. This is related to the presence of nitrogen, naturally contained in plants in the form of free amino acids, proteins, some heterocyclic structures (including pyrroline or pyridine), and chlorophyll. The results of different studies on burning herbaceous and arboraceous samples, discussed comparatively by Coggon et al.,[13] show that a fraction of emitted nitrogen-containing VOCs (up to even 15%) is clearly correlated to the nitrogen content in fuel. Nevertheless, burning conditions substantially influence the fraction of nitrogen-containing VOCs. The important factor is also the distribution of proteins in the combusted biomass; for instance, under the same conditions the burning of poly-L-proline gives an HCN/NH_3 ratio of 9.5 and poly-L-leucine - a ratio of 2.2. On the other hand, the combustion of biomass with a high nitrogen content results in the formation of higher carbon nitriles (mainly acetonitrile and acrylonitrile), whereas a nitrogen-lean fuel produces predominantly hydrogen cyanide.

An increasing significance of biomass as a fuel for domestic heating has been observed since the mid-90s. This is caused by growing fossil fuel prices (including taxes), ecological reasons (suitable legislation related to the CO_2 balance), and facilitated accessibility to renewable energy sources. Suffice to say that Europe, North America, and Asia are considered as quickly expending markets for consumption of wood fuels (such as wood pellets and logs) for various types of stoves and boilers. Unfortunately, the domestic biomass burning emits relatively high amounts of air pollutants (mainly $PM_{2.5}$, VOCs, semi-VOCs, CO, and NO_x). From this point of view, continuously fired pellet boilers are more environmentally friendly compared with various wood log stoves, but they still have significantly higher $PM_{2.5}$ emission than oil boilers, which sometimes work alternatively in a single-house heating system.[14] A beneficial effect of wood pellet fuel in relation to wood log burning is usually accounted for by a lower content of moisture, uniform shape, as well as enhanced porosity obtained by compression of chips and sawdust. As a consequence, pyrolytic gases are released fast from the burned fuel, and the heat transfer is highly effective. During burning of wood pellets in a 32-kW boiler at 10% O_2 content, Szyszlak-Bargłowicz et al.[15] found a concentration of 16 PAHs of 323.8 $\mu g/m^3$ and benzo(a)pyrene of 4.8 $\mu g/m^3$ in the emitted gases. The biomass pellet combustion also results in the emission of high amounts of VOCs (mainly methane, ethylene, n-butane, but with identified contents of furan and benzene).

2.3 Emission From Mobile Fuel Combustion

One could expect that the problem of emission of VOCs by mobile fuel combustion has been solved due to the implementation of highly effective catalysts used in a huge majority of on-road vehicles. Indeed, as shown in Fig. 1, the application of two-way (diesel engines) and three-way catalysts (gasoline engines) has strongly limited the amounts of particulate matters and VOCs released by transportation. However, some issues are still actually making motorization one of the strongest VOCs emission sources. Two main reasons are responsible for this situation, viz., fuel evaporation and incomplete combustion. It is obvious that the profile of VOCs emitted due to fuel evaporation is very similar to the fuel composition. On the other hand, the incomplete combustion gives mainly light products, which depend on various factors including type of fuel, engine construction, exhaust treatment system, age of engine and extent of its exploitation, as well as driving conditions. Generally, diesel vehicles emit more alkenes compared with gasoline vehicles. Heavy-duty vehicles release mainly

C_9-C_{11} straight-chain alkanes, whereas light-duty gasoline vehicles release C_5-C_8 branched alkanes.[16]

Some regulations were implemented that cause increasing significance of biofuels. For example, the European Union has declared achieving a minimum 10% share of renewable fuels in base transportation energy consumption by 2020 in each member country.[17,18] Commonly, bioethanol, widely produced in Brazil and the United States, is used as an easily blended fuel component characterized by lower emission of particulate matter and VOCs compared with the traditional petroleum-based fuels.[19] Nevertheless, it should be kept in mind that the utilization of ethanol can cause some air quality problems. In spite of its low capacity for ozone formation in oxidation processes in the atmosphere, it is converted to highly reactive formaldehyde, acetaldehyde, and peroxyacetyl nitrates.[20]

The most serious problem with VOCs emission from mobile fuel combustion is related to the so-called cold-start driving cycle, when from one side the formation of organic pollutants is very intense due to incomplete combustion, and from the other side the catalyst does not work perfectly. Unfortunately, such driving mode is very often identified in the case of motorcycles, which are in addition most frequently not equipped with a catalytic converter. It is estimated that about 65%–75% of on-road vehicles are motorcycles, for which the average travel distance is less than 10 km and the average travel time is shorter than 20 min.[18,21,22]

3. IMPACT OF VOLATILE ORGANIC COMPOUNDS ON ENVIRONMENT AND HUMAN HEALTH

3.1 Photochemical Smog

VOCs emitted to the atmosphere are usually fairly reactive and can be involved in various photochemical reactions with other compounds forming even more hazardous products. First of them is ground-level ozone, which is recognized as a major constituent of photochemical smog. Even short exposure to ambient ground-level ozone has a significant effect on the human respiratory system (especially on the lung function). Generally, tropospheric O_3 is formed by photolysis of NO_2 followed by the reaction of atomic oxygen with molecular O_2:

$$NO_2 + h\nu \rightarrow NO + O(^3P) \quad (1)$$

$$O(^3P) + O_2 \rightarrow O_3 \quad (2)$$

The presence of VOCs drastically enhances the NO_2 formation when $OH^{\bullet}$ radicals (produced by photolysis of aldehydes, hydroperoxides, or nitrous acid) are in addition available in the atmosphere. Simply, these radicals preferentially oxidize VOCs (including CH_4) and CO giving alkyl, alkoxy, or alkylperoxy radicals, which react with NO to give NO_2, as is illustrated below for the example of methane:

$$CH_4 + OH^{\bullet} \rightarrow CH_3^{\bullet} + H_2O \quad (3)$$

$$CH_3^{\bullet} + O_2 \rightarrow CH_3O_2^{\bullet} \quad (4)$$

$$CH_3O_2^{\bullet} + NO \rightarrow CH_3O^{\bullet} + NO_2 \quad (5)$$

VOCs emitted to the atmosphere can also undergo oxidation, followed by condensation resulting in semi-VOCs and low-volatility products (so-called secondary organic aerosol). The role of the oxidant can be played by $OH^{\bullet}$ radicals (active during daylight hours) or nitrate ($NO_3^{\bullet}$) radicals and O_3 (active during evening and night hours; Fig. 2). During these processes, oxygen-containing products are created, including ketones, aldehydes, alcohols, nitrates, and carboxylic acids, with vapor pressures from 10 to 10,000 times lower than those characteristics for primary VOCs.[23,24] They condense on existing soot particles to form secondary organic aerosol. Fortunately, the most abundant volatile organic compound, namely methane, has a rather low contribution to secondary organic aerosols, estimated to be about 0.13% of the total atmospheric aerosol carbon mass.[3] However, aromatic hydrocarbons (especially

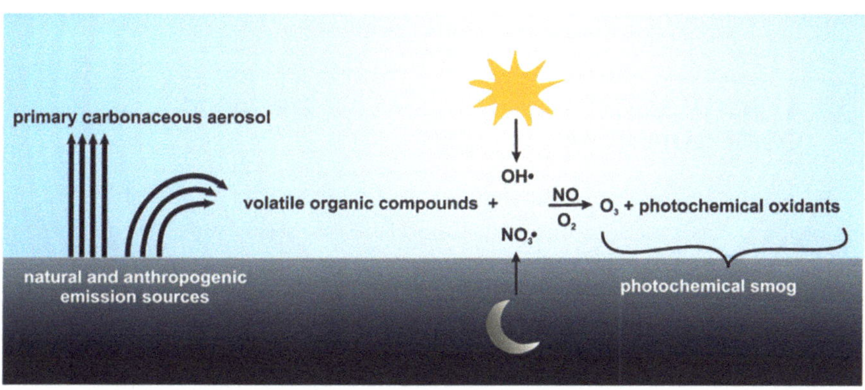

Figure 2 Formation of primary organic aerosol and photochemical smog.

benzene, toluene, xylenes, or ethylbenzene) are significantly more reactive in the formation of secondary organic aerosols.[25]

3.2 Health Hazard

In addition to the environmental risks, VOCs are harmful for human health. A huge number of VOCs (including aromatic compounds, aldehydes, ketones, and alcohols) are recognized to be highly toxic and carcinogenic. The negative health effects, which appear at a given VOCs concentration recognized by the US National Institute of Occupational Safety and Health, have been reviewed in detail elsewhere.[2] The maximum safe limit of VOCs in air should not exceed a few hundred parts per million (for example, in the case of aromatic compounds it is estimated as 200 ppm). However, for an occupational exposure this content should not be higher than a fraction of parts per million. Even at such a low VOCs concentration in indoor air, their long-term contact with the human body can lead to various neuropsychoimmunological manifestations (so-called "sick building syndrome"), which is the most often affliction related to the exposure to VOCs. Furthermore, a low concentration of VOCs in air (even below 1 ppm) usually causes confusion, tiredness, and nausea. Throat and eyes irritation, shortness of breath, and chest tightness appear as well. An increase in the VOCs concentration can result in acute poisoning, unconsciousness, dizziness, and even death (exposure for 5−10 min to air containing 2 vol.% of benzene could be fatal). Some more serious health problems (such as central nerve system depression and irritation of the respiratory tract, skin, or mucous membranes) after a long-term exposure to VOCs have also been confirmed. The contact with benzene vapor is a known reason of leukemia and lymphomas, whereas aldehydes and ketones cause nasopharyngeal cancer.

4. MAIN METHODS OF VOLATILE ORGANIC COMPOUNDS ABATEMENT

Taking into account the deleterious VOCs effects, their content in the atmosphere has to be monitored. Therefore, rigorous regulations limiting the permissible level of VOCs emission have been enunciated in many countries. The current Göteborg Protocol (1999) sets a decrease in the national emission ceilings of VOCs in the EU member countries by nearly half by 2020, as compared with the base year 2000.[26−28] In 2015, the US Environmental Protection Agency strengthened the National Ambient

Air Quality Standards for ground-level ozone to 70 ppb. To reduce its formation, the US Clean Air Act regulates the emission of VOCs from anthropogenic sources. Under the Clean Air Act, areas in the United States that do not meet the national standards are required to reduce the VOCs emission within their boundaries (excluding emission from mobile sources) by 3% each year until the national standard will be achieved.[29]

4.1 Variety of Volatile Organic Compounds Elimination Techniques

The emission of VOCs can be limited using primary and secondary methods (Fig. 3). The primary methods consist of an improvement of existing technologies and reduction in the number of emission sources. However, such approach is insufficient to achieve ambitious goal for the VOCs abatement defined by the strict legislation. Therefore, many secondary techniques have been developed.

The secondary techniques for the elimination of VOCs can be classified in two groups: (1) recovery methods (e.g., absorption, adsorption, condensation, and membrane separation), which open the perspective of recycling of valuable organic compounds, and (2) destructive methods (e.g., biofiltration, thermal oxidation, and catalytic oxidation), which result in an irreversible elimination of undesirable compounds.[2,26,30–32] The choice of suitable removal techniques depends mainly on the kind and concentration of emitted VOCs.

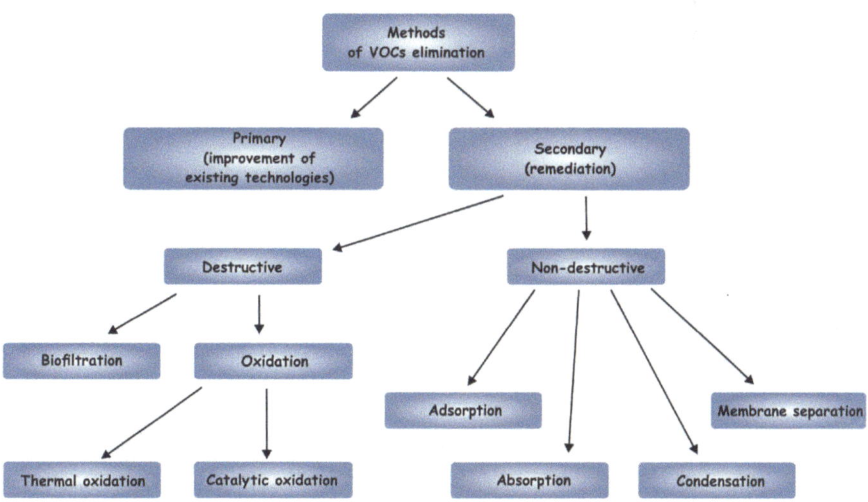

Figure 3 Classification of methods used for VOCs abatement.

When the content of VOCs in a contaminated gas stream is relatively high (between 500 and 5000 ppm), the organic pollutants can viably be removed by absorption in an appropriately chosen solvent (for instance, H_2O in the case of water-soluble compounds). The process is conducted in an absorber designed in a way to enable easy mass transfer.[30,31] The main disadvantage of this method is, however, the need for separation of dissolved VOCs from the spent solvent. Alternatively, adsorption can be applied as the recovery technique. In this approach, solid adsorbents with expanded surface areas (e.g., activated carbons, zeolites, and resins) interact physically and/or chemically with organic molecules. This solution is only reasonable for the removal of valuable VOCs highly diluted in a gas stream. Nevertheless, a high cost of production and necessity of frequent regeneration are serious limitations of the adsorption technology.[2,30,31,33,34] VOCs can also be eliminated by condensation induced by an increased pressure at a constant temperature or by a lowered temperature at a given pressure. This method is the most efficient during the abatement of VOCs characterized by a boiling point higher than 38°C and their high concentration (preferably exceeding 5000 ppm). The removal of low-boiling VOCs can demand intense cooling, which results in high costs for this operation.[30] Another shortcoming of the condensation process is the need for the utilization of huge quantities of cooling fluids, which due to their energy-consuming character, limits the applicability of this technology to only highly volatile solvents.[33] The last important nondestructive method of VOCs elimination is membrane separation. In this technique, both polymer and silicone membranes are tested. The selected baffle leaks VOC molecules due to a difference in concentration (gas permeation) or pressures (reverse osmosis) between both zones situated on the opposite sides of a membrane. It should, however, be kept in mind that commercialization of this solution is questionable because the developed membranes are relatively expensive, less stable, and less efficient compared to the expected results.[30]

The destructive methods of VOCs elimination are usually more effective. After all, the technologies based on total oxidation (both thermal and catalytic), which are described in detail below, seem to be very flexible. Furthermore, biofiltration was proposed in which VOCs are removed by passing gas-containing VOCs through a biofilter being a support of a thin water film filled with living microorganisms. After adsorption and/or absorption of VOCs on the biofilter, they are decomposed by the biologically active material.[31,32,35]

Thinking about the elimination of VOCs present in flue gases emitted mainly from stationery sources, special attention should be given to the presence of particulate matter, which usually disturbs the operation of classical methods of VOCs elimination. Therefore, before the fundamental step of VOCs removal, a pollutant stream has to be cleaned by particle precipitators or filters. Such solutions, besides using dedicated constructions of heat exchangers, are commercially available for the reduction of particulate matter released even by domestic wood log stoves and boilers combusting different kinds of wood fuel (e.g., pellets, logs, or chips).[36] Moreover, it should be kept in mind that an essential part of biomass-derived $PM_{2.5}$ consists of an inorganic core (constituted mainly by potassium compounds, such as K_2SO_4) surrounded by a collapsed graphitic structure.[25] An interesting approach to the elimination of alkali-rich particles together with soot and VOCs was proposed by Chen et al.[37] who applied hollandite manganese oxides with a one-dimensional tunnel structure (diameter of about 0.46 nm) to trap K^+ ions with a very high efficiency. The trapping capacity of hollandite manganese oxides was calculated to be 1450 μmol/g according to a theoretical equation

$$\square Mn_8O_{16} + K^+ \rightarrow KMn_8O_{16}, \tag{6}$$

where $\square$ represents a vacant site in the material structure. The role of this material was, first, to disintegrate the potassium-containing core and subsequent burning of the organic part of the pollutants. It was found that the presence of K (potassium) in the hollandite manganese oxide structure accelerated the rate of $PM_{2.5}$ and VOCs combustion due to a high electron transport ability enabling easy dissociation of O_2 molecules.

5. CATALYTIC COMBUSTION OF VOLATILE ORGANIC COMPOUNDS

5.1 Thermal Oxidation Versus Catalytic Oxidation

Many different studies presented in the literature clearly show that total oxidation of VOCs in the presence of solid catalysts is one of the most efficient methods of disposal of these harmful pollutants. The best results in the case of this approach are obtained by collection of VOCs at relatively low concentration from a contaminated gas stream (up to 1 vol.%; however, usually less than 1000 ppm).[33,38] Nevertheless, the catalytic process should be performed selectively toward products, which are environmentally friendly and less toxic compared to the converted VOCs. It should be stressed that the catalytic combustion exhibits some advantages compared to the corresponding thermal oxidation; primarily, the operating temperatures are lower (200–500°C vs. about 1000°C). In consequence, the use of reactors resistant to high temperature is not required, implying economic benefits (due to energy saving) and reduction in the amount of other harmful products formed (such as NO_x or dioxins).[26,39] It is obvious that the role of the catalyst is to increase the reaction rate by a change in the reaction mechanism, resulting in reducing the activation energy between reactants and products. Moreover, desirable reaction products can be formed over an appropriately selected catalyst. In the case of the combustion of VOCs, which is an exothermic reaction, thermodynamic equilibrium is strongly shifted toward the products of deep oxidation (mainly CO_2 and H_2O):

$$C_xH_y + \left(x + \frac{1}{4}y\right)O_2 \rightarrow xCO_2 + \frac{y}{2}H_2O. \tag{7}$$

However, the kinetic aspect requires that the process has to be carried out at elevated temperatures. From this point of view, stability of the catalyst becomes a very important parameter. The life time of the catalyst is mostly influenced by the resistance to poisoning by various components of the flue gases. On the other hand, a decrease in the reaction temperature could result in the appearance of higher amounts of by-products (kind and concentration depend on the converted substrate) formed by partial oxidation favored under such conditions.

5.2 Demands for Volatile Organic Compounds Combustion Catalysts

In some cases, catalysts of VOCs combustion gradually lose their initial activity and selectivity with time-on-stream. The main reason for this phenomenon is accumulation of various impurities on the catalyst surface. Compounds containing halogens, nitrogen, phosphorous, or sulfur atoms are known to gather on active sites and block the catalytic activity. Alternatively, by-products formed through side reactions can permanently be adsorbed and most frequently form carbon deposits (so-called inactive coke).

Catalysts used for the elimination of chlorinated VOCs should be characterized by an enhanced resistance to chlorine and its derivatives, and a high selectivity toward CO_2, H_2O, and HCl during the catalytic performance at relatively low temperatures. The latter issue is crucial because unsatisfactory selectivity to HCl can result in the generation of other by-products (such as $COCl_2$ or Cl_2), which are even more toxic compared to primary VOCs.[40] Dai et al.[41] demonstrated that the main products of trichloroethene combustion over a CeO_2 catalyst were CO_2, HCl, Cl_2, and small amounts of strongly undesirable products, including CO and chloroethyne. It was found that the introduction of water vapor to the stream of contaminated gases (at a level of 3 vol.%) suppressed the catalytic activity but simultaneously improved the selectivity to HCl. The authors showed that the catalytic elimination of chloroalkenes is essentially more difficult than that of chloroalkanes due to the higher C—Cl bond energy in unsaturated hydrocarbons.

The development of properly active catalysts for the total oxidation of organic pollutants containing sulfur is a real challenge for scientists working in the field of environmental catalysis. The main problem is to find a catalyst, which could be appropriately resistant to S compounds exhibiting high affinity to various inorganic surfaces. Furthermore, low selectivity for the expected products (namely CO_2, SO_2, and H_2O) is achieved. Unfortunately, sulfur oxides appear among the oxidation products. It should, however, be kept in mind that both SO_2 and SO_3 are strong, volatile oxidants, which in contact with moisture present in the atmosphere form sulfur acids. These compounds after condensation return to the earth surface in the form of so-called acid rain (very dangerous for plants and animals).

5.3 Mechanism of Catalytic Total Oxidation

A variety of organic pollutants classified as VOCs, which present sometimes very different properties, makes it impossible to postulate one general catalytic way for their elimination. O'Malley and Hodnett[42] studied the combustion of VOCs over typical Pt catalysts supported on β zeolite, mordenite, SiO_2, and Al_2O_3. The following tendency of VOCs to be oxidized was found:

alcohols > aromatic hydrocarbons > ketones > carboxylic acids > alkanes

However, more reactive alkanes were more easily combusted compared to carboxylic acid. The bond energies in the VOC molecules are considered to be the most important factor influencing the stability of organic molecules during the total oxidation process. A similar sequence of VOCs reactivity proposed by Tichenol and Palazzolo[43] is as follows:

alcohols > aldehydes > aromatic hydrocarbons > ketones > acetates > alkanes > chlorinated compounds.

According to Hermia and Vigneron,[44] when the molar mass of VOCs is higher, its total oxidation is harder. Furthermore, due to higher reactivity of unsaturated hydrocarbons in relation to their saturated analogues, the former ones undergo combustion easier.

The catalytic process involving a solid catalyst is very complex in which some distinct steps can be distinguished. First of all, reactant molecules have to migrate to an external surface of the catalyst from the gas or liquid phase. Because the catalyst is very often a porous material, they diffuse later into the pores (inner surface) and are chemisorbed (at least one of the reactants taking part in the catalytic process) on active sites. Subsequently, the surface reaction occurs in case the activation energy is achieved, and the molecules are in a suitable intermediate state and contact each other. Finally, the formed products desorb and migrate from the pore system back to the gas phase. For the VOCs combustion, three different mechanisms have been discussed in the open literature: (1) Mars-van Krevelen, (2) Langmuir–Hinshelwood, and (3) Eley-Rideal.[38,45–47] In the Mars-van Krevelen model, the reaction between adsorbed VOC molecules and oxygen located in the catalyst lattice is postulated. The combustion of VOCs occurs in two steps as illustrated schematically in Fig. 4.

First, the adsorbed molecule reacts with the lattice oxygen giving gaseous products and results in the reduction of the active phase. The number of oxygen vacancies distributed on the catalyst surface grows with the reaction time. Subsequently, the reduced surface is reoxidized by molecular oxygen available in the reaction mixture, which is adsorbed in the neighborhood of the reduced site. Alternatively, O^{2-} ions can migrate from the bulk to the reduced sites, and oxygen molecules can be adsorbed on the surface in a place located far from the formed vacancy. Under steady-state conditions, the rate of reduction and reoxidation should be the same.

The Langmuir–Hinshelwood mechanism postulates a reaction between the adsorbed VOC molecules and the chemisorbed oxygen (Fig. 5). The

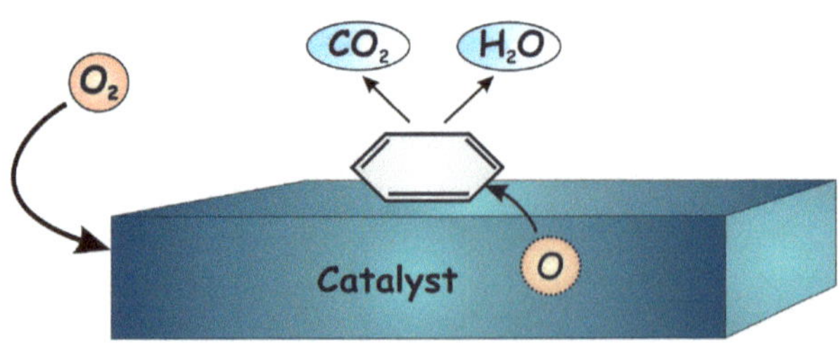

Figure 4 Mechanism of VOCs combustion according to the Mars-van Krevelen theory.

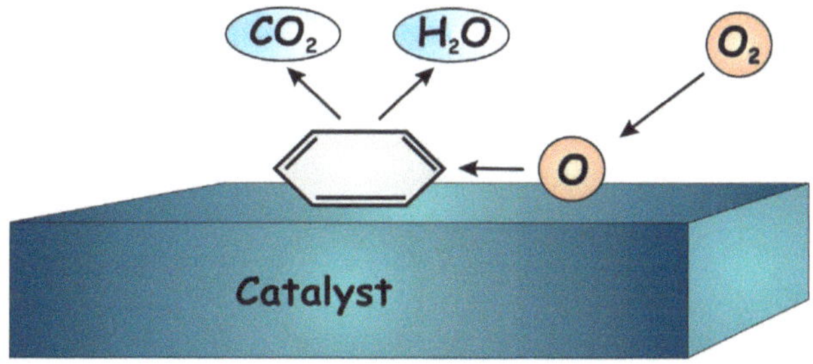

Figure 5 Mechanism of VOCs combustion according to the Langmuir–Hinshelwood theory.

VOC and O_2 molecules can be activated on the same type of surface sites (single-site model) or on different surface sites (dual-site model). The rate determining step in this process is the surface reaction. Therefore, homogeneous dispersion of adsorbed reactants is a crucial parameter.

The last considered mechanism for VOCs transformation corresponds to the Eley-Rideal theory, which assumes the reaction of chemisorbed oxygen species and VOC molecules from the gas phase (Fig. 6).

The total oxidation of VOCs can proceed on the same catalyst according to different mechanisms depending on the reaction conditions. Generally, at low temperatures, the Langmuir–Hinshelwood approach dominates, whereas an increase in the process temperature usually results in a changeover in mechanism to that proposed by Mars-van Krevelen. This effect is explained by a significant mobility of lattice oxygen at high temperatures (400–500°C). Nonetheless, the Langmuir–Hinshelwood mechanism describes more often the total oxidation of VOCs, opposite to the redox mechanism being more suitable to account for the kinetics of the selective

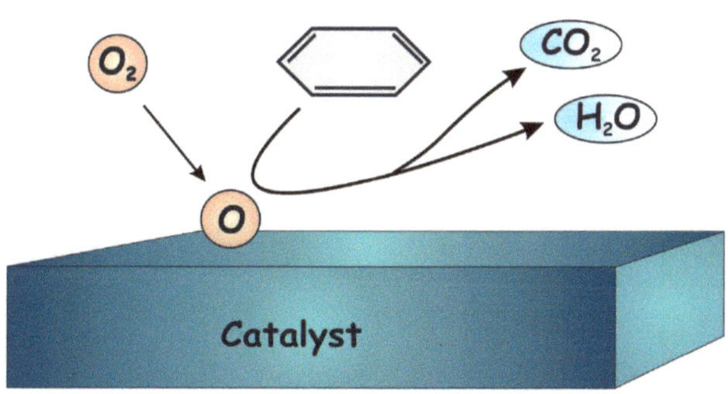

Figure 6 Mechanism of VOCs combustion according to the Eley-Rideal theory.

oxidation of organic compounds. To determine the reaction mechanism, isotopic studies are very often applied, for example, ^{16}O is exchanged by ^{18}O, ^{12}C by ^{13}C, or H by D, and the reaction products are analyzed online by mass spectrometry using the detection of the intensity of selected m/z lines.[48,49] Behar et al.[46] investigated the kinetics of toluene combustion in the presence of a $Cu_{1.5}Mn_{1.5}O_{4.0}$ catalyst. The performed studies revealed that the catalytic oxidation of toluene proceeds according to the Mars-van Krevelen redox mechanism. This mechanism was confirmed by other authors[49] who proposed the following consecutive steps: (1) adsorption of toluene on a catalyst surface, (2) simultaneous elimination of hydrogen from methyl and phenyl groups, (3) removal of carbon from the methyl group, and finally, (4) destruction of the aromatic ring.

5.4 Types of Volatile Organic Compounds Combustion Catalysts Used

The performance of studied VOCs combustion catalysts depends on many different factors, including physicochemical characteristics of the oxidizing organic molecules, active phase (its concentration, chemical state and dispersion, preparation method, and kind of precursors used), support (porosity, acid-base properties, hydrophilicity, particle size, and morphology), and reaction procedure (concentration of VOCs and O_2 in the reactant mixture, flow rate, and type of reactor).[26,50–52]

The easy deactivation by heteroatom-containing compounds is typical for noble metal–based catalysts, which are the largest group of materials used as catalysts in the VOCs combustion. These catalysts usually contain a small amount (between 0.1 and 0.5 wt.%) of metal (typically Pt or Pd, and rarely Au, Rh, or Ag) and are characterized by extremely high activity and selectivity toward CO_2 production.[26,40,45,53–56] Due to high sensitivity to deactivation and relatively high cost of production, the use of noble metal–modified catalysts is undesirable on a commercial scale, and alternative catalytically active materials have been developed. Among noble metal-free catalysts, materials containing CuO, CeO_2, Co_3O_4, MnO_x, V_2O_5, Cr_2O_3 and Fe_2O_3 are usually used.[39,57] The production of these heteroatom-resistant catalysts is more economic compared to Pt- or Pd-loaded materials.[58,59] However, the oxide catalysts are generally less active, and higher reaction temperatures have to be used. Regardless of the nature of the active phase, an increase in the accessibility of VOC and O_2 molecules to the surface sites is required. Thus, the catalysts are produced using methods generating enhanced porosity, for example, by decomposition of hydrotalcite-like[60–81] or alginate precursors,[46] as well as polymer-assisted,[82–85] glycothermal,[86] hydrothermal,[87–89] autocombustion,[90–93] or sol-gel routes.[92,94–100] Furthermore, the active phase is very often deposited on a support characterized by an expanded surface area. The support in catalysts for VOCs combustion can play an important role: Al_2O_3,[49,59,101–106] TiO_2,[107–115] ZrO_2,[48,53,116–118] CeO_2,[41,119–124] Nb_2O_5,[125] SnO_2,[126,127] Cr_2O_3,[128,129] MnO_2,[130,131] mesoporous silica,[57,132–137] zeolites,[137–148] clays,[149–154] or even various carbons.[39,50,155–157] The deposition of the active phase on a porous support not only increases its dispersion and limits consumption of expensive ingredients used in the catalyst production but also results in the appearance of specific properties related to synergetic effects of the system components. Some different techniques are used for the introduction of the catalytically active phase.[158,159] The most popular methods are wet impregnation and incipient wetness impregnation with an aqueous solution containing a precursor of the active phase (usually metal nitrate, acetate, or chloride). Apart from these approaches, also ion exchange, chemical vapor deposition, plasma deposition, and precipitation are often utilized for the introduction of the active phase onto the support.

In commercial applications, catalysts containing small amounts of noble metals (mainly Pt or Pd) dispersed on a moderately acidic support (e.g., γ-Al_2O_3), which is formed in monoliths or shaped particles, are encountered. The choice of suitable shape and size of catalyst particles is essentially determined by the type of reactor used. Besides desirable activity and selectivity, and low cost of production, the developed catalysts should not cause pressure drops, diffusion limitations, and temperature gradients as well as should be characterized by mechanical and thermal stability.

It is obvious that dispersion of noble metal particles on a support should have a significant impact on the catalytic activity in the total oxidation of VOCs. However, for Pt–V catalysts synthesized for the combustion of short-chain hydrocarbons (e.g., propane), no effect of particle size and oxidation state of Pt on the catalytic performance was found.[101] The main factor influencing the efficiency of the catalyst was the interaction between Pt and V leading to changes in reducibility. Nevertheless, in the case of typical supported noble metal–containing catalysts, particle size of the active phase is strongly correlated with the catalytic activity. Such observation was described, for example, by Abdelouahab-Reddam et al.[39] for Pt introduced by impregnation on commercial activated carbon doped with Ce (5–20 wt.%). The conversion of ethanol over these catalysts resulted in the formation of CH_3CHO, CO_2, and H_2O. The best results were obtained for a sample modified with 10 wt.% of Ce (total conversion of ethanol to CO_2 at 160°C). Pd and Pt dispersion can be improved by the addition of oleic acid to an aqueous solution of metal precursors (H_2PdCl_6 and H_2PtCl_6) used during incipient wetness impregnation.[160] Xie et al.[161] proposed to enclose Pt particles inside Pd@Co core-shell structures dispersed on the surface of three-dimensionally ordered macroporous CeO_2 as a way to stabilize noble metal nanoparticles during the highly exothermic methane combustion, which usually migrate at elevated temperatures.[162] From this point of view, an important parameter influencing the catalytic performance of noble metal–containing catalysts is a strong metal–support interaction enhanced by the application of various oxide-type carriers, including monolith-like TiO_2 nanotube arrays,[163] Co_3O_4,[164] or CeO_2.[165]

Water vapor, which is one of the VOCs combustion products and always present in exhaust gases, is simultaneously identified as an important inhibitor.[39,50,166,167] Hydrophilic inorganic oxides, contained as supports or active phases in some catalysts, facilitate adsorption of H_2O molecules on a catalyst surface and results in gradual deactivation. To avoid this effect, which is especially distinct at low temperatures, an active phase can be deposited on carbonaceous or organic hydrophobic supports, such as activated carbons,[39,168] nanofibers,[166] nanodiamonds,[169] mesoporous replicas, polymers, and aerogels.[158] These materials can be formed in different ways (including powders, granulates, pellets, films, and monoliths) and are characterized by high surface areas, expanded porosity, and open opportunities for further physical and chemical surface modifications.[50,155,156] Nevertheless, a serious risk in the application of carbon-containing catalysts is the possibility of their continuing gasification at elevated temperatures in an oxidizing atmosphere, promoted in addition by the presence of noble metals and transition metal oxides.[50,155,156] Pt/carbon aerogel catalysts were examined in the combustion of toluene and acetone.[158] Because of the limited thermal stability of the carbon support, the highest applied temperature of the developed systems was restricted to about 300–400°C depending on pretreatment conditions (higher stability of carbon supports is favored by higher pyrolysis temperature). It should, however, be kept in mind that during the strong exothermic reaction of VOCs burning, a huge amount of heat is released, which can result in local overheating and destruction/gasification of the support.

A decrease in the reaction temperature protects against decomposition but can simultaneously lead to an enhancement of adsorption effects over oxide- and carbon-based materials. Matejova et al.[170] investigated the performance of catalysts (containing 0.05–0.13 wt.% of Pt–Pd and Cu(3.34 wt.%)–Mn(5.44 wt.%)–O supported on Al_2O_3 or Al_2O_3–CeO_2) in the total oxidation of ethanol, dichloromethane, and toluene, chosen as representatives of various groups of VOC oxygenates, as well as halogen-containing and aromatic compounds. The catalytic activity decreased in the following sequence:

$$Pd/Al_2O_3 > Pt{-}Pd/Al_2O_3{-}CeO_2 > Cu{-}Mn{-}O/Al_2O_3$$

The temperatures of 50% (T_{50}) toluene conversion were 214°C, 270°C, and 282°C, whereas T_{90} were 223°C, 285°C, and 301°C, respectively. At lower temperatures (140–260°C), a partial disappearance of toluene was observed over the Cu–Mn–O/Al_2O_3 catalyst, but no adequate amounts of CO_2 were produced. A significant increase in CO_2 concentration was found at temperature as high as 275°C. The observed effect was explained by chemisorption of toluene, which played a dominant role at lower temperatures. After heating the system to 275°C, chemisorbed VOC molecules began to be oxidized to CO_2.

Cerium oxide (CeO_2), manganese oxides (MnO_x), and chromium oxides (CrO_x) show very promising catalytic properties in the total oxidation of VOCs due to their specific reducibility, oxygen storage capacity, and oxygen mobility.[171] Therefore, these oxide systems are often used as components of catalysts for the VOCs removal.[38,39,101] Sinha and Suzuki[108] synthesized novel 3D mesoporous chromium oxide, being an analogue of mesoporous silica sieves. The catalytic measurements in toluene combustion revealed that at a temperature of 85°C the specific activity of meso-CrO_x related to 1 m^2 of its surface was 35 times higher compared with meso-SiO_2. The proposed type of the active phase is, however, undesirable due to the carcinogenic potential of Cr(VI) species, which are predominantly formed under the oxidizing atmosphere at elevated temperatures.[57]

The cryptomelane-type octahedral MnO_2 molecular sieve with a one-dimensional tunnel structure composed of 2 × 2 edge-shared MnO_6 octahedral chains was recognized as a very attractive VOCs combustion catalyst due to bulk oxygen vacancy, high reducibility, and reoxidation ability.[172] From this point of view, MnO_2 containing Mn^{4+} with tremendous reducibility is the most frequently studied manganese oxide–based catalyst.[173–175] However, Kim and Shim[176] reported the following order of decreasing activity of manganese oxides in the total oxidation of aromatic VOCs (benzene and toluene):

$$Mn_3O_4 > Mn_2O_3 > MnO_2.$$

For the best Mn_3O_4 catalyst, T_{50} and T_{90} in the toluene combustion were achieved at 245°C and 270°C, respectively. Mn_3O_4 exhibits the presence of the highest amounts of surface electrophilic oxygen, as well as Brønsted acid sites (corresponding to Mn–OH species) beneficial for the total oxidation of VOCs and their initial adsorption/activation.[177] In addition, an introduction of structural promoters (0.5 wt.% of K, Ca, and Mg) to MnO_x clearly suppressed the temperature of toluene conversion.[176]

Because a wide spectrum of various forms of reactive oxygen (O_2^-, O_2^{2-} and O^-) appears at the interface as a result of strong interaction between MnO_x and CeO_2, mixtures of these phases were extensively applied in catalysts for the total oxidation of VOCs.[178] Manganese ions are partially incorporated into the CeO_2 structure to give enhanced oxygen mobility and oxygen storage capacity. Furthermore, the appearance of Mn–O–Ce bonds greatly lowers the energy of oxygen vacancy formation.[179–184] It was shown that the catalytic performance of these systems can be amended by, e.g., synthesis based on a hydrolysis-driven redox coprecipitation method,[185] as well as an acid treatment, resulting in an improvement of textural and redox properties.[186]

Zirconium oxide (ZrO_2) is very often investigated as support of various active phases. ZrO_2 is characterized by excellent ionic conductivity, which is connected with diffusion of O^{2-} ions through oxygen vacancies present in the lattice. It was confirmed that catalysts supported on tetragonal ZrO_2 exhibit higher catalytic activity in the total oxidation of aromatic VOCs compared to materials based on γ-Al_2O_3.[118] Stabilization of the required tetragonal phase of ZrO_2 is attained by the addition of foreign ions, such as Y^{3+}, Mg^{2+}, and Ca^{2+}. The introduced cations partially exchange Zr^{4+} ions in the lattice. The generated charge is compensated by the appearance of additional oxygen vacancies. The best stabilization effect was found after the introduction of 5 mol% of Y_2O_3.[117] The ZrO_2–Y_2O_3 supports were synthesized for instance by the coprecipitation and sol-gel methods.[116] The selected preparation route influenced the porosity of the final support. The intraparticle porosity dominated in the coprecipitated material, whereas the interparticle voids appeared in the catalysts obtained by the sol-gel approach. The deposition of CuO (2.5–7.5 wt.% of Cu) caused a partial blocking of the pore system manifested by a decrease in specific surface area and pore volume, especially distinct for the sol-gel samples. Simultaneously, these catalysts showed higher catalytic activity in the toluene combustion compared with the corresponding coprecipitated samples at the same CuO contents. This phenomenon was explained by a smaller CuO crystallite size. The best catalytic results were collected for the sample containing 5 wt.% of Cu supported on $ZrO_2-Y_2O_3$ prepared by the coprecipitation technique ($T_{50} = 264°C$).

Porous ZrO_2, TiO_2, and ZrO_2-TiO_2 containing Pd were tested in the combustion of toluene and chlorobenzene.[187] The highest efficiency in toluene elimination was observed for Pd/TiO_2, which gave 100% conversion at a temperature as low as 190°C. It was postulated that in the initial step of the reaction, Pd^0 is oxidized by gaseous O_2. The formed $Pd^{2+}O^{2-}$ species are subsequently active in the combustion of VOC molecules and return to the starting Pd^0 state.

Unfortunately, a comparison of literature data related to the catalytic activity of the studied materials in the VOCs elimination is not easy due to various reaction conditions used by different scientific groups working on this topic. Nevertheless, we analyzed the available papers with respect to activity in toluene combustion, by which a coherent view was obtained as presented in Fig. 7, where T_{50} achieved over different catalysts containing noble metals[53,107,187–190] and transition metal oxides[60,116,150,151,191] under the same reaction conditions (total flow rate = 100 mL/min, toluene vapor concentration = 1000 ppm, and catalyst mass = 100 mg) was compared.

Co_3O_4 spinel was recognized as one of the most active oxide-type phases in the VOCs combustion. Its dispersion was increased by the generation of mesoporosity,[192–194] colloidal dispersion,[195] as well as using various supports, including CeO_2,[123] clay (halloysite),[149] and mesoporous silicas,[132] which resulted in higher specific surface areas, weak acid sites, and higher abundances of surface Co^{2+} and adsorbed O species, being crucial parameters in the catalytic combustion of VOCs. The Co_3O_4 phase was also deposited on zeolites, viz., ZSM-5[196] and BEA.[138] The latter one was previously found to be ideal in the dispersion of the MnO_x active phase for toluene combustion.[197] First of all, influence of acidity of catalyst surface on the activity and selectivity in toluene combustion was investigated over Co_3O_4/BEA. It was shown that the complete elimination of Bronsted acid sites (related to Al^{3+} framework cations) from the zeolite structure by dealumination did not change the catalytic activity but improved significantly the selectivity for desired reaction products (H_2O and CO_2). No

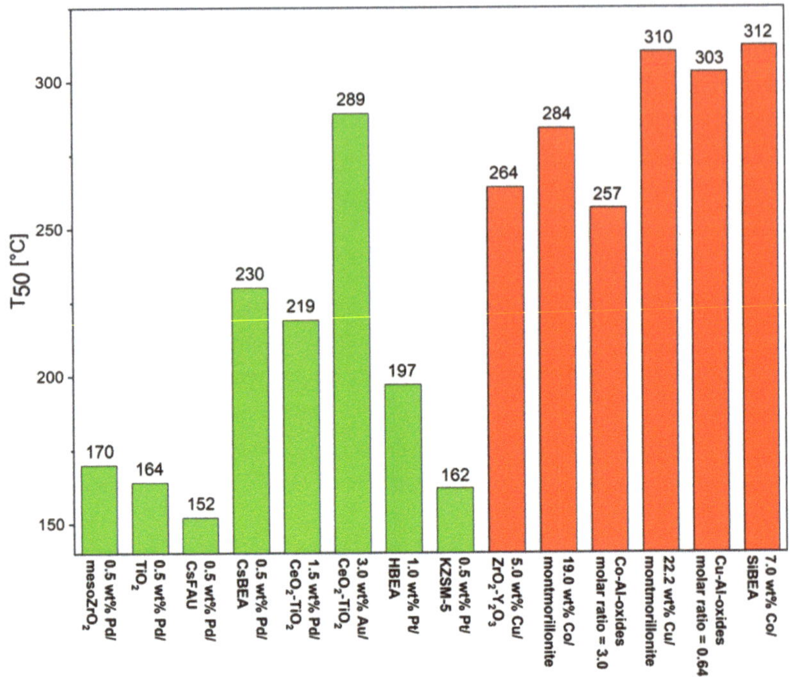

Figure 7 Temperature of 50% toluene conversion observed over noble metal– (green bars) and transition metal oxide–containing catalysts (red bars) under identical reaction conditions (total flow rate = 100 mL/min, toluene vapor concentration = 1000 ppm, and catalyst mass = 100 mg).

CO and benzene observed over the β zeolite with the Si/Al molar ratio of 14 were found for the catalyst based on the support with Si/Al → ∞. Co_3O_4 deposited on the surface in the form of clusters and nanocrystallites showed considerably higher activity (working according to the Mars-van Krevelen mechanism) compared to framework Co^{2+}.

Rokicińska et al.[151] demonstrated a very innovative idea of formation of a Co_3O_4 phase dispersed in the structure of partially delaminated aluminosilicate (e.g., montmorillonite). This approach was presented for an example of production of hydrogel–clay composites by intercalation polymerization of acrylic acid in an aqueous slurry of montmorillonite, resulting in partial exfoliation of its structure. In the subsequent step, Co^{2+} cations were deposited in the composite by adsorption in an aqueous solution of cobalt(II) nitrate at controlled pH within the range of 3.3–9.0. The presence of adsorption sites on the clay layers and intercalated polymer chains facilitated the Co^{2+} adsorption, which was most efficient at higher pH. Under such conditions, Co^{2+} ions easily interact with dissociated COO^- groups and can be distributed homogeneously over the whole composite structure. Finally, after high-temperature calcination at 500°C, the hydrogel template is removed, and the introduced Co^{2+} forms the highly dispersed Co_3O_4 spinel nanophase. What is very important, the Co_3O_4 nanocrystals are well protected against sintering by the exfoliated clay layers.

The amount of deposited Co^{2+} can be controlled not only by pH but also by the volume and concentration of the metal cation solution used for adsorption. In Fig. 8, the catalysts containing various Co loadings (from 5 to 25 wt.%) are compared. The samples were prepared by adsorption at a constant pH = 6.0 and varying volume (100–700 mL) of 0.01 M solution of $Co(NO_3)_2$. The formed Co_3O_4 nanoparticles pillared the exfoliated aluminosilicate layers, forming mesopores and increasing specific surface area (under optimal preparation conditions exceeding 100 m^2/g). As shown in Fig. 8, reducibility (measured as a temperature at which reduction of Co_3O_4 begins) and catalytic activity (expressed as conversion of toluene determined at 300°C) of the prepared catalysts depend on the Co content. In the case of the catalysts with low Co loadings, the active Co_3O_4 phase is strongly stabilized on the clay surface and its reduction demands temperatures higher than 350°C. This results in low catalytic activity of these materials. An increase in the Co content enables to obtain

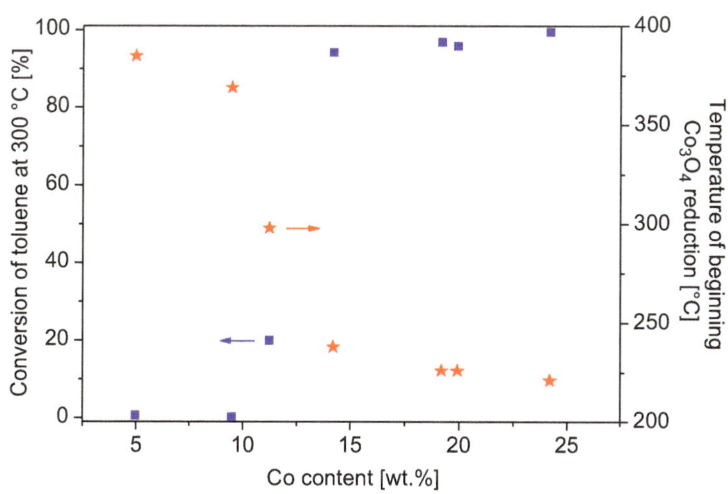

Figure 8 Correlation between content of Co_3O_4 spinel phase dispersed in partially delaminated clay and its reducibility (stars) as well as catalytic activity in toluene combustion (squares).

larger Co_3O_4 nanoparticles characterized by easier reducibility and higher activity. The presented tendency clearly confirms that the studied reaction proceeds according to the Mars-van Krevelen mechanism, and the presence of labile lattice oxygen in the used catalyst is obviously required.

5.5 Perspective Combination of Adsorption and Catalytic Combustion

The process of VOCs combustion can be executed without external heating and addition of fuel when the concentration of organic pollutants is 0.6–0.8 g/m^3 and higher.[198] For lower contents of VOCs, energy needs to be introduced into a converter. A great challenge from an economic point of view is especially the removal of VOCs from exhaust gases at concentrations lower than 0.1 g/m^3. In such case, an excellent solution is a combination of adsorption and catalytic oxidation. Initially VOC molecules are accumulated inside the pores of the adsorbent and then desorbed after a significant condensation and transferred to a catalytic converter.[199] Both processes can also be performed within one material, being simultaneously an efficient adsorbent and catalyst. The material can work sequentially in cycles, including adsorption and combustion, for a long time.

Zazhigalov et al.[198] studied commercially available alumocopper–chromium oxide catalyst (in the form of 10.5 × 10.5 × 2.5 mm Raschig rings) in the adsorption–catalytic processes for VOCs elimination. They proposed separation of the catalyst bed into several independent sections, with regeneration in a sequential manner. The developed flowsheet resulted in reduction of maximum concentrations in flash emissions of desorbed VOCs and peak outlet temperatures during regeneration (temperature of the gases could be reduced from 275 to 65°C).

In the role of adsorbent catalysts, Pt nanoparticles (1 wt%) and/or Bi_2O_3, CeO_2, Fe_2O_3, NiO, RuO_2, and ZrO_2 (1, 5 or 10 wt%) supported on mesoporous Al_2O_3 were tested.[200] Such systems can be used as VOC sensors. In the first step, VOCs were adsorbed at room temperature, 100–150°C, and next pulsed temperature heating to 250–450°C resulting in the occurrence of catalytic destruction. Characteristic sensor-signal profiles were recorded with one large dynamic response (due to flash catalytic combustion of adsorbed molecules) and subsequent static response (due to general catalytic combustion) during the pulse heating. The highest responses were observed for the sensor containing 1 wt% of Pt and 10 wt% of CeO_2, which exhibited the largest surface area. It was found that the integrated dynamic response is

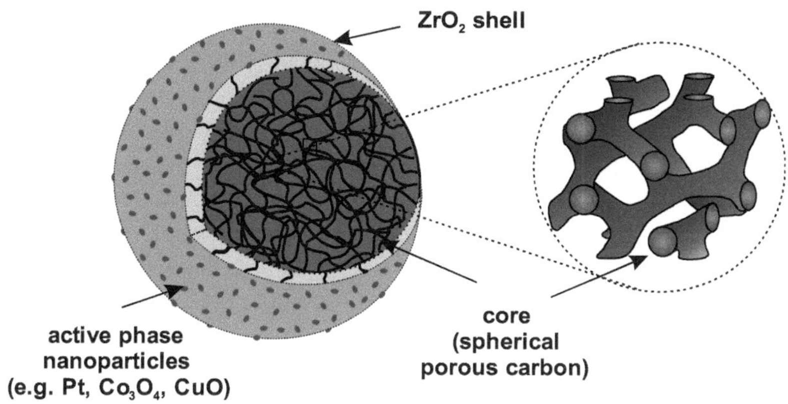

Figure 9 Model of bifunctional core-shell materials for the elimination of VOCs using an adsorption—catalytic method.

the most important response for quantitative detection of VOCs for the adsorption-/combustion-type gas sensors.[201]

Recently, we have demonstrated a new idea for the construction of dual active materials working in adsorption—combustion cycles.[202] We proposed core shell—type carbon-inorganic composites containing simultaneously components functionalized toward adsorption of VOCs and their subsequent catalytic combustion (Fig. 9).

The developed materials were synthesized in a subsequent step procedure including (1) preparation of spherical, mesoporous silica (e.g., MCM-48 sieve) being the structure-directing template; (2) introduction of furfuryl alcohol to the silica mesopores by precipitation polycondensation in an aqueous slurry; (3) deposition of a thin layer of ZrO_2 precursor on the surface of the formed polymer-silica composite; (4) thermal treatment to obtain the ZrO_2@carbon-silica core-shell material; (5) removal of the silica template; and (6) decoration of the external surface with a metal or oxide phase catalytically active in the VOCs combustion. Our preliminary studies revealed optimal conditions for the synthesis. It was shown that crystallization of tetragonal ZrO_2 (the best form of zirconia support) demands calcination at temperatures as high as 850°C. Furthermore, effective extraction of the silica template is possible by etching with a 3M NaOH solution at 50°C for 24 h. The final ZrO_2@carbon materials showed high surface areas ($S_{BET} \sim 500$ m^2/g) and pore volumes ($V_{total} \sim 0.6$ cm^3/g) with a dominant role of mesoporosity (contribution of about 60% in total porosity). Using the incipient wetness technique, the surface of core-shell samples was modified with the CuO or Co_3O_4 active phase (content of 2.5–15.0 wt %). Moreover, deposition of Pt nanoparticles on the ZrO_2@carbon materials was performed. The introduced Pt nanoparticles were obtained by reduction of $PtCl_4$ with NaOH in water-free ethylene glycol.

The spherical, porous materials containing the carbon core surrounded by the zirconia shell (important as support for the active phase, resulting in its high dispersion and a lack of contact with carbon, which could be easily oxidized together with VOC molecules), decorated with the active phase, were tested in the adsorption and catalytic combustion of VOCs. In the case of toluene, the prepared composites had the adsorption capacity of about 250 mg/g (including ca. 50 mg/g of chemisorbed VOC molecules). During the catalytic step, the most active materials containing CuO or Co_3O_4 exhibited total conversion of toluene at temperature below 300°C. The Pt-doped (1 wt.%) catalysts were significantly more active, viz., over these samples, toluene disappeared completely even below 200°C. Taking into account such promising results, one can expect that further studies on the proposed core-shell systems may result in the development of very attractive commercial materials for the removal of various VOCs present in air at low concentrations.

6. CONCLUSIONS AND PERSPECTIVES

VOCs are released to the atmosphere with increasing intensity from various natural and anthropogenic sources. In many places their concentration exceeds acceptable limits and becomes to be dangerous for the environment and living organisms. Therefore, continuously tightening VOCs emission regulations have been introduced, bringing new challenges to the developed VOCs abatement technologies. Because the overwhelming majority of the emission sources discharges VOCs at low concentrations, some typical methods, such as thermal combustion, condensation, and absorption, are impractical. Thus, the best known approach to the VOCs elimination is catalytic oxidation, which enables to transform VOCs into CO_2, H_2O, and other relatively less harmful products at significantly lower temperatures compared with thermal combustion. Nevertheless, suitable catalysts for this process are still being searched for. The highly active noble metal–containing materials are expensive and sensitive to poisoning. On the other hand, more stable and cheaper catalytic systems based on transition metal oxides have been proposed. Unfortunately, these catalysts are usually less active compared to the supported noble metals, and higher temperatures must be used to achieve the expected yields. Consequently, further efforts have to be made toward improving the activity of transition metal oxides in the VOCs combustion. Undoubtedly, the redox properties of such materials are extremely important due to the Mars-van Krevelen mechanism of the VOCs combustion. Furthermore, taking into account the low concentration of organic pollutants, various multiple remediation technologies can be considered. Among them, adsorption followed by catalytic oxidation seems to be a very promising solution.

ACKNOWLEDGMENTS

This work was supported by the National Science Centre under the grant no. 2013/11/B/ST5/01550.

REFERENCES

1. Moulijn, J. A.; Makkee, M.; van Diepen, A. E. *Chemical Process Technology*, 2nd ed.; John Wiley & Sons Ltd, 2013.
2. Zhanga, X.; Gaoc, B.; Creamerc, A. E.; Caoa, C.; Li, Y. J. *Hazard. Mater.* **2017,** *338,* 102–123.
3. Bianchi, F.; Barmet, P.; Stirnweis, L.; El Haddad, I.; Platt, S. M.; Saurer, M.; Lötscher, C.; Siegwolf, R.; Bigi, A.; Hoyle, C. R.; DeCarlo, P. F.; Slowik, J. G.; Prévôt, A. S. H.; Baltensperger, U.; Dommen, J. *Atmos. Environ.* **2016,** *141,* 41–47.
4. https://www.epa.gov/air-emissions-inventories/air-pollutant-emissions-trends-data.
5. Dinh, T. V.; Choi, I. Y.; Son, Y. S.; Song, K. Y.; Sunwoo, Y.; Kim, J. C. J. *Environ. Manage.* **2016,** *168,* 157–164.
6. Fiedrich, M.; Kurtenbach, R.; Wiesen, P.; Kleffmann, J. *Atmos. Environ.* **2017,** *165,* 57–61.
7. Wang, L.; Xiang, Z.; Stevanovic, S.; Ristovski, Z.; Salimi, F.; Gao, J.; Wang, H.; Li, L. *Sci. Total Environ.* **2017,** *589,* 173–181.
8. Oanh, N. T. K.; Tipayarom, A.; Bich, T. L.; Tipayarom, D.; Simpson, C. D.; Hardie, D.; Liu, L. J. S. *Atmos. Environ.* **2015,** *119,* 182–191.
9. Inomata, S.; Tanimoto, H.; Pan, X.; Taketani, F.; Komazaki, Y.; Miyakawa, T.; Kanaya, Y.; Wang, Z. J. *Geophys. Res. Atmos.* **2015,** *120,* 5237–5252.
10. Holder, A. L.; Gullett, B. K.; Urbanski, S. P.; Elleman, R.; O'Neill, S.; Tabor, D.; Mitchell, W.; Baker, K. R. *Atmos. Environ.* **2017,** *166,* 22–33.
11. Aurell, J.; Gullett, B. K.; Tabor, D.; Yonker, N. *Atmos. Environ.* **2017,** *150,* 395–406.
12. Coudour, B.; Chetehouna, K.; Halter, F.; Mounaïm-Rousselle, C.; Garo, J. P. *Combust. Sci. Technol.* **2016,** *188,* 2128–2136.
13. Coggon, M. M.; Veres, P. R.; Yuan, B.; Koss, A.; Warneke, C.; Gilman, J. B.; Lerner, B. M.; Peischl, J.; Aikin, K. C.; Stockwell, C. E.; Hatch, L. E.; Ryerson, T. B.; Roberts, J. M.; Yokelson, R. J.; de Gouw, J. A. *Geophys. Res. Lett.* **2016,** *43,* 9903–9912.
14. Czech, H.; Pieber, S. M.; Tiitta, P.; Sippula, O.; Kortelainen, M.; Lamberg, H.; Grigonyte, J.; Streibel, T.; Prévôt, A. S. H.; Jokiniemi, J.; Zimmermann, R. *Atmos. Environ.* **2017,** *158,* 236–245.

15. Szyszlak-Bargłowicz, J.; Zając, G.; Słowik, T. *Pol. J. Environ. Stud.* **2015,** *24,* 1349–1354.
16. Hong-li, W.; Sheng-ao, J.; Sheng-rong, L.; Qing-yao, H.; Li, L.; Shi-kang, T.; Cheng, H.; Li-ping, Q.; Chang hong, C. *Sci. Total Environ.* **2017,** *607–608,* 253–261.
17. Lim, C.; Kim, D.; Song, C.; Kim, J.; Han, J.; Cha, J. S. *Appl. Energy* **2015,** *139,* 17–29.
18. Costagliola, M. A.; Prati, M. V.; Florio, S.; Scorletti, P.; Terna, D.; Iodice, P.; Buono, D.; Senatore, A. *Fuel* **2016,** *183,* 470–477.
19. Jin, D.; Choi, K.; Myung, C. L.; Lim, Y.; Lee, J.; Park, S. *Fuel* **2017,** *209,* 702–712.
20. Nogueira, T.; Dominutti, P. A.; Fornaro, A.; de Fatima Andrade, M. *Atmosphere* **2017,** *8,* 144.
21. Tsai, J. H.; Yao, Y. C.; Huang, P. H.; Chiang, H. L. *Aerosol. Air Qual. Res.* **2017,** *17,* 1214–1223.
22. Li, L.; Ge, Y.; Wang, M.; Peng, Z.; Song, Y.; Zhang, L.; Yuan, W. *Sci. Total Environ.* **2015,** *502,* 627–631.
23. Gilman, J. B.; Lerner, B. M.; Kuster, W. C.; Goldan, P. D.; Warneke, C.; Veres, P. R.; Roberts, J. M.; De Gouw, J. A.; Burling, I. R.; Yokelson, R. J. *Atmos. Chem. Phys.* **2015,** *15,* 13915–13938.
24. Tiitta, P.; Leskinen, A.; Hao, L.; Yli-Pirilä, P.; Kortelainen, M.; Grigonyte, J.; Tissari, J.; Lamberg, H.; Hartikainen, A.; Kuuspalo, K.; Kortelainen, A. M.; Virtanen, A.; Lehtinen, K. E. J.; Komppula, M.; Pieber, S.; Prévôt, A. S. H.; Onasch, T. B.; Worsnop, D. R.; Czech, H.; Zimmermann, R.; Jokiniemi, J.; Sippula, O. *Atmos. Chem. Phys.* **2016,** *16,* 13251–13269.
25. Yao, L.; Yang, L.; Chen, J.; Wang, X.; Xue, L.; Li, W.; Sui, X.; Wen, L.; Chi, J.; Zhu, Y.; Zhang, J.; Xu, C.; Zhu, T.; Wang, W. *Sci. Total Environ.* **2016,** *557–558,* 520–530.
26. Liotta, L. F. *Appl. Catal. B* **2010,** *100,* 403–412.
27. Scire, S.; Liotta, L. F. *Appl. Catal. B* **2012,** *125,* 222–246.
28. Environmental Fact Sheet No 19. *Produced by The Swedish NGO Secretariat on Acid Rain, Goteborg, Sweden,* January 2006.
29. https://www.issa.com/data/moxiestorage/regulatory_education/voc_limits_summary_12-6-16.pdf.
30. Khan, F. I.; Kr.Ghoshal, A. *J. Loss Prev. Process. Ind.* **2000,** *13,* 527–545.
31. Hariz, R.; del Rio Sanz, J. I.; Mercier, C.; Valentin, R.; Dietrich, N.; Mouloungui, Z.; Hébrard, G. *Chem. Eng. Sci.* **2017,** *157,* 264–271.
32. Malakar, S.; Das Saha, P.; Baskaran, D.; Rajamanickam, R. *Environ. Technol. Innov.* **2017,** *8,* 441–461.
33. Kamal, M. S.; Razzak, S. A.; Hossain, M. M. *Atmos. Environ.* **2016,** *140,* 117–134.
34. Everaert, K.; Baeyens, J. J. *Hazard. Mater.* **2004,** *109,* 113–139.
35. Chen, H.; Yang, C.; Zeng, G.; Luo, S.; Yu, G. *Bioresour. Technol.* **2012,** *121,* 199–204.
36. Matthes, M.; Hartmann, I. *Chem. Eng. Technol.* **2017,** *40,* 340–350.
37. Chen, Y.; Tian, G.; Zhou, M.; Huang, Z.; Lu, C.; Hu, P.; Gao, J.; Zhang, Z.; Tang, X. *Environ. Sci. Technol.* **2016,** *50,* 5825–5831.
38. Huang, H.; Xu, Y.; Fenga, Q.; Leung, D. Y. C. *Catal. Sci. Technol.* **2015,** *5,* 2649–2669.
39. Abdelouahab-Reddam, Z.; El Mail, R.; Coloma, F.; Sepulveda-Escribano, A. *Catal. Today* **2015,** *249,* 109–116.
40. Aranzabal, A.; Pereda-Ayo, B.; Gonzalez-Marcos, M. P.; Gonzalez-Marcos, J. A.; Lopez-Fonseca, R.; Gonzalez-Velasco, J. R. *Chem. Pap.* **2014,** *68,* 1169–1186.
41. Dai, Q.; Wang, X.; Lu, G. *Catal. Commun.* **2007,** *8,* 1645–1649.
42. O'Malley, A.; Hodnett, B. K. *Catal. Today* **1999,** *54,* 31–38.
43. Tichenor, B. A.; Palazzolo, M. A. *Environ. Prog.* **1987,** *6,* 172–176.
44. Hermia, J.; Vigneron, S. *Catal. Today* **1993,** *17,* 349–358.
45. Spivey, J. J. *Ind. Eng. Chem. Res.* **1987,** *26,* 2165–2180.

46. Behar, S.; Gomez-Mendoza, N. A.; Gomez-Garcia, M. A.; Świerczyński, D.; Quignard, F.; Tanchoux, N. *Appl. Catal. A* **2015,** *504,* 203–210.
47. Ojala, S.; Pitkaaho, S.; Laitinen, T.; Koivikko, N.; Brahmi, R.; Gaalova, J.; Matejova, L.; Kucherov, A.; Paivarinta, S.; Hirschmann, C.; Nevanpera, T.; Riihimaki, M.; Pirila, M.; Keiski, R. *Top. Catal.* **2011,** *54,* 1224–1256.
48. Dongare, M. K.; Ramaswamy, V.; Gopinath, C. S.; Ramaswamy, A. V.; Scheurell, S.; Brueckner, M.; Kemnitz, E. *J. Catal.* **2001,** *199,* 209–216.
49. Menon, U.; Galvita, V. V.; Marin, G. B. *J. Catal.* **2011,** *283,* 1–9.
50. Morales-Torres, S.; Carrasco-Marin, F.; Perez-Cadenas, A. F.; Maldonado-Hodar, F. *J. Catal.* **2015,** *5,* 774–799.
51. Du, C.; Lu, S.; Wang, Q.; Buekens, A. G.; Ni, M.; Debecker, D. P. *Chem. Eng. J.* **2018,** *334,* 519–544.
52. Zhang, Z.; Jiang, Z.; Shangguan, W. *Catal. Today* **2016,** *264,* 270–278.
53. Tidahy, H. L.; Hosseni, M.; Siffert, S.; Cousin, R.; Lamonier, J. F.; Aboukais, A.; Su, B. L.; Giraudon, J. M.; Leclercq, G. *Catal. Today* **2008,** *137,* 335–339.
54. Chlala, D.; Labaki, M.; Giraudon, J. M.; Gardoll, O.; Denicourt-Nowicki, A.; Roucoux, A.; Lamonier, J. F. *C R Chim.* **2016,** *19,* 525–537.
55. Guo, Y.; Zhang, S.; Mu, W.; Li, X.; Li, Z. J. *Mol. Catal. A: Inside Chem.* **2017,** *429,* 18–26.
56. Zhu, J.; Mu, W.; Su, L.; Li, X.; Guo, Y.; Zhang, S.; Li, Z. J. *Solid State Chem.* **2017,** *248,* 142–149.
57. Somekawa, S.; Watanabe, H.; Oaki, Y.; Imai, H. *Catal. Commun.* **2015,** *72,* 161–164.
58. Li, W. B.; Wang, J. X.; Gong, H. *Catal. Today* **2009,** *148,* 81–87.
59. Larsson, P. O.; Andersson, A. *Appl. Catal. B* **2000,** *24,* 175–192.
60. Białas, A.; Mazur, M.; Natkański, P.; Dudek, B.; Kozak, M.; Wach, A.; Kuśtrowski, P. *Appl. Surf. Sci.* **2016,** *362,* 297–303.
61. Basąg, S.; Kovanda, F.; Piwowarska, Z.; Kowalczyk, A.; Pamin, K.; Chmielarz, L. J. *Therm. Anal. Calorim.* **2017,** *129,* 1301–1311.
62. Castano, M. H.; Molina, R.; Moreno, S. *Appl. Catal. A* **2015,** *492,* 48–59.
63. Mo, S.; Li, S.; Li, W.; Li, J.; Chen, J.; Chen, Y. J. *Mater. Chem. A* **2016,** *4,* 8113–8122.
64. Jirátová, K.; Kovanda, F.; Ludvíková, J.; Balabánová, J.; Klempa, J. *Catal. Today* **2016,** *277,* 61–67.
65. Zhao, S.; Yi, H.; Tang, X.; Kang, D.; Gao, F.; Wang, J.; Huang, Y.; Yang, Z. J. *Hazard. Mater.* **2018,** *344,* 797–810.
66. Brunet, J.; Genty, E.; Bugnon, L.; De Weireld, G.; Thomas, D.; Decroly, A.; Siffert, S.; Cousin, R. *Mater. Today* **2016,** *3,* 188–193.
67. Qu, Z.; Gao, K.; Fu, Q.; Qin, Y. *Catal. Commun.* **2014,** *52,* 31–35.
68. Mo, S.; Li, S.; Li, J.; Peng, S.; Chen, J.; Chen, Y. *Catal. Commun.* **2016,** *87,* 102–105.
69. Jirátová, K.; Kovanda, F.; Balabánová, J.; Kšírová, P. *Catal. Today* **2018,** *304,* 165–171.
80. Zhao, S.; Li, K.; Jiang, S.; Li, J. *Appl. Catal. B* **2016,** *181,* 236–248.
81. Mrad, R.; Cousin, R.; Saliba, N. A.; Tidahy, L.; Siffert, S. *C R Chim.* **2015,** *18,* 351–357.
82. Xie, S.; Deng, J.; Zang, S.; Yang, H.; Guo, G.; Arandiyan, H.; Dai, H. *J. Catal.* **2015,** *322,* 38–48.
83. Jiang, Y.; Xie, S.; Yang, H.; Deng, J.; Liu, Y.; Dai, H. *Catal. Today* **2017,** *281,* 437–446.
84. Yang, H.; Deng, J.; Liu, Y.; Xie, S.; Wu, Z.; Dai, H. J. *Mol. Catal. A: Inside Chem.* **2016,** *414,* 9–18.
85. Xie, S.; Liu, Y.; Deng, J.; Zhao, X.; Yang, J.; Zhang, K.; Han, Z.; Arandiyan, H.; Dai, H. *Appl. Catal. B* **2017,** *206,* 221–232.
86. Miniajluka, N.; Trawczyński, J.; Zawadzki, M. *Appl. Catal. A* **2017,** *531,* 119–128.
87. Suárez-Vázquez, S. I.; Gil, S.; García-Vargas, J. M.; Cruz-López, A.; Giroir-Fendler, A. *Appl. Catal. B* **2018,** *223,* 201–208.

88. Hu, Z.; Qiu, S.; You, Y.; Guo, Y.; Guo, Y.; Wang, L.; Zhan, W.; Lu, G. *Appl. Catal. B* **2018,** *225,* 110–120.
89. Tsoncheva, T.; Ivanova, R.; Henych, J.; Dimitrov, M.; Kormunda, M.; Kovacheva, D.; Scotti, N.; Santo, V. D.; Stengl, V. *Appl. Catal. A* **2015,** *502,* 418–432.
90. Castaño, M. H.; Molina, R.; Moreno, S. *Catalysts* **2015,** *5,* 905–925.
91. Castaño, M. H.; Molina, R.; Moreno, S. J. *Mol. Catal. A Chem.* **2015,** *398,* 358–367.
92. Wang, Y.; Xue, R.; Zhao, C.; Liu, F.; Liu, C.; Han, F. *Colloids Surf. A* **2018,** *540,* 90–97.
93. Piumetti, M.; Bensaid, S.; Andana, T.; Russo, N.; Pirone, R.; Fino, D. *Appl. Catal. B* **2017,** *205,* 455–468.
94. Rezlescu, N.; Rezlescu, E.; Popa, P. D.; Doroftei, C.; Ignat, M. *Ceram. Int.* **2015,** *41,* 4430–4437.
95. Chen, J.; He, Z.; Li, G.; An, T.; Shi, H.; Li, Y. *Appl. Catal. B* **2017,** *209,* 146–154.
96. Lu, H.; Kong, X.; Huang, H.; Zhou, Y.; Chen, Y. J. *Environ. Sci.* **2015,** *32,* 102–107.
97. Chen, D. L.; Pan, K. L.; Chang, M. B. J. *Environ. Sci.* **2017,** *56,* 131–139.
98. Sihaib, Z.; Puleo, F.; Garcia-Vargas, J. M.; Retailleau, L.; Descorme, C.; Liotta, L. F.; Valverde, J. L.; Gil, S.; Giroir-Fendler, A. *Appl. Catal. B* **2017,** *209,* 689–700.
99. Giroir-Fendler, A.; Alves-Fortunato, M.; Richard, M.; Wang, C.; Díaza, J. A.; Gil, S.; Zhang, C.; Can, F.; Bion, N.; Guo, Y. *Appl. Catal. B* **2016,** *180,* 29–37.
100. Wang, C.; Zhang, C.; Hua, W.; Guo, Y.; Lu, G.; Gil, S.; Giroir-Fendler, A. *Chem. Eng. J.* **2017,** *315,* 392–402.
101. Garcia, T.; Agouram, S.; Taylor, S. H.; Morgan, D.; Dejoz, A.; Vázquez, I.; Solsona, B. *Catal. Today* **2015,** *254,* 12–20.
102. Baranowska, K.; Okal, J. *Catal. Lett.* **2016,** *146,* 72–81.
103. Yang, P.; Li, J.; Cheng, Z.; Zuo, S. *Appl. Catal. A* **2017,** *542,* 38–46.
104. Wang, H.; Lu, Y.; Han, Y. X.; Lu, C.; Wan, H.; Xu, Z.; Zheng, S. *Appl. Surf. Sci.* **2017,** *420,* 260–266.
105. Kim, I. H.; Park, E. J.; Park, C. H.; Han, S. W.; Seo, H. O.; Kim, Y. D. *Catal. Today* **2017,** *295,* 56–64.
106. Norsic, C.; Tatibouët, J. M.; Batiot-Dupeyrat, C.; Fourré, E. *Chem. Eng. J.* **2016,** *304,* 563–572.
107. Barakat, T.; Rooke, J. C.; Tidahy, H. L.; Hosseini, M.; Cousin, R.; Lamonier, J. F.; Giraudon, J. M.; De Weireld, G.; Su, B. L.; Siffert, S. *ChemSusChem* **2011,** *4,* 1420–1430.
108. Sinha, A. K.; Suzuki, K. *Angew. Chem. Int. Ed.* **2005,** *44,* 271–273.
109. Cao, S.; Fei, X.; Wen, Y.; Sun, Z.; Wang, H.; Wu, Z. *Appl. Catal. A* **2018,** *550,* 20–27.
110. Calzada, L. A.; Collins, S. E.; Han, C. W.; Ortalan, V.; Zanella, R. *Appl. Catal. B* **2017,** *207,* 79–92.
111. Chen, H.; Tang, M.; Rui, Z.; Wang, X.; Ji, H. *Catal. Today* **2016,** *264,* 23–30.
112. Tsoncheva, T.; Ivanova, R.; Henych, J.; Velinov, N.; Kormunda, M.; Dimitrov, M.; Paneva, D.; Slušná, M.; Mitov, I.; Štengl, V. *Catal. Commun.* **2016,** *81,* 14–19.
113. Wang, J.; Liu, X.; Zeng, J.; Zhu, T. *Catal. Commun.* **2016,** *76,* 13–18.
114. Guo, Y.; Zhang, S.; Zhu, J.; Su, L.; Xie, X.; Li, Z. *Appl. Surf. Sci.* **2017,** *416,* 358–364.
115. Zeng, J.; Liu, X.; Wang, J.; Lv, H.; Zhu, T. J. *Mol. Catal. A: Inside Chem.* **2015,** *408,* 221–227.
116. Białas, A.; Kondratowicz, T.; Drozdek, M.; Kuśtrowski, P. *Catal. Today* **2015,** *257,* 144–149.
117. Labaki, M.; Siffert, S.; Lamonier, J. F.; Zhilinskaya, E. A.; Aboukais, A. *Appl. Catal. B* **2003,** *43,* 261–271.

118. Dole, H. A. E.; Isaifan, R. J.; Sapountzi, F. M.; Lizarraga, L.; Aubert, D.; Princivalle, A.; Vernoux, P.; Baranova, E. A. *Catal. Lett.* **2013,** *143,* 996–1002.
119. Aboukaïs, A.; Skaf, M.; Hany, S.; Cousin, R.; Aouad, S.; Labaki, M.; Abi-Aad, E. *Mater. Chem. Phys.* **2016,** *177,* 570–576.
120. Wang, Y.; Xue, Y.; Zhao, C.; Zhao, D.; Liu, F.; Wang, K.; Dionysiou, D. D. *Chem. Eng. J.* **2016,** *300,* 300–305.
121. Peng, R.; Li, S.; Sun, X.; Ren, Q.; Chen, L.; Fu, M.; Wu, J.; Ye, D. *Appl. Catal. B* **2018,** *220,* 462–470.
122. Peng, R.; Sun, X.; Li, S.; Chen, L.; Fu, M.; Wu, J.; Ye, D. *Chem. Eng. J.* **2016,** *306,* 1234–1246.
123. Mei, J.; Ke, Y.; Yu, Z.; Hu, X.; Qu, Z.; Yan, N. *Chem. Eng. J.* **2017,** *320,* 124–134.
124. Fiorenza, R.; Crisafulli, C.; Condorelli, G. G.; Lupo, F.; Scire, S. *Catal. Lett.* **2015,** *145,* 1691–1702.
125. Wolski, L.; Sobczak, I.; Ziolek, M. *J. Catal.* **2017,** *354,* 100–112.
126. Jiang, W.; Pang, Y.; Gu, L.; Yao, Y.; Su, Q.; Ji, W.; Au, C. T. *J. Catal.* **2017,** *349,* 183–196.
127. Liu, Y.; Guo, Y.; Liu, Y.; Xu, X.; Peng, H.; Fang, X.; Wang, X. *Appl. Surf. Sci.* **2017,** *420,* 186–195.
128. Chen, X.; Chen, X.; Cai, S.; Chen, J.; Xu, W.; Jia, H.; Chen, J. *Chem. Eng. J.* **2018,** *334,* 768–779.
129. Wu, Z.; Deng, J.; Xie, S.; Yang, H.; Zhao, X.; Zhang, K.; Lin, H.; Dai, H.; Guo, G. *Microporous Mesoporous Mater.* **2016,** *224,* 311–322.
130. Li, J.; Qu, Z.; Qin, Y.; Wang, H. *Appl. Surf. Sci.* **2016,** *385,* 234–240.
131. Wu, Y.; Yuan, S.; Feng, R.; Ma, Z.; Gao, Y.; Xing, S. *Mol. Catal.* **2017,** *442,* 164–172.
132. Gonzalez-Prior, J.; Gutierrez-Ortiz, J. I.; Lopez-Fonseca, R.; Busca, G.; Finocchio, E.; de Rivas, B. *Catal. Sci. Technol.* **2016,** *6,* 5618–5630.
133. Ivanova, R.; Genova, I.; Kovacheva, D.; Atanasova, G.; Tsoncheva, T. *J. Porous Mater.* **2016,** *23,* 1005–1013.
134. Wang, C.; Tian, C.; Guo, Y.; Zhang, Z.; Hua, W.; Zhan, W.; Guo, Y.; Wang, L.; Lu, G. J. *Hazard. Mater.* **2018,** *342,* 290–296.
135. Fu, X.; Liu, Y.; Yao, W.; Wu, Z. *Catal. Commun.* **2016,** *83,* 22–26.
136. Sabour, S.; Especel, C.; Fontaine, C.; Bidaoui, M.; Benatallah, L.; Saib-Bouchenafa, N.; Barbier, J., Jr.; Mohammedi, O. *J. Mol. Catal. A: Chem.* **2016,** *420,* 50–55.
137. Chen, D.; Qu, Z.; Lv, Y.; Lu, X.; Chen, W.; Gao, X. *J. Mol. Catal. A: Chem.* **2015,** *404–405,* 98–105.
138. Rokicińska, A.; Drozdek, M.; Dudek, B.; Gil, B.; Michorczyk, P.; Brouri, D.; Dzwigaj, S.; Kuśtrowski, P. *Appl. Catal. B* **2017,** *212,* 59–67.
139. Chen, C.; Wang, X.; Zhang, J.; Bian, C.; Pan, S.; Chen, F.; Meng, X.; Zheng, X.; Gao, X.; Xiao, F. S. *Catal. Today* **2015,** *258,* 190–195.
140. Yosefi, L.; Haghighi, M.; Allahyari, S.; Ashkriz, S. *J. Chem. Technol. Biotechnol.* **2015,** *90,* 765–774.
141. Romero, D.; Chlala, D.; Labaki, M.; Royer, S.; Bellat, J. P.; Bezverkhyy, I.; Giraudon, J. M.; Lamonier, J. F. *Catalysts* **2015,** *5,* 1479–1497.
142. Hosseini, M.; Haghighi, M.; Kahforoushan, D.; Zarrabi, M. *Process Saf. Environ. Prot.* **2017,** *106,* 284–293.
143. Wang, Y.; Yang, D.; Li, S.; Chen, M.; Guo, L.; Zhou, J. *Microporous Mesoporous Mater.* **2018,** *258,* 17–25.
144. Zhang, J.; Rao, C.; Peng, H.; Peng, C.; Zhang, L.; Xu, X.; Liu, W.; Wang, Z.; Zhang, N.; Wang, X. *Chem. Eng. J.* **2018,** *334,* 10–18.
145. Yang, P.; Li, J.; Zuo, S. *Chem. Eng. Sci.* **2017,** *162,* 218–226.

146. Zhang, Y.; Zhang, H.; Yan, Y. *Microporous Mesoporous Mater.* **2018,** *261,* 244–251.
147. Romero-Sáez, M.; Divakar, D.; Aranzabal, A.; González-Velasco, J. R.; González-Marcos, J. A. *Appl. Catal. B* **2016,** *180,* 210–218.
148. Li, S.; Hao, Q.; Zhao, R.; Liu, D.; Duan, H.; Dou, B. *Chem. Eng. J.* **2016,** *285,* 536–543.
149. Carrillo, A. M.; Carriazo, J. G. *Appl. Catal. B* **2015,** *164,* 443–452.
150. Natkański, P.; Rokicińska, A.; Wach, A.; Drozdek, M.; Dudek, B.; Lityńska-Dobrzyńska, L.; Kuśtrowski, P. *RSC Adv.* **2016,** *6,* 100373–100382.
151. Rokicińska, A.; Natkański, P.; Dudek, B.; Drozdek, M.; Lityńska-Dobrzyńska, L.; Kuśtrowski, P. *Appl. Catal. B* **2016,** *195,* 59–68.
152. Gatica, J. M.; Castiglioni, J.; de los Santos, C.; Yeste, M. P.; Cifredo, G.; Torres, M.; Vidal, H. *Catal. Today* **2017,** *296,* 84–94.
153. Sanchis, R.; Cecilia, J. A.; Soriano, M. D.; Vázquez, M. I.; Dejoz, A.; López Nieto, J. M.; Castellón, E. R.; Solsona, B. *Chem. Eng. J.* **2018,** *334,* 1159–1168.
154. Liang, X.; Qi, F.; Liu, P.; Wei, G.; Su, X.; Ma, L.; He, H.; Lin, X.; Xi, Y.; Zhu, J.; Zhu, R. *Appl. Clay Sci.* **2016,** *132–133,* 96–104.
155. Abdelouahab-Reddam, Z.; El Mail, R.; Coloma, F.; Sepúlveda-Escribano, A. *Appl. Catal. A* **2015,** *494,* 87–94.
156. Choun, M.; Ham, K.; Shin, D.; Lee, J. K.; Lee, J. *Catal. Today* **2017,** *295,* 26–31.
157. Li, W.; Zhang, Z.; Wang, J.; Qiao, W.; Long, D.; Ling, L. *Chem. Eng. J.* **2016,** *293,* 243–251.
158. Maldonado-Hodar, F. J. *J. Hazard Mater.* **2011,** *194,* 216–222.
159. Campanati, M.; Fornasari, G.; Vaccari, A. *Catal. Today* **2003,** *77,* 299–314.
160. Wang, H.; Yang, W.; Tian, P.; Zhou, J.; Tang, R.; Wu, S. *Appl. Catal. A* **2017,** *529,* 60–67.
161. Xie, S.; Liu, Y.; Deng, J.; Zhao, X.; Yang, J.; Zhang, K.; Han, Z.; Dai, H. *J. Catal.* **2016,** *342,* 17–26.
162. Kaminski, P.; Ziolek, M. *Appl. Catal. B* **2016,** *187,* 328–341.
163. Rui, Z.; Tang, M.; Ji, W.; Ding, J.; Ji, H. *Catal. Today* **2017,** *297,* 159–166.
164. Wang, Z.; Liu, Y.; Yang, T.; Deng, J.; Xie, S.; Dai, H. *Chinese J. Catal.* **2017,** *38,* 207–216.
165. Fiorenza, R.; Bellardita, M.; Palmisano, L.; Scirè, S. *J. Mol. Catal. A: Chem.* **2016,** *415,* 56–64.
166. Morales-Torres, S.; Perez-Cadenas, A. F.; Kapteijn, F.; Carrasco-Marin, F.; Maldonado-Hodar, F. J.; Moulijn, J. A. *Appl. Catal. B* **2009,** *89,* 411–419.
167. Chen, J.; Chen, X.; Xu, W.; Xu, Z.; Chen, J.; Jia, H.; Chen, J. *Chem. Eng. J.* **2017,** *330,* 281–293.
168. Zhou, G.; He, X.; Liu, S.; Xie, H.; Fu, M. *J. Ind. Eng. Chem.* **2015,** *21,* 932–941.
169. Park, E. J.; Seo, H. O.; Kim, Y. D. *Catal. Today* **2017,** *295,* 3–13.
170. Matejova, L.; Topka, P.; Jiratova, K.; Solcova, O. *Appl. Catal. A* **2012,** *443–444,* 40–49.
171. Hu, F.; Chen, J.; Peng, Y.; Song, H.; Li, K.; Li, J. *Chem. Eng. J.* **2018,** *331,* 425–434.
172. Wu, Y.; Feng, R.; Song, C.; Xing, S.; Gao, Y.; Ma, Z. *Catal. Today* **2017,** *281,* 500–506.
173. Liu, P.; He, H.; Wei, G.; Liu, D.; Liang, X.; Chen, T.; Zhu, J.; Zhu, R. *Microporous Mesoporous Mater.* **2017,** *239,* 101–110.
174. Wei, Y.; Ni, L.; Li, M.; Zhao, J. *Catal. Today* **2017,** *297,* 188–192.
175. Morales, M. R.; Yeste, M. P.; Vidal, H.; Gatica, J. M.; Cadus, L. E. *Fuel* **2017,** *208,* 637–646.
176. Kim, S. C.; Shim, W. G. *Appl. Catal. B* **2010,** *98,* 180–185.
177. Piumetti, M.; Fino, D.; Russo, N. *Appl. Catal. B* **2015,** *163,* 277–287.
178. Deng, L.; Ding, Y.; Duan, B.; Chen, Y.; Li, P.; Zhu, S.; Shen, S. *Mol. Catal.* **2018,** *446,* 72–80.

179. Hu, F.; Chen, J.; Zhao, S.; Li, K.; Si, W.; Song, H.; Li, J. *Appl. Catal. A* **2017,** *540,* 57–67.
180. Li, L.; Jing, F.; Yan, J.; Jing, J.; Chu, W. *Catal. Today* **2017,** *297,* 167–172.
181. Tang, W.; Wu, X.; Liu, G.; Li, S.; Li, D.; Li, W.; Chen, Y. J. *Rare Earth* **2015,** *33,* 62.
182. Colman-Lerner, E.; Peluso, M. A.; Sambeth, J.; Thomas, H. J. *Rare Earth* **2016,** *34,* 675.
183. Lin, X.; Li, S.; He, H.; Wu, Z.; Wu, J.; Chen, L.; Ye, D.; Fu, M. *Appl. Catal. B* **2018,** *223,* 91–102.
184. Kan, J.; Deng, L.; Li, B.; Huang, Q.; Zhu, S.; Shen, S.; Chen, Y. *Appl. Catal. A* **2017,** *530,* 21–29.
185. Chen, J.; Chen, X.; Chen, X.; Xu, W.; Xu, Z.; Jia, H.; Chen, J. *Appl. Catal. B* **2018,** *224,* 825–835.
186. Quiroz, J.; Giraudon, J. M.; Gervasini, A.; Dujardin, C.; Lancelot, C.; Trentesaux, M.; Lamonier, J. F. *ACS Catal.* **2015,** *5,* 2260–2269.
187. Tidahy, H. L.; Siffert, S.; Lamonier, J. F.; Zhilinskaya, E. A.; Aboukais, A.; Yuan, Z. Y.; Vantomme, A.; Su, B. L.; Canet, X.; De Weireld, G.; Frere, M.; N'Guyen, T. B.; Giraudon, J. M.; Leclercq, G. *Appl. Catal. A* **2006,** *310,* 61–69.
188. Barakat, T.; Idakiev, V.; Cousin, R.; Shao, G. S.; Yuan, Z. Y.; Tabakova, T.; Siffert, S. *Appl. Catal. B* **2014,** *146,* 138–146.
189. Chen, C.; Wang, X.; Zhang, J.; Pan, S.; Bian, C.; Wang, L.; Chen, F.; Meng, X.; Zheng, X.; Gao, X.; Xiao, F. S. *Catal. Lett.* **2014,** *144,* 1851–1859.
190. Chen, C.; Zhu, J.; Chen, F.; Meng, X.; Zheng, X.; Gao, X.; Xiao, F. S. *Appl. Catal. B* **2013,** *140–141,* 199–205.
191. Białas, A.; Kuśtrowski, P.; Dudek, B.; Piwowarska, Z.; Wach, A.; Michalik, M.; Kozak, M. *Thermochim. Acta* **2014,** *590,* 191–197.
192. Li, K.; Chen, J.; Bai, B.; Zhao, S.; Hu, F.; Li, J. *Catal. Today* **2017,** *297,* 173–181.
193. Li, G.; Zhang, C.; Wang, Z.; Huang, H.; Peng, H.; Li, X. *Appl. Catal. A* **2018,** *550,* 67–76.
194. Xie, S.; Liu, Y.; Deng, J.; Yang, J.; Zhao, X.; Han, Z.; Zhang, K.; Dai, H. *J. Catal.* **2017,** *352,* 282–292.
195. Zhang, W.; Wu, F.; Li, J.; You, Z. *Appl. Surf. Sci.* **2017,** *411,* 136–143.
196. Wang, T.; Zhang, H.; Yan, Y. *J. Solid State Chem.* **2017,** *251,* 55–60.
197. Peng, Y.; Zhang, L.; Chen, L.; Yuan, D.; Wang, G.; Meng, X.; Xiao, F. S. *Catal. Today* **2017,** *297,* 182–187.
198. Zazhigalov, S. V.; Mikenin, P. E.; Lopatin, S. A.; Baranov, D. V.; Pisarev, D. A.; Chumakova, N. A.; Zagoruiko, A. N. *Catal. Ind.* **2016,** *8,* 231–241.
199. Campesi, M. A.; Luzi, C. D.; Barreto, G. F.; Martínez, O. M. *J. Environ. Manage.* **2015,** *154,* 216–224.
200. Nagae, K.; Ueda, T.; Sasahara, T.; Nakagoe, O.; Tanabe, S.; Hyodo, T.; Shimizu, Y. *ECS Trans.* **2016,** *75,* 23–29.
201. Hyodo, T.; Hashimoto, T.; Ueda, T.; Nakagoe, O.; Kamada, K.; Sasahara, T.; Tanabe, S.; Shimizu, Y. *Sens. Actuators B* **2015,** *220,* 1091–1104.
202. Kuśtrowski, P.; Kondratowicz, T.; Machowski, K.; Natkański, P. *Polish Patent Application.* P.423615, 2017.

INDEX

'*Note*: Page numbers followed by "f" indicate figures, "t" indicate tables.'

E

F

L

N

O

U

V

W

X

Y

Z

Papier | Fördert gute Waldnutzung
FSC® C083411

Kolibri 360 GmbH
Ferdinand-Jühlke-Straße 7
99095 Erfurt, Deutschland
produktsicherheit@kolibri360.de